SONS OF MARTHA

*Civil Engineering
Readings in
Modern Literature*

Collected & Edited
by Augustine J. Fredrich

Published by the
American Society of Civil Engineers
345 East 47th Street
New York, New York 10017-2398

ABSTRACT

Sons of Martha: Civil Engineering Readings in Modern Literature explores the history and heritage, the people, the projects and the ethics involved in the development of the civil engineering profession. The first part of this anthology discusses such topics as the art of engineering, the role of the engineer in society, and the development of engineering education. The second and third sections bring to life some of the important people and projects in the civil engineering field. Finally, the fourth part examines the civil engineering profession and its relationship with the rest of society. In all, this collection of writings present an intriguing look at the history and development of the civil engineering field.

Dedicated to:

Cecelia, my wife

Barbara, my mother

and

Gerald E. Thomas and Leo R. Beard, Sons of Martha

PREFACE

Almost every civil engineer, at some time or another, is faced with the necessity of responding to the question "What *is* a civil engineer" or "What does a civil engineer *do*?" Often those questions are posed in the course of an informal discussion, when the questioner neither wants nor needs a definitive or complete answer. A response based on personal experience or a general response indicating that a civil engineer is a person who designs and builds things will frequently satisfy the low level of curiosity that stimulates the question under those circumstances.

However, when the inquiry comes from a student interested in the civil engineering career field, or from his or her parents, (or when the casual response based on the presumption of limited interest elicits further inquiries) a more complete kind of response is needed, one that conveys the nature and scope of the civil engineer's activities and the sense of pride, service and accomplishment that most civil engineers feel about their profession. This book has been prepared to fulfill the need for information that would be the basis for such a response.

Civil engineering is a unusual profession. It has many dimensions, provides many opportunities, offers many challenges and promises many rewards. There are rewarding careers for those who prefer to work at a desk as well as for those who enjoy working outdoors; for those whose are interested in working with things as well as for those who prefer to work with people; and for those who want the security of employment as a salaried professional as well as for those who seek the challenge of entrepreneurship as self-employed professionals. There are positions for those who are most comfortable working within the framework prescribed by handbooks, standards and manuals as well as for those who thrive on the stimulus of problems for which there are no textbook solutions; for those who want to apply their creative genius to the development of concepts and alternatives in project planning (where the results of one's efforts may not be seen for years) as well as for those who prefer the more routine and familiar tasks inherent in operating and maintaining existing projects (where there's a new problem every day and the results of one's efforts are visible almost immediately).

The variety doesn't end with the many different types of work in which civil engineers are engaged. There is also an almost unlimited variety of locales where civil engineers may find employment opportunities. The adventure and excitement of civil engineering projects in foreign nations is a possibility for those who dream of a life different from what they have known. Construction projects in wilderness areas provide challenges totally different from those experienced by persons engaged in civil engineering work in the offices of one of the many city, county, state or federal government agencies involved in the planning, design, construction, operation and maintenance of civil engineering projects. Those attracted to the hustle and bustle of cosmopolitan centers can seek employment with the large national and international civil engineering design and construction firms or with government agencies located in any major metropolitan area while those who prefer the more pedestrian lifestyle of a smaller communities will likely find more satisfaction in municipal engineering in smaller communities or in consulting firms with a local or regional practice.

For those who seek fame, who enjoy being in the public eye, and who thrive on public acclaim and recognition, civil engineering offers elected and appointive engineering positions at almost every level of government. These positions lead some civil engineers to elected political office as mayors, governors, state legislators, and U.S. congressmen and senators. For those who are most comfortable working in relative anonymity and who find fulfillment primarily in the personal satisfaction that comes from an important job well done there are rewarding staff jobs in government, private industry and civil engineering consulting. And for those who aspire to careers that involve developing the knowledge needed to solve tomorrow's problems and communicating that knowledge to young people who will be tomorrow's problem solvers there are careers in civil engineering research and education.

For almost any conceivable human motive or aspiration--fame, fortune, service, security--there is a niche in the civil engineering profession that provides the opportunity to find happiness and personal fulfillment.

Communicating all of that to a person who really seeks to understand the civil engineering profession, especially to prospective students who wish to make career decisions, is a formidable task. Where law, medicine, architecture, advertising, commerce, law enforcement and a host of other professions and occupations have been widely publicized in movies, television series, newspapers, magazines, novels and other forms of popular communications, civil engineering work and civil engineers have not. Thus, there is a real need to provide young people with a source of information that will assist them in understanding the challenges and rewards of a career in civil engineering.

The purpose of this collection is to provide in a single source a representative selection of readings that provide for the interested reader a reasonably accurate portrayal of the nature and scope of the civil engineering profession. A second objective is to stimulate interest in civil engineering and civil engineers on the part of those who benefit from the work of the profession but have heretofore thought little about the role of civil engineering in modern society.

Finally, many civil engineers themselves seek to become more conversant with the many facets of their profession, particularly with the history and heritage of the profession, an aspect of engineering education that has not received enough attention in the recent past. In this collection the interested civil engineer will find a wealth of relevant material--not only relevant but interesting, stimulating and well worth reading.

ACKNOWLEDGMENTS

The editor and publisher gratefully acknowledge the cooperation of the authors and publishers for granting the permission to reprint this material:

History and Heritage

"I Deem It of the First and Greatest Importance to Familiarize Myself with the Business I Have Undertaken" from **Strategems and Aqueducts** by Sextus Julius Frontinus. Copyright © 1961 by the President and Fellows of Harvard College. Excerpted by permission of Harvard University Press.

"Engineering: The Heroic Art" from **Great Engineers and Pioneers in Technology**. Copyright © 1981 by St. Martin's Press, Inc. Reprinted with permission of St. Martin's Press, Inc.

"Michelangelo and the Dome of St. Peter's in Rome" from **Engineers and Engineering in the Renaissance** by William B. Parsons. Copyright © 1967 by The Williams and Wilkins Company. Reproduced by permission of The Williams and Wilkins Company.

"Taking the Measure of a New Nation" by Neal FitzSimons reprinted from Civil Engineering with the permission of the American Society of Civil Engineers. Copyright © 1967 by the American Society of Civil Engineers.

"The First School of Engineering" by Sidney Forman reprinted from The Military Engineer, March-April, 1952. Copyright © by the Society of American Military Engineers. Reprinted with permission of the Society of American Military Engineers.

"Early Life" from **The Reminiscences of John B. Jervis: Engineer of the Old Croton** edited by Neal FitzSimons. Copyright © 1971 by Syracuse University Press. Reprinted by permission of the publisher.

"Champions of the Public Interest" from **The American Civil Engineer: 1852-1974** by William H. Wisely. Copyright © 1974 by the American Society of Civil Engineers. Excerpted with permission of the publisher.

"The Functional Intellectual" from **Engineering in American Society, 1850-1875** by Raymond H. Merritt. Copyright © 1969 by the University Press of Kentucky. Reprinted by permission of the publisher.

"What Was Troy to This?" from **The Existential Pleasures of Engineering** by Samuel C. Florman. Copyright © 1976 by Samuel C. Florman. Reprinted by permission of the author and St. Martin's Press, Inc.

"Sons of Martha" by Rudyard Kipling. Copyright © 1907 by Rudyard Kipling. Reprinted with permission of Doubleday, a division of Bantam, Doubleday, Dell Publishing Group, Inc.

Civil Engineering Projects

In addition to the cooperation of the authors and publishers listed above, the assistance, advice and support of many people were helpful in making this book become a reality. The time required to convert a hodgepodge of articles, books, newspaper clippings and other readings into a coherent anthology was made possible by a sabbatical leave granted by the Board of Trustees of the University of Southern Indiana. The leave would not have been possible without the encouragement of Dr. Robert Reid, Vice President for Academic Affairs at USI, the cooperation of Professor George Eadie who fulfilled my administrative responsibilities during the leave, and the support of my faculty colleagues who assumed responsibilities for my teaching assignments.

In return for a modest amount of assistance and consultation, the U. S. Army Corps of Engineers provided the financial support which made the sabbatical leave possible. James R. Hanchey, Director of the Corps' Institute for Water Resources, and Bill S. Eichert, Director of the Corps' Hydrologic Engineering Center, provided funding for the sabbatical leave from within their budgets and made working space and facilities available for completion of the final version of the manuscript. Kyle Schilling at the Institute for Water Resources arranged my workload for the Corps in a way that made it possible for me to fulfill commitments to both the Corps and the publisher.

Roxanna Nolt and Courtney Klingelhoefer, secretaries at the University of Southern Indiana, typed a substantial portion of the first version of the manuscript. Al Onodera and Penni Baker of the Hydrologic Engineering Center staff and Mike Walsh, David Moser, Martin Titus and Bob Swett of the Institute for Water Resources staff provided assistance in the use of facilities for preparation of the manuscript and its illustrations.

Bill Johnson and Jerry Delli Priscoli, of the Hydrologic Engineering Center and Institute for Water Resources staffs respectively, were advocates for this book long before I actually began work on it, and they both offered support and encouragement throughout its preparation. Leo R. Beard, James Flynn, and Robyn Colosimo read each selection while the manuscript was in preparation and provided helpful suggestions which shaped the content of the final manuscript.

However, the one person who contributed the most to this work was my wife Cecelia who read every version of every selection (and in the process learned more about civil engineering than she ever wanted to know). She offered advice on which selections were most interesting; she suggested places where definitions of terms and explanations were needed; she served as editor and proofreader; and, like all good spouses, she was a constant source of encouragement, optimism and enthusiasm. Without her, **Sons of Martha** would be just another idea that never really got off the ground.

Illustration Sources and Credits

American Society of Civil Engineers; New York, NY: Pages 34, 47, 53, 55, 95, 96, 280, 285, 318, 321, 390, 404 and 442

Bechtel Group; San Francisco, CA: Page 436

Empire State Building Associates; New York, NY: Page 401

Neal FitzSimons Collection; Kensington, MD: Pages 2 (top) and 108

National Aeronautics and Space Administration; Washington, DC: Pages 2 (bottom) and 274

Smithsonian Institution; Washington, DC: Pages 178 and 315

U. S. Army Corps of Engineers, Office of History; Washington, DC: Pages 30, 37, 49, 184, 244, 249, 250 and 336

U. S. Department of the Interior, Bureau of Reclamation; Denver, CO: Pages 383 and 547

U. S. Military Academy Archives; West Point, NY: Page 32 (top and bottom)

Steinman Boynton Gronquist & Birdsall; New York, NY: Page 438

The cover illustration was drawn by Lili Rethi, F.R.S.A., during the construction of the Verrazano-Narrows Bridge. Used courtesy of the Triborough Bridge and Tunnel Authority; New York, NY.

TABLE OF CONTENTS

HISTORY AND HERITAGE

The readings in this section are not intended to provide a comprehensive history of the civil engineering profession. Neither are they intended to dramatize the effects that civil engineers and the civil engineering profession have had on the course of civilization. Rather, they have been chosen with a view toward providing a collection of "snapshots" which give the reader a series of glimpses of civil engineers playing out their roles on the stage of history. They are intended to provoke the reader to ponder the role of civil engineering in the evolution of modern society in general and in the development of this nation in particular, and to reflect on the manner in which our modern perceptions of the civil engineer have been influenced by the needs and desires of human society through the ages.

Civil engineers have unquestionably played a crucial role in the evolution of civilization, but the profession has also been shaped by that evolution. In a sense, today's civil engineers have become the creatures of their own creation. The very existence of the profession depends on recognition of a strange difference between it and most of the other professions. Like most other professions, the status of civil engineers as professionals depends on societal perception of them as individuals motivated by their contributions to the good of humanity rather than by advancement of their personal well-being. However, unlike other professionals, civil engineers cannot instill trust and confidence in themselves or their motives through person-to-person contacts with the general public as attorneys, physicians, clergymen and other professionals do. Instead, they have to rely on the public perception of their work and the extent to which it provides beneficial service to society as a whole.

In the public view there is no such thing as a "good" civil engineer or a "bad" civil engineer, since society does not generally know civil engineers as individuals; there is only a "good" or a "bad" civil engineering profession, and that perception is based on the value society ascribes to the work accomplished by the members of the profession in aggregate. If the products of the labors of civil engineers are generally regarded as "good" then individual members of the profession will be individually regarded as "good." "Bad" products will result in a degradation of the status of the profession and its individual members regardless of their individual intelligence, integrity and ability.

Examination of the history and heritage of the civil engineering profession has two objectives: to develop an appreciation of the personal and professional standards that have produced professional status for the modern civil engineer; and to create an awareness of the necessity for maintaining those standards in order to preserve that status in the future.

British engineer John Smeaton - first to employ the title Civil Engineer

American civil engineer James van Hoften - first civil engineer in space

I Deem It of the First and Greatest Importance to Familiarize Myself with the Business I Have Undertaken

Excerpts from **The Aqueducts of Rome**
Sextus Julius Frontinus

Julius Sextus Frontinus, was born about 35 A.D. He was sent to Britain as provincial governor around 75 A.D. In 97 A.D. he was appointed water commissioner for Rome. Other than these few facts, not much is known about the details of the life of Frontinus. However, in addition to the writings from which this selection was drawn, he is known to have produced writings on surveying and military strategy.

In the introduction to **The Aqueducts of Rome** *Professor C.E. Bennett writes: "It {***The Aqueducts of Rome***} gives us a picture of the faithful public servant, charged with immense responsibility, called suddenly to an office that had long been a sinecure and wretchedly mismanaged, confronted with abuses and corruption of long standing, and yet administering his charge with an eye only to the public service and an economical use of public funds. . . Were one asked to point out, in all Roman history, another such example of civic virtue and conscientious performance of simple duty, it would be difficult to know where to find it. . . Rich and valuable as is his treatise . . . in facts relating to the administration of ancient aqueducts, it is the personality of the writer that one loves to contemplate, his sturdy honesty, his conscientious devotion to the duties of his office, his patient attention to details, his loyal attachment to the sovereign whom he delighted to serve, his willing labours in behalf of the people whose convenience, comfort and safety he aimed to promote. We sympathize with him in his proud boast that by his reforms he has not only made the city cleaner, but the air purer, and has removed the causes of pestilence that had formerly given the city such a bad repute. . ."*

Those who would aspire to be employed as engineers in the public service, especially as municipal engineers, need look no farther for a written code of conduct than this nineteen-hundred-year-old document.

Inasmuch as every task assigned by the Emperor demands especial attention; and inasmuch as I am incited, not merely to diligence, but also to devotion, when any matter is entrusted to me, be it as a consequence of my natural sense of responsibility or of my fidelity; and inasmuch as Nerva Augustus (an emperor of whom I am at a loss to say whether he devotes more industry or love to the State) has laid upon me the duties of water commissioner, an office which concerns not merely the convenience but also the health and even the safety of the City, and which has always been administered by the most eminent men of our State; now therefore I deem it of the first and greatest importance to familiarize myself with the business I have undertaken, a policy which I have always made a principle in other affairs.

For I believe that there is no surer foundation for any business than this, and that it would be otherwise impossible to determine what ought to be done, what ought to be avoided; likewise that there is nothing so disgraceful for a decent man as to conduct an office delegated to him, according to the instructions of assistants. Yet precisely this is inevitable whenever a person inexperienced in the matter in hand has to have recourse to the practical knowledge of subordinates . . . For though the latter play a necessary role in the way of rendering assistance, yet they are, as it were, but the hands and tools of the directing head. Observing, therefore, the practice which I have followed in many offices, I have gathered in this sketch (into one systematic body, so to speak) such facts, hitherto scattered, as I have been able to get together, which bear on the general subject, and which might serve to guide me in my administration. Now in the case of other books which I have written after practical experience, I consulted the interests of my successors. The present treatise also may be found useful by my successor, but it will serve especially for my own instruction and guidance, being prepared, as it is, at the beginning of my administration.

And lest I seem to have omitted anything requisite to a familiarity with the entire subject, I will first set down the names of the waters which enter the City of Rome; then I will tell by whom, under what consuls, and in what year after the founding of the City each one was brought in; then at what point and at what milestone each water was taken; how far each is carried in a subterranean channel, how far on substructures, how far on arches. Then I will give the elevation of each, [the plan] of the taps, and the distributions that are made from them; how much each aqueduct brings to points outside the City, what proportion to each quarter within the City; how many public reservoirs there are, and from these how much is delivered to public works, how much to ornamental fountains (munera, as the more polite call them), how much to the waterbasins; how much is granted in the name of Caesar; how much for private uses by favour of the Emperor; what is the law with regard to the construction and maintenance of the aqueducts, what penalties enforce it, whether established by resolutions of the Senate or by edicts of the Emperors.

There follows a lengthy description of the various aqueducts providing water to the city. The descriptions include details such as the size and length of the pipes and the quantity of water delivered. Most of the information is of limited interest today, even for engineers. However, the descriptions also include information on the relationship of aqueduct location to major roads, the quality of the water delivered by each aqueduct, the rationale for naming some of the aqueducts and the political conditions for the construction of the various aqueducts. Then the treatise resumes.

With such an array of indispensable structures carrying so many waters, compare, if you will, the idle Pyramids or the useless, though famous works of the Greeks!

It has seemed to me not inappropriate to include also a statement of the lengths of the channels of the several aqueducts, according to the kinds of construction. For since the chief function of this office of water commissioner lies in their upkeep, the man in charge of them ought to know which of them demand the heavier outlay. My zeal was not satisfied with submitting details to examination; I also had plans made of the aqueducts, on which it is shown where there are valleys and how great these are; where rivers are crossed; and where conduits laid on hillsides demand more particular constant care for their maintenance and repair. By this provision, one reaps the advantage of being able to have the works before one's eyes, so to speak, at a moment's notice, and to consider then as though standing by their side. . . .

4

Since I have given in detail the builders of the several aqueducts, their dates, and, in addition, their sources, the lengths of their channels, and their elevations in sequence, it seems to me not out of keeping to add also some separate details, and to show how great is the supply which suffices not only for public and private uses and purposes, but also for the satisfaction of luxury; by how many reservoirs it is distributed and in what wards; how much water is delivered outside the City; how much in the City itself; how much of this latter amount is used for water-basins, how much for fountains, how much for public buildings, how much in the name of Caesar, how much for private consumption.

A description of the techniques, devices and units of measurement for measuring the quantities of water delivered follows at this point.

I will now set down what discharge each aqueduct, according to the imperial records, was thought to have up to the time of my administration, and also how much it actually did deliver; then the true measure, which I reached by careful investigation, acting on the suggestion of that best and most industrious emperor, Nerva. Now there were, in the aggregate, 12,755 quinariae* set down in the records, but 14,018 quinariae actually delivered; that is, 1,263 more quinariae were reported as delivered than were reckoned as received. Since I considered it the most important function of my office to determine the facts concerning the water-supply, my astonishment at this state of affairs stirred me profoundly and led me to investigate how it happened that more was being delivered than belonged to the property, so to speak. . . .

A description follows of the amount of water actually entering each aqueduct; the amount of water available to be delivered according to imperial records; the amount of water actually delivered; and the amount of water reported as received. Then Frontinus sums up his findings.

I do not doubt that many will be surprised that according to our gaugings, the quantity of water was found to be much greater than that given in the imperial records. The reason for this is to be found in the blunders of those who carelessly computed each of these waters at the outset. Moreover, I am prevented from believing that it was from fear of droughts in the summer that they deviated so far from the truth, for the reason that I myself made my gaugings in the month of July, and found the above-recorded supply of each [aqueduct] remaining constant throughout the remainder of the summer. But whatever the reason may be, it has at any rate been discovered that 10,000 quinariae were intercepted** . . .

----

* A quinaria {plural: quinariae} was a Roman unit for measuring flow capacity or rate {volume per unit of time}. A quinaria is estimated to be equivalent to about 5,000 gallons per 24 hours {about 4 gallons per minute}.

** That is, there was a discrepancy of 10,000 quinariae between the amount of water measured at the entrance to the aqueducts and the amount reported to have been received.

The cause of this is the dishonesty of the water-men[*], whom we have detected diverting water from the public conduits for private use. But a large number of landed proprietors also, past whose fields the aqueducts run, tap the conduits; whence it comes that the public water-courses are actually brought to a standstill by private citizens, just to water their gardens. . . .

We have found irrigated fields, shops, garrets even, and lastly all disorderly houses fitted up with fixtures through which a constant supply of flowing water might be assured. . . .

With this, enough has been said about the volume of each aqueduct, and, if I may so express it, about a new way of acquiring water: about frauds and about offences committed in connection with all this. It remains to account in detail for the supply delivered . . .

Frontinus provides a detailed listing of the distribution of water from each of the city's aqueducts--public and private uses, inside and outside the city. He then concludes this summary with an assessment of the effects of these detailed accountings and the changes which result from them.

This is the schedule of the amount of water as reckoned up to the time of the emperor Nerva and this is the way in which it was distributed. But now, by the foresight of the most painstaking of sovereigns, whatever was unlawfully drawn by the water-men, or was wasted as the result of negligence, has been added to our supply; just as though new sources had been discovered. And in fact the supply has been almost doubled, and has been distributed with such careful allotment that wards which were previously supplied by only one aqueduct now receive the water of several. . . In all parts of the City also, the basins, new and old alike, have for the most part been connected with the different aqueducts by two pipes each, so that if accident should put either of the two out of commission, the other may serve and the service may not be interrupted.

The effect of this care . . . is felt from day to day . . . and will be felt still more in the improved health of the city, as a result of the increase in the number of the works, reservoirs, fountains and water basins. . . . The appearance of the City is clean and altered; the air is purer; and the causes of the unwholesome atmosphere, which gave the air of the City so bad a name with the ancients, are now removed.

----

[*] The water-men were the men who received the water from the State and delivered it to the consumers.

[**] Namely, by discovering and correcting frauds.

6

Engineering, The Heroic Art
Arnold Pacey

Great Engineers and Pioneers in Technology
Volume I: From Antiquity Through the Industrial Revolution
Editors Roland Turner and Steven L. Goulden

Arnold Pacey's essay sweeps through the ages providing an historian's view of the engineer in various cultures and eras since the dawn of recorded history. His portrayals provide an inspirational, yet sobering, view of the heritage of the modern civil engineer.

Engineers, Kings and Gods

The "great engineer" of a century ago was a figure who personified the way many people then felt about technology. He was a folk-hero who built railroads through the Rockies, canals through deserts, and bridges of incredible span across wild gorges. Gustave Eiffel, the French railroad engineer, caught the spirit of the age in the great tower which bears his name. As a contemporary Frenchman remarked, astonished Paris saw a "new shape of a new adventure" rising in its midst.[1] It was the adventuring spirit of engineering as well as the individualism of its practitioners which made this an age of great engineers.

Every major civilization has produced important works of engineering, often on an heroic scale. Among the best-known are the pyramids of Egypt, the aqueducts of Rome, and the Great Wall and Grand Canal of China. There were also many large-scale irrigation and water projects in India, Mesopotamia and Palestine. But though history tells us much about the emperors and warriors of those times, and a little about the writers and priests, only fragments of information have come to us about the craftsmen and designers who undertook engineering work. Engineers in ancient times did not rise to personal fame--but their achievements, even so, were appreciated, and cherished in legend. The most popular stories linked engineering and metallurgy; they described marvelous feats by mythical blacksmiths whose activities were dramatized by the flaming furnaces they mastered. One such figure was Hephaistos, the Greek artisan-god whose fire was stolen by Prometheus and given to men. The Egyptians honored Ptah, another blacksmith-god who was said to have been related to Imhotep, the builder of the first pyramids. And the Israelites remembered Tubal-Cain, the ancient Biblical ancestor of all metalworkers and other craftsmen.[2]

The legends about these characters not only show that technical achievement was recognized and valued--they also suggest that invention was feared as well. Indeed, it may have been more congenial to celebrate craftsmen as legendary heroes rather than come to terms with real men whose genius, skill and power was very disconcerting. The Greeks of Homer's time certainly appreciated what craftsmen could do, even if they did not commemorate craftsmen's names. Thus Homer's poetry lingers lovingly over every description of craft skill and construction, and is full of "ramparts, harbors, and causeways ... sailing ships, hawsers, oars and ropes ... axes, adzes, augers and dowels."[3] But there was always a certain ambivalence about such things, which is clearly seen in the character of the god Hephaistos. His skill was truly fabulous and Homer never tired of praising it. But Hephaistos was represented as an ugly, lame man with a hairy chest--a figure of fun among the other gods. The engineer, then, was a man to be valued for his skill, but not as a member of polite society.

One other deity the Greeks linked with craft skill was the goddess Pallas Athena. Homer wrote of an expert carpenter who was "well versed in all his craft's subtlety" through Athena's inspiration, and he mentioned a goldsmith who was taught his trade jointly by Athena and Hephaistos. This reflects a respect for the practicality of women, whose craft skills were admired more wholeheartedly than those of men, and who could teach the craftsman about the utility of what he made.[4]

In most early civilizations, then, the achievements of engineers and craftsmen were regarded with wonder and even awe, and were celebrated in verse and legend. But if it was the names of gods rather than of men which express this heroic view of engineering most clearly in Homer's Greece, in other civilizations it was kings, emperors and even civil servants whose names became most strongly connected with daring feats of construction. Among the people of Old Testament Israel, for example, engineering works brought fame to the kings who commissioned them, not to artisans who designed them. King Uzziah was remembered for the catapults and other engines of war he had installed for the defense of Jerusalem, and Hezekiah for an improved water supply with a tunnel cut through rock.[5] Solomon, the great predecessor of these two kings, employed a craftsman from Tyre to work on the Temple. This man's achievements are described at length, perhaps because he was closely connected with the king of Tyre. He was a skilled metal-worker who was also an inventor, able "to find out every device."[6]

In Israel, as in Greece, it is only very occasionally that a technician is identified by name. Hezekiah's water supply tunnel of about 700 B.C. still exists, with an inscription describing how its builders started at both ends and met at the midpoint, but there is no account of its designer's career. It is all the more unusual, then, that Eupalinos is recorded as the engineer of a similar tunnel, also for water supply, which was made on the island of Samos about 530 B.C.

Engineers and Bureaucrats

The great engineer does not become a folkhero to the bureaucrat, no matter how much his services are appreciated, and this twentieth-century commonplace was also true of the Roman Empire and of dynastic China. In these two very different empires, the majority of big engineering projects were government undertakings, and engineers worked for large organizations and not as individual consultants. They thus remained almost as anonymous as the constructors of earlier times.

For example, the book on Rome's aqueducts written by Sextus Julius Frontinus is the work of a professional administrator, not an engineer, and reflects an official point of view. When Frontinus names the man responsible for building an aqueduct, this was often the commissioner under whom the aqueduct was planned. The commissioner's office employed several hundred people--plumbers and surveyors, masons and engineers--and no effort is made to say which members of this staff designed any given aqueduct. Thus even an imaginative planner of new works, such as Appius Claudius Crassus, was an administrator and politician, not an engineer. And while Frontinus himself had an excellent technical knowledge of aqueducts, his job was primarily that of a manager.[7] By contrast, Marcus Vitruvius Pollio had worked as a military engineer before he wrote his "ten books" on subjects connected with architecture and engineering. Another man with a strong practical bent was Gaius Sergius Orata, who is credited with the invention of the Roman system for heating houses. But he was a businessman, not an engineer. He lived near Naples around 80 B.C., and made a good income from buying up houses and installing heated bathrooms before reselling them. Pliny thought him unscrupulous, but his enterprise caused him to be remembered when the engineers who worked for the imperial bureaucracy were not.

In China, too, many of the best-known names associated with engineering achievements are officials who planned and managed them. One such is Sung-li, the government minister who supervised construction on the summit level of the Ming Grand Canal in 1411. This had an elaborate water supply, and boats ascended to it through locks of a distinctly modern type.[8] Other officials became known through their connection with Chinese government policies for distributing technical books. The policy was based on the wood-block printing process introduced around 870 A.D. Many books were produced in editions of a thousand or more copies to inform provincial officials of improved techniques in agriculture, medicine or construction. In the eleventh century, one civil servant who became well-known as an author[9] was Su Sung, who wrote one book about medicinal herbs and another about an elaborate water clock he designed. Contemporary with these developments, the publication of books on civil engineering marks "a very rich period" of achievement in China,[10] extending from about a century after the introduction of printing up to about 1270. Bridge-building gives one indication of this, with the erection of several iron-chain suspension bridges and the building of some daringly flat masonry arches. These times of heroic construction and technical advance are reflected in the present volume by a cluster of Chinese names from the three centuries between 1000 and 1300.

The frustration we may feel at the anonymity of most early engineers is eased somewhat when we turn from the great empires of China and Rome to the small kingdoms of medieval Europe. Here the records of municipalities, abbeys and royal households mention many names of craftsmen involved in engineering construction. One comprehensive analysis of such documents in England identifies no less than twenty-six men who, prior to 1550, described themselves as "engineers."[11] Most of them were designers of siege engines or carpenters who constructed military works, but many undertook civil projects also: one built a wharf by the River Thames, another erected a windmill. Indeed, from about 1200 onwards, the records of a tiny, nascent bureaucracy, the Office of the King's Works in England, give us the kind of information which is so singularly lacking for the aqueducts of Rome: almost for the first time, it is possible to distinguish clearly between administrators, designers and craftsmen. One example is Geoffrey Chaucer, the poet, who was at one time an administrator in the Office of Works. Henry Yevele stands out as one of its greatest architects: he extended the work done by William of Sens at Canterbury Cathedral and designed part of Westminster Abbey. Another outstanding figure in the history of the Office was the military designer James St. George, who worked on the construction of a castle in Switzerland[12] before building the major castles of North Wales for the English King.

More typical, though, than any of these was an individual who took the word "engineer" into his surname, Richard L'enginour (fl. 1277-1315). He worked as an assistant to James St. George, building bridges and pontoons for river crossings in North Wales. He also constructed a siege engine known as a "springald," and supervised building work on the Conway town walls. Later, he managed a group of water-mills at Chester, where he eventually became mayor. He was so well thought of in the Office of Works that he was given a pension of 12 pence per day for life (over 18 pounds per year). James St. George himself got three times as much, with separate provision for his wife Ambrosia should she be widowed. But then, James was an engineer of unique standing: there was no other craftsman in medieval England who was given such important assignments.[13]

Although military engineering and the construction of castles had reached a high level of sophistication by the thirteenth century, it was the great cathedrals of that time which showed the most daring and adventurous structural design, and particularly the cathedrals of northern France. With vaulted ceilings 130 or more feet (40 meters) above the ground and steep roofs rising considerably higher, they required flying buttresses of remarkable complexity to support the vaults and take up wind pressures. Soaring to still greater heights were towers and spires, many of which proved too ambitious and were left unfinished. The culmination of this development is to be seen in the cathedrals at Chartres (begun in 1194), Amiens (1220) and Beauvals (1247).

Of course, cathedral-building was more than engineering. It was an art form, expressing religious beliefs through symbolism, and religious emotion through mass and space and soaring verticals. It expressed other things as well, not least the competitive spirit of the prosperous commercial cities in which many of the cathedrals were built. It was an art in which aesthetic and technical innovation was continuously pursued, as if the builders, always unsatisfied with their handiwork, were constantly driven to explore and experiment and to attempt the untried.

The Artist-Engineer

The engineering structure of a cathedral was not just a framework on which works of art could hang, but an integral part of the medium with which the artist worked. Indeed, the artist who designed the cathedral was also the engineer responsible for its structure--and often also designed any special hoists or other equipment needed while building was under way. William of Sens was not untypical in these respects. But there is a deeper point to appreciate here: all works of engineering--mechanical or structural--have some of the same qualities as a work of art, whether their designers think of them that way or not. As one civil engineer has put it, "every manmade structure, no matter how mundane, has a little bit of cathedral in it, since man cannot help but transcend himself as soon as he begins to design and construct."[14] This is true of railroad structures in the nineteenth century as well as buildings in the Middle Ages; all were products of the culture of their time and expressive of human striving and aspiration. And in their daring as well as in scale, engineering structures were often the most heroic of all artistic creations.

But did the heroic quality of a work demand that its designer be celebrated as a great artist or even a "great engineer"? With the cathedrals, there was probably a strong feeling that the greatness of the work far surpassed the greatness of any man, whether designer or craftsman. Thus it was left for later generations possessing the humanistic values of the Renaissance to focus on the creativity of individuals--which meant creativity in both engineering and art. So marked was this combination, indeed, that a number of historians have spoken of the Renaissance as an age of "artist-engineers"; Leonardo da Vinci illustrates what is meant by this very clearly.

There was also at this time great curiosity about the process of invention, and three especially notable inventions widely discussed were printing, the magnetic compass and firearms. In due course, Johannes Gutenberg became known as the inventor of printing by moveable type. This emphasis on a single, creative individual was, of course, misleading in that it overlooked other pioneers, not to mention the earlier Chinese printing technique. But it supported the contemporary idea that certain individuals had a special creative genius. These were men with access to divine inspiration, among whom the deeply spiritual and profoundly gifted Michelangelo was taken as the outstanding example. Another much-discussed inventor was the artist-engineer Filippo Brunelleschi, one of the pioneers of Renaissance architecture. He invented hoists for lifting stones on building sites and devised special techniques for constructing the dome of Florence Cathedral. His inventions were rewarded by prizes, and in one case by a patent.

But the manifold heritage of the Renaissance came not only from this stress on invention and on the individuality of engineers. There was also a more general acceptance that some practical arts were fitting occupations for men of learning or noble birth. This applied particularly to the artist-engineers previously mentioned and also to military engineers. Among such men, a select few had something approaching professional status, and were no longer regarded simply as rather accomplished craftsmen. There was also an injection of book-learning into their work, with a growing interest in scholarly aspects of engineering, including its mathematical basis and its literary background in the writings of ancient authors.

The mathematical emphasis, already evident in Leonardo's work, came to fruition with Galileo, who had himself learned much through the study of ancient authors, particularly Archimedes. Technical problems that came Galileo's way included land drainage, pumps and some aspects of military engineering. Books he wrote covered the elements of three major technological subjects: ballistics, machines and the use of scale models in structural design. His writings provided a wide range of new ideas which were taken up most strongly a generation later in France, particularly by a succession of teachers of surveying, architecture and military engineering.[15] Edme Mariotte (1620-84) is perhaps the best known of them. Recent historical studies have shown that these teachers--many working as private tutors--contributed much to the early development of engineering education in France; it is even suggested that they helped bring about a "galilean revolution" in the conceptual framework of engineering. More specifically, they applied Galileo's ideas to a wide range of problems, particularly related to hydraulics. Mariotte, for example, made calculations and tests to help the designers of elaborate machines which pumped water to the Palace of Versailles, and one of his pupils (Sebastien Truchet, 1657-1729) became consulting engineer for the Orleans Canal. Other studies were made of the efficiency of water wheels and the flow of water in pipes. Daniel Bernoulli, remembered for the Bernoulli theorem in hydraulics (1738), had a loose affinity with this group.

Among the small number of mathematics teachers and elite engineers who did this work, something of the heritage of the Renaissance artist-engineer lingered on. In many practical projects architectural work was also involved, and fountains for the gardens at Versailles or other big mansions presented problems in hydraulic design. Thus the type of engineering in which the work of Galileo and Mariotte was most actively developed became known in France as *architecture hydraulique*.[16] These words, indeed, formed the title of the most important general book on engineering to be produced in the first half of the eighteenth century. Its author was Bernard Forest de Belidor, and in its extensive illustrations it shows the artistry of engineering as well as the technical achievements of the period. It is also noteworthy for its extended account of Thomas Newcomen's steam-powered pumping engine.

Engineering Archetypes

Recognition of the engineer as a species of artist is a Renaissance idea which seems a long way removed from twentieth-century attitudes. Yet there have always been a few engineers whose work has an obvious affinity with artistic creation. Robert Maillart, pioneer designer of reinforced concrete bridges, has been hailed as an "artist-engineer" of the present century, and detailed studies have been made to identify "the connection between the utilitarian and the artistic features" of his work.[17] One of the best British writers on nineteenth century engineers also suggests that an appreciation of the engineer as artist is essential for understanding two of the greatest figures of that time, Marc Brunel and his more famous son, Isambard Kingdom Brunel.[18] Designs made by Isambard Brunel for important parts of the London-Bristol railroad include "exquisite sketches ... of tunnel mouth, bridge, pediment or balustrade" which reveal "not Brunel the engineer, but Brunel the artist."[19] His sketchbooks remind one "of the notebooks of Leonardo da Vinci" in their imagination and scope.

Near the end of his life, during two visits to Rome, Isambard Brunel spent hours alone in St. Peter's. He was not a religious man, and these contemplations provoked comment from his travelling companions. His biographer suggests that Brunel felt rather particularly "at one" with that building. A great deal of the spirit in which Brunel himself worked was expressed there; and he felt a strong resonance with his own aims under the dome conceived by Michelangelo and modelled partly on Brunelleschi's earlier dome at Florence.

To suggest that Brunel saw St. Peter's as an archetype for the artistic impulse that had gone into his own work is to suggest no more than a number of modern scientists and technologists have openly said about their own attitudes. In 1961, Alvin Weinberg, Director of Oak Ridge National Laboratory, commented that when historians of the future look back at the twentieth century, they will find in "the huge rockets, the high-energy accelerators, the high-flux research reactors, symbols of our time just as surely as ... Notre Dame [is] a symbol of the Middle Ages."[20] Other writers have taken up the image of a space probe, which "like a cathedral is a symbol of aspiration to higher things."

For many people, though, it is not St. Peter's or Notre Dame, not the aqueducts or pyramids which best demonstrate the cultural significance of engineering. Rather, it is the attitudes and achievements of individual engineers. That is why the image of the "great engineer" is so important. From the time when Andrea Palladio and his contemporaries sought to explain their work in terms of the writings of Vitruvius, the great engineer of the past has been regularly used as an archetype of what modern engineering should be about. Galileo saw his work in terms of what Archimedes had achieved; many British engineers today seem haunted by the near-mythic, Promethean figure of Isambard Brunel.

For an engineer from the past to be appreciated in this way, it is not sufficient only that he should be identified with major constructional achievements. Many engineers well-known to historians remain in obscurity, among them two previously mentioned: Richard L'enginour and Edme Mariotte. An even better example is Joseph Karl Hell (1718-85), a contemporary of John Smeaton who built and improved Newcomen-type steam engines at mines in the Austrian Empire, and who was a brilliant inventor of water-pressure engines and pneumatic devices. In terms of technical achievement, he was as great an engineer as Smeaton, but the industry in which he worked was declining, and he had few successors with whom to leave his heritage. Men such as Joseph Hell remain unrecognized because circumstances did not allow them to enhance the significance of engineering or extend its cultural meaning.

Another reason why some engineers of high achievement remain unknown is that the whole significance of what they did was expressed in what they designed, not in their lives. For this reason, the builders of medieval cathedrals known to historians by name are very rarely revered as "great engineers" or even great architects.[21] It is only occasionally that one of them is a significant cultural figure as an individual--as, for example, William of Sens because of his personal heroism and ingenuity. For much the same reasons, engineers in the twentieth century also tend to remain anonymous; it is their achievements that have meaning, and the organizations for which they work, not their own individuality. So in recent years we have celebrated the Apollo moon rockets, the CERN [European Commission on Nuclear Research] nuclear accelerator and the Concorde airliner without ever hearing the names of those who led design teams or dreamed up key ideas.

Myths and Responsibilities

It should by now be clear that the "great engineer" is, in one sense, part of the mythology of technology. His archetypal image serves us rather as the image of Hephaistos and other gods served the Greeks--as a symbol of what engineering is all about. We use his memory in the same way as we use the image of the pyramid or the cathedral--as an archetype for the cultural meaning of technology.

Biographers and other writers bear an important responsibility here, because it is very often their interpretations of technology which canonize men as

13

great engineers. Brunelleschi's reputation was greatly enhanced for later generations by the way Giorgio Vasari wrote about him in a collection of biographies published in 1550. In a similar way, many British engineers were first immortalized by Samuel Smiles in books which commented on the "dramatic interest," the heroism, and the "noble efforts" of these men. The element of myth-making in all this becomes even clearer when we notice that the study of great engineers or inventors as isolated individuals flies in the face of modern trends in historical scholarship. As one eminent historian of the Industrial Revolution has remarked, it is misleading to present invention "as the achievement of individual genius, and not as a social process."[22] Similarly, to concentrate on "great engineers" may mean ignoring modern research on more representative engineers and on the social history of technology.

When historians do make comments of this kind, one answer they can be given is that the use of symbols and archetypes for understanding many human endeavors is persistent and inevitable; biographers of the great engineers can serve the community by ensuring that the symbols we use are clearly understood. And this present volume does much to enlarge such understanding, for although it accepts the canon we have inherited from the Renaissance and the nineteenth century, it also adds enormously to that canon in the international dimension and in relation to the scientific background of engineering.

But there is another, more serious bias in our accepted appreciation of the "great engineer." The men who are most often celebrated in this way are those whose careers express a daring inventiveness or a quasi-artistic creativity. Only rarely do "great engineers" stand out as men whose lives were dedicated to humanitarian ends, and some great engineers, placed on pedestals in their own lifetimes, became arrogant and careless even of their technical responsibilities. In England, where engineering training was almost nonexistent for much of the nineteenth century, engineers would not always admit to the limitations imposed by their ignorance of theory. Thus ships they built were unseaworthy, sewers caved in, and bridges collapsed due to miscalculation (as did Robert Stephenson's Dee Bridge at Chester in 1847). Even the great Brunel made serious mistakes about the locomotives on his railroad, and regularly saddled his financial backers with projects that were too grandiose and expensive. And when reformers did at length do careful scientific tests on the design of sewers and began to install improved drainage in London, the "great engineers" refused to accept that improvements had been made, and even resorted to sabotage, digging up streets at night in order to tamper with the new drains.[23] The historian who describes these episodes notes that the great railroad engineers of the time were "the folk-heroes of mid-Victorian England," but that they "showed a swashbuckling disdain for the social evils around them."

So at the same time as we recognize the cultural significance of the engineer's endeavors and of the monuments he erects, we should also note that the vocation of the engineer is not simply that of an artist on an heroic scale, building cathedrals, railroads and space vehicles to demonstrate the adventuring spirit of man. Engineering is also a social and humanitarian vocation. Thus the great figures we remember in the history of engineering should also include men who dedicated themselves to reform of the profession, or to humanitarian service. Such men include Georgius Agricola, much-loved physician and sympathetic observer of mine engineering; John Smeaton, who emphasized "civil" engineering as opposed to the military branch; William Strutt who attempted to create a technology of social welfare[24] applicable in hospitals and homes; John Roe, whose careful, detailed experiments in the 1830s and '40s created improved designs of sewers; and Edwin Chadwick, a reformer prepared to challenge the professional arrogance of many nineteenth century engineers in his efforts to create effective forms of public health engineering.[25]

These men are all great figures in the history of engineering, though not all are recognized as such. For it is a strange thing that while so many engineers are remembered for their creativity and daring, few indeed are known primarily as benefactors of mankind. That, rather than any objection from the academic historian, seems the greatest limitation in the popular concept of the "great engineer."

Notes

1. Samuel C. Florman, **The Existential Pleasures of Engineering**, St. Martin's Press, New York, 1976, p. 124.

2. Genesis, 4, 22.

3. Florman, op. cit., p. 105.

4. M. I. Finley, **The World of Odysseus**, Chatto & Windus, London, 1964.

5. 2 Chronicles, 26, 15 and 2 Kings 20, 20.

6. 2 Chronicles, 2, 14.

7. L. Sprague de Camp, **The Ancient Engineers**, Doubleday, New York, 1960.

8. Mark Elvin, **The Pattern of the Chinese Past**, Eyre Methuen, London, 1973, p. 104.

9. Ibid, pp. 180, 184.

10. Joseph Needham, **Science and Civilization in China**, Cambridge University Press, Cambridge, England, 1954-71, Vol. 4 (3), p. 325.

11. John Harvey, **English Medieval Architects: a Biographical Dictionary**, Batsford, London, 1954.

12. This is the castle at Yverdon, on Lake Neuchatel, which was then in Savoy: James took his name from St. Georges d'Esperanche, also in Savoy.

13. R. A. Brown, H. M. Colvin and A. J. Taylor, **The History of the King's Works**, Her Majesty's Stationery Office, London, 1963, Vol. 1, pp. 204-6.

14. Florman, op. cit., page 126.

15. Arnold Pacey, **The Maze of Ingenuity**, MIT Press, Cambridge, Mass., pp. 107-115 (on Galileo) and pp. 123-5 (on his French successors).

16. Jacques Grinevald, "Revolution industrielle ... et Revolutions scientifiques." in **La Fin des Ourils: Technologie et Domination**, Cahiers de l'IUED Geneve, No. 5, Presses Universitaire de France, Paris, pp. 164 and 182-3.

17. David P. Billington, **Robert Maillart's Bridges: The Art of Engineering**, Princeton University Press, 1979.

18. L. T. C. Rolt, **Isambard Kingdom Brunel**, Longman, London, 1957, pp. 5-6.

19. Ibid, pp. 141, 325.

20. Alvin M. Weinberg, "The Impact of Large-scale Science on the United States," *Science*, 134 (1961), p. 161 (21 July 1961).

21. See, for example, the biographical dictionary by Harvey, op. cit.

22. T. S. Ashton, **The Industrial Revolution 1760-1830**, revised edition, Oxford University Press, 1968, p. 11.

23. S. E. Finer, **The Life and Times of Sir Edwin Chadwick**, Methuen, London, 1952, reprinted 1970, pp. 439 and 448-9.

24. Pacey, op. cit., pp. 250, 255-7.

25. Finer, op. cit., see pp. 220-2 on John Roe.

Michelangelo and the Dome of St. Peter's

An Excerpt from **Engineers and Engineering in the Renaissance**
William B. Parsons

The works of Michelangelo the artist are well known. That he was also an engineer of considerable repute is not so well known today, even by engineers. In his time there was no distinction between civil engineer and military engineer. Nor was there a clear delineation between engineer and architect. In the absence of labels that might tend to restrict one to a single field of endeavor as well as barriers that might preclude a person demonstrating genius in a variety of undertakings, Michelangelo employed his talents in all of these areas, producing works that in their own way are as awesome today as his painting and sculpture, although not as widely recognized.

This excerpt from Parsons' classic study of medieval engineering shows aspects of Michelangelo's talents and personality that are somewhat different from the conventional perspective.

There are two buildings that stand out above all others erected during the period of the Renaissance, and whose design and manner of construction place them in the field of higher engineering. Both structures are domes.

The first, that of Santa Maria del Fiore, was described in the preceding chapter. The second, although it is smaller than Santa Maria and inferior to it in boldness of plan and execution, has, on account of its historic importance and striking setting, obtained a far greater hold on popular imagination, if not on popular understanding.

It is not the purpose of this chapter to describe the architectural and decorative features of the second of these great domes, that of St. Peter's at Rome, because these have been exhaustively treated by many authors from both the critical and the historical standpoints. It will, however, aid in considering the dome as an element of engineering construction to give a brief resume of the genesis of the idea, of the successive steps that were taken and of the men who directed them until finally Michelangelo came upon the scene and designed the great culminating member.

St. Peter's was long in building; a century and a half elapsed between the beginning of its construction and its completion as we see it today.

The present church is not the first to stand on the site, an earlier St. Peter's having been erected in the year 334, during the reign of Constantine the Great, and in which the remains of the Apostle St. Peter had been interred. This was a simple structure of the T-cross plan. The exterior length of the main stem was about 370 feet and that of the arms 303 feet. The nave, 213 feet wide between the side walls, was composed of five bays. At the beginning of the Renaissance the old building was in a bad condition, due more probably to decay and insufficient repairs than to faulty original construction, because a building that has successfully endured for more than eleven hundred years can claim to have justified the work of its constructors.

Pope Nicholas V (1447-1455) was ambitious to have a building that would be more in keeping with the new spirit just coming to life than one that marked the time of Constantine. He referred the question of saving the old church to Leon Battista Alberti, the famous engineer, who reported that the walls had become so seriously out of plumb that repairs were inadvisable. On receiving this report Nicholas ordered the demolition of the ancient edifice, the first Christian church erected in Rome. One cannot help feeling that the "modernist" ambition of Nicholas had as much to do with the destruction as the decayed state of the structure itself. He appointed an architect named Rossellino to take charge of the new construction, but it had not progressed much beyond the foundation stage when the death of Nicholas caused the work to be suspended. It was resumed for a while, but only on a small scale, during the pontificate of Paulus II (1464-1471).

On October 21, 1503, Giuliano, Cardinal della Rovere, was elected pope and took the title of Julius II. Julius was an outstanding character and churchman. He was a man of vision, indomitable will and unlimited energy. His ambition was to restore to the Church and the papacy their pristine glory, which had been so sadly tarnished by the scandals of the Borgias, Alexander VI and his brother Cesare. He planned first to construct an outward sign that a new regime in ecclesiastical affairs had been established, and obviously the most conspicuous sign of the change would be an enlarged papal palace and, next, a great cathedral to which the name of St. Peter could be restored. At that time the Vatican was an unworthy residence for a powerful sovereign, and the unfinished building begun nearly fifty years before was a disgrace to the Church. To carry out these projects, or at least to put them on paper, Julius called on Bramante, the leading architect of the day.

Of Bramante personally and of his antecedents little is definitely known. In fact, there is question as to his first name, and neither the place nor date of his birth can be stated with certainty, although for the last two Castel Durante in Urbino and the year 1444 are usually accepted. It appears that he lived in Milan from about 1472 to 1500. During this time he executed with great success plans for churches and other buildings in Milan and neighboring cities. In the year 1500 he established himself in Rome and was, therefore, at hand when Julius II suggested his program of large construction.

Bramante's conception for a new church of St. Peter was brilliantly magnificent. He brushed aside the remains of Rossellino's undertaking in order that something really monumental could be erected that would be both striking in character and thoroughly expressive of the mightiness of the Revival. As a fitting canopy for the tomb of the great Apostle he imagined a great dome. In order that this dome might be the one feature of the building, he decided to rest it on a Greek cross with equal arms. The structure would, therefore, be symmetrical from any point of view, with the dome as the central and dominating member to which all the other parts of the building would be equally subservient.

*   *   *

The dome was circular in plan and semicircular in elevation with a radius of 92.5 palmi (67 feet, 7 inches). Bramante departed from the precedent set by

18

Brunelleschi at Florence*, by adopting the circle, not the octagon, in plan and a full-centered arch in place of a segmental one for the rise. This he did in obedience to his esthetic judgment, but his selection of curves modified to a great extent the development of stresses in the dome . . . There was, in addition, a third difference between the two plans in which Bramante showed himself to be inferior to Brunelleschi as an engineer, however sound an architect he may have been. But Bramante was one of the few architects of this period who did not claim to be also an engineer. Nor was he a painter. He had a true artistic sense but he confined it to the production of buildings.

The third difference was the designing of the shell of the dome as a solid instead of in two parts. Although Bramante was not an engineer, it is singular that he did not appreciate the great constructive advantages offered by the dome of Santa Maria. The dome of the Pantheon was nearer by, and he followed it rather than the dome of Florence as his model.

The drum on which the dome was to rest consisted of a double colonnade, and if placed, as was proposed, in the center of a Greek cross, it would have made a splendid monument.

Bramante died on March 11, 1514. Towards the end of the year 1513, Fra Giocondo, the engineer of the Pont Notre-Dame was appointed assistant architect, and in January, 1514, Giuliano da Sangallo received a similar appointment, because of the advanced years of Giocondo, who died a few months after Bramante. Bramante, on dying, expressed the wish that he should be succeeded as architect-in-chief of St. Peter's by Raphael Sanzio da Urbino, his nephew, a selection which was confirmed officially April 1, 1515.

Bramante was an architect but not an engineer, Raphael was a painter of great repute but neither an architect nor an engineer. In fact, there are only three or four architectural drawings that can be attributed to him. Raphael held the office for just five years, and practically all that he did was to consider substituting a Latin for the Greek cross, a radical change to Bramante's excellent plan. Raphael died in 1520, and Antonio da Sangallo, a nephew of Giuliano, with Baldassare Peruzzi as colleague, was appointed architect of St. Peter's.

Under the administrations of Raphael and Sangallo, the foundations were extended and the four great columns which now support the dome were strengthened, as Bramante had made them of insufficient size to carry so great a load. . . . No work was done on the dome, except that Sangallo prepared a new design.

Sangallo's design differed in several respects from Bramante's. Externally, he placed the colonnades on the drum in two tiers and offset the lower tier so as to act as buttresses. This was an improvement, as it added strength at a point of weakness. Then he omitted the stepped-up rings around the dome, a suggestion that came from the Pantheon, and substituted longitudinal ribs. Within, he changed the semi-circular form . . . This increased the height from spring to lantern opening by about 30 feet. He retained Bramante's

----

* Brunelleschi's precedent was the dome of Santa Maria mentioned earlier.

single shell. . . . Neither Raphael nor Sangallo left specifications as to the materials of which their domes were to consist or a description of the methods to be employed in construction.

Peruzzi's contribution was to insist on the rejection of Raphael's proposal to use a Latin cross for the ground plan and to return to Bramante's plan. This practically eliminated Raphael's influence on the structural design of the great edifice.

Pope Leo X died in 1521 and with him all driving force to complete the church. In fact, a period of absolute stagnation in construction followed. It was to last for many years, as neither of his immediate successors, Hadrianus VI nor Clement VII, had any interest in art. Clement's lack of appreciation is the more remarkable as he was a Medici. But unlike the other members of his great family, he was weak and incompetent. He became embroiled in European politics, was defeated by the Spanish troops who sacked Rome and made him a prisoner. The schisms that followed almost disrupted the Church, and under these conditions no one thought of continuing the construction of St. Peter's.

All was not lost, however. Clement died in 1534, and Paulus III, a man of entirely different mold--strong, resolute and religious--ascended the papal throne. His reign, which lasted until 1540, was marked at the outset by his directing Peruzzi to resume work. Unfortunately, Peruzzi died in 1537 without having accomplished anything of moment, leaving his colleague, Sangallo in charge. The latter remained in office until his death in 1546.

Not only was St. Peter's still unfinished, but no design for its completion had been officially adopted. Pope Paulus then summoned to his aid the only man in Italy who seemed capable of bringing order out of the chaos caused by a delay of forty years under the supervision of many architects who held radically divergent views. Michelangelo was at the time in Rome, engaged on various projects, the most important being the decoration of the Sistine Chapel. Before describing Michelangelo's contribution to the design of St. Peter's, it will be well to give an account of the man himself.

Michelangelo, or Michelangelo Buonarroti, his full name in Italian, was another of those combinations of several types of genius of which the Renaissance produced so many examples. He was a painter, sculptor, architect, poet and engineer, excelling in each capacity, and his engineering work was in the fields both of military and of civil practice, a fact that is frequently overlooked.

The Buonarroti family was of old and highly respectable Florentine stock. Michelangelo, the second of five sons, was born at Caprese in the valley of the Arno a few miles east of Florence, on March 6, 1474, where his father was exercising the functions of the local magistrate. From his childhood his inclination was towards art, both painting and sculpture. When still a boy he was noticed by Lorenzo de Medici, who placed him in the school for the better teaching of artists, of which he was the patron. It is not necessary to follow Michelangelo's subsequent and brilliant career in art: this history is concerned chiefly with what he did in engineering.

In 1503, Julius II was elected pope. One item of his ambitious building program was a gigantic sepulchral monument. Michelangelo's fame had reached Rome, and from what Julius had heard he decided that the young but promising artist possessed the necessary talent to design and construct what he had in mind. Consequently, he summoned Michelangelo to Rome.

This sepulchre was to be no ordinary monument, but something far more grandiose than had yet been attempted. It was to be surmounted with colossal statues which, with other parts of the structure, would demand special stones of unusual size. As this constructive detail was manifestly one of prime importance, and as it presented considerable difficulty in execution,

Julius sent Michelangelo to Carrara to supervise the quarrying of the required blocks of marble and to devise means to transport them to Rome. To raise and move great single masses of stone is in itself a serious mechanical matter, but in this instance there were added the no less serious problems of providing for their transportation to the sea, for their loading on vessels and for their discharge and subsequent rehandling at Rome. This was all successfully accomplished and was probably Michelangelo's first experience in engineering.

After attending to this task Michelangelo returned to Rome only to find that his patron had begun to transfer his interest from a single monument to the completion of a new St. Peter's. In this he was probably influenced by Bramante, whom Julius, as stated above, had appointed Vatican architect. Between Bramante and Michelangelo there was more than ordinary professional jealousy. The former saw in the latter a coming rival who might some day contest his supremacy. Michelangelo on his side was far from tactful. As Julius was domineering, the situation easily developed to the disadvantage of Michelangelo, who, in a fit of temper, suddenly left Rome for Florence. His departure coincided with the ceremony of laying the first stone in Bramante's plan for the rebuilding of St. Peter's, so that some attribute his leaving not so much to a break in his relations with the Pope as to an unwillingness to be present at Bramante's triumph.

But, whatever the cause, no sooner was he gone than Julius regretted his loss and sought to induce him to return. After long-drawn-out negotiations and interviews, the accounts of which are amusing to read, the Pope succeeded.

One of the influencing factors was a suggestion by Bramante that Michelangelo should be given the commission to paint the ceiling of the Sistine Chapel in the Vatican. This was no friendly act; on the contrary, it was a hostile one, because Bramante thought it would involve the complete downfall of Michelangelo, who had had no experience in fresco work.

It is an interesting fact that the task forced on Michelangelo with a view to his undoing became his best-known masterpiece, and a further consequence was that through an unforeseen turn of the wheel of fortune it led to his name being associated with the design of the dome of St. Peter's, rather than Bramante's.

The death of Julius in 1513 brought temporarily to an end Michelangelo's influence in Rome, so he decided to return to Florence with the intention of making that city his permanent residence.

Soon afterward he began his engineering work. In those times any city, particularly one that was wealthy, was almost always exposed to attack, and Florence was no exception to the rule. During the decade of 1520, it became necessary, or at least desirable, to reconstruct and improve the permanent defenses. This work was entrusted to Michelangelo and so well did he do it that the Council of Ten of War gave him complete charge by a commission dated April 6, 1529. This commission reads in part:--

> "Considering the genius and practical attainments of Michelangelo di Lodovico Buonarroti, our citizen, and knowing how excellent he is in architecture, beside his other most singular talents in the liberal arts, by virtue whereof the common consent of men regards him as unsurpassed by any master of our times; and, moreover, being assured that in love and affection towards the country he is the equal of any other good and loyal burgher; bearing in mind too, the labor he has undergone and the diligence he has displayed, gratis and of his free will, in the said work [fortification] up to this day; and wishing to employ his industry and energies to the like effect in future, we of our motion and initiative do appoint him to be governor and procurator-general over the construction and fortification of the city walls, as well as every other sort of defensive operation and munition for the town of Florence."

The salary was fixed at one gold florin ($2.39) per diem. In his new capacity he laid out an extensive plan, the main feature of which was the conversion of the hill of San Miniato into a fort. In the attack on Florence in 1529, San Miniato was subjected to a heavy artillery bombardment, and it is recorded that Michelangelo saved the masonry by covering it with sacks of wool and other mattresses which absorbed the shock of the cannon balls. This is the same device that Leonardo had previously recorded in his note-books. As Leonardo's notes were not published, Michelangelo may have invented the method independently. However successful the engineer may have been in building structures to resist attack, his works could not withstand incompetence in the high military command, and the capture of Florence by the Medicean allies was seen to be inevitable. When Michelangelo realized this, he showed that he could be a wise man as well as a skillful engineer, for under cover of night he succeeded, with two friends, in passing the opposing lines and in taking with him a large sum of money.

After the bitterness of the Florentine war had subsided, Pope Clement VII (1523-1534) invited Michelangelo to return to Rome to continue the unfinished frescoes in the Sistine Chapel. This commission was one of the very few acts of Clement that showed any recognition or appreciation of art. During the pontificate of his successor, Paulus III (1534-1549), war threatened, and Paulus, recalling Michelangelo's success in constructing fortifications for Florence, commanded him to do the same for Rome, even at the cost of interrupting his work on the Sistine Chapel. Antonio de Sangallo was already engaged in the studies for these works and, resenting Michelangelo's being called in even for advice, informed him bluntly "that sculpture and painting were his arts and not fortification." According to

Vasari,[*] Michelangelo replied, "Of sculpture and painting he knew but little, of fortification on the contrary the much he had thought of it, with what he had accomplished, had taught him more than had ever been known by Sangallo and all his house put together."

Michelangelo had also considerable experience in bridge-building. It is reported but not confirmed that in about 1529, when he fled from Florence to Venice, he prepared a plan for a new Rialto bridge. He made a design for a bridge across the Tiber and was directed by Paulus III to repair the foundations of the bridge of Santa Maria, which had been weakened by scour. Michelangelo placed cofferdams in the river, but when he had nearly finished restoring the foundations, the authorities without notification transferred the control to an architect named Nanni di Baccio Bigio. The latter undid much of what Michelangelo had completed and even replaced masonry with gravel. The result was that the bridge collapsed in 1557 during a heavy flood. This incident, as will be seen, had several repercussions, since it made Nanni and his friends bitter enemies of Michelangelo.

During the pontificate of Paulus IV (1555-1559) Michelangelo had another, and this time unopposed, opportunity to plan fortifications for Rome. The Pope, who had embroiled himself in war with Spain, placed Michelangelo in full charge as engineer to reconstruct the defenses. But when in 1557 the enemy advanced, Michelangelo repeated at Rome his previous conduct at Florence and fled secretly from the city to the neighborhood of Spoleto. It can be said in partial defense that at this time he was eighty-four years old, but it certainly seems that however skillful he may have been in constructing means of defense, he was not enthusiastic about sacrificing himself in their service.

Michelangelo died at Rome on February 17, 1563, just before the close of his eighty-ninth year. He was continuously engaged in many fields of art and construction to the very end of his life. He never married. Art, he said, was his wife, and his works were his children. He was buried with great pomp in the church of Santa Croce in Florence.

The salient traits in his character were an insatiable capacity for work, a strange aloofness that kept him from making intimate friends, and unexplainable affectation of poverty and a deep antagonism to all forms of graft. The last trait did not harmonize with the accepted moral standards of the time. His poverty was a pose. When he died 7000 gold ducats are reported to have been found in his bedroom, and the large sum in cash ($16,460) was in addition to investments in Florence. In spite of all this he wrote in various letters: "I have no time to eat." "For the past twelve years I have been ruining my body with fatigue." "I suffer a thousand ills." I live in a state of poverty and suffering." His health was never good, owing in part to his abstemious habits, which, in turn, aggrandized his normal inclination to withdraw from the society of other men. All these traits made him pessimistic, morose and irascible, and were the causes of much of the friction that existed between him and his associates, superiors and assistants.

----

* Giorgio Vasari was a contemporary biographer of Michelangelo.

In Michelangelo, the artist surpassed the engineer. He did not possess the inventive boldness of Brunelleschi, his predecessor, nor the scientific reasoning of Leonardo, his contemporary. Nevertheless, he proved himself to be an engineer through the creation of his greatest piece of construction, the dome of St. Peter's. As Symonds states in his **Life of Michelangelo Buonarroti** (1893): "Michelangelo's reputation not only as an imaginative builder, but also as a practical engineer in architecture, depends in a very large measure upon the cupola of St. Peter's."

It will be remembered that the post of Vatican architect became vacant in the autumn of 1546, through the death of Antonio de Sangallo. On January 1, 1547, Pope Paulus named Michelangelo the "Commissary, prefect, surveyor of the works and architect, with full authority to change the model, form and structure of the church at pleasure, and to dismiss and remove the workingmen and foremen employed upon same."

These were plenary powers, but under the existing circumstances such powers were needed.

Two situations that called for firmness and correct judgment confronted Michelangelo when he took charge. The first was a matter of men, the second a matter of plans.

The animosity towards Michelangelo that had been fostered, if not actually created, by Bramante continued and perhaps was increased by Michelangelo's unfortunate temperament, the asperity of which had been sharpened by the course of time.

For nearly thirty years Antonio da Sangallo had been in control of design and construction, and during this long period he had built up a complete organization whose members owed their appointments to him and who looked to him not only as their technical leader but as their patron. It is hardly likely that this group, so firmly intrenched in power, made any serious effort to refrain from the custom of diverting funds to private ends, a custom that was not strongly disapproved by the superior powers even when they were high officials of the Church. To be freed of the hampering influence of Sangallo, who was dead, and to succeed in carrying out his own plan, Michelangelo had to remove the whole Sangallo entourage and appoint men whom he could trust. This he did under the last sentence in his commission, but his action, while necessary, served to make more bitter the existing hostility. For, although dismissed, the Sangallo adherents carried on organized warfare in which Nanni di Baccio Bigio was a leader. They had great influence with certain cardinals and through them they sought to overthrow Michelangelo, but Pope Paulus faithfully supported him.

The following letter (quoted by Symonds) from Michelangelo to one of his own staff, a superintendent, shows how he was called on to fight the practice of peculation:--

> "You know that I told Balduccio not to send his lime unless it were good. He has sent bad quality, and does not seem to think he will be forced to take it back; which proves that he is in collusion with the person who accepted it. This gives great encouragement to the men I have dismissed for similar transgressions. One who

24

accepts bad goods needed for the fabric, when I have forbidden them, is doing nothing else than make [sic] friends of people whom I have turned into enemies against myself. I believe there will be a new conspiracy."

The making of a plan was surrounded with many difficulties. In spite of the Pope's permission "to change the form and structure of the Church at pleasure," Michelangelo did not have a free hand. He had to accept much of what had been done because it was too extensive to be entirely rejected. In the first place, the plan of a Greek cross had been accepted, and although Raphael had thought of adopting the Latin cross, Sangallo had reverted to Bramante's plan and had laid the foundations. He had also completed the four piers at the crossing, with their connecting arches. After mature consideration, Michelangelo decided to retain the Greek cross as best adapted to a monumental structure and to use the four main piers. The latter decision automatically fixed the diameter of the surmounting dome and at the same time demonstrated the greatness of Michelangelo's character: it showed that he placed the dictates of his artistic and constructive judgment over what must have been a sore temptation to discredit the work of two men who had deeply wronged him. . . .

[The manuscript ended here.]

Although Parsons' book ends here, other sources provide considerable information on St. Peter's dome and Michelangelo's role in its completion. One of the most informative of these is the catalog **Michelangelo Architect** *written by Henry A. Millon and Craig Hugh Smyth to accompany an exhibit of Michelangelo's model of the dome (and related drawings) at the National Gallery of Art in the fall of 1988.*

According to the catalog, work on the base of the dome was completed in 1552 and construction of the drum supporting the dome was initiated in 1554. By 1556 work on the drum was terminated because of the hostilities with Spain, and little was done until 1561. In the interim, Michelangelo was considering an invitation to leave Rome and return to Florence. Because of the slow progress on the dome and the possibility that he might leave (and probably also because of his advanced age), his friends requested that he prepare a model of the drum and dome. In 1557 he completed a terra cotta model. A larger scale wood model was started in November, 1558, and was not completed until November, 1561. This wood model is the model exhibited at the National Gallery of Art in 1988.

According to the authors of the catalog most experts agree that Michelangelo envisioned a hemispherical dome, although there is evidence that he considered an elevated dome (a dome that is higher than a hemispherical dome of the same diameter). Drawings by Michelangelo (which were included in the exhibit) and a description of Michelangelo's model written by Vasari shortly after Michelangelo's death indicate that the model originally depicted a hemispherical dome. However, the model now has an elevated dome, and the dome, as constructed, is elevated.

Work on the drum supporting the dome was almost complete at the time of Michelangelo's death in 1563 but work on the dome itself had not yet started. Work on the drum continued for five years under two successors to

Michelangelo until in 1568 all work was suspended. After a twenty-year period during which nothing further was done, Giacomo Della Porta was appointed architect. Under his supervision the dome was completed in 1593.

Vatican records suggest that Della Porta may have worked with Michelangelo at some time, but there is no supporting evidence that this is so. Similarly, there is no indication that Michelangelo reconsidered the shape of the dome after the model was completed, despite the evidence that he considered an elevated dome as an alternative to the hemispherical shape before initiating work on the model. Hence, one is led to the conclusion that the shape of the dome as we see it today reflects a modification of Michelangelo's design--a change in shape from hemispherical to elevated--and Della Porta is generally believed to be responsible for the change. But the engineering of the dome, its conformance to the existing piers and the design and construction of the supporting drum are unquestionably the work of the master engineer, Michelangelo.

In 1694 Italian architect Carlo Fontana wrote: "Everyone consequently ceased to celebrate the most renowned buildings, both ancient and modern, in Rome and elsewhere in the world, saving praise only for the vast construction of the dome of the unsurpassable Vatican Temple. They accorded greater praise to the excellent merit of [Michelangelo] Buonarroti, who gave birth to a marvelous work, indeed stupendous. Thus, that great builder will deserve to live eternally, having demonstrated a more than human genius in the invention of a building that shows such miraculous understanding, never before equalled in the world." Michelangelo, we now know, will live forever--not for his works of engineering as Fortana believed, but for his works of art. He should, however, live forever in the memories of civil engineers as one of their most illustrious professional ancestors.

Taking the Measure of a New Nation

Civil Engineering Magazine - May and June, 1967
Neal FitzSimons

This selection includes the last two articles from a three-article series published in Civil Engineering magazine as part of Mr. FitzSimons' series of columns on civil engineering history. The series continued over a period of several years, providing civil engineers with a wealth of information on the history and heritage of their profession.

In modern times we seldom think about the important role played by the civil engineer/surveyor in the early days of the nation's history. The fact that states had an office of Surveyor General, unheard of in modern times, is an indication of the importance of this function to early governments. When almost everything was wilderness and when there were virtually no maps, the person who could measure a tract of land and describe and document its boundaries was an individual whose talents were not only useful but absolutely essential if there was to be orderly development of uncharted regions.

The transition of American surveying from the frontier art of the early 18th Century to the sophisticated engineering science of the new republic was largely due to the efforts of David Rittenhouse (1732-1796). The son of a Pennsylvania farmer of modest means, he had very little formal schooling. His brilliant mind and tenacious spirit, however, enabled him to advance himself intellectually so that before he was twenty he is said to have mastered Newton's *Principia*. He began his surveying career on the oft-disputed boundary between the holdings of Lord Baltimore and the Penns. His work was accepted and incorporated into the survey of Mason and Dixon.* Much of Rittenhouse's life was devoted to astronomy, mathematics and instrument making; and although his contributions to surveying were great, he probably would be best characterized as an astronomer. He did, however, run accurate boundary surveys in every state from Massachusetts to Virginia and conducted some of the early canal surveys. David Rittenhouse and his brother, Benjamin (1740-1820, also had an important impact on American surveying as makers of high-quality instruments. Several of Andrew Ellicott's surveys were conducted with these instruments, and George Washington himself is believed to have used a Rittenhouse surveying compass when he made his last complete survey of Mount Vernon.

A colorful contemporary of Rittenhouse was the Dutch-American engineer Bernard Romans (1720-1784). Trained in Britain, Romans came to the Southern colonies at the time of the French and Indian Wars. For more than a decade he explored, surveyed and botanically investigated the area from South Carolina to Louisiana. When the Revolution broke out, Romans

----

* The Mason-Dixon Line.

** Ellicott, together with the French engineer, Pierre L'Enfant, laid out the city of Washington in the newly established District of Columbia.

happened to be in the North and immediately offered his services to the rebel cause. While David Rittenhouse was helping the Continental Army by supervising the manufacture of munitions, Romans was engaged in constructing fortifications. After serving with General Gates in New York, he took an extended leave in Wethersfield, Connecticut, where he married and his son was born. Recalled to active duty in 1780, Romans was sent to join [General Nathanael] Greene's Southern army, but unfortunately his ship was captured and he spent the rest of the war a prisoner in the West Indies. He died at sea en route home soon after the war was over.

An interesting sidelight to the surveying history of this period is the career of Roger Sherman (1721-1793) who began his public life as a county surveyor in Connecticut from 1745 to 1758. A pillar of the new republic, Sherman was the only person whose signature appears on all four of our founding documents-- the Articles of Association (1774), the Declaration of Independence (1776), the Articles of Confederation (1781) and the Constitution (1787). Also among the signers of the Declaration of Independence was Abraham Clark (1726-1794) of New Jersey, who began his public career as a surveyor but later became engrossed in politics.

Another surveyor of historical interest was Thomas Marshall (1730-1802), colleague of George Washington on the Fairfax surveys and father of John Marshall (1755-1835), the famed Chief Justice of the United States. A pair of frontier surveyors who, like Thomas Marshall, continued to use the crude techniques of their colonial predecessors, was the father-son team of John Jenkins Sr. (1728-1785) and Jr. (1751-1827). Typical of their kind, they spent as much time fighting Indians as they did chaining in the wilds of the Wyoming Valley of northeastern Pennsylvania, then claimed by the state of Connecticut.

About a three-day ride to the south was the Montgomery County (Pennsylvania) birthplace of Andrew Porter (1743-1813), a gallant soldier in the Revolutionary Army. Unlike the Jenkins team, Porter was a surveyor of the "Rittenhouse school" and did precision work in running the boundaries of his home state. Later (1800) he was commissioned to settle the Wyoming Valley claims and made State Surveyor-General in 1809.

A topographical surveyor whose untimely death occurred just before the birth of the new republic was the Scotsman, Robert Erskine (1735-1780). Educated in Scotland, he spend most of his life in England. Although not successful in business, Erskine did make contributions to hydraulic engineering significant enough to gain his election as a Fellow of the Royal Society in 1771. About that time he was sent to manage the mines of the American Iron Company in northern New Jersey. Soon after the outbreak of the Revolution Erskine joined the local militia, but upon learning of his skills in engineering, Washington personally asked him to serve as Geographer and Surveyor-General of the Continental Army. For the next three years, Erskine made topographic maps of the forested Hudson-Highlands area, which were the

----

* The device surveyors use for measuring linear distances is called a chain.

basis of the successful defense plans (including the fortifications at West Point by Kosciuszko) that maintained American control of the region and prevented the isolation of New England.

Erskine's successor, who bore the title "Geographer to the United States" was the remarkable patriot, Thomas Hutchins (1730-1789). Born in New Jersey, he began his military career in the Pennsylvania colonial troops during the French and Indian War, later gaining a regular commission in the British Army. As a British engineer officer, he traveled extensively in the Colonies engaging in making maps and fortifications. For his geographical descriptions of the Middle Atlantic and Southern colonies, he was elected to the American Philosophical Society. In 1789, when the personable scoundrel Abel Buell compiled and published the first American map of the United States, he relied greatly on the work of Hutchins. When the Revolution erupted, Captain Hutchins was in London where, after declining a major's commission, he was imprisoned for treason. In 1780 he was released from prison and made his way to America and, the following year, was appointed to Erskine's position. When peace returned, Hutchins continued this work as a civilian and was the prime mover in the surveys resulting from the famous Land Ordinance of 1785. He was responsible for the survey of the entire Northwest Territory, for which he was paid about $2.00 per mile. This famous survey, with the "Seven Ranges of Townships," based on a "Geographer's Line," was the foundation of the Public Land System in the United States.

Thomas Hutchins died on April 28, 1789. Two days later in New York, George Washington (1732-1799) took office as president of the United States under the new Constitution. Hutchins' work in surveying the old Northwest Territory had, perhaps, been the most significant project under the [Articles of] Confederation. Now, a new project arose--the establishment of the Constitutional capital, the city of Washington. A political agreement between [Alexander] Hamilton and [Thomas] Jefferson moved the capital from New York City to Philadelphia in 1790 for a period of ten years, after which it would be located in the 100-square-mile tract near the confluence of the Potomac River and its eastern branch, the Anacostia River.

Washington himself is credited with the general location of the capital, but Andrew Ellicott was the first man on the job. In February, 1791, at Washington's suggestion, Secretary of State Jefferson requested Ellicott to "proceed by the first stage to the Federal Territory on the Potomac for the purpose of making a survey of it." The following months Jefferson (apparently again at Washington's suggestion) wrote the peppery French military engineer, Major Pierre L'Enfant (1754-1825). He asked the Major's aid ". . . to have drawings of the particular grounds most likely to be approved for the site of the Federal Town and buildings . . . connecting the whole [from Georgetown to the eastern branch] with certain fixed points on the map Mr. Ellicott is preparing."

By March 26, the Gazette of the United States announced, "Mr. Ellicott and Major L'Enfant are now engaged in layout of the ground on the Patowmac (sic) on which the Federal buildings are to be erected." This team must have been very busy indeed for less than nine months later L'Enfant had submitted his plan for the city to Congress even though Washington appeared to have had some misgivings about it. The plan was rebuffed by Congress and returned to L'Enfant, who refused to permit any further use of it. When L'Enfant was dismissed by Jefferson the following March (1792) after repeated antagonisms with the three Commissioners (General Thomas Johnson, Daniel Carroll Esq., and Dr. David Stuart, Washington's physician), it remained for Ellicott to finish the job. Relying on his knowledge of L'Enfant's work and adding some of his own ideas, Ellicott prepared a revised plan which, with Washington's approbation, was accepted by Congress the following spring. It seems, however, that the Commissioners also rankled Ellicott and he was happy to leave the final details of the project to his assistants, among them his two brothers Joseph and Benjamin.

Andrew had one other noteworthy assistant whose name was Benjamin Banneker (1734-1806). Until he was, perhaps, fifty years old, this son of an African slave farmed in the area around Baltimore. His bright mind and scholarly disposition came to the attention of the Ellicott family who had opened their mills on the Patapsco River just outside Baltimore. With their encouragement and help (especially from Joseph), Banneker became an accomplished surveyor. This is how he came to work with Andrew Ellicott on the National Capital. During this same period, Banneker had, largely through his own calculations, produced an astronomical almanac for the Pennsylvania, Maryland, Delaware and Virginia area. It was so well received by the Ellicotts that their cousin, George, aided its publication. This remarkable work, which was annually revised by Banneker until 1802, received such international attention that his death was noted by the French Academy, the English Parliament, and distinguished men like Jefferson in the United States.

Isaac Roberdean

Isaac Roberdean (1763-1829) was another famous surveyor and engineer who began his career working with Ellicott and L'Enfant on the first surveys of Washington City. The son of a Revolutionary patriot, Roberdean was born and raised in Philadelphia. During his early twenties he traveled extensively and studied engineering in London. A few years after returning to the United States, Roberdean joined L'Enfant in Washington and upon his [L'Enfant's] dismissal followed him to Paterson, New Jersey. It was there that Alexander Hamilton planned to have the ideal American Industrial City (N.B. Historically, Hamilton favored an industrial-bank economy for the United States while his political rival, Jefferson, sought to maintain an agrarian nation.)

30

When L'Enfant left Paterson to work on Fort Miflin in the Delaware River, authorized under the Fortifications Act of 1794, Roberdean returned to his home city and opened his own engineering practice. His major work at that time was the famous canal scheme to connect the Schuykill [River] with the Susquehanna River. First proposed by William Penn (1644-1718) about 1690, the route was later surveyed by the great David Rittenhouse and the vociferous Loyalist, Dr. William Smith (1727-1803). With Roberdean's help, the canal progressed about 15 miles before its funds were exhausted. It was not until 1828, the year before Roberdean's death, that the Union Canal Company finally completed this project.

About ten months after the opening of the War of 1812, the United States Army established the Corps of Topographical Engineers. Among the first four majors appointed to the Corps (this was the highest rank authorized) was Isaac Roberdean. Through a Congressional oversight the Corps was dis-established in 1815, but Majors Roberdean and Anderson were retained on active duty to complete surveys they had started on the northern frontier and Lake Champlain. Later when the Corps was re-established (1816), such outstanding men as John Abist, James Kearney and Stephen Long became associated with Roberdean in surveys for seacoast fortifications and later, for the famous explorations of the West.

In 1818, Roberdean was appointed to head the newly established "Topographic Bureau" and he moved back to Washington, D.C. It was during his tenure that the famous Board of Internal Improvements was established, and much of the survey work for the country's expanding canal and road network was done by the military and civilian engineers under his overall direction.

Alden Partridge, the first Professor of Engineering in the United States

Claudius Crozet, Professor of Engineering at West Point 1816-1823

The First School Of Engineering
Sidney Forman

The Military Engineer - March-April 1952

The U. S. Military Academy at West Point was the first school of engineering in the English-speaking world. This article from The Military Engineer, the publication of the Society of American Military Engineers, recounts some of the early history of the academy. It also details the influential role the academy and its graduates had on the development of the civil engineering profession in the United States.

When the United States Military Academy was founded by Act of Congress on March 16, 1802, its chief purpose was to train young men in ordnance and artillery manufacture and in engineering. The engineer of that period was expected to be qualified in both the military and civil applications of his science.

An indication of how the word engineer was defined may be seen in Alexander Hamilton's proposal that the Military Academy teach "the principles of construction, with particular reference to aqueducts, canals and bridges." The Secretary of War, James McHenry, in advising the Congress as to the advantages of establishing a military academy, wrote of both civil and military training:

> We must not conclude . . . that the service of the engineer
> is limited to constructing, connecting, consolidating, and
> keeping in repair fortifications. This is but a single
> branch of their profession, though, indeed, a most important
> one. Their utility extends to almost every department of
> war, and every description of general officers, besides
> embracing whatever respects public buildings, roads, bridges
> canals, and all such works of a civil nature.

The law establishing the Military Academy was not very specific as to the details of academic organization. In the wording of the act, "That the said Corps of Engineers, when so organized, shall be stationed at West Point in the State of New York, and shall constitute a military academy," the Congress seems to have envisioned some kind of apprentice school or classical college in which the engineer officers were the masters, or teachers, and the cadets were the apprentices or students. A curriculum was not specified in the law. Nevertheless, it was a school of, by, and for the engineers.

The level of training was dependent on the conscientiousness and capability of the superintendent and the assigned instructional staff. As principal engineer in the Corps of Engineers when Congress established the Military Academy, Jonathan Williams automatically became the first academy superintendent. Born in Boston on May 26, 1750, Williams was educated there and at Harvard. During the Revolution, he served as a representative of the American Government in France, where he met Thomas Jefferson in 1784. A grand-nephew of Benjamin Franklin, Williams expressed an early interest in science and in military art, and had helped the War Department by translating French works on artillery and fortifications.

33

West Point

As soon as the cadets reported for duty, Williams issued books, organized classes, and established a routine for his staff. He gave lectures on fortifications and led the cadets in surveys of the country around West Point, training them in the use of surveying instruments.

Williams' assistants were Captain William A. Barron and Captain Jared Mansfield. Mansfield was well-equipped for the assignment of teaching geometry and also served as acting professor of natural and experimental philosophy. He had been a teacher of mathematics, navigation and the classics at Yale University, and in Philadelphia. He had also written a volume of original *Essays, Mathematical and Physical*. The bulk of his studies dealt with astronomy and the theory of gunnery, subjects of basic importance to the academy. Captain Barron, a Harvard graduate who had tutored in mathematics at Cambridge, taught algebra.

The texts used were new in America. The cadets were given Williams' translation of H. O. de Scheel's *Treatise of Artillery* and Charles Hutton's *Mathematics*, a two-volume work especially designed for the Royal Military Academy at Woolrich (*ed.: the British Military Academy*). The title of the last-named work was not at all descriptive of the contents which included a compilation of mathematics and natural philosophy with a considerable section on surveying, practical exercises in mechanics, statics, sound, motion, use of the ballistic pendulum, and the application of mensuration to artificer's work such as bricklaying masonry, tiling, plastering and painting. The cadets also received William Enfield's *Natural Philosophy*, a full elementary course which was used also at Harvard, and Vauban's classic *Trait de Fortifications* in the original French.

34

Teachers of French and of drawing were appointed according to an act of February 28, 1803. These subjects were important additions to the original courses at the Academy. Francis D. Masson, appointed as teacher of French on July 12, 1803, was a scholar in all fields of engineering and was engaged in translating into English "all that is known in Europe" on the subject of engineering. Superintendent Jonathan Williams soon realized that the curriculum, heavily weighted in the favor of mathematics, was too narrow in scope for an engineering school, and in 1808 he was instrumental in having Masson assigned as temporary professor of engineering. Christian E. Zoeller, a Swiss, replaced Masson as teacher of drawing in September, 1808. The instruction offered by Zoeller was largely confined to drawing military fortifications and bridges, and to practical surveying.

Professor Barron resigned in 1807, and was succeeded by Ferdinand R. Hassler, a Swiss by birth, with an excellent reputation as a mathematician. He remained at West Point until 1809. Hassler had studied geodesy and mathematics at the University of Bern. He introduced analytical trigonometry into the West Point course--the first time the subject was taught in this country. Alden Partridge, a Dartmouth graduate, who was commissioned from West Point in 1806, was appointed assistant professor of mathematics. These men made up the faculty which, in the first ten years of the academy's existence, educated eighty-nine graduates of whom at least 10 per cent were engaged in civil engineering work after graduation from West Point.

The threat of war in 1812 forced expansion of the Army, the Corps of Engineers, and the Military Academy. The law of April 29, 1812 increased the corps of cadets to two hundred and fifty and authorized a full academic staff, including "one professor of engineering in all its branches," the first such professorship in any institution of higher education in the United States. Secretary of War Eustis offered the position to Pierre Charles L'Enfant, the planner of Washington, D. C. After L'Enfant declined. Alden Partridge was appointed to the new professorship. He served as professor of engineering from September 1, 1813 to December 31, 1816. Partridge concentrated his instruction on the building of fortifications.

Professor Claudius Crozet, who had been assistant professor of engineering since October 1, 1816, followed Captain Partridge into the professorship on March 16, 1817. A Frenchman, Crozet had received his education at the Ecole Polytechnique. He introduced descriptive geometry into the West Point curriculum as a necessary preliminary to the proper study of engineering, and made other improvements in the course of study. By 1819, the course based on Gay de Vernon's Ecole Polytechnique text, *A Treatise on the Science of War and Fortification* in the translation prepared for the academy by Captain John M. O'Connor, was supplemented by a study of building construction, arches, canals, bridges, and other public works.

By 1820, Crozet had developed the course to the extent that the Military Academy at West Point was the only institution in the United States:

> where a complete system of instruction in mathematics, descriptive geometry, mechanics, astronomy, and civil and military architecture (*civil architecture and civil engineering were then used synonymously*) is procured and which afterwards places the pupils in situations, where they

have an opportunity of exercising the different civil and
military avocations, to which the application of the exact
sciences is indispensable.

Crozet introduced new textbooks in the original French, *Program D'un Course de Construction* by Sganzin and *Trait des Machines* by Hachette, and published his own *Treatise on Descriptive Geometry* "for the Use of the Cadets," the first English work of importance on the subject. Perhaps his greatest contribution was his introduction in 1823 of M. I. Sganzin's *An Elementary Course of Civil Engineering*, an Ecole Polytechnique text which dealt with construction materials, canals, roads, bridges, railways and harbors. The Board of Visitors for 1824 reported that the cadets also studied the construction of "public edifices." Because of the severity of the climate, Crozet resigned on April 28, 1823, to continue his career in Virginia as a civil engineer, and at the Virginia Military Institute as a teacher.

David B. Douglass, who succeeded Crozet, extended and further improved the course in civil engineering during his incumbency from 1823 to 1831, and changed his title from professor of engineering to professor of civil and military engineering in 1827. Professor Douglass introduced his practical engineering experience into his lectures, and as the utility of his course increased, more and more graduates were assigned to the non-military engineering projects. The achievement of the graduates in this field led the President, John Quincy Adams, to say in his First Annual Message to Congress (December 6, 1825):

> The Military Academy . . . recommends itself more and more
> to the patronage of the nation, and the numbers of
> meritorious officers which it forms and introduces to the
> public service furnishes the means of multiplying the
> undertakings of public improvements to which their
> acquirements at that institution are particularly adapted.

Successive Boards of Visitors also expressed their approval and praise. George Ticknor, a Harvard professor serving on the West Point Board of Visitors in 1826, attended an examination in civil engineering and wrote in a letter to his wife: "The examination of the upper class in civil engineering was very beautiful, and I think I picked up some knowledge of roadmaking, canals, tunneling, et cetera, which I should be sorry to forget." The Board of 1830 fully accepted the idea that "a large portion of the cadets are destined to act as civil engineers . . . capable of giving wholesome direction to the spirit of enterprise which pervades our country."

The library at the academy also reflected the heightened interest in civil engineering. Under the heading of "Engineering and Fortification" the first library catalog of 1822 lists 64 titles; under "Architecture, Bridges, Canals, Perspective and Topography" there are 59 titles. There are 729 titles in all. The library included Belidor's *Architecture Hydraulique*, a widely used four-volume text on hydraulic engineering; the writings of Perronet who perfected the stone masonry arch in the Pont de la Concorde at Paris; and of John Smeaton, builder of the Eddystone Lighthouse, who first used the title of civil engineer. Of course, Vauban was represented.

By 1830, the Military Academy had the largest engineering library in the United States. Approximately 692 titles dealing with various phases of engineering were then known to Americans. The catalog of the academy library of that year listed 431 engineering works; 123 in the military section, and 308 in the civil engineering list. Included in the latter group were works in Latin, Italian, German, French, and English. Among the English works were the publications of such contemporary engineering giants as Telford, Rennie, Tredgold and Macadam.

Professor Dennis Mahan

Civil engineering reached its most extensive development under Dennis Hart Mahan, a graduate of the academy in the class of 1824. A four-year tour of professional duty in Europe, where he studied public works and military institutions, established Mahan as an authority in this field. He spent more than a year at the Military School of Application for Engineers and Artillerists at Metz. When he returned to the Military Academy as acting professor of civil and military engineering, on September 1, 1830, he expanded the course and prepared a complete set of new textbooks. Instruction in the department under his direction included the properties, preparations and use of materials for construction; elementary parts of buildings and the art of construction generally, including decorative architecture; the manner of laying out and constructing roads; construction of various kinds of bridges; general principles which regulate the removal of obstructions that impede the navigation of rivers; and survey, location, and construction of canals and railroads; and the formation of artificial harbors and the improvement of natural harbors.

The description of Mahan's course of civil engineering, in the report of the Board of Visitors for 1835, significantly preceded the details of the course in military engineering. Clearly, in the eighteen thirties, civil engineering dominated the curriculum. Parents encouraged their sons at West Point to qualify as engineers, and some of the cadets came to the academy solely for that purpose. The class of 1832 formed themselves into a society for the purpose of improving their knowledge of civil engineering. The group held meetings every other Saturday evening to read essays on the subject and enjoyed the extra privilege of using the library two hours each day.

Professor Mahan lectured from notes gathered while touring France and Italy. From his notes and continued study, he prepared a series of lithographic texts which were published at the academy for the use of the cadets. His most important work was the *Course of Civil Engineering*, first published in 1837 and continually enlarged and improved. It found widespread use in America, was reproduced in England, and was translated, in whole or in part, into several foreign languages.

Francis Weyland, President of Brown University and student of American higher education at the time that the Military Academy was so largely concerned with civil engineering, recognized West Point's role in technical education and reported in 1850, that of the more than one hundred twenty colleges in the United States, the academy did more "to build up the system of internal improvements in the United States than all the colleges combined."

Among the academy graduates who won distinction in civil engineering was the first graduate of 1802, Joseph G. Swift; Joseph G. Totten (Class of 1805) was a leading technician in river and harbor improvement; James J. Abert (Class of 1811), headed the Topographical Engineers for more than thirty-two years.* William G. McNeill, (Class of 1817) and George W. Whistler (Class of 1819), achieved even greater renown in canal and road construction. There were few works undertaken before 1850 in connection with which their names did not appear. George S. Greene (Class of 1823), who worked on the Croton Water Supply of New York City, was consulting engineer for many important municipal enterprises. The wings and the dome of the National Capitol were built under the superintendence of Montgomery C. Meigs (Class of 1836), also of Washington Aqueduct fame. When the American Society of Civil Engineers was formed in 1852, Military Academy graduates James Barnes, George S. Greene, William W. Morris, and William H. Sidell were listed as charter members. Four of the six honorary members on the same list were West Point graduates: John J. Abert, Alexander D. Bache, Dennis H. Mahan, and Joseph G. Totten.

By 1870 there were nineteen technical schools in the United States, of which at least ten had direct West Point pedagogical affiliation. For example, the Columbia University School of Mines called upon three West Point graduates: Professors Francis L. Vinton, William G. Peck, and later Petit W. Trowbridge, to give it "a thorough mathematical foundation for engineering studies characteristic of West Point."

Before the close of the century more than forty-five schools were giving degrees in engineering, and their ever-increasing number of graduates, well-endowed laboratories, advanced courses, and graduate schools indicated the great progress made since the foundation of the first American school of engineering.

----

* During the first half of the nineteenth century the Corps of Topograph-ical Engineers and the Corps of Engineers were separate branches of the Army).

38

Early Life
John B. Jervis

An Excerpt from The Reminiscences of John B. Jervis
Edited by Neal FitzSimons

John Jervis was one of the most prominent and most versatile of the nineteenth century American civil engineers. He was involved in the planning, design, construction and operation of canals, railroads and aqueducts over a period of more than fifty years. After he had retired from the active practice of civil engineering he prepared a series of reflections on his career. Mr. FitzSimons collected and edited these reminiscences for publication. This selection describes Jervis' introduction to the civil engineering career field.

At one time I entertained the idea of obtaining an education of a higher order than I could have in common school. But my father was not able to help me, and in fact, he needed the service I could render him. I thought to study and work at the same time, and made arrangement with an educated teacher to hear my recitations, and began the study of Latin. The time I had to study was-- after setting the log, I could study until admonished of the time to reset the log. After making what I considered a fair experiment in this method of obtaining an education, I gave it up. The demands of the work I had to do absorbed so much of my time and thought that I felt compelled to give up the attempt.

I may here remark that I regard it a great error to bring up a boy without giving him the training of a regular vocation. But my home was pleasant, and my father needed my services; and though he manifested anxiety in regard to my future, he could not overcome the urgency of my usefulness with him.

As an instance of the times, I mention that in the winter season I engaged to transport a load of wheat for a merchant in town from Rome to Albany [about one hundred and ten miles]. It was good sleighing until I passed Schenectady when I found the snow rather light. I took with me my provisions and oats for my horses. It required seven-days' journey. I received for the service about eleven dollars. My expenses were about four dollars, besides provisions and oats for my horses. The latter at that day were cheap, worth together about three dollars, leaving me for myself and team, about four dollars for seven-days' work. This is the way we transported in those days, a circumstance that impressed upon me the importance of the canal, which soon after was presented to the public mind.

Though but a youth at that time, I regarded the canal project with much interest. I had no definite idea as to whether it would pay for construction, but somehow I thought it a great work. I had not the least idea of anything personal, more than to supply such demands as it might make for lumber. The subject was very warmly discussed, and many intelligent men regarded it as a measure that would involve the state in inevitable pecuniary ruin.

In 1816 the state made an appropriation for an instrumental survey and estimate of the Erie and also the Champlain Canal. With some previous surveys the estimates were made by the Honorable Benjamin Wright[4] of Rome, the Honorable James Geddes[5] of Onondaga, and John Broadhead[6] Esq. of Utica for the Erie Canal, and by Mr. G. Lewis Garin for the Champlain Canal. With the information furnished by these engineers, the state entered on the construction of the works. Mr. Wright was appointed chief engineer of the Erie and Mr. Geddes of the Champlain Canal.

Several sections of the canal between Rome and Utica were put under contract, and the formal ceremony of breaking ground occurred on July 4, 1817. There had been a difference of opinion on the question of location through Rome, in consequence of which it was delayed until late in the autumn. The general principle of location having been settled, a party of engineers was sent on to make the survey through the Rome swamp.

Mr. Wright, the chief engineer, as above noticed, resided in Rome. When the party came to make the survey, it was composed of only the experts, depending upon obtaining axemen at the locality. These latter were an important adjunct in carrying a line through a dense cedar swamp. To supply this defect in the engineering party, Mr. Wright, the chief engineer, knowing that my father had in his employ men working as axemen, called upon him to engage two for a few days on this service. My father referred the matter to me; he was induced to do this from an intention of making proposals for a contract for this section of the canal, and thought I would be able to obtain useful information as to the character of the work. I assented to his proposition to go as one of the axemen myself, taking with me one of my father's men.

Having been brought up in a new country, I had acquired a very handy use of the axe, and my associate was a first-rate chopper. As a brace of axemen, we entered on this new field with good spirits, not doubting we could do anything required in this line; and beyond this I had no thoughts except to notice the prospect of a contract for this swamp section of about two miles. We had had no experience in cutting lines for such a purpose. The cedar brush was very thick, and the ground very soft, and to keep on a line was a thing we had not practiced. But we soon got the run of the work, and having myself a pretty good eye for a line, we worked our way in what appeared a satisfactory method. The swamp was so soft, it was necessary to drive stakes to set the leveling instrument on. In our progress (which was in November) a fall of snow occurred, loading the cedar brush so as to render the work for the axemen very disagreeable. It was necessary that they should go forward, and by cutting shake the snow from the brush and trees, and so open a path for the engineers.

I was of a slender frame, rarely weighing over one hundred and twenty-five pounds; my associate was tall and stout, and we pushed through the brush and snow with the zeal of a new enterprise. Our main ambition was to satisfy the chief of the party, N.S. Roberts Esq.,[7] who was a man of austere manners who did not hesitate to speak plainly.

Though my occupation in this service was that of an axeman in which I exerted myself to cut stakes, pegs, and line in a satisfactory manner, I had not proceeded long before my attention was attracted to the operations of the men using the instruments. These instruments and the method of using them appeared to me a profound mystery, and I at first had no idea, considering my education, that I could learn their mysteries. The two target men, for whom I drove the pegs on which they rested their instrument for an observation from the level, came often to my position, and after driving their pegs, I had some leisure until the observation was made to notice their manipulations. On such occasions I was led, at first from mere curiosity, to watch their movements and to ask some questions, which I did with great modesty, as they quite outranked me in the service. After a time, as I became more familiar, and, having some approving words for my expertness as an axeman, I ventured at some leisure time to ask the privilege of taking the target and of examining its movements, which was kindly conceded. After a day or two, I ventured to ask some questions in relation to this mysterious instrument and the mode of its operation. I learned in these conversations that the target man was regarded as occupying the first step in the science of engineering.[8] On reflection it appeared to me I could do that service if I could have a little practice.

The chief of the party, Mr. Roberts, was, as I have observed, a man of austere manners, and the party generally regarded him with a reverence that did not allow familiarity. In the low rank of my place in the party, I did not feel at liberty to ask of my principal any questions, and all my communications were with the target men, who were both young. For a time I did not see my way clear to bring out the working of my own mind; but by and by the principal seemed to unbend a bit, and on an occasion expressed his approbation of our dexterity as axemen, and very frequently stated that he had never had a pair of axemen so skillful and efficient as we were; and at the same time took occasion to manifest a partiality to men who were not afraid to work. This encouraged me to think that the same application of industry might enable me to master an art that seemed then so mysterious. Gathering strength by reflection, the idea I had long indulged so pressed on my mind that on the last day of this service, while the party gathered in a huddle in the swamp and were eating our dinners, I mustered courage to make a proposition, partly in joke and partly in earnest. I put it in the modest form of a question to Mr. Roberts: "What will you give me to go with you next summer and carry one of those targets?" He very promptly replied, "Twelve dollars per month."[9] I as promptly assented and said I would go.

As this seemed to settle the question, I began to consider seriously what it was that I had given my assent to. It was no object to engage my services for twelve dollars per month. I must see something beyond this. It was natural the question should arise, could I ever become an engineer? The thing appeared a mystery to me, of which I had only seen a very small edge of the great field. In this course of reflection I considered first that I only had a common school education with no view to the science of engineering. So far I had only looked at the target; the mystery of the level, the taking of sights, its adjustment, and the computations of the observations were all dark to me. As I pondered these matters, I began to think that after all, it might be a mere joke and that Mr. Roberts had no idea of my accepting his prompt assent. But the idea held fast to my mind, and finally I determined to ascertain if the parties were in earnest. I say parties, for Judge Wright[10] was the principal, and I had intimated to him Mr. Roberts' acceptance, to which he did not object. Having a personal acquaintance with Judge Wright, the chief engineer, after reflecting several weeks on the subject, I called on him and requested

41

him to inform me if they certainly accepted my proposition. He replied very promptly, "If you say you will go, I say you shall go."

Now a new occupation suddenly burst on my view. How shall I prepare for it? Only a few months and I shall be called upon to enter a party, all probably better educated than myself. I had no fancy for appearing more ignorant than those engaged in the same field. The whole aspect appeared to me a profound mystery. But I had voluntarily put on the harness, and had faith to believe that what others had learned, I could learn.

On inquiry I found that land surveyors were considered best fitted for engineers. So, at the recommendation of Mr. Roberts, I purchased two books on surveying.[11] These I studied in the evenings, and at such odd times as the weather did not allow team work, until 10th of April 1818, when I started with my target on my shoulder, with a party of about a dozen men for Geddesburg (now Syracuse). Except N. S. Roberts, engineer, the principal of the party, we all made the journey on foot. A baggage wagon accompanied us, with our baggage and camp furniture. We started in the afternoon and only made seven miles that day. The roads were excessively muddy, and we reached Geddesburg in the afternoon of the third day and then pitched our tents among hemlock logs and bushes. On the journey, I occupied such opportunity as I found in studying the target, making myself familiar with the movements, as I had seen in the swamp, and especially with the reading of the figures. The target rod used at that time was ten and a half feet long.[12]

My neighbors at home laughed at my venture, and predicted I would soon return home to sleep.

On arriving at Geddesburg, the party immediately organized for work, Mr. Roberts having arrived about the same time and joined the camp. The party then proceeded on the work of locating the canal from this point to Montezuma[13] on the Seneca River, a distance of about thirty-six miles.

A few days after we began work, a severe snow storm came on, falling to a depth of six inches, and followed by high winds not agreeable for tent life. As the season advanced, the warm weather brought out large quantities of mosquitos,[14] which proved the most serious annoyance to our comfort, especially in their night visits.

In our operations it was common to walk from one to five miles to and from our camp to work. Our principal walked with us, and the task was regarded a necessary and proper thing. At the present day this would be thought rather severe service, especially as the day's work on the line was a continuous labor with half an hour for dinner.

At times, when in the opinion of the chief of the party it was necessary, our young aspirants for the profession of engineering were required to take an axe and cut pegs and stakes and clear line. Mostly they did not relish this, regarding it an infringement on their dignity, doing their work in a hesitating and slovenly manner. As I had been brought up to work, and did not regard any honest labor a degradation, I performed my part on such occasions with the same alacrity as my regular work. My readiness on all such occasions gave me favor with my principal, which I improved to obtain knowledge in the art I sought.

42

The target men were allowed and requested to keep a field book of the observations and work out the levels, so we had opportunity to become familiar with the method of taking the observations and the principle of computing the levels. No one was deemed competent to work the levels except Principal Roberts, and we could do this only for our own practice, but after a time, when Mr. Roberts was occupied in camp on his plans and calculations, he would allow us to practice with the level. I eagerly embraced such opportunities.

I had prepared myself with a very plain set of drawing instruments and on rainy days employed myself in plotting the line and the profile, making such a map as my limited skill allowed me. These were of no business value except for my own improvement. Mr. Roberts made all the maps that were regarded as necessary for the business. This made me familiar with the art of plotting lines and profiles, which then appeared to me the great things to be learned.

There were sometimes errors by the target men in calling off the observation. This would be discovered on running the proof line, which was always carefully done. But the discovery often led to the necessity of making the second proof. An error of this kind did not fail to call out severe reproof from the Principal. As there were two target men, it was ordinarily impossible to decide which committed the error, and so both were compelled to feel the reproof. Sometimes I thought my associate was rather punished over my shoulder; but it could not be certain on which the fault properly rested and we were both cautioned against similar error. Sometimes it was finally ascertained the fault was not with the target men at all. All this was good discipline to impress on our minds the strict care required in conducting levels--much more important on canals than on rail or other roads. Principal Roberts had a profound sense of his responsibility in establishing a correct level for the contemplated canal, and this very naturally impressed itself on my mind.

The location for the canal was completed to the east shore of the Seneca River, at Montezuma, about the tenth of July, three months from the day we left Rome, traveling on foot about eighty miles.

At Rome, the party was broken up and reorganized into several small parties that were assigned to the charge of staking out and superintending the works. At this time, or very soon after, the whole of the middle section was placed under contract. In this new organization, one target man was considered sufficient for a party. The party to which I was assigned was placed under David S. Bates as principal, usually known as Judge Bates.[15] I had seen Judge Bates, but had no special acquaintance with him. His chief qualification was as a land surveyor of good standing. I was the only one of the old party that was attached to Judge Bates, to whom was assigned the position of resident engineer of a division extending from Canastota in Madison County to Limestone Creek in Onondaga County, about seventeen miles in length.

This opened quite a new field of experience to me. I soon found that Judge Bates had very little experience in the use of leveling instruments, or the computation involved in conducting levels, and was very ready to avail himself of what I had learned.[16] He was a man of very pleasant manner, and rather to my surprise, was ready to learn, even from me. This was extremely favorable to my progress, as he allowed me to use the instruments freely, and very soon came to depend on me for taking the levels and as a check to the computations. I very readily availed myself of this privilege, by which I became very familiar with the use of all the instruments, and the calculations required in laying out and computing the contents of the work. The latter, as the surface of the country was pretty level, did not require high mathematical science, though more perhaps in some cases than was applied.

The work we had to do consisted mainly in laying out curves on the angles of the line and setting the side stakes to guide the workmen. We went along and for several days had no side stakes to set, except when the surface of the ground was above the bottom level of the canal. This was simple, as we only had to add the slope to the half-width of the bottom of the canal, the measure being from center line. After a few days we came to a piece of work where the surface of the ground was about two feet below bottom level. Mr. Bates was at the level, and I was carrying the target. The observation and calculation showed the ground below level. Considering that the same slope would be carried in from bottom level, I deducted the slope from the distance to center, and so called out my calculation to Judge Bates as showing where the stake should be set. On this, Judge Bates directed the stake set at fourteen feet from center, not deducting the slope. As it was a new case, I carefully looked over the matter on which, it appeared to me, I was right. Judge Bates could not see the propriety of putting any stake less than fourteen feet from center, and so we were at issue. We pondered over the question for a while, and then left for dinner. At dinner, we were joined by Canvass White,[17] the principal assistant engineer, who was reviewing the line generally. We submitted the question to him, and after dinner he went out with us, and taking a seat on a log, appeared to us as making an examination of the question. Whether he was or not, we did not know, for after sitting about an hour, he left without giving us his opinions. We thought he had concluded it was better to let us work it out for ourselves, thinking that one of us was right and that we should ultimately reach a correct decision. I then drew sketches to illustrate my position, and toward evening we came to what we agreed was the true solution of the question. I mentioned these circumstances as historic of the science as held at that day, though it must not be understood there were none to comprehend such questions.

By the preceding, it will be noticed my situation was favorably advanced, especially as to acquiring knowledge. I received no increase of compensation, for I was technically the target man at twelve dollars per month. I was, however, well satisfied to bear all the responsibility for the privilege I had of gaining a better knowledge for the profession, which I now looked upon with great interest. In this new position, I derived great benefit from the discipline I had had under Mr. Roberts, by which I was impressed with the necessity of great care in observations and computations. Our party concluded its season's work in December and returned to Rome, the headquarters of the chief engineer. Here we were disbanded for the season. I may remark, very little work was done at that day on the canal in winter. I had no idea of any employment for the remainder of the winter, but I soon found I was assigned to the duty of weighing stone for locks.[18] For this purpose, I took up my winter quarters about midway between Onondaga Hollow and Syracuse, the latter then known as Crossets Corners. This service I completed about the first of April. As a feature of the times, I will state how young aspirants for engineering traveled in those days. I accidentally met at Crossets Corners two of my associates of Judge Roberts' party of the previous summer, who were residents of Rome and who had spent the winter in the same work of weighing stone further west. There were Messrs. Barrett and Tibbits.[19] We were forty miles from Rome; it was late in the afternoon, and we decided to go on to Orville, a village four miles east, and lodge there, with the view of making Rome, thirty-six miles from Orville, the evening of the next day. Accordingly, we started from Orville the next morning at sunrise, each carrying the bundle of clothes that had served our changes for the winter. The morning was bright and the ground hard frozen, and we walked fifteen miles to Canaseraga and stopped for breakfast. After breakfast, at about ten o'clock, we resumed our journey. The clear sun had now so far dissolved the frost as to make about an inch of soft mud, materially injuring the walking. Shortly after we had resumed our journey, a stagecoach came along. There was plenty of room in the coach for us all, but not the least intimation of a desire to avail oneself of its benefits was made by any of us, and we allowed it to pass on as a thing in no way suitable for us, at least not in accordance with our independent method of traveling. When we arrived at a point twelve miles from Rome (Oneida Castle) we found the ice ridge made by the sleigh track was firm, and the valleys between nearly filled with water from the melting snow. The condition of the road, as we found it after breakfast, made our traveling heavy and disagreeable. We reached Rome about nine o'clock in the evening, pretty well fatigued, but hardly felt the worse for it the next day. Not much of this sort of service is undertaken at the present day by young engineers.

NOTES

4. Benjamin Wright was Jervis' mentor and life-long friend. Born in Wethersfield, Connecticut, of old New England stock, Wright rose from a local surveyor in Fort Stanwix to chief engineer of the Erie Canal. Later he directed major canals in New York, Maryland, and Virginia, railroads in New York and Cuba, and many other public works. He has been called the Father of American Civil Engineering.

5. James Geddes came to New York from Pennsylvania. He studied law and was a judge and surveyor who, as early as 1808, made surveys for the Erie Canal. Later he was chief engineer on the Champlain Canal. He also worked on canals in Ohio, Maine, and Maryland.

6. Charles C. Brodhead, also Broadhead, came to Utica from Kingston, where he had been a local surveyor. He had surveyed the Black River area in 1793. Why Jervis calls him John is uncertain.

7. Nathan S. Roberts was of Pilgrim stock. A part-time school teacher from Plainfield, New Jersey, he arrived at Oriskany in 1806 to practice this profession. It was Wright who persuaded Roberts to be a surveyor and engineer on the Erie. Later Roberts worked on the Chesapeake and Ohio Canal and was chief engineer of the Pennsylvania State Canal.

8. The survey party was and still is generally organized as follows: party chief, instrument man, chainmen (front and rear), targetmen (or rodmen), axemen, and helpers. A potential surveyor would advance through each of these steps to party chief and thence perhaps to resident engineer.

9. The engineers, by comparison, received from about one hundred to one hundred and fifty dollars per month.

10. Wright was active in local politics. He served as a state legislator and, during the War of 1812, was appointed an Oneida County judge.

11. J. Day's **Principles of Navigation and Surveying** (New Haven, Connecticut) became available in 1817. Other common books on surveying were Abel Flint's **System of Geometry and Trigonometry, Together with a Treatise on Surveying** (Hartford, Connecticut, 1804), and S. Morris' **An Accurate System of Surveying** (Litchfield, Connecticut, 1796). It is not known, however, which books Jervis bought at this time.

12. Modern rods are about six and a half feet long, and can be extended to twelve to thirteen feet.

13. In the 1790's possibly the longest wooden trestle bridge in the world (over two hundred spans at twenty-five feet) was built over the Seneca River near here. The swampy areas which necessitated the long bridge also caused many problems for the canal builders. It is surprising that Jervis did not mention this remarkable bridge.

14. Later, many construction workers were to die from fevers carried by mosquitoes in this area. In 1820, for example, about a thousand workers were disabled from sickness between July and October.

15. David Stanhope Bates, son of a Revolutionary War officer, was trained as a minister but took up surveying because he loved mathematics. In 1810, he left his native Parsippan, New Jersey, for Oneida County, New York, where he was first a surveyor, then superintendent of the local iron works, and later, county judge. He joined Wright on the Erie in 1817 as an assistant engineer and advanced to division engineer. Later he was the principal engineer of the canal system for the State of Ohio.

16. Most early surveyors were trained for compass and chain type surveys and not in leveling. That is, they knew how to make a "plan" of a piece of property but could not make a topographical map of the land as this required measuring differences in elevations ("levels").

17. Canvass White was, like Jervis, a local farm boy who became an engineer under Wright's mentorship. He joined the canal staff in 1816 and was selected to tour English canals in 1817. Returning to New York, he became a close engineering aide to Wright and achieved much fame for discovering a natural cement in Chittenango, New York, near the route of the Erie. Later White worked on canals in Connecticut, Pennsylvania, and New Jersey.

18. Masonry contractors were paid on the basis of the amount of stone actually placed. The stone-weigher determined the basis for these payments.

19. Presumably Jervis is referring to Alfred Barrett, who was later chief engineer on the enlargement of the Erie, and Hiram Tibbits, a surveyor who served the Erie at least until 1823.

Champions of the Public Interest

An Excerpt from Chapter 1 -**The American Civil Engineer**
William H. Wisely

William Wisely was Executive Director of the American Society of Civil Engineers from 1955 to 1972. His book, **The American Civil Engineer,** *published in 1974, documents the origins of the profession in America, the establishment of the American Society of Civil Engineers (in 1852) and the role of the Society in the affairs of the profession from its beginning until the time of the book's completion.*

In this selection Wisely briefly describes the important role engineers played in the early development of the nation.

The first half of the 19th century was a crucial period in the youth of America. Migration to the west from the eastern seaboard extended to the Mississippi by 1820. This development was largely rural in character, with 90 percent of the national population of 10 million living in settlements of less than 2,500. The fertile western farms were devoted to food crops, while agriculture in the South was dedicated to the production of cotton. Manufacturing was making a modest beginning in New England and some larger western cities.

Facilities for communication and transport were sorely needed to integrate these regional elements into a national structure for trade and commerce. Fortunately, these services were to be forthcoming through many bold and imaginative enterprises in which engineers played important roles.

The earliest non-military engineering art was applied mainly in the domain of transportation. The 1787 laws establishing the Northwest Territory decreed that 2 percent of revenues from land sales would be allocated to construction of the National Road or "Cumberland Pike" from Washington to the Ohio

The National Road

47

River. Begun in 1806, the first 130 miles to Wheeling, Virginia (now West Virginia), were opened in 1820, and the road was eventually extended to Columbus and St. Louis. The project was significant in that it represented the first use of federal funds for major civil works construction.
The 9,000 miles of rock-and-gravel-surfaced roads in 1820 grew to 88,000 miles in the next four decades.

Equally significant was the Canal Era, even though it was to be shortlived. Beginning with the South Hadley and Middlesex Canals in Massachusetts (1793 and 1804), the period was highlighted by the greatly successful Erie Canal in New York (1825) and followed by the Union Canal in Pennsylvania (1829), the Morris Canal in New Jersey (1831), and the Chesapeake and Ohio Canal (1851). By mid-century the steam engine was in general use for water and land transportation. About 3,000 miles of canals were in operation in 1840, by which time the waterway was rapidly yielding its leading role in transportation to the railroad.

The South Carolina Railroad was the first in the United States. It was soon followed by the Baltimore and Ohio; the Pennsylvania; the Pittsburgh, Fort Wayne and Chicago; the Rock Island; and the Erie railroads. It is noteworthy that the Erie was to be a training ground for seven engineers who were later to become presidents of the American Society of Civil Engineers.

By 1852, the experimental and early development phase of the American railroad industry was accomplished, with about 9,000 miles of operating trackage. This increased ten-fold in the next thirty years.

The railroad brought new emphasis to the art of bridge building, which until the early 19th century was largely confined to timber and stone masonry. The middle years of the century, however, brought a renaissance in bridge building, both in the transition from wood to cast and wrought iron as well as in the advancement from the early empirical trussed arches to the proprietary trusses of Howe, Pratt, Warren, Whipple, Bollman, Fink and others. The systematic analysis of stress in the truss members and early efforts toward rational design soon followed. The first metal truss bridge (cast and wrought iron) was designed and built on the Reading Railroad in 1845. By mid-century wrought iron was in general use as a structural material, and the arrival of the Bessemer converter in 1856 introduced the Steel Age.

In 1850, there were 83 municipal water supplies in the United States, thanks largely to pioneering efforts in Boston, New York City and Bethlehem, Pennsylvania. The supplies provided untreated water from wells, springs and surface sources. The earliest American work on water treatment was initiated at St. Louis in 1866, when slow sand filters were utilized.

Although some sewer lines in Boston were in operation by 1800, the earliest public system of sewers was built in Chicago in 1855 under the direction of City Engineer E. S. Chesborough. By 1860, public sewer systems were serving about a million people in ten of the largest cities. Sewage was discharged without treatment to the nearest watercourse.

What manner of men were these "civil engineers" who were assuming such a significant leadership role in planning, building and managing the bold public works projects that were shaping America in the mid-19th century? The term

"Civil Engineer" was first adopted by John Smeaton, builder of early roads, structures and canals in England, who about 1782 signed himself under that title in presenting expert testimony in the courts.

In America the Continental Congress legislated the appointment of engineer officers in the army, most of these positions being filled by Europeans. Although this "Corps of Engineers" was disbanded at the end of the Revolutionary War, it was reconstituted in 1794 and has prevailed ever since. The United States Military Academy was created by act of Congress in 1802, with the intent that it would function as an arm of the Corps of Engineers. This administration prevailed for more than 60 years. In 1821 the Congress enacted legislation directing the Corps of Engineers to make surveys of major roads and canals, and prescribed that this work be performed by office and field parties under the direction of a supervisory board, all to be jointly constituted of "engineer officers" and "civil engineers." This is certainly one of the earlier distinctions in America between the military engineer and the civilian or "civil" engineer.

Sylvanus Thayer

The importance of the United States Military Academy as a source of trained engineers is attested by the fact that of 572 graduates in the period 1802-1829, 49 had been appointed chief or resident engineers on railroad or canal projects by 1840.

Thus, most of the formally educated engineers of this period were graduates of West Point, the remainder having supplemented their general academic training by scientific study and field experience. A large segment of engineers in the mid-1800's, however, had little or no formal education, acquiring their technical knowledge through self-study and apprenticeship, often as axmen or rodmen in surveying parties. The roads, canals and railroads on which they worked served as their "universities."

Engineering education in America began in the United States Military Academy, which produced its first graduate in 1802. The curriculum was greatly strengthened during the tenure of Sylvanus Thayer as Superintendent from 1817-1833. The first civil engineering course outside of West Point was offered in 1821 by the American Literary, Scientific and Military Academy, later to be known as Lewis College and then, in 1834, renamed Norwich University.* The first civil engineering degree was conferred by Rensselaer Polytechnic Institute in 1835. By mid-century engineering courses were being offered by Union College (1845), Harvard College (1846), and Yale College (1846). In the next twenty years about seventy institutions of higher learning initiated engineering programs.

The national census of 1850 counted only 512 civil engineers, of whom two-thirds were resident in the states of Massachusetts (68), New York (62), Ohio (59), Pennsylvania (55), Connecticut (46) and Wisconsin (44). Although few in numbers, they were an elite group in stature. Regardless of the pathway of their careers, the civil engineers of this era were supremely confident of their capabilities, and they jealously guarded their independence of professional decision and action. As champions of the public interest, they were outspoken in their criticism of questionable political and industrial management. Their performance earned them prestige, public respect and financial remuneration to a degree unexcelled by any other profession at the time.

----

* This institution had been founded in 1818 by Alden Partridge, himself a graduate of West Point {1806} and the first professor of engineering in the United States {at West Point from 1813 to 1817}.

The Functional Intellectual

An Excerpt from **Engineering in American Society, 1850-1875**
Raymond H. Merritt

The maturing of the nation in the second half of the nineteenth century, with its accompanying increases in industrialization and urbanization, brought about the need for a more broadly educated civil engineer to deal with an increasingly complex set of nation-building problems. However, the young profession was not ready to surrender the benefits realized from the practical education inherent in the apprenticeship system which had produced the majority of its members in the first half of the century. The solution to the dilemma was the emergence of an educational system that combined the theoretical and practical aspects of engineering.

This selection from Raymond Merritt's study of the engineering profession in the middle of the nineteenth century describes the roots of the modern civil engineering educational system. In this selection Merritt documents another important dimension of the profession which materialized during this period-- the emergence of the professional society as the primary advocate for continuing professional education.

Knowledge was the critical agent in the growth of the engineering profession. Private and public corporations hired engineers because they were informed on the latest technological developments and had the administrative skill to organize and carry out technical projects efficiently. Formal academic study and an extensive apprenticeship-training qualified an engineer to direct industrial enterprises. Once in the field, he kept abreast of his practice by reading technical journals and by participating in professional societies. These national and regional groups arranged formal discussion and lectures, supplied reference and library services, and encouraged their members to publish accounts of their own successes and failures.

Publishing houses and professional associations fulfilled an educational purpose and sought to make general knowledge functional, rather than cultivate a special esoteric know-how. Most engineers did not achieve professional status merely by mastering a quantity of technical data. They wanted to be known as men with inquiring minds who were professional problem-solvers. Their success depended upon continual study and observation. In a debate over Thomas Clarke's paper "The Education of Civil Engineers," Francis Collingwood told the American Society of Civil Engineers that education was a lifelong process and that the critical task of the schools was to teach a youth how to learn. Thinking aimed thus at innovation and deviation was a prerequisite for America's technological growth. The technician who merely imitated others, or who made only small improvements in the details of basic design, was not functioning as an engineer but as a craftsman. Such a person who understood accepted building methods could be called a construction foreman or chief mechanic, not a civil or mechanical engineer. Herein lay the distinction between shopmen and those with a thorough academic training. William J. McAlpine, president of the American Society of Civil Engineers, reminded his colleagues in 1868 that the man who neglected his daily study "may rest assured that sooner or later he will be shelved, and his place supplied by one of those who by closer study and better acquaintance with modern developments" would fill his position.

Engineers could perhaps be called functional intellectuals. They were men who employed the methods and discipline of the scholar but to whom ideas were tools of cultural change rather than aesthetic experiences. During the period preceding and following the Civil War, American society called upon such men to fulfill needs and solve problems for which the past provided little help. Improving the methods of building bridges, sewers, canals, railroads, harbors, tunnels, and levees required alert and critical minds. The new materials and the new systems of construction were so complex that society could no longer rely upon craftsmen to provide estimates, supervise contractors, and efficiently manage expanding operations. Moreover, in the cities traditional solutions to the problems of transportation, communication, and sanitation were no longer adequate. The nineteenth-century engineer committed himself to an organic concept of learning much as the twentieth-century architect espoused organic architecture. A functional and expanding knowledge became the expected means of improving society.

*   *   *

The growth of industrial education was quite remarkable when one considers that there were only two institutions, Rensselaer and West Point, preparing men for careers in applied science in 1840 and that by 1870 over seventy institutions of higher learning offered students an engineering curriculum with courses in mathematics, geology, physics, chemistry, hydraulics, and mechanics. Due to the growth of new schools, engineering graduates increased from less than 900 in the decade before 1870 to more than 3,800 in the 1880s.

The growth of engineering curriculums introduced two tendencies into American higher education: an emphasis upon learning through personal experience and a corresponding concern for developing professional attitudes. Rensselaer Polytechnic Institute, which pioneered in extending a student's education beyond the confines of the lecture and classroom through laboratory assignments, industrial visits, fieldwork, and a program of graduate level research, emphasized that the purpose of college education was the "discipline of the mental facilities as working forces."

*   *   *

An engineering graduate, however, would soon fall behind if he relied entirely upon his formal education. He needed a continuing program of learning to keep pace with a rapidly expanding fund of technical knowledge. Professional societies provided this constant exchange of technical information. The American Society of Civil Engineers was reorganized in 1868 under determined leaders who emphasized "the advancement of science and practice in Engineering, the acquisition and dissemination of experimental knowledge, the comparison of professional experience, and the encouragement of social intercourse among its members." The Society not only arranged for the discussion and publication of formal papers and a review of current literature but also developed a reference library around the personal collection of William McAlpine. Members were asked to contribute old and new copies of government, municipal, railway, canal, and other reports, including specifications, profiles, maps, and photographs, so as to facilitate comparative studies and to preserve the historical record of important public works. Each of the Society's monthly *Proceedings* contained a section entitled "Library and Museum," listing about a hundred acquisitions.

William McAlpine

The early presidents of the Society believed that knowledge was the principle means by which engineers could maintain their status and authority. In his presidential address of 1868, William McAlpine boasted that engineering was the noblest profession because its collective intellect had breadth and depth as well as the capacity to put its thought to immediate use for the benefit of all mankind. He noted that the world had become too educated to tolerate charlatans. The Society discussed the possibility of requiring each new candidate to present a thesis on some professional subject and to require from all members the annual contribution of a paper, description, or drawing of some work of interest.

Though the officers never enforced this policy, they did succeed in making the Society an educational center for the profession.

*   *   *

. . . Many men actively engaged in engineering practice in the years 1850-1875 remembered the time when the engineer was a craftsman or a tradesman. William McAlpine, whose father was a millwright, noted that the "men of the old school" had become engineers "from a natural bent of mind and an intense love of the developments of the physical sciences." Formal training in mathematics and science, according to McAlpine, was the major factor in lifting engineering "from a trade to the dignity of one of the liberal professions." McAlpine had himself been apprenticed to John B. Jervis for eight years after which he served as chief engineer of the Erie Canal, chief engineer of the Brooklyn Dry Dock, state engineer of New York (1852-1857), chief engineer of the Erie Railway (1852-1857), the Chicago and Galena

Railroad (1857), and the Ohio and Mississippi Railway (1861-1864). Despite his lack of formal education, McAlpine became a consulting engineer for many of the major public works of the period. He helped plan city water systems for Chicago, Brooklyn, Buffalo, Montreal, Philadelphia, San Francisco, and Toronto; bridges for St. Louis, Niagara, and New York; the Manchester Ship Canal in England; a railway in India; and navigation on the Danube River.

McAlpine's mentor, John B. Jervis, likewise urged the American Society of Civil Engineers to fuse the recently developed program of formal scholastic study with the apprenticeship system. Jervis, who was one of America's early canal and railroad engineers, addressed the Society in 1869 on the transition of engineering from a craft to a profession. He suggested that future engineers study mathematics, mechanical philosophy, hydraulics, and existing structures, then enter the field under the direction of an experienced professional. This was similar to the advice that William Gillespie gave his students at Union College. He suggested that after graduation they "take the lowest position in some engineering corps, that of rodman, and only then expect to work up gradually to the rank of chief engineer."

Some exceptional engineers in the period 1850-1875 attained professional status through the apprenticeship system without attending a college. Isham Randolph, for example, son of a notable Virginia family, reached college age during the Civil War, when schools were closed. In 1868 he decided to apprentice as an engineer; five years later he became a resident engineer for the Baltimore and Ohio Railroad. In the 1880s he was chief engineer for the Illinois Central and in 1893 was appointed chief engineer of the Sanitary District of Chicago. He thus attained a national reputation through an effective apprenticeship training, combined with continual study and persistent work.

Most engineers advanced through four stages of apprenticeship training. The first level was that of axman or rodman, the second, transitman or levelman, the third was assistant, resident, or division engineer, and the fourth, chief engineer. Many such as Thomas N. McNair and Albert Robinson, who grew up during the early days of professionalization, attended college first, then passed through the apprenticeship system. Robinson graduated from the University of Michigan with a degree in civil engineering, remained to earn a Master of Science before he was hired in 1871 by the Atchison, Topeka and Santa Fe. He stayed with the same company for most of his career, working his way through the jobs as axman, rodman, levelman, assistant engineer, and chief engineer. After twenty-two years he became vice president and general manager of the railroad.

*   *   *

Advancement to the position of chief engineer was never a routine matter. Such a man had to demonstrate the capacity to work under political and economic pressure, as well as to display technical competence. One newly appointed chief engineer wrote, "The responsibility came upon me like a thunder bolt, which caused a rather curious sensation . . . a sharp sudden pain on the upper left side of my head, that never ceased for weeks until I got thoroughly acquainted with the work of my new position."

The apprenticeship system trained young men in careful habits of observation and note-taking. The mobility of the profession demanded that notebooks should be so precise and clear that at any moment they could be turned over to a successor who would be able to understand them as well as the author. As many literary men recorded their ideas and experiences in notebooks, so also leading engineers documented their intellectual growth by systematically noting new information.

The apprenticeship system selectively advanced and rewarded only those blessed with an alert mind and administrative capacities. The apprenticeship system taught executive skills, but turned back students who were ineffective leaders.

The "new education" with its emphasis upon discovery and research fostered a great expansion of publication. The idea that publication was a necessary part of academic life also prevailed among engineers. To print up formally the results of a project or the proposal for a new design not only prepared the ground for prospective capital but also eliminated the necessity of answering countless letters from other engineers throughout the world who wished to accumulate technical descriptions, cost analyses, production methods, and other details for similar projects. The publishing house thus served as an institution of higher learning where engineers registered for a lifelong curriculum of reading and writing without expecting ever to graduate.

* * *

Octave Chanute

An examination of the personal correspondence of Octave Chanute, chief engineer of the Kansas City Bridge (1867-1868), reveals how the printing press became an important tool in the work of a construction engineer. Whenever Chanute confronted a new problem, he requested Van Nostrand's to send any available technical materials dealing with his specific needs. He also ordered, at premium prices, complete sets of periodicals such as *Engineering* and meanwhile subscribed to *Van Nostrand's Eclectic Engineering Magazine*, the *Journal of the Franklin Institute*, the London-published *Engineer*, *The Nation*, *The Week*, *Putnam's Monthly*, and *The Railroad Times*. In 1868 Chanute suggested to Henry Morton that the *Journal of the Franklin Institute* publish the formal papers being prepared for discussion by the reactivated American Society of Civil Engineers, but the society made arrangements instead to publish their own *Transactions*.

Chanute's work on the Kansas City Bridge evoked much interest in the profession, part of it flowing from his argument with William McAlpine in the *Franklin Journal* over proper foundation construction. After completion of the project Chanute and his assistant George Morison wrote **The Kansas City Bridge**, a book describing the technical aspects of spanning the Missouri River. Chanute continued his research and reading even in retirement when he became interested in the gliding experiments of Otto and Gustav Lilienthal.

One might expect an extensive search for new ideas and a willingness to discuss formally new techniques from an immigrant engineer such as Octave Chanute; however, Elmer Corthell, a native American engineer trained in a liberal arts college, also boasted a large cosmopolitan library. Corthell accumulated over twelve hundred volumes, about three-fourths of them connected with his work on the Syn Island Levee; Eads' South Pass Jetties; the New York, East Shore and Buffalo Railroad; Eads' Tehuantepec Ship Railroad scheme; the St. Louis Merchants Bridge; the Chicago, Madison and Northern Railroad; harbor improvements at Tampico, Mexico; public works in Buenos Aires; and commercial projects in Para, Brazil. He published many articles in foreign and domestic newspapers, popular journals, and technical periodicals, and also issued in 1880 his **History of the Jetties at the Mouth of the Mississippi River** (1880).

*   *   *

At the turn of the century George Morison, a Harvard graduate who entered the profession as an assistant engineer on Octave Chanute's Kansas City Bridge in 1865, observed while looking back over his career, that the age of "Yankee ingenuity" based on a general "knowledge of what has been done" had been replaced by persistent study and observation. Simple honesty and hard work would no longer assure success. The ability to organize, skill in communication, knowledge of a broad array of scientific principles, and the capacity to analyze and modify technical problems had become indispensable. Engineers acquired these through a whole community of schools, professional associations, and apprenticeship experiences. Classical training had given way to the new education which was concerned with changing the future rather than knowing the past. Specialization and utilitarianism were its benchmarks. The nation's colleges trained young men not simply because they wished to be scholars, but because scholars were necessary to "the good of the community." Specialization had made the professional and academic communities more dependent upon one another. Polytechnic education, like technological growth, fostered an urban world and compelled men to expand their knowledge and experience beyond a study of esoteric subjects.

The education of engineers then, rested upon a combination of formal learning and practical experience. Members of the new profession were aware that their prestige and status were dependent upon an ability to assimilate, organize and communicate useful information. Consequently, engineers became intellectuals who combined disciplined study with practical experience. Scientific discoveries, professional advancements, public criticism, and administrative innovations constantly tempered their knowledge and understanding. The learning process of the engineer was never complete.

"What Was Troy to This?"

Chapter 1, **The Existential Pleasures of Engineering**
Samuel C. Florman

Samuel Florman, a civil engineer, introduces us to what he calls "The Golden Age of Engineering"--the period between 1850 and 1950--in this first chapter of his widely acclaimed little book, **The Existential Pleasures of Engineering**. *In subsequent chapters of the book Florman describes the decline in prestige of the engineering profession--a decline which he dates as beginning on January 31, 1950, the date that President Truman announced initiation of work on development of the hydrogen bomb--and provides arguments to support his contention that engineering is still a profession in which an individual may find fulfillment and self-actualization. Few other engineers have written as clearly and as scholarly of the joys of their profession as Florman does in this tribute to his profession.*

In May 1902 the fifty-year-old American Society of Civil Engineers held its annual convention in Washington, D.C. Robert Moore, the newly elected president, gave a welcoming address entitled, "The Engineer of the Twentieth Century." He began by eulogizing the engineers of the past for making human life "not only longer, but richer and better worth living." Then he acclaimed the achievements of his contemporaries and fellow members. Finally he warmed to his chosen topic, the engineer of the coming era:

> And in the future, even more than in the present, will the
> secrets of power be in his keeping, and more and more will he
> be a leader and benefactor of men. That his place in the esteem
> of his fellows and of the world will keep pace with his growing
> capacity and widening achievement is as certain as that effect
> will follow cause.

What a flush of pleasure they must have felt, those engineers of 1902, to hear themselves described as benefactors of mankind. What a quickening of the pulse there must have been as they listened to their leader predict success and glory for them in the years ahead. Doubtless they sat quietly, looking solemn in their starched collars and frock coats, the way we see them in faded photographs. But beneath those sedate facades they could not have helped but feel the stirrings of a fierce joy.

To be an engineer in 1902, or at any time between 1850 and 1950, was to be a participant in a great adventure, a leader in a great crusade. Technology, as everyone could see, was making miraculous advances, and, as a natural consequence, the prospects for mankind were becoming increasingly bright.

Every few months, it seemed, some new technological marvel was unveiled and greeted with wild public enthusiasm. There were marvels of transportation--trains, ocean liners, trolley cars, subways, automobiles, dirigibles, and airplanes; marvels of communication--telegraph, telephone, phonograph, movies, radio, and television; marvels of construction--bridges, tunnels, dams, and skyscrapers; miraculous new sources of power--steam engines, gasoline engines, Diesel engines, electric dynamos; wondrous new materials--steel, petroleum, aluminum, rayon, and plastics; machinery to save labor and expand production--reapers, looms, presses, derricks, and lathes; perhaps the biggest thrills of all--sewing machines, toilets, typewriters, bicycles, cameras, watches, electric lights, refrigerators, air conditioners, and so forth.

The completion of sizable technological undertakings was marked with celebrations fitting for an armistice or a coronation. Twenty-five thousand people paraded around the streets of New York for sixteen hours when the first cablegram was sent from America to Europe. At the moment that the Union Pacific and the Central Pacific railways were joined at Promontory in 1869, all over America bells were rung, cannons were fired, and bonfires were lit. That same year a glittering array of royalty and a procession of flag-bedecked ships marked the inauguration of the Suez Canal. As late as 1937, the opening of the Golden Gate Bridge was celebrated by a "Fiesta" attended by 200,000 persons, followed by a week of pageants and other festivities.

At the great international exhibitions, starting with London's Crystal Palace in 1851, technology was literally idolized. Six million people visited the Crystal Palace in Hyde Park to gape at the miraculous new machines on display. At the 1876 Centennial Exhibition in Philadelphia, in the glass and iron Machine Hall, the Corliss steam engine dominated the scene like a gargantuan icon. Alexandre Eiffel's one-thousand-foot tower was the sensation of the 1889 Universal Exposition in Paris, attended by more than thirty-two-million visitors. Again in Paris, in 1900, Henry Adams tells of how he "haunted" the machinery exhibits, mesmerized particularly by the "occult mechanism" of the dynamo. Every few years--at St. Louis (1904), Brussels (1910), San Francisco (1915), Wembley (1924), Philadelphia (1926), Chicago (1933), New York (1939)--the engineers and inventors presented their new magic show to a dazzled and appreciative public.

In the late 1800s engineers began to appear regularly as heroes in novels and short stories. Rudyard Kipling's **Bridge-Builders** were depicted as robust creators of the British empire. August Strindberg eulogized in fiction the builders of the St. Gotthard Tunnel. The engineer hero of Zane Grey's **The U.P. Trail** was "wild for adventure, keen for achievement, eager, ardent, bronze-faced, and keen-eyed," a man who "had been seized by the spirit of some great thing to be." A whole series of best sellers glorified the American civil engineer of the early 1900's--**The Iron Trail, End of Steel, The Trail of the Lonesome Pine, Whispering Smith, The Fight on the Standing Stone, The Fire Bringers, Empire Builders**. When one of this genre, **The Winning of Barbara Worth**, was made into a movie, Ronald Colman played the part of the handsome engineer who gets the girl. Jules Verne's more than fifty fantastic tales were perennial favorites; so were the futuristic visions of H. G. Wells. As for the poets, in 1855 Walt Whitman was "Singing the great achievements of the present,/Singing the strong light works of engineers," a hymn of affirmation that Carl Sandburg was to echo sixty years later. On the occasion of the completion of the transcontinental railway, Joaquin Miller, a popular poet of the day, wrote, "There is more poetry in the railway that crosses the continent than in all the history of the Trojan War." Robert Louis Stevenson, in his journal, **Across the Plains**, used the same epic image to describe the building of the railroad. "If it be romance," he wrote, "if it be contrast, if it be heroism we require, what was Troy to this?"

A few querulous voices were raised in alarm against the coming of the new machines and deplored the worship of material progress. Thoreau comes first to mind, of course, then Samuel Butler, Matthew Arnold, Nathaniel Hawthorne, and others, up to Aldous Huxley whose **Brave New World** appeared in 1932. Some of those writers who most admired technology-- Whitman, Henry Adams, and H. G. Wells, for example--also feared it greatly. Charlie Chaplin's <u>Modern Times</u> expressed a wry but deep-felt protest. Yet the doubts and objections of a handful of artists and intellectuals could not stem the tides of public opinion. The conventional wisdom was the technological progress brought with it real progress--good progress--for all of humanity, and that the men responsible for this progress had reason to consider themselves heroes.

The years between 1850 and 1950 were indeed good years for engineering, "the Golden Age of Engineering," one is tempted to call them, in today's language of nostalgia. Before 1850 there had been many fine engineers and many outstanding engineering works. But engineering itself had been rather a craft than a profession, relying more on common sense and time-honored experience than on the application of scientific principles, and lacking those essentials of true professionalism--professional schools and professional societies. After 1950, as we shall see, engineering entered into a dark age of criticism and self-doubt. But during the hundred years between, the profession flourished. Its mighty schools and societies proliferated. And as its members grew in numbers (from two thousand to more than half a million in the United States), they also grew in prestige, power, accomplishment, and self-satisfaction.

The self-satisfaction came from many different sources. First of all, there was the elementary pleasure of solving technical problems and successfully completing constructive projects. This pleasure was as old as the human race. What was new about engineers as they started to develop as a profession was the delight they took in thinking of themselves as saviors of mankind. Since the lot of the common man had traditionally been one of unrelenting hardship, engineers looked upon their works as man's "redeemer from despairing drudgery and burdensome labor." Once the common man was released from drudgery, the engineers reasoned, he would inevitably become educated, cultured and ennobled, and this improvement in the race would also be to the credit of the engineering profession. Improved human beings, of course, would have to be happier human beings.

Next, elevation of the common man would tend to make all men more nearly equal, thus making the engineer an agent in the realization of the democratic dream, "an apostle of democracy," as one engineer orator put it in 1905. Comfort, leisure, and equality would imbue men with confidence in themselves and in the objectives of their society, making for the growth of patriotism and domestic tranquillity. Jealousies and class hostilities would diminish, a sense of brotherhood would spread, and the cause of peace would be served.

The engineer's works would also contribute to brotherhood by literally bringing men closer together. In the earliest days of the American republic, both Jefferson and Hamilton had remarked on how roads and canals would serve this end. Through the years, each advance in transportation and communication evoked new commentary on the theme. Walt Whitman rhapsodized:

> Lo, soul, seest thou not God's purpose from the first?
> The earth to be spann'd, connected by network,
> The races, neighbors, to marry and be given in marriage,
> The oceans to be cross'd, the distant brought near,
> The lands to be welded together.

As the Panama Canal neared completion, poet Percy MacKaye exulted over this wondrous work:

> Where the tribes of man are led toward peace
> By the prophet-engineer.

But all of this was only a small part--the most obvious part--of the satisfaction that engineers found in their work. They felt that they were improving the world, not only by their deeds, but also by their way of thinking. If engineers could solve problems by being open-minded and free of preconceptions and prejudices--by applying scientific methods -- could not all men learn to think in this mode, and then would not ignorance, superstition and bigotry vanish? "We are the priests of the new epoch," an engineering leader told his colleagues in 1895, "without superstitions."

Ever since the days of the eighteenth-century Enlightenment, men had been tantalized by the thought that social and moral truths existed, just like scientific truths, and could be discovered by the application of "right reason." This idea appealed mightily to engineers as their profession came of age. They fancied themselves the group best qualified to show the way in this great enterprise. Were they not, after all, the appliers of scientific method to practical problems?

"There is evil and plenty of it, the world over," wrote John Augustus Roebling, designer of the Brooklyn Bridge, "but all this evil may be traced in its origin to man's transgression of the laws of his own being." If the laws of man's being were discoverable, like the laws of thermodynamics, then indeed the blueprint for Utopia might be close at hand. In his native Germany, Roebling had been a student of the philosopher, Hegel. He brought with him to America the Hegelian concept that man might eventually free himself from the irrationalities of history by mastering nature. Not all engineers studied Hegel, of course, but many were influenced by Herbert Spencer (himself an engineer), who, during the 1880s and 1890s, expounded his version of Darwinism. In Darwin's theory of evolution, Spencer saw proof that man and society were governed by scientific laws, just as nature was. The first president of the American Society of Mechanical Engineers cited Spencer in declaring that the engineering profession should concern itself with general problems of politics and economics, since they might well prove susceptible to engineering solutions.

Frederick W. Taylor, the mechanical engineer who, around the turn of the century, "discovered" the rationalization of work (time and motion studies, quotas, unit costs--what came to be called efficiency engineering) considered himself a prophet whose "scientific principles" would settle all social conflicts.

The principles of scientific management, he averred, "can be applied with equal force to all social activities: to the management of our homes; the management of our farms; the management of the business of our tradesmen large and small; of our churches, our philanthropic institutions, our universities, and our governmental departments." Henry L. Gantt, a disciple of Taylor's, went so far as to form an organization in 1916 called the New Machine. Society, declared Gantt, "must accord leadership to him who knows what to do and how to do it for the benefit of the community. This man is the engineer."

For those engineers who were neither formal philosophers nor social visionaries, there was always support to be found in that most venerated of sources--the Bible. For had not God given the earth to man and ordered him to "subdue it"? Many engineers saw their work as the carrying out of the Christian mission of subduing the earth for the benefit of man and for the greater glory of God.

In short, engineers of the Golden Age were not at a loss for intellectual, philosophical, and spiritual gratifications to go along with their often not inconsiderable tangible rewards.

There were a few frustrations, to be sure. Although they were generally admired and occasionally lionized, engineers never seemed to be satisfied with the "status" accorded them. There was a social basis for this: in Europe, engineering had been a career traditionally shunned by the upper classes. (When Herbert Hoover told a lady he was an engineer, she replied, "Why, I thought you were a gentleman!") There was a professional basis as well: engineering was the youngest of the professions, with none of the established traditions of medicine, divinity, and law. And engineers, jealous of their new identity, were repeatedly irked by being confused with scientists, or rather for having scientists get all the credit for engineering accomplishments.

But the greatest irritant by far was the domination of so many engineering activities by the business community. For all their declarations of professional independence and moral purity, engineers again and again found themselves subservient to financiers and businessmen. Indeed, serious cleavages developed between those engineers who pledged allegiance to the business community, for whom so many of them worked, and those who attempted to build an aloof and independent profession. These cleavages were largely responsible for the splintering of the profession through the proliferation of diverse engineering societies.

Another embarrassment was the failure of the engineer's proclaimed social programs to really take hold. Compared to all the talk of applying engineering methods to social problems, results were exceedingly meager. Henry Gantt's organization, the New Machine, came to naught, as did all similar ventures. During the 1920s Thorstein Veblen aroused some public interest in "technocracy," a proposed takeover of power by "a soviet of technicians." But when the idea was revived briefly in 1932, most engineers considered it to be a maverick movement controlled by left-wing impostors.

These few annoyances and embarrassments paled, however, in contrast to the real accomplishments, the popular acclaim, and the promising dreams to which the engineering profession could point. Whatever the engineering profession had not achieved between 1850 and 1950, it certainly expected to achieve in the very near future. During World War II, the development of radar, sonar, and other techniques promised new engineering marvels for the postwar years. Even the horror of Hiroshima was quickly forgotten in talk of atomic power for peaceful purposes. "The engineer," wrote a noted historian of the profession, "has faith in his work and in the ultimate beneficence of the forces he serves." "It is a great profession," reminisced Herbert Hoover as the twentieth century reached its midpoint. Few engineers of the previous hundred years would have disagreed.

If we had asked an engineer of the Golden Age whether he found engineering existentially satisfying, he probably would not have understood the question. For the word *existential* achieved currency only with the popularity of the work of the French existentialist philosophers after World War II. But engineers did, I believe, find their work thrilling in a deep-down, elemental way that we think of when the word *existential* is used today. They felt fulfilled as men. They felt a part of the flow of history. They loved their work and believed it was inherently good.

At least, that is how it appears to me as I look back upon that age. And that is how it appeared to me as a boy growing up in the 1930s thinking about becoming an engineer. I remember well those old newsreels showing the dedication of yet another TVA dam. Instead of the floods and dust bowls of which we had seen so much, here were peaceful scenes of tamed rivers and humming transmission wires. We applauded F.D.R. as he promised more of such wonders in the years ahead. When I visited the General Motors Futurama Exhibit at the 1939 New York World's Fair, I believed that I was truly looking at "The World of Tomorrow." It promised to be a better world, too. It would have to be, with its superhighways, its sleek cars, and its sparkling cities. When I graduated as an engineer in the midst of World War II, the dreams of the engineering profession were still intact. The war was a temporary nastiness, after which the building of a better world would be resumed. In the Seabees--Navy construction battalions led by civil engineer officers--the proud boast was "Can Do!" "The difficult we do immediately," we used to say. "The impossible takes a little longer."

The Sons of Martha
Rudyard Kipling

Kipling lived in a milieu in which great value was placed on the accomplishments of engineers. Great engineering works were not yet so commonplace as to be taken for granted by the general populace. Leading engineers were regarded as heroes. The Sons of Martha is Kipling's tribute to the engineering profession. Drawing the analogy between Martha in the biblical story of Mary and Martha (Luke 10:38-42) and engineers, he portrays engineers as those who make it possible for the rest of society to "choose the better part."

Kipling includes in the poem recognition of the engineer's reliance on his own watchfulness as a safeguard against things about to go wrong; his tendency to analyze rather than trust in faith alone; his willingness to exercise particular care with his work on behalf of his fellow man; his inclination to work until the job is done rather than until the end of the prescribed workday; and his commitment to service to his fellow man.

Those searching for recognition of the value of civil engineering to society or for motivation to continue pursuit of their professional goals might find their search fulfilled in a thoughtful reading of this poem.

The Sons of Mary seldom bother, for they have
    inherited that good part;
But the Sons of Martha favour their Mother of the
    careful soul and the troubled heart.
And because she lost her temper once, and because
    she was rude to the Lord her Guest,
Her Sons must wait upon Mary's Sons, world with-
    out end, reprieve, or rest.

It is their care in all the ages to take the buffet
    and cushion the shock.
It is their care that the gear engages; it is their
    care that the switches lock.
It is their care that the wheels run truly; it is their
    care to embark and entrain,
Tally, transport, and deliver duly the Sons of Mary
    by land and main.

They say to the mountains, 'Be ye removed.' They
    say to the lesser floods 'Be dry.'
Under their rods are the rocks reproved--they are
    not afraid of that which is high.
Then do the hill-tops shake to the summit--then
    is the bed of the deep laid bare,
That the Sons of Mary may overcome it, pleasantly
    sleeping and unaware.

They finger death at their gloves' end where they
    piece and repiece the living wires.
He rears against the gates they tend: they feed him hungry
    behind their fires.
Early at dawn, ere men see clear, they stumble into
    his terrible stall,
And hale him forth like a haltered steer, and goad
    and turn him till evenfall.

To these from birth is Belief forbidden; from these
    till death is Relief afar.
They are concerned with matter hidden--under
    the earth-line their altars are:
The secret fountains to follow up, waters withdrawn
    to restore to the mouth,
And gather the floods as in a cup, and pour them
    again at a city's drought.

They do not preach that their God will rouse them
    a little before the nuts work loose.
They do not teach that His Pity allows them to
    leave their work when they damn-well choose.
As in the thronged and the lighted ways, so in the
    dark and the desert they stand,
Wary and watchful all their days, that their breth-
    ren's days may be long in the land.

Raise ye the stone or cleave the wood to make a
    path more fair or flat;
Lo, it is black already with blood some Son of
    Martha spilled for that!
Not as a ladder from earth to Heaven, not as a wit-
    ness to any creed,
But simple service simply given to his own kind in
    their common need.

And the Sons of Mary smile and are blessed--they
    know the angels are on their side.
They know in them is the Grace confessed, and for
    them are the Mercies multiplied.
They sit at the feet--they hear the Word--they
    see how truly the Promise runs:
They have cast their burden upon the Lord, and--
    the Lord He lays it on Martha's Sons!

The Evolution of a Profession

Excerpts from **The Revolt of the Engineers**
Edwin T. Layton

Engineers today, like other professionals, have little reason to reflect on the evolution of their profession--the change in status from a semi-skilled vocation based largely on on-the-job training rather than education to the status of licensed professionals with proscribed standards of professional education and experience. Only a few engineers ever think about how their profession came to be recognized as a profession or about the important role professional societies played in that recognition.

Edwin Layton's book, **The Revolt of the Engineers,** *is subtitled "Social Responsibility and the American Engineering Profession." It records the emergence of the American engineering profession in the nineteenth century. It explains the role of the professional societies in that emergence; the conflicts between engineers and their employers; and the influences of political and societal concerns.*

The balance between business and professionalism has been one of the most important forces in the formation and evolution of engineering societies in America. Most American engineering societies represent compromises between business and professionalism. A purely professional association might fail to win business support. Employer approval is needed, however, to permit employees to attend meetings or read papers. A wholly commercial organization, on the other hand, would tend to alienate creative professionals. And without their participation a technical society would fail of its fundamental purpose: to increase knowledge. Although the support of both professionals and businessmen is needed for a successful engineering society, the degree of influence accorded to each has varied widely; almost no two engineering societies are alike in this respect. However, the particular compromise adopted will have widespread influence on almost every aspect of the society's policies.

There are two fundamental ways of altering this balance: by defining the field to be covered and by professional standards of membership. Those who think of engineering as an independent profession have favored a comprehensive body that would represent and unite all engineers of whatever specialty. In contrast, engineers who think of themselves as businessmen have sometimes preferred technical societies built around a single industry and devoted to its interests. Strongly professional societies set high standards of membership that restrict effective power to initiates in a body of esoteric knowledge. Because such qualifications tend to exclude businessmen, management-oriented engineers have favored societies with lower membership requirements, which could bring together on an equal basis the businessmen, managers, and engineers affiliated with a given industry. As one engineer sympathetic to management put it, membership standards should be "inclusive and not too rigidly graded, granting to engineers who pass into administrative duties the full fellowship of the profession."

Most engineers agree that their profession is a "vaguely bounded nucleus within a large body of technical workers." Engineering societies mark off this professional nucleus by means of membership standards. At one time or another four different tests for full membership in engineering societies have been suggested: technical creativity, the ability to design, being in "responsible charge" of engineering work, and company or industrial affiliation. The crucial element in each is the degree to which businessmen--including managers--would be excluded from full membership. By the tests of technical creativity and design, almost all businessmen would be relegated to a class of associates; officeholding and power would be effectively restricted to professional engineers. The criterion of "responsible charge" would admit businessmen who performed some technical functions, but would exclude some others from full membership. A company or industrial standard would admit all businessmen who might be interested in full membership, but would effectively exclude most professional engineers. It is not uncommon for engineering societies to apply more than one of these tests.

Engineers closely allied with science have sometimes favored restricting full membership to those having done creative technical work. This is, in fact, the test that scientists have applied, though often informally, in determining who is a professional "insider." Such a standard would exclude from power all but a tiny minority of businessmen and managers. It would also exclude most of those who think of themselves as engineers. The problem is that the number of creative innovators is small. At the beginning of the twentieth century electrical technology was perhaps the most rapidly developing field of engineering; yet at that time only about 2 percent of the membership of the American Institute of Electrical Engineers was contributing papers in a given year. A society restricted to the creative would have few full members. Nevertheless, creativity has been important, at least informally; the presidency of engineering societies is usually reserved for such men.

The test that most clearly distinguishes the professional engineer from all other groups is his ability to design engineering works. Placing the emphasis on the application of professional knowledge, rather than on its creation, distinguishes between the scientist and the engineer. Stressing the ability to design as well as to direct or construct engineering works draws a sharp line between professional engineers and managers and businessmen. Design has several drawbacks as a criterion of full membership, however. In actual fact, engineers perform a diversity of roles; they are not all designers. As a result, the requirement has been phrased as "qualified to design." This, in turn, is rather hypothetical and creates problems of administration. But however difficult of application, the intent is clear: to restrict full membership to professional engineers. It has, therefore, been adopted by the most strongly professional societies. Such organizations limit full membership to a professional elite; they exclude many persons who perform engineering work as well as most businessmen and managers.

A more inclusive test for full membership is that of being in "responsible charge" of engineering work. It is, in essence, a bureaucratic definition; it applies to engineers who do not have to refer their plans or drawings to a higher authority for technical review. Such a criterion of membership accepts the reality that the engineer in America is only partly differentiated from management. It is adapted to the heterogeneous tasks actually performed by engineers, who might be variously engaged in management, sales, production, testing, or any of a number of other functions. From the professional standpoint this criterion has a serious disadvantage: bureaucratic standing does not guarantee technical excellence. Despite this, "responsible charge" has been widely used as a professional qualification for membership in American engineering societies. The reason for its popularity is that it represents a compromise between professionalism and business. Those in charge of engineering work are in some sense professional engineers, but this test is broad enough to include most businessmen or managers who might be interested in membership. It is a common denominator between the engineer and the businessman.

Engineering societies that define their membership in terms of industrial affiliation represent a further dilution of professionalism. Such membership standards may be variously phrased. Some societies admit anyone interested in a given field of practice. If this field is defined narrowly enough, it will effectively limit membership to those "practically" engaged in the field. A further limitation is implied by a membership based on employment by a specific company or group of companies. The extreme of this tendency is reached when the technical society becomes a trade association, the litmus-paper test of this being the adoption of company rather than individual memberships. There is, however, an important borderland of societies that have both company and individual members. Societies with company members usually drop the term "engineering" from their titles; for example, the American Ceramic Society, the American Concrete Institute, and the American Waterworks Association.

Each of these four tests of full membership constitutes, in effect, a definition of the term "engineer." A membership qualification based on design implies that engineering is an independent profession, but one based on industrial affiliation suggests a basic solidarity between the engineer and the businessman. "Responsible charge" is a compromise which orients the engineer toward both his profession and his employer. "Creativity" tends to ally engineers with science rather than with either business or engineering. As recent sociological studies have indicated, these represent four possible orientations for the individual engineer in industry. In practice, however, societies are seldom pure types. Most of them combine two or more of these tests of membership, in order to appeal to as many members as possible. By applying a diversity of qualifications, some perhaps informally, almost any desired combination of business and professionalism is possible. Consequently, engineering societies represent not four sharply defined types, but a continuous spectrum.

Engineering professionalism first appeared in America as an offshoot of the scientific variety. From 1829 to 1836, Alexander Dallas Bache utilized the Franklin Institute in an effort to advance professional science; a significant feature of these early activities was the attempt to gain patronage and prestige for science by linking it to practical activities, such as the investigation of steam-boiler explosions. Thus, when Bache and his scientific friends refashioned the institute's *Journal*, one of the three professional subdivisions they created was that of civil engineering. In this way, the institute became the first American center for the encouragement of a professional spirit in engineering.

The first effort to form an engineering society in America occurred in February, 1839, when some forty engineers met at Baltimore. They appointed a committee to draft a constitution. This committee, logically enough, proposed to graft the new organization upon the Franklin Institute. The membership standards proposed would have pleased Bache: each member would be required to contribute at least one communication annually. British precedent was also important. A subcommittee quoted Thomas Telford, the first president of the British Institution of Civil Engineers, on the dangers of "too easy and promiscuous admission," which, he had warned, led to "unavoidable, and not infrequently incurable, inconveniences."

The committee's proposals did not win the support of American civil engineers. There were several reasons for this failure. American engineers may not have relished such a close alliance with science; in any case, the membership standards were too high. Another reason may have been the desire of Pennsylvania engineers recently dismissed from their positions on state works to utilize the projected society for their own welfare. Sectional jealousies and indifference were, however, probably the most important factors. A counterproposal by Mr. Edward Miller, in 1840, was based on the idea of four independent regional societies. In fact the earliest engineering societies to endure were of precisely this character. The first was the Boston Society of Civil Engineers, founded in 1848, followed by the American Society of Civil Engineers, in 1852, which, despite its name, functioned initially as a local society for New York. Other early societies were the Western Society of Engineers, established at Chicago in 1868, and the Engineers Club of St. Louis, which appeared in 1868. Although local associations of engineers continued to multiply, their importance as carriers of professionalism was soon overshadowed by the emergence of national engineering societies.

Two engineering societies dominated the national scene in the 1870s; they represented something like thesis and antithesis in the dialectic between professionalism and business. The American Society of Civil Engineers claimed to represent all American engineers not in military service; it maintained high standards of membership that drew a sharp line between an elite of professional engineers and all others. But its ideals were challenged by the American Institute of Mining Engineers. The AIME did not attempt to represent the engineering profession. Its fundamental aim was service to the mining and metals industries. The AIME did not restrict its membership to professional engineers. The civil engineers were attempting to separate business and engineering, the mining engineers to merge the two. Fundamentally, the ASCE stood for the ideal of engineering as an independent profession, and the AIME embodied the notion that engineering was an integral part of business.

The ASCE was the first national engineering society to endure, and it set the pace for the professional development of engineering in America. Founded in 1852, the ASCE soon lapsed into a moribund condition, but it was revived in 1867. In its earliest days the ASCE had admitted all persons professionally interested in the advancement of engineering. Slightly more than two years after its revival, however, the ASCE created a new grade of associate, identical to its previous rank of member. Members were now required to have been in active professional practice for five years and to be in charge of engineering work. In 1891, an even sharper upgrading took place. A new rank of associate member was created between member and associate, with requirements identical to those previously specified for members. The full member now had to be over thirty, have been in active practice of his profession for ten years, and in "responsible charge" of engineering work for five years. Even more

68

significant was the requirement that a full member must be qualified to design as well as to direct engineering work.

The ability to design was the crux of the ASCE's membership standards. Spokesmen for the ASCE maintained that the difference between civil and other engineers was not that between coordinate branches of engineering, but rather between professionals and nonprofessionals. As one ASCE president put it in 1895,

> Any man who is thoroughly capable of understanding and handling a machine may be called a mechanical engineer, but only he who knows the principles behind that machine so thoroughly that he would be able to design it or to adapt it to a new purpose . . . can be classed as a civil engineer.

Thus, ability to design was rooted in esoteric knowledge. Without this knowledge, the engineer was a workman rather than a professional.

The ASCE's high membership standards marked off an elite of full professionals, established a graded hierarchy of professional excellence, and protected the autonomy of the professional society. Realistically, the chief threat to professional independence lay with the great railroad corporations. A purely bureaucratic test for membership might have given a decisive position in the society to railroad employees holding managerial and supervisory positions. Although not without influence, major business interests, such as the railroads, were balanced by the substantial number of consultants and government employees among members. According to a tabulation in 1909, railroad employees, manufacturers, and contractors together constituted less than a third of all members. Consultants and government employees totaled about one-fifth each. The consultants, in particular, constituted the cutting edge of professionalism. They could be assumed to be lifelong practitioners. They were self-employed and presumably independent. High membership standards enabled such professionals to dominate the society.

After its revival in 1867, the ASCE evolved from a local to a national society. In 1870, the society adopted the principle of holding its annual meetings at various population centers across the country, which enabled more of its members to attend and participate. Equally important were measures to decentralize the society's governmental machinery. In 1873, a letter-ballot was adopted to allow members distant from New York to vote in all elections. Five years later, the ASCE created a nominating committee to maintain geographical balance in the selection of officers. The society was divided into five geographical districts, and one member of the committee was selected from each district. These measures were not without success. By 1873, residents of the New York area constituted 30 percent of the society; by 1897, the proportion of nonresident members had climbed to 80 percent.

The ASCE, however, paid a heavy price for its exclusive standards of membership. Its elitist tendencies alienated at least three important groups of engineers: the local groupings of civil engineers, the engineering specialists emerging in industry, and the younger engineers. Efforts by the ASCE to incorporate the local societies of civil engineers as branches failed because of the high membership standards of the national society. City engineers and others who were locally important did not relish the prospect of being relegated to inferior grades of membership in a national society.

In the case of the engineers in industry, it was probably the ASCE's attempt to draw a sharp line between engineers and managers that was the source of difficulty. The typical mechanical engineer in the 1870s and 1880s was simply a plant superintendent or small manufacturer. One of the most widely respected mechanical engineers of this era was John Fritz. His early training had been as blacksmith and machinist; he worked his way up to the position of plant superintendent by self-training. He was active in the American Institute of Mining Engineers from its inception in 1871, but did not join the ASCE until 1893, a year after his retirement.

The ASCE benefited the senior and successful men who qualified as full or associate members. Younger engineers who did not meet these standards found little to attract them to the ASCE; they were not even allowed to vote in society elections. Young civil engineers were to constitute the core of the protest organizations that arose to challenge the ASCE in the early years of the twentieth century.

The leaders of the ASCE do not appear to have regretted the failure of their society to become all inclusive. In fact, the society developed an elaborate set of procedures to maintain professional standards. Publications offer an important example. All papers submitted were first passed on by a committee. If accepted, the paper was distributed to the membership for comments; these monthly issues became the *Proceedings*. After further scrutiny, the paper then appeared in the annual *Transactions*, along with the written comments of other members. Committees were vested with the nomination of officers, the admission of new members, meetings, standardization, prizes, and similar functions. It was vital that this control machinery be kept in the hands of professionals. A more heterogeneous membership might allow nonprofessionals to take control and misuse these powers. The ASCE preferred to maintain its autonomy rather than to extend its influence.

The professional autonomy sought by the ASCE implied more than freedom from external control; its more profound meaning was the group's desire for moral independence. This required that practitioners look to colleagues for praise or disapproval, rather than to employers or the general public. The ASCE developed several mechanisms to enhance the importance of colleagual opinion. Informal contacts at society meetings were probably the most important. But they were supplemented by the practice of soliciting written comments on papers. Another way of expressing colleagual spirit was by awarding medals. In 1872, the ASCE established the Norman medal, the first such prize for American engineering. It was followed by several other prizes. Although never as significant as the major scientific prizes, these awards were designed to serve much the same purpose. The ASCE also encouraged colleagual spirit by participating in international professional gatherings. The ASCE was represented at the U.S. Centennial of 1876, the Paris Exposition of 1878, and the International Engineering Congress held in connection with the Columbian Exposition in Chicago in 1893.

In relation to public policy questions, as with membership, the ASCE chose the safer course of guarding its independence rather than of attempting to expand its influence. It avoided antagonizing major interests. Although the railroads and other industrial groups were unable to control the society, they constituted powerful minorities that had to be appeased. Apart from the internal strife such groups could produce, there was the further danger that members might transfer their allegiance to rival societies. The formation of the American Railway Engineering Association in 1899 demonstrated how real this threat was. Thus, the ASCE largely ignored the delicate matter of railroad mismanagement. By confining its activities in sensitive areas such as standardization and safety to innocuous topics, the society avoided internal dissension and external pressures. But the result of the ASCE's prudence was that the society came close to abdicating its social and professional responsibilities.

The ASCE's lack of aggressiveness extended to its relations with the army engineers. In 1872, the ASCE appointed a special committee to persuade the federal government to undertake tests of the strength of American iron and steel. The first effort along this line met with success, and Congress created a board of seven engineers: two from the army, two from the navy, and three from ASCE. But in 1878, Congress ended the board, and the testing machine that had been built was turned over to the Army Ordnance Department. The entire program seems to have become a casualty in the continuing hostilities between military and civilian engineers in America. Despite a congressional admonition in 1881 to "give attention to such programme of tests as may be submitted by the American Society of Civil Engineers," the army maintained that funds were insufficient to perform the tests desired by the civilian engineers. The leaders of the ASCE avoided open conflict with the army and confined themselves to mild suggestions that a new testing machine be constructed for the use of civilian engineers. They left it to engineers affiliated with local and state societies to continue the battle.

By the end of the nineteenth century, leaders of the ASCE viewed the progress of their society with a satisfaction bordering on complacency. The success of the ASCE rested on its very high professional standards of membership and the diversity of occupational roles open to civil engineers. Corporation employees were not in ascendency; nor was any single industry dominant. Presidents of the ASCE took a high degree of professional autonomy for granted. They depreciated the disagreements between civil engineers serving as expert witnesses before the courts and suggested, instead, that the professional engineer ought to serve the courts themselves rather than the litigants. Such a role implied that engineers should be independent of specific commercial interests.

But the ASCE's professionalism was not without its drawbacks; the society did not serve the needs of the bulk of those who considered themselves professional engineers. It demanded a higher level of professionalism than the heterogeneous character of American engineering warranted; and, in consequence, rival societies sprang up that offered different balances between professionalism and business.

At the turn of the century, the American Society of Civil Engineers was smugly complacent. It had long since achieved the professional goals that electrical and mining engineers were struggling to gain. It had austere

71

entrance requirements that excluded all but professional engineers from its higher ranks of membership. It embodied, at least in theory, the principle of professional unity; civil engineers were all engineers not engaged in military service. Its members, therefore, rejected [Andrew] Carnegie's offer in 1903 of a unified headquarters for the four major engineering societies.* They viewed with disdain the lower entrance requirements of other societies and saw the proposed building as a scheme by the other societies to bask in the ASCE's accumulated prestige. Those civil engineers who had been able to secure for themselves an occupational role as independent consultants looked down on the employed engineers in other societies as unworthy of being considered truly professional.

But behind the placid exterior of the ASCE, the profession was in a ferment. The society was living on the prestige of the past, when civil engineering had been the queen of applied sciences. Those days were over. Much of the glamour of engineering had passed on to the newer fields, such as electrical engineering. But the schools continued to turn out great numbers of civil engineers, who found that the profession was overcrowded. Economic pressures, especially on the younger civil engineers, added to a pervasive discontent in the early twentieth century. Other factors influenced even those engineers who were financially comfortable. Set in a pattern of an older age, the ASCE gave little support to the new aspirations of engineers. It did little to support engineers in government service or to encourage a wider application of engineering to social problems. Above all, the society did nothing to assuage the mounting status anxieties of American engineers. It refused, for example, to adopt a code of ethics--a step advocated for its prestige value, rather than for any serious concern with self-policing. It showed no interest in a policy of militant professionalism, at a time when drastic action appeared to many engineers to be the only course that might salvage the status of their profession.

"We see, then, that the businessman is the master; the engineer is his good slave," commented one engineer. This statement, published in the *Engineering News* in 1904, precipitated a flurry of letters and editorials on the subject of the status of the engineer. Two points emerged from this discussion with stark clarity: the engineer was poorly rewarded and lacked power in relation to his employers. Some engineers thought the only remedy lay in the engineer becoming a businessman. Others vehemently dissented. The engineer was a scientific man and should pursue his profession even if it did not make him wealthy, they argued. The debate over the role and status of the engineer was intensified by an editorial in the New York *Evening Post*, in 1907, entitled "Commercialized Engineering." The editors pointed out that the alliance of engineers with predatory monopolies was contrary to the interests of the public as well as the profession. They went on to suggest that professional societies should determine "what the relations of special talent should be to employing corporations, what the obligations of a man to the severe demands of science when they are in conflict with mere money-making." *Engineering News* condemned commercialism and urged

----

* American Institute of Electrical Engineers, American Institute of Mining Engineers, American Society of Civil Engineers and American Society of Mechanical Engineers.

engineering societies to adopt codes of ethics to check the "business view" of engineering. Charles Whiting Baker, the editor, summed up the progressive viewpoint in a signed article. "And is the engineer a mere passive looker on in this struggle to establish anew our liberties?" he asked. He suggested that the engineer was more than a "cog in the industrial machine," and that by vigorous action he could establish his rightful position as "more than a faithful slave, or a faithful overseer."

Of all civil engineers, it was the younger men who were the most alienated by the ASCE's conservatism. They had virtually no say in the society's policies, since junior men could not vote in society elections. The society was geared to the interests of an elite of senior and successful men; it showed no real concern for the plight of the younger men. Indeed, the senior men sometimes exploited the younger engineers. As one younger engineer bitterly commented, "we can expect very little encouragement from the generals of the profession who are mainly responsible for the conditions in which young engineers find themselves."

In the spring of 1909, the young men took matters into their own hands by forming a protest organization, the Technical League. Its membership, confined to civil engineers, was at first secret, since its leaders feared retaliation on the part of employers. Largely centered in New York, the Technical League's principal aim was the "closing" of the engineering profession. This they proposed to achieve by a licensing law that would restrict the practice of engineering to those with four years of college education. Other aims included publicity, the inculcation of professional ethics, and the elevation of the dignity and *esprit de corps* of civil engineers. The president of the Technical League denounced the ASCE as "a mere private corporation ruled by a handful, mostly New Yorkers."

The new organization met hostility and criticism from older engineers. President Onward Bates of the ASCE, while acknowledging that the engineer's lot was not satisfactory, insisted, "there is no room in the profession for trade unionism." A. L. Dabney noted that he used to employ uneducated persons as rodmen and chainmen, but that he had found college-trained engineers were willing to accept such jobs for the same compensation. He criticized the Technical League on the ground that there would be less demand for young engineers at higher prices. Charles Whiting Baker expressed a more sophisticated opposition, which was typical of the attitude of many progressive engineers. He blamed the over-rapid expansion of engineering education for the overcrowding of the profession. He thought the remedy lay in broadening the training of engineers, presumably by lengthening the course of study. This would serve the twofold purpose of limiting the supply of engineers and preparing engineers to serve as leaders of public affairs.

The young men were not content to wait for a long run solution. The Technical League got a licensing law introduced before the New York Legislature in 1910. To get a license, a candidate would be required to have four years of college. The bill was intended to cover all engineers, subordinates as well as those in responsible charge of work. This measure met the united opposition of senior civil engineers, progressive and conservative alike. A number of prominent engineers testified against the bill, including the president of the ASCE, John A. Bensel. J.L. Raldiris, the

73

sponsor of the licensing law, suggested a compromise: that both parties agree on a model law. The ASCE went along reluctantly with this suggestion. In 1910, a special committee of the society on licensing helped to draft the model code. The ASCE's bill watered down the Technical League's proposals. This measure did not close the engineering profession; it applied only to senior men and it did not require a college degree.

The Technical League appears to have collapsed soon after the failure of its licensing bill. But its very existence and its attraction of approximately one thousand members in New York suggested that discontent among the younger members was widespread. It had a permanent impact on changing the ASCE's policy of isolation. The ASCE's special committee on licensing concluded the society should assist in the formulation of all laws affecting civil engineering. . . . The committee concerned itself with such issues as the tenure of public officials, a suggested department of public works, and the regulation of public utilities.

The younger engineers were not the only ones discontented with the ASCE's policies; progressives were unhappy because of the lack of genuine democracy within the society. Only members who held the two highest grades could vote at all. . . . Contests for society offices were rare. The society opposed, at least informally, any publicity of its internal affairs. . . .

Pressure for the reform of ASCE came from a rather loose alliance of progressive civil engineers. Their most vocal spokesman was probably Charles Whiting Baker, editor of *Engineering News*, the leading technical journal in civil engineering. The progressive's greatest strength lay outside of New York, where ASCE members sometimes felt left out by the dominant group. Many such members were active in local engineering societies. These organizations had lower entrance requirements than the ASCE, but they were often more involved in advancing professionalism than was the national society. The reformers, therefore, favored a drastic decentralization of the ASCE, with the creation of autonomous sections that would cooperate in professional matters with local societies. They hoped to make these sections the basic building blocks of the society: each holding its own elections and governing its own affairs. Above all, progressive engineers wanted their profession to act positively to advance the welfare and status of engineers; they rejected the "fallacious idea" that the only functions of engineering societies were holding meetings, discussing papers, and producing technical literature. But progressives insisted that before organizations like the ASCE could undertake new responsibilities, they should become "truly representative of the membership at large."

. . . The leaders of the ASCE were willing to offer three substantial concessions to the dissidents. First, the society in 1914 adopted a code of ethics. Second, the ASCE abandoned its outdated claim to represent the entire profession, and it began to cooperate with other major engineering societies. . . . Finally, the society accepted a significant measure of decentralization.

The Building of a Theory

Chapter 1 of **Men and Ideas in Engineering**
R. A. Kingery, R. D. Berg and E. H. Schillinger

*This selection from a collection of twelve stories about technological
innovations which were rooted in the University of Illinois College of
Engineering describes the evolution of technical expertise in the structural
discipline of civil engineering. Although this particular example occurs in an
academic setting, it is representative of what is required to build a team of
experts in any technical organization--government, private corporation or
university. The coming together of two or more individuals with a common
bond of knowledge in an environment that allows (and perhaps even requires)
them to challenge, stimulate, reinforce, support and encourage one another
creates the kind of creative intellectual atmosphere that fosters innovation.*

*Despite significant differences in personality and teaching styles, all of the
engineer-teachers who established and nurtured the tradition of excellence in
structural engineering at the University of Illinois were recognized as good
teachers--men who could communicate knowledge and inspire students. Their
teaching methods, as described in this reading, would not endear them to
today's students or administrators, but they clearly commanded the respect of
their peers at the time, and their contributions to the profession have stood the
test of time.*

On July 28, 1957, Mexico's worst earthquake in 50 years devastated Mexico
City. A dozen buildings collapsed and many more were so greatly damaged
that they would have to be demolished. In nearly every large building there
were shattered windows and broken plaster; sprung elevator shafts and
broken plumbing were commonplace, sidewalks were tilted as much as 45
degrees, and buildings that had been standing side by side were learning apart
at odd angles. The city was shattered.

Amid the rubble the city's tallest building, the 43-story Latino-Americana
Tower, stood unharmed. Not a window or a partition in the giant building was
cracked. The building's survival was not an accident. The story of its
dramatic success had begun 80 years earlier with the arrival of a twenty-year-
old engineering student on the campus of the Illinois Industrial University.

Arthur Newell Talbot came to the campus in 1877 on a horse-drawn streetcar,
the only public transportation connecting the University and the neighboring
towns of Champaign and Urbana. This tiny, isolated outpost, consisting of a
handful of mismatched buildings in the middle of the prairie, would eventually
become the University of Illinois. Except for a four-year period, Talbot would
spend the rest of his life here.

After graduating in 1881, he followed a boyhood interest in railroading and
went to Colorado to survey for a narrow-gauge line through the mountains. In
a letter to a friend in Urbana, he described the job and some of his co-workers:
"The workmen of our party are mountain roughs; gamblers and jailbirds some
of them are. When the engineer of the party took us up to the bar on our
arrival, and I called for lemonade, there was an audible smile which went over
the boss' face as well as the rest of the crowd, ten in all, who all took their

whiskey." Such attitudes set the tone for Talbot's relationship with people all his life.

In 1885 Talbot, now twenty-eight years old, came back to his alma mater as an assistant professor of engineering and mathematics. These were depression years and the 28 faculty members earned average salaries of $2,000 (not including Regent Peabody, who made $4,000). Salaries had been declining along with student enrollments. It was hardly an auspicious start for a career.

Talbot's first assignment was in the Civil Engineering Department. There he began teaching courses in the new subject of engineering mechanics, and in 1890 the University administration recognized its popularity and importance by organizing the Theoretical and Applied Mechanics Department. Talbot was placed in charge of this department, a position he would occupy during the rest of his career at the University. His jaw-breaking title was "Professor of Municipal and Sanitary Engineering in Charge of Theoretical and Applied Mechanics."

The new department head prided himself on his ability to learn the names and faces of all his students in a single class meeting; his lectures, however, were impersonal and always seemed to be addressed to the back wall of the room. In his course TAM 6, Engineering Materials, he would enter the classroom and lecture through the period without ever looking at the class. Many of the students called this course "Sleep 6" and him "Stoneface."

Talbot once decided to illustrate his lectures with slides. The lights were turned off during the first lecture. When they were switched on again, the room was empty except for the projectionist and Talbot.

The "Stoneface" label might have applied to his family life as well--he was a 24-hours-a-day engineer who spent his own time reading technical journals, even during meals. His vacations were often spent reading and writing technical articles with the help of a secretary who sometimes accompanied him and his family on these holidays.

Talbot's accomplishments in research brought him recognition early in his career. His formulas for relating the rates of rainfall runoff to the sizes of waterways became standards for engineering practice. He developed the "Talbot Spiral" or "railroad transition spiral," a flexible method of laying out gradual, jolt-free railroad curves. At his urging, facilities for hydraulic studies and the testing of construction materials were established at the University. He pioneered in sewage treatment with septic tanks, testing of roadway materials, the design and use of reinforced concrete structures, and research on stresses in railroad rails.

In spite of the importance of these accomplishments, Talbot could never quite understand the fame he achieved. He once said that he thought the many awards and honors he received could have been "accidents," many of the later ones coming because "once you get one, you get others." He felt that he lived in a time of opportunity when little had been done, and that his accomplishments were commonplace. "It has been surprise after surprise to find that my own work has been widely accepted, generally commended, and has received such continuing recognition."

Talbot did achieve an international reputation, but apparently neither he nor any of his contemporaries realized that his greatest claim to fame was the number of other outstanding engineers he attracted to his profession and the tradition of excellence he established at Illinois.

The first of the men who came to work with him was a dark-haired, gaunt-faced engineer named Wilbur Wilson. Wilson was thirty-two when he received a one-year appointment as an assistant professor in 1913. During that year, however, the Western Society of Engineers awarded him the Chanute Medal for the best paper of the year presented before the Society, and he was invited to stay at Illinois as a permanent staff member. Although he was popular with his colleagues and students, Wilson did share Talbot's Calvinistic attitude toward life: their co-workers were afraid to smoke, chew, or swear near either of them.

Early in his career Wilson became a proponent of a newly developed structure, the rigid-frame bridge. Cheaper to construct and maintain, as well as stronger than older style bridges, the rigid-frame bridge was a simple structure with top and sides of one solid piece of reinforced concrete. The design was so novel that it took many years to gain acceptance, but eventually it became the most popular kind of small bridge.

Wilson was as interested in the deterioration of bridges and other structures as he was in their design. His work with metal fatigue enabled engineers to determine the most likely locations of fatigue cracks, so that the flaws could be detected before they endangered the life of the structure.

One winter day, while testing a bridge to learn how temperature was affecting it, Wilson worked so long in the cold that he froze part of one hand. Because of this experience and because he disliked waiting for years to observe fatigue developing naturally in buildings and bridges, he joined Herbert F. Moore, famed for his work with metal fatigue, in designing and building testing machines that would give the same results in the laboratory in much shorter periods of time. Many of these testing machines were used for decades on the Illinois campus, and similar machines were used in other laboratories around the world.

In the last years before he retired in 1949, Wilbur Wilson made his greatest contribution to engineering. In designing, constructing, and testing structural members, he had become interested in how they were fastened together. Joints were made primarily by two means, welding and riveting, and Wilson had done a great deal of research on both. Rivets were expensive to install and difficult to replace when they failed. Wilson was particularly concerned because in many structures, particularly bridges, rivets were failing at an alarming rate. If bolts were strong enough, he felt, many of these problems would be solved. His research indicated that joints fastened with high-strength bolts could be stronger than riveted joints.

Many engineers thought Wilson was crazy: "If rivets, which fill the holes completely, can work loose, how can you think that bolts won't?" But rivet and bolt manufacturers had more faith in Wilson, and they performed tests that agreed with his. As it became clear that bolts were not only stronger, but also easier and cheaper to replace than rivets, engineers began to realize the fasteners' potential. Wilson was instrumental in the formation of the

Research Council on Riveted and Bolted Structural Joints, and through that organization carried on further research. Eventually such structures as the Pan American Building in New York City, which used over four million high-strength bolts, and the Verrazano Narrows Bridge, with three million, would testify to the value of Wilson's idea.

The next man after Wilson to join the structural engineering team was Harald M. Westergaard. "The Great Dane," as he was called, had been studying in Germany when World War I started. He found it necessary as a foreign national to leave quickly. "Go to Illinois and study under Talbot," a friend advised, and Westergaard followed the advice. He was one of the first students to qualify for a Ph.D. in theoretical and applied mechanics at Illinois, and in 1916 he became a member of the staff.

Westergaard was an unusual man. He was a scholar with a wide range of interests, an ability to speak and read a number of languages, and an exceptional talent for theoretical and mathematical problems. Unlike Wilson, he disliked working with his hands. He had one known affectation: he carried a cane because "one should have an eccentricity to be remembered by." He was as likely to be remembered for his absent-mindedness. Sometimes he drove his car to work, forgot it, and walked home.

Westergaard fit the picture of an aloof, erudite professor, but his reputation as a "loner" was not fully deserved. Slightly deaf, he was often preoccupied with his thoughts. "If Doc ever snubbed you," one associate said, "it was for one of three reasons: he didn't hear you, he was thinking about something else, or you were such a hopeless fool he couldn't tolerate you."

He was not an easy man to meet. Because he disliked coming to work early, he sandwiched his regular office hours into the ten-minute periods between his morning classes. A little sign on his office door read: "Office hours--10:50 to 11:00 a. m., Monday, Wednesday and Friday."

But the Great Dane was an outstanding speaker and writer. To many of his students his insistence on precision and style seemed to border on fanaticism. He was seldom appreciated by undergraduates, but he was a remarkable graduate instructor. His exams sometimes contained trick questions that his students either saw through and answered immediately or never managed to answer at all.

Westergaard's meticulousness and originality were unmixed virtues for research. He was an expert in structural mechanics, the analysis of pavement slabs, and the effect of earthquakes on structures. His studies of Japanese earthquakes with Mikishi Abe, a Ph.D. student of Professor Talbot, had an important influence on Japanese building design.

Westergaard became the special analyst for the Stevenson Creek experimental arch dam in California, and was later hired as chief mathematician on the specialized design problems of Hoover (Boulder) Dam. Part of his work was to determine the degree of earthquake resistance necessary for intake towers in Lake Mead. Another part involved calculating the stresses that could be imposed on the ground by the world's largest man-made lake.

While Westergaard was establishing his reputation, another man joined this brilliant circle of structural engineers. Hardy Cross, an engineer philosopher whose specialty was structural analysis and design, came to Illinois in 1921.

The man who would become known as the greatest teacher of structural engineering of all time looked to his students like a bad-tempered, sarcastic perfectionist. Cross, like Westergaard, suffered from deafness, and he used this handicap to his own advantage both in and out of the classroom. Students found it difficult to improvise answers at the tops of their voices, and soon learned either to be explicit or to admit that they couldn't answer his questions.

In the classroom Cross lectured without notes, but his performance was always calculated to create the atmosphere he wanted. Occasionally he would stomp out of the classroom early because no one had attempted a certain problem, then later ask someone who had observed the exit, "How do you think they took it?" If everything had gone well in a class, Cross would walk to the door at the end of the period, smile at the students, and then rush out of the room. After observing one of these performances, Wilson remarked, "It's just like the Cheshire Cat's smile in *Alice in Wonderland*--it's still hanging there after he's gone!"

Sometimes he chose to play an intellectual devil's advocate. Once a student named Alford told Cross that he thought one of the problem solutions in their text was wrong. Cross paced back and forth, staring hard at the student, and pointing at him fiercely. "Can you, a graduate student, actually have the temerity to accuse the internationally known engineer who wrote this book of MAKING A MISTAKE? Can you really believe that the publishers would allow such an alleged error to be printed? Can you show us the error?"

Alford seemed unable to answer.

Still pacing, Cross said, "Can anyone help Mr. Alford? Do any of you see a mistake in problem four?"

The class was silent.

"Well, Mr. Alford," Cross said sternly, "would you care to retract your accusation?"

"It's just that I can't . . ."

"Speak up!" Cross thundered.

"I still believe it's wrong!" Alford shouted, his face red with embarrassment.

"Then kindly come to the board and prove it to us," Cross taunted. "We shall be pleased to see the proof of your unfounded allegation."

Alford labored at the board without success for the rest of the period.

Cross began his next lecture by saying, "In our last meeting Mr. Alford raised a serious and unfounded charge against the author of our text." Staring at Alford, he said, "Have you reconsidered your accusation?"

"No, sir," Alford replied. "I still believe he is wrong."

"To the board, then. We still await your proof."

Alford's labors were again unsuccessful.

The third time the class met, Cross said, "Mr. Alford, are you ready to withdraw your ill-considered accusation about problem four?"

Moments later Alford was at the board. Within a few minutes he managed to show that the solution to the problem in the book was incorrect, and he returned to his seat. Cross's pleasure was evident from his expression. "You must always have the courage of your convictions," he said. "Mr. Alford does; apparently the rest of you do not, or you are not yet sufficiently well educated to realize that authority--the authority of a reputation or the authority of a printed page--means very little. All of you should hope to someday develop as much insight and persistence as Mr. Alford."

Cross believed that the classroom was the place to develop the student's personal ingenuity and self-confidence. "The Univesity is a place to get into as much intellectual trouble as possible, a place to make mistakes, many mistakes, and to rectify them."

Dean Milo Ketchum resented Cross's reputation as a great teacher. During his term as dean he had prevented Cross from getting salary increases, and had suggested a number of times that Cross might be better off working elsewhere. One of the charges Ketchum leveled against Cross was that he failed to publish enough papers, although Cross's publications between 1925 and the early thirties were among the most important and far-reaching of any written in the college during that period.

Finally, in answer to Ketchum's threats, Cross published a ten-page paper in the *Proceedings of the American Society of Civil Engineers* that was a classic. The paper, entitled "Analysis of Continuous Frames by Distributing Fixed-End Moments," set forth a new method of analyzing building frames. Cross won the Norman Medal of the ASCE for it, and the technique became known the world over as the "Hardy Cross method" or the "moment distribution method."

Cross later quipped that there were "no further discussions with the Dean on the subject of my being a Cross the College wouldn't bear."

Thus by the early 1920s structural engineering had become an important part of the University of Illinois. The first string team consisted of Talbot, founder of the tradition; Wilson, the experimentalist; Westergaard, the theoretician; and Cross, the philosopher. All of them drew other outstanding men to Illinois.

In 1930 the man who was to inherit the expertise accumulated over the years at Illinois came to the University to do graduate work in structural engineering. Nathan M. Newmark was a brilliant student who had earned a bachelor's degree from Rutgers University before he was twenty years old. Like many others, he came to study under the giants of the field at Illinois.

In Newmark's first encounter with the engineer-philosopher Hardy Cross, Cross asked where each student had studied. When Newmark answered, "Rutgers," Cross looked down his nose and answered, "You've got a lot of things to unlearn."

But in time Newmark and Cross developed a mutual admiration and a broad spectrum of interests. Their relationship was based on the interplay and exchange of ideas--not only in engineering, but in every conceivable subject. They discussed with equal relish politics, philosophy, art, or anything else that interested them. As Newmark put it, his part in these discussions "must have been audible for blocks" because Cross was so deaf.

Newmark's initial appointment was as a research assistant working under Professor Wilson. In 1934 he received the third Ph.D. ever awarded by the Civil Engineering Department. Because of his outstanding work, Wilson recommended that a permanent position be created for him on the civil engineering staff. In a letter of recommendation for this appointment Cross said, "He is, I think, the ablest man we have ever had in graduate work. He has a great power for originating methods and viewpoints." In another letter Westergaard said, "I believe that his appointment will be a contribution to the solution of the difficult problems, a source of future distinction for the University. His intellectual capacity is rare, his personality attractive. He is among the few who can be rated as truly brilliant."

The Dean and the head of the Department felt that Newmark's great potentialities could best be developed by allowing him to work on whatever interested him most. The first project he undertook was an extended, detailed study of stresses in building foundations.

The team had begun to break up before Newmark joined it. Talbot, endowed with an international reputation, had retired in 1926. His pioneering work had touched on nearly every aspect of civil, hydraulic, structural, sanitary, and railway engineering, and his name was solidly linked with the diverse uses of concrete as an engineering material.

The second man to leave the team, Hardy Cross, resigned to take the position of Chairman of Civil Engineering at Yale in 1936. The following year Harald Westergaard resigned to become the Dean of the Harvard Graduate School of Engineering.

With Wilson's retirement in 1949 all four of the great engineers were gone from Illinois. They had been rewarded handsomely by the profession they had served. Between them they had accumulated ten honorary degrees, fifteen medals, four plaques, and a certificate of appreciation for determining the causes of ship plate fracture during World War II. Talbot had been especially honored by Illinois: the Materials Testing Laboratory had been renamed the Arthur Newell Talbot Laboratory. The engineers' monuments were everywhere: in buildings, railroads, highways, dams, culverts, and bridges. They left a tradition of excellence for the structural engineers who would follow them at Illinois--and their protege, Nathan Newmark.

Newmark by now had had fifteen years to develop his own ideas. Deeply influenced by his predecessors, he was left to carry on the work the others had started.

In 1950 Newmark was given an opportunity to put into practice the knowledge that had accumulated during the last 70 years. The project would demand knowledge of all the aspects of structural engineering which had been the strongest interests of Newmark and his teachers: reinforced concrete, foundations, steel frame connections, and factors related to earthquake resistance. He was invited to serve as a consultant on the design of the Latino-Americana Tower in Mexico City.

It was a complex problem. The tower would be built in the heart of an area plagued by earthquakes, on the poorest subsoil imaginable for a tall building. Two of Newmark's former students, Leonardo Zeevaert and Emilio Rosenblueth, worked with him on the project. The steel for a 28-story building had been purchased and was on the site when Newmark arrived to discuss the project with the young Mexican engineers. They concluded that the building would be too weak for the earth motions it might be forced to endure.

Newmark suggested that with a better design the building could go as high as 43 stories. The redesigning began. The ground itself created problems: to a depth of 100 feet or more there was watery clay; below this were layers of fine sand, gravel, and more clay. Zeevaert and Newmark designed the Tower to stand on 361 one-hundred-foot concrete pilings that were driven down to the level of the sand, 117 feet below street level. The foundation was planned to behave as a floating box, sitting on these roots in such a way that the load of the building was carried by the piles to the thin, hard stratum of sand. The building was to be as light and flexible as possible. The tops and bottoms of all windows were anchored with single bolts, and small spaces were left between the tops of partitions and the ceilings. The building was designed to withstand earthquake tremors three times stronger than any previously recorded in Mexico City.

As completed, the Tower stood 43 stories high, with a 138-foot television antenna towering more than 600 feet above the city. Instruments were mounted at various elevations to record motions of the floor due to winds or earthquakes. Newmark's part of the work was concluded in 1954, and the structure became another part of his past.

After the earthquake of 1957, engineers went into the Tower and examined the recording instruments. The top of the building, they learned, had whipped back and forth approximately a foot. Cushioned by its special foundation, the gigantic structure had undergone almost exactly the forces it had been designed to withstand.

Many sightseers, walking or driving through Mexico City after the earthquake, were amazed to see the city's tallest building standing undamaged. Four of the men whose work had made the Tower possible would not have been surprised at all--but none of them were ever to see it.

"If You Go Home Tonight, Don't Come Back Tomorrow"

Excerpts from Chapters 19 and 20, **The Power Broker**
Robert A. Caro

*During periods such as the past twenty years or so, when those who are
responsible for "getting things done" often find themselves handcuffed by rules
and regulations which--while designed to promote safety, or efficiency, or the
public welfare--seem to produce interminable delays and insurmountable
obstacles, it is difficult to remember that there were times when such
restrictions did not exist (or were easily circumvented). Many of the public
works projects which we look upon now as essential elements of modern
American life were produced in such times. It is doubtful whether such projects
could be duplicated today, even though technological advances have
theoretically made it possible to do more with less.*

*This selection from Robert Caro's biography of Robert Moses reminds us of the
way things were in a different time. It graphically displays the role that
"vision" plays in accomplishing great undertakings. It also reminds us of some
of the hardships men endured to produce works of "engineering art" that we
take for granted today. Finally, it reminds us that the rules and regulations
which so often hamstring us in our efforts to get things done are, in fact,
necessary to prevent unscrupulous people from abusing the human condition
and quashing due process.*

There were now [1934] seven separate governmental agencies concerned with
parks and major roads in the New York metropolitan area. They were the
Long Island State Park Commission, the New York State Council of Parks, the
Jones Beach State Park Authority, the Bethpage State Park Authority, the
New York City Park Department, the Triborough Bridge Authority and the
Marine Parkway Authority. Robert Moses was in charge of all of them.

The New Deal's attempts to combat the Depression had apparently already
given New York an opportunity to refurbish existing city parks. Harry
Hopkins' federal Civil Works Administration (CWA), set up in November 1933,
had 68,000 men working on park clean-up projects in the city by Christmas.
But Moses and his top Long Island park administrators, driving around to the
parks to see what these men were doing, found that the city had given them
neither adequate tools, materials, supervision nor instruction. Crews were
laying asphalt roads and paths without adequate foundations--and even as
they laid one section, another, completed a week earlier, was already heaving
and cracking behind them from frost action. Six thousand men, assigned to
"move ash dumps" in Riverside Park, were standing on the banks of the
Hudson pecking at frozen cinders; two thousand were standing on truck beds
on a little reef off Staten Island "building up" the reef by dumping out sand--
which was washed away, at a cost of five dollars per cubic yard, almost as fast
as they could dump it. Fifty-four hundred more were assigned to Brooklyn's
Marine Park . . . Moses' engineers sneaked into the cupola of an old mansion
in the park so that they could watch the work unobserved--and found that
there was nothing to watch. Spread out over expanses of sand more
reminiscent of a French bivouac during the retreat from Moscow than a park
reclamation project, all but a handful of the fifty-four hundred sat huddled
around small fires built against the freezing wind whipping out of Jamaica

Bay. Some were passing around wine bottles held in brown paper bags. Others were throwing dice. Most had no tools--and Moses' men understood why when they saw men chopping up shovels and using their handles as firewood. Adding a poignant detail to the scene were a few men who had kept their tools and who obviously wanted to work; they spent hours "raking" the frozen ground or building little fences out of stone they found in the area, "just so," as one of them was later to recall, "I could feel I was doing something to earn my money."

Moses dispatched teams of engineers to "inventory" New York City's parks--their acreage (incredibly no one knew their exact size), the buildings, paths, roadways, statues and equipment in them, the condition of these items and the type and amount of labor and materials that would be required to renovate them. He filed this information in a loose-leaf notebook kept atop his desk. By the time he was sworn in as Park Commissioner, the notebook was more than a foot thick, and he had a list of 1800 urgent renovation projects on which 80,000 men could immediately be put to work.

By late December [1933], the outline of his ideas for large-scale park construction projects was ready . . . One crew [of engineers] would drive with him to certain parks--describing these trips, one engineer echoed Frances Perkins' words of two decades before, "Everything he saw made him think of some way it could be better"--and then that crew would go back to Babylon and translate his ideas into general engineering plans while another would head out with him to other parks.

Relays were needed to keep up with him. "His orders just poured out," recalled the engineer. "Bam! Bam! Bam! So fast that we used to all try to take them down at once so that when we got out to Babylon we could put them together and maybe get one complete list of everything he wanted. You'd start at dawn--hell, sometimes we'd start before dawn; I remember driving around Manhattan when everybody was still sleeping except the milkmen, maybe, and the cops on the beats . . . and by late afternoon, I can tell you, your head would be absolutely spinning. But he'd still be firing things at you." Didn't they break for lunch? "You didn't break for lunch when you were out driving around with Robert Moses."

Soon the engineers' concepts of his ideas were being presented for his approval.

On January 19, 1934, Governor Lehman signed the legislation allowing Robert Moses to become the first commissioner of a citywide Park Department. At 5 p.m. that same day, Mayor La Guardia swore Moses into office at City Hall. As soon as his right hand came down from the oath, Moses turned to the assembled reporters and told them he had an announcement to make: the five borough park commissioners and the five borough park superintendents, along with the commissioners' personal secretaries and stenographers, and assorted deputy commissioners and other top-level park aides, were fired--"as of now."

The next morning at 9 a.m., a fleet of chauffeured black Packards roared up to the curb on Fifth Avenue in front of the Arsenal. Out of them stepped Moses and a squad of his top Long Island aides. These men would be running the Park Department from now on, Moses announced. Leading them up the steps of the Arsenal, which he announced would henceforth be departmental headquarters, he assigned them offices.

Moses had given these men their orders. They were to weed out--immediately--those headquarters employees who would not or could not work at the pace he demanded. The weeding out would be accomplished by making all employees work at that pace--immediately.

Unlike the commissioners and their personal secretaries, most headquarters employees were protected by civil service, but that didn't help them much. Men who lived in the Bronx were told that henceforth they would be working in Staten Island; men who lived in Staten Island were assigned to the Bronx. Or they were given tasks so disagreeable they couldn't stomach them. Women were treated no better. One ancient biddy, accustomed to spending her days at the Arsenal knitting in a rocking chair, refused to admit she was over retirement age and gracefully accept a pension. When a search failed to produce a birth certificate to disprove her story, she was ordered to work overtime--all night. Every time she tried to rest, she was ordered to keep working. She retired at 2 a.m.

Down at his State Parks Council office at 80 Centre Street--Moses was never to have an office at the Arsenal and was to visit the building infrequently during his twenty-six years as Park Commissioner because he didn't want to be accessible to departmental employees--Moses was confronting the CWA. Its officials were worried themselves about the demoralization of the 68,000 relief workers in the parks. Moses told them that the first requirement for getting those men working on worthwhile projects was to provide them with plans. Blueprints in volume were needed, he said, and they were needed immediately. He must be allowed to hire the best architects and engineers available and he must be allowed to hire them fast. The CWA must forget about its policy of keeping expenditures for plans small to keep as much money as possible for salaries for men in the field. The agency must forget its policy that only unemployed men could be hired so that Moses could hire a good architect even if he was working as a ditch digger or was being kept on by his firm at partial salary. And the agency must drop its rule that no worker could be paid more than thirty dollars per week. The CWA refused: rules were rules, it said. I quit, Moses said. La Guardia hastily intervened. After seven days of haggling, the CWA surrendered. Moses was given permission to hire 600 architects and engineers without regard to present job status, and to pay them up to eighty dollars a week. So that he could hire them as fast as possible, he was given an emergency allocation to summon them to interviews not by letter but by telegram.

The CWA capitulated on the morning of January 27. By noon, 1,300 telegrams were being delivered to carefully selected architects and engineers all over New York State, telling each of them that if he was interested in a job, he should report to the Arsenal the next day.

No profession had been hit harder than architecture and engineering. Engineers were particularly reluctant to accept relief. "I simply had to murder my pride," one said. "We'd lived on bread and water for three weeks before I could make myself do it." But The Nation estimated that fully half of all engineers were out of work--and six out of seven architects.

85

These men had been hiding out in public libraries to avoid meeting anyone they knew, or trampling the streets carrying their customary attache cases although those cases contained only a sandwich. Although the Park Department interviews weren't supposed to begin on January 28 until 2 p.m., on that day, with the temperature below freezing, when dawn broke over the city, it disclosed a line of shivering men outside the Arsenal. The line began at the front door. It wound down the steps, out to Fifth Avenue at Sixty-fourth Street, and along the avenue to Seventy-second Street.

The interviewing went on all day; some of the men who had been waiting on line before dawn didn't get into the Arsenal until late afternoon. But for 600 of them the wait was worth it.

"When you got inside, nobody asked you how much money you had in the bank or what was the maiden name of your great-grandmother," one architect recalls. "All they asked you was: "What are your qualifications?" Those whose qualifications satisfied Moses' men were hired on the spot, shown to the Arsenal's garage, in which drafting tables had been set up, handed assignments and told to get to work. In the evening, some started to go home. Those whose assignments didn't have to be finished for a few days were allowed to do so; several hundred whose plans were needed faster were told flatly, "If you go home tonight, don't come back tomorrow." Without exception, these men stayed, catching naps on cots Moses had had set up in the Arsenal's corridors.

Out in the parks, the ragtag ranks of the CWA workers were being shaped up.

Moses had found the men to do the shaping. Out of fear of losing them permanently to rivals, idle construction contractors were struggling to keep on salary their "field superintendents," the foremen or "ramrods" whose special gift for whipping tough Irish laborers into line made them an almost irreplaceable asset. Pointing out that the CWA was not a rival and would probably go out of existence when business improved, Moses persuaded contractors throughout New York, New Jersey, Pennsylvania and New England to give him their best ramrods, "the toughest you've got." And when the new men arrived, 300 in the first batch, 450 more within two months, his instructions to them were equally explicit. CWA workers, he was to say later, "were not accustomed to work under people who drove them. I see to it that my men do drive them." Arriving at Marine Park on January 31, new superintendent Percy H. Kenah ordered the men away from their bonfires, and when some moved too slowly for him, fired sixty-six on the spot. The men refused to leave. They moved threateningly toward him--and then they noticed that patrol cars crammed with policemen had quietly driven up behind them. On the same day, with patrol cars backing them up, new superintendents fired hundreds of other workers in other parks.

If this method disposed of malingerers and malcontents, it nonetheless proved difficult to whip the relief workers who remained into an efficient work force. Most wanted to work at the pace demanded but were unable to do so. The suits, overcoats and fedoras many wore while wielding shovels were testimony not only to their inability to afford warmer work clothes but also to their lack of experience in performing hard physical labor outdoors. Even in mild weather, they would have had difficulty. In winter, they suffered bitterly from exposure. Since their pay was $13.44 per week, many were still scrimping on their own food so that their children could have more. "I remember guys just

86

keeling over on the job," one laborer recalls, "and you always knew that it was just that they had come to work without anything to eat." And they were scrimping in other ways, too. A worker at Dyker Beach Park at the southern edge of Brooklyn, who had caught a reporter's eye because "the side of his neck is swollen and his breath is from a sick throat," told the reporter that he lived in Manhattan. Shivering in a thin overcoat, he said that to get to work "I walk over four hours. I set the alarm clock for half past two and start walking quarter of four. The reporter asked him why he didn't take a trolley, "Carfare is twenty cents a day," he replied.

Some of the new superintendents quietly handed quarters to laborers whose inability to keep up was due to hunger or frostbite; others fired them. But none of the ramrods stopped driving. If they did, they knew, they would be fired themselves. They were, after all, working for a boss, who when questioned about a new wave of firing that almost touched off riots in several parks, said, "The government and the taxpayers have a right to demand an adequate return in good work, faithfully performed, for the money that is being spent. . . . We inherited men who were working without plan and without supervision. The plans have now been made, the supervision is being supplied, and we expect the men to work."

The winter of 1934 was the first of five of the most severe in New York's history. The temperature dropped below zero on five different days--on one day it hit fourteen below--and a steady succession of heavy storms dumped a total of fifty-two inches of snow on the city. The mean temperature for the entire month of February was 11.5 degrees. But all through that winter, the residents of the tall apartment houses rimming Central Park could look down into the park and see, in the snow, thousands of men swinging pickaxes and shovels, climbing ladders set against trees, swarming over scaffolding erected around older structures and building new ones. From behind the park's granite-block walls came the pounding of pneumatic drills, the rumble of concrete mixers, the dull roar of steam shovels and the sharp rapping of hammers.

And to the consternation of those apartment-house residents, this clangor did not stop at five o'clock. At dusk, thousands of men filed into Central Park to replace those who had been working during the day, and when the watchers in the apartment houses retired for the night--for nights that they complained were made restless by the noise--they could take a last look out their windows and see the pickaxes still swinging in the harsh glare of hundreds of high-powered carbide lamps. When they awoke in the morning, the pickaxes were still swinging--and they realized that a third shift had filed into the park during the night. The work was going on twenty-four hours a day. . . .

Sometimes, now, the laborers were even performing the construction phenomenon known as "working ahead of plans." By February, there were more than 800 architects and engineers in the Arsenal and they had become accustomed to working fourteen-hour days. But often, after they had finished a blueprint and it had been approved . . . and they rushed it themselves out to the project site, they would find that the work crews had already begun, or finished, digging ditches for pipes and foundations, or other preliminary work, and they would have to sit down on the spot and draw new plans to fit in with the work that had already been completed. The team of fifteen architects working at the Arsenal . . . to design a new Central Park Zoo . . . were,

Latham[*] recalls, "working [while] looking out the window to see what had already been done." These men . . . had to work "with little equipment, crowded together two or three to the table, and moved about from one place to another every few days." They completed the plans for the entire new zoo in sixteen days.

By March, the economy was beginning to recover and optimism was rising-- along with demands from the nation's press, heavily anti-New Deal, that the government begin phasing out the spending of "taxpayers' money" on such "socialistic" practices as work relief. Moses had been led to expect an extension of the act creating CWA, but at the last moment Congress changed its mind, and the agency went out of existence on March 31, on forty-eight hours' notice. With only a limited amount of funds from the Federal Emergency Relief Administration available for park work, half of Moses' men were abruptly dismissed. But he kept the remaining half working. The harshness of the winter persisted into April, and every weekend was either cold or rainy. But on Saturday, May 1, 1934, the weather turned balmy, and, as they do on the first warm Saturday of every spring, New Yorkers poured into their parks.

Seventeen hundred of the eighteen hundred renovation projects had been completed.

Every structure in every park in the city had been repainted. Every tennis court had been resurfaced. Every lawn had been reseeded. Eight antiquated golf courses had been reshaped, eleven miles of bridle paths rebuilt, thirty-eight miles of walks repaved, 145 comfort stations renovated, 284 statues refurbished, 678 drinking fountains repaired, 7,000 wastepaper baskets replaced, 22,500 benches reslatted, 7,000 dead trees removed, 11,000 new ones planted in their place and 62,000 others pruned, eighty-six miles of fencing, most of it unnecessary, torn down and nineteen miles of new fencing installed in its place. Every playground in the city had been resurfaced, not with cinders but with a new type of asphalt that Moses' engineers assured him would prevent skinned knees, and every playground had been re-equipped with jungle gyms, slides and sandboxes for children and benches for their mothers. And around each playground had been planted trees for shade.

"Generations of New Yorkers," as the Times put it, "have grown up in the firm belief that park benches are green by law of nature, like the grass itself." But now, as New Yorkers strolled through their parks, they saw that the benches had been painted a cool cafe au lait. Generations of New Yorkers had believed that the six miles of granite walls around Central Park were a grimy blackish gray. Now they saw that sand blasting had restored them to their original color, a handsome dark cream. . . . And a thousand plots in the parks, plots which as long as New Yorkers could remember had contained nothing but dirt and weeds, were gay with spring-blooming flowers.

----

* William Latham, an engineer who worked for Robert Moses for more than 40 years

88

By midsummer, new construction projects in the parks were being completed. Ten new golf courses, six new golf houses, 240 new tennis courts, three new tennis houses and 51 new baseball diamonds were to be opened to the public before Labor Day. The Prospect Park Zoo was completely rebuilt and a new zoo erected at Barrett Park on Staten Island. Complete reconstruction jobs were done on St. James, Crotona and Macombs Dam parks in the Bronx; Owl's Head, McCarren and Fort Greene parks in Brooklyn; Crocheron, Chisholm and Kissena parks in Queens; and Mount Morris, Manhattan Square and Carl Schurz parks in Manhattan.

On a sunny Saturday, the fence around Bryant Park came down and thousands of spectators in a reviewing stand set up behind the Lowell Fountain saw that the weed-filled lot had been transformed into a magnificent formal garden. Two hundred large plane trees, grown in Moses' Long Island Park Commission nurseries, trucked to the city and then lifted over the fence and lowered into prepared holes by giant cranes, had been planted along its edges, and their broad leaves shaded graceful benches and long flower beds bordered by low, neat hedges. The four acres they surrounded were four acres of lush and neatly trimmed grass, set off by long, low stone balustrades and flower-bordered flagstone walks, that looked all the greener against the grayness of the masses of concrete stores and office buildings around it. As a newly formed sixty-six-piece Park Department band, outfitted in white duck trousers, forest-green jackets with white belts and white caps trimmed with green and gold braid, blew a fanfare, the great-granddaughter of William Cullen Bryant, the poet and journalist for whom the park had been named, and the sister of Mrs. Josephine S. Lowell, in whose memory the fountain had been built, walked together from the reviewing stand to the fountain, escorted by twenty youthful pages and Park Department attendants in uniform, and flung handfuls of petals into it. At that signal, water gushed from the fountain's five dolphin spouts for the first time in a decade, and a speaker said that Robert Moses had outdone his biblical namesake because while the Moses of the Israelites had smote a rock in the desert and brought forth water, Moses of New York had "smote the city's parks" and brought forth not only water but trees, grass and flowers.

In Central Park, Moses' men restored Olmsted's long-defaced buildings, replanted the Shakespeare garden, placing next to every flower a quotation from the Bard in which it was mentioned, and exterminated herds of rats; 230,000 dead ones were counted in a single week at the zoo site alone. While seven hundred men were working night and day to build a new zoo, another thousand were transforming the dried-up reservoir bed that had been called "Hoover Valley"--Moses had torn down the shanty town there--into a verdant, thirty-acre "Great Lawn," were laying flagstone walks around it and planting along them hundreds of Japanese cherry trees. Then, having satisfied those who objected to use of the reservoir bed entirely for active play, Moses constructed a playground and wading pool in the northeast corner of the bed, outside the lawn's borders, for small children and a game field in the northwest corner for older children. On the North Meadow he built handball courts, wading pools and thirteen baseball diamonds. He deported the deformed sheep and turned the old sheepfold into a "Tavern-on-the-Green," an old English inn-in-a-park complete with doormen wearing riding boots and hunting coats and top hats and cigarette girls in court costumes complete with bustles--and with the added touch of an outdoor flagstone terrace on which couples could dance among tables shaded by gaily colored umbrellas to the music of a twelve-piece orchestra costumed in forest green.

And Moses was not merely beautifying the city's parks. He was doing what generations of reformers had despaired of doing: he was creating new ones--in the areas that needed them.

In his first flush of enthusiasm following La Guardia's offer of the park commissionership, Moses had believed that by forcing landlords to dump real estate on the market at a fraction of its former value, the Depression had given the city at last a chance to acquire and tear down slum tenements and use the space thus gained for play space for the slum children who so badly needed it. But then La Guardia disclosed to him the extent of the city's financial crisis and told him that, because of the Depression, even fractions were beyond the city's ability to pay.

"I remember one time he came back from talking to La Guardia and he told us this," said Bill Latham. "And I remember that he said then--don't remember the words, really, but the idea was: 'All right, then, god-dammit, we'll get land without money.'"

Moses instructed Latham to set his surveyors to making an "inventory" of every piece of publicly owned land in New York City, every tract or parcel owned by any city department, and to determine, not by asking departmental officials but by personal inspection, whether every piece of that land was actually being used. Within a month, he had learned that on the Lower East Side there were nine long-vacant strips of land along Houston Street that had been acquired by the Board of Transportation to store equipment during subway construction but that had been lying idle ever since the construction was completed, ten elementary schools so old that they had been abandoned by the Board of Education for years and five vacant lots that were owned by the Park Department itself but that the Park Department had somehow not been aware it owned. . . .In the Red Hook tenement slums, Brooklyn's version of Manhattan's "Hell's Kitchen," thirty-eight acres of land had been purchased for a public housing project, but no such project had yet begun.
    . . . And throughout all the city's slums were scores of small triangular "gores," where streets angled together or bits of land had been left over from street-widening condemnation proceedings, that were now just unnoticed pieces of dirt or concrete and that were too small to be used for play but that were, if planted with grass and a tree or two, large enough to add a touch of green to the drabness around them. Moses asked La Guardia to direct the city Sinking Fund Commission, the body which, under existing charter provisions, held the actual title to all city-owned land, to turn this land over to the Park Department. Often, the other departments involved objected to such incursions into their jealously guarded empires--the Tenement House Commission hastily began drawing up plans for the Red Hook housing project to prove that construction on it was imminent--and sometimes, as in the Red Hook case, La Guardia sided with them. But generally the new mayor backed Moses. Within four months after taking office, the new Park Commissioner had obtained, in slum areas in which there had been no significant park or playground development for at least half a century, no fewer than sixty-nine separate small park and playground sites.

Survival of the Fittest

Chapter 3, **The Bridge**
Gay Talese

*Gay Talese captures the romance and the tragedy of more than a century of
North American bridge building in this brief chapter from his book on the
construction of the Verrazano-Narrows Bridge. The book is illustrated with
beautiful sketches completed by artist Lili Rethi while the bridge was under
construction.*

The bridge began as bridges always begin--silently. It began with underwater
investigations and soil studies and survey sheets; and when the noise finally
started, on January 16, 1959, nobody in Brooklyn or Staten Island heard it.

It started with the sound of a steam pile driver ramming a pipe thirty-six
inches in diameter into the silt of a small island off the Brooklyn shore. The
island held an old battered bastion called Fort Lafayette, which had been a
prison during the Civil War, but now it was about to be demolished, and the
island would only serve as a base for one of the bridge's two gigantic towers.

Nobody heard the first sounds of the bridge because they were soft and
because the island was six hundred feet off the Brooklyn shore; but even if it
had been closer, the sounds would not have risen above the rancor and clamor
of the people, for when the drilling began, the people still were protesting, still
were hopeful that the bridge would never be built. They were aware that the
city had not yet formally condemned their property--but that came three
months later. On April 30, 1959, in Brooklyn Supreme Court, Justice J.
Vincent Keogh--who would later go to jail on charges of sharing in a bribe to
fix another case--signed the acquisition papers, and four hundred Bay Ridge
residents suddenly stopped protesting and submitted in silence.

The next new noise was the spirited, high-stepping sound of a marching band
and the blaring platitudes of politicians echoing over a sun-baked parade
ground on August 14, 1959--it was ground-breaking day for the bridge, with
the ceremony held, wisely, on the Staten Island side. Over in Brooklyn, when
a reporter asked State Senator William T. Conklin for a reaction, the Bay
Ridge representative snapped, "It is not a ground-breaking--to many it will be
heartbreaking." And then, slowly and more emotionally, he continued: "Any
public official attending should always be identified in the future with the
cruelty that has been inflicted on the community in the name of progress."

Governor Nelson Rockefeller of New York had been invited to attend the
ceremony in Staten Island, but he sent a telegram expressing regret that a
prior engagement made it impossible for him to be there. He designated
Assembly Speaker Joseph Carlino to read his message. But Mr. Carlino did
not show up. Robert Moses had to read it.

As Mr. Moses expressed all the grand hope of the future, a small airplane
chartered by the Staten Island Chamber of Commerce circled overhead with
an advertising banner that urged: "Name it the Staten Island Bridge." Many
people opposed the name Verrazano--which had been loudly recommended by
the Italian Historical Society of America and its founder, John N. La Corte--

because they could not spell it. Others, many of them Irish, did not want a bridge named after an Italian, and they took to calling it the "Guinea Gangplank." Still others advocated simpler names--"The Gateway Bridge," "Freedom Bridge," "Neptune Bridge," "New World Bridge," "The Narrows Bridge." One of the last things ever written by Ludwig Bemelmans was a letter to the <u>New York Times</u> expressing the hope that the name "Verrazano" be dropped in favor of a more "romantic" and "tremendous" name, and he suggested calling it the "Commissioner Moses Bridge." But the Italian Historical Society, boasting a large membership of emotional voters, was not about to knuckle under, and finally after months of debate and threats, a compromise was reached in the name "Verrazano-Narrows Bridge."

Verrazano-Narrows Bridge

The person making the least amount of noise about the bridge all this time was the man who was creating it--Othmar H. Ammann, a lean, elderly proper man in a high starched collar, who now, in his eightieth year, was recognized as probably the greatest bridge engineer in the world. His monumental achievement so far, the one that soared above dozens of others, was the George Washington Bridge, the sight of which had quietly thrilled him since its completion in 1931. Since then, when he and his wife drove down along the Hudson River from upstate New York and suddenly saw the bridge looming in the distance, stretching like a silver rainbow over the river between New York and New Jersey, they often gently bowed and saluted it.

"That bridge is his firstborn, and it was a difficult birth," his wife once explained. "He'll always love it best." And Othmar Ammann, though reluctant to reveal any sentimentality, nevertheless once described its effect upon him. "It is as if you have a beautiful daughter," he said, "and you are the father."

But now the Verrazano-Narrows Bridge presented Ammann with an even larger task. And to master its gigantic design he would even have to take into account the curvature of the earth. The two 693-foot towers, though exactly perpendicular to the earth's surface, would have to be one and five-eighths inches farther apart at their summits than at their bases.

Though the Verrazano-Narrows Bridge would require 188,000 tons of steel -- three times the amount used in the Empire State Building--Ammann knew that it would be an ever restless structure, would always sway slightly in the wind. Its steel cables would swell when hot and contract when cold, and its roadway would be twelve feet closer to the water in summer than in winter. Sometimes, on long hot summer days, the sun would beat down on one side of the structure with such intensity that it might warp the steel slightly, making the bridge a fraction lower on its hot side than on its shady side. So, Ammann knew, any precision measuring to be done during the bridge's construction would have to be done at night.

From the start of a career that began in 1902, when he graduated from the Swiss Federal Polytechnic Institute with a degree in civil engineering, Ammann had made few mistakes. He had been a careful student, a perfectionist. He had witnessed the rise and fall of other men's creations, had seen how one flaw in mathematics could ruin an engineer's reputation for life--and he was determined it would not happen to him.

Othmar Hermann Ammann had been born on March 26, 1879, in Schaffhausen, Switzerland, into a family that had been established in Schaffhausen since the twelfth century. His father had been a prominent manufacturer and his forebears had been physicians, clergymen, lawyers, government leaders, but none had been engineers, and few had shared his enthusiasm for bridges.

There had always been a wooden bridge stretching from the village of Schaffhausen across the Rhine, the most famous of them being built at a length of 364 feet in the 1700's by a Swiss named Hans Ulrich Grubenmmann. It had been destroyed by the French in 1799, but had been replaced by others, and as a boy Othmar Ammann saw bridges as a symbol of challenge and a monument to beauty.

In 1904, after working for a time in Germany as a design engineer, Ammann came to the United States--which, after slumbering for many decades in a kind of dark age of bridge design, was now finally experiencing a renaissance. American bridges were getting bigger and safer; American engineers were now bolder than any in the world.

There were still disasters, but it was nothing like it had been in the middle 1800's, when as many as forty bridges might collapse in a single year, a figure that meant that for every four bridges put up one would fall down. Usually it was a case of engineers not knowing precisely the stress and strain a bridge could withstand, and also there were cases of contractors being too cost-conscious and willing to use inferior building materials. Many bridges in those days, even some railroad bridges, were made of timber. Others were made of a new material, wrought iron, and nobody knew exactly how it would hold up until two disasters--one in Ohio, the other in Scotland--proved its weakness.

The first occurred on a snowy December night in 1877 when a train from New York going west over the Ashtabula Bridge in Ohio suddenly crumbled the bridge's iron beams and then, one by one, the rail cars fell into the icy waters, killing ninety-two people. Two years later, the Firth of Tay Bridge in Scotland collapsed under the strain of a locomotive pulling six coaches and a brakeman's van. It had been a windy Sunday night, and seventy-five people were killed, and religious extremists blamed the railroad for running trains on Sunday. But engineers realized that it was the wrought iron that was wrong, and these two bridge failures hastened the acceptance of steel--which has a working strength twenty-five percent greater than wrought iron--and thus began the great era that would influence young Othmar Ammann.

This era drew its confidence from two spectacular events--the completion in 1874 of the world's first steel bridge, a triple arch over the Mississippi River at St. Louis designed and built by James Buchanan Eads; and the completion in 1883 of the Brooklyn Bridge, first steel cable suspension span, designed by John Roebling and, upon his tragic death, completed by his son, Washington Roebling. Both structures would shape the future course in American bridge-building, and would establish a foundation of knowledge, a link of trial and error, that would guide every engineer through the twentieth century. The Roeblings and James Buchanan Eads were America's first heroes in high steel.

James B. Eads was a flamboyant and cocky Indiana boy whose first engineering work was raising sunken steamers from the bottom of the Mississippi. He also was among the first to explore the river's bed in a diver's suit, and he realized, when it came time for him to start constructing the foundations for his St. Louis bridge, that he could not rely on the Mississippi River soil for firmness, because it had a peculiar and powerful shifting movement.

So he introduced to America the European pneumatic caisson--an air-tight enclosure that would allow men to work underwater without being hindered by the shifting tides. Eventually, as the caisson sank deeper and deeper and the men dug up more and more of the river bed below, the bridge's foundation could penetrate the soft sand and silt and could settle solidly on the hard rock beneath the Mississippi. Part of this delicate operation was helped by Eads' invention--a sand pump that could lift and eject gravel, silt and sand from the caisson's chamber.

Before Ead's bridge would be finished, however, 352 workmen would suffer from a strange new ailment--caisson's disease or "the bends"--and twelve men would die from it, and two would remain crippled for life. But from the experience and observations made by James Eads' physician, Dr. Jaminet, who spent time in the caisson with the men and became temporarily paralyzed himself, sufficient knowledge was obtained to greatly reduce the occurrence of the ailment on future jobs.

Eads Bridge at St. Louis under construction

When the St. Louis steel bridge was finished, James Eads, to show its strength, ran fourteen locomotives across each of the bridge's three arches. Later a fifteen-mile parade marched across it, President Grant applauded from the reviewing stand, General Sherman drove in the last spike on the Illinois side, and Andrew Carnegie, who had been selling bonds for the project, made his first fortune. The bridge was suddenly instrumental in the development of St. Louis as the most important city on the Mississippi River, and it helped develop the transcontinental railroad systems. It was credited with "the winning of the West" and was pictured on a United States stamp in 1898; and in 1920 James Buchanan Eads became the first engineer elected to the American Hall of Fame.

He died an unhappy man. A project he envisioned across the Isthmus of Tehuantepec did not work out.

John Augustus Roebling was a studious German youth born in 1806, in a small town called Muhlhausen, to a tobacco merchant who smoked more than he sold and to a mother who prayed he would someday amount to more than his father. Largely through her ambition and thrift he received a fine education in architecture and engineering in Berlin, and later he worked for the Prussian government building roads and bridges.

But there was little opportunity for originality, and so at the age of twenty-five he came to America and soon, in Pennsylvania, he was working as a surveyor for the railroads and canals. And one day, while observing how the hemp rope that hauled canal boats often broke, John Roebling began to experiment with a more durable fiber, and soon he was twisting iron wire into the hemp--an idea that would eventually lead him and his family into a prosperous industry that today, in Trenton, N.J., is the basis for the Roebling Company--world's largest manufacturers of wire rope and cable.

But in those days it led John Roebling toward his more immediate goal, the construction of suspension bridges. He had seen smaller suspensions, hung with iron chains, during his student days in Germany, and he wondered if the suspension bridge might not be more graceful, longer and stronger with iron wire rope, maybe even strong enough to support rail cars.

He had his chance to find out when, in 1851, he received a commission to build a suspension bridge over Niagara Falls. This opportunity arose only because the original engineer had abandoned the project after a financial dispute with the bridge company--this engineer being a brilliant but wholly unpredictable and daring man named Charles Ellet. Ellet, when confronted with the problem of getting the first rope across Niagara, found the solution by offering five dollars to any boy who could fly a kite across it. Ellet later had a basket carrier made and he pulled himself over the rushing waters of Niagara to the other side; and next he did the same thing accompanied by his horse, as crowds standing on the cliffs screamed and some women fainted.

John Roebling

Things quieted down when Ellet left Niagara, but John Roebling, in his methodical way, got the job done. "Engineering," as Joseph Gies, an editor and bridge historian, wrote, "is the art of the efficient, and the success of an engineering project often may be measured by the absence of any dramatic history."

In 1855, Roebling's 821-foot single span was finished, and on March 6 of that year a 368-ton train crossed it--the first train in history to cross a span sustained by wire cables. The success quickly led Roebling to other bridge commissions, and in 1867 he started his greatest task, the Brooklyn Bridge.

It would take thirteen years years to complete the Brooklyn Bridge, and both John Roebling and his son would be its victims.

One summer morning in 1869, while standing on a pier off Manhattan, surveying the location of one of the towers, and paying no attention to the docking ferryboat that was about to bump into the pier, John Roebling suddenly had his foot caught and crushed between the pier floor and piles; tetanus set in, and two weeks later, at the age of sixty-three, he died.

At the death of his father, Washington Roebling, then thirty-two years old and the chief engineering assistant for the bridge, took over the job. Roebling had previously supervised the construction of other bridges that his father had designed, and had served as an engineering officer for the Union Army during the Civil War. During the war he had been one of General Grant's airborne spies, ascending in a balloon to watch the movement of Lee's army during its invasion of Pennsylvania.

When he took over the building of the Brooklyn Bridge, Washington Roebling decided that since the bridge's tower foundations would have to be sunk forty-four feet into the East River on the Brooklyn side and seventy-six feet on the New York side, he would use pneumatic caissons--as James Eads had done a few years before with his bridge over the Mississippi. Roebling drove himself relentlessly, working in the caisson day and night and he finally collapsed. When he was carried up, he was paralyzed for life. He was thirty-five years old.

But Washington Roebling, assisted by his wife Emily, continued to direct the building of the bridge from his sickbed; he would watch the construction through field glasses while sitting at the window of his home on the Brooklyn shore; and then his wife--to whom he had taught the engineer's language, and who understood the problems involved--would carry his instructions to the superintendents on the bridge itself.

Washington Roebling was the first bridge engineer to use steel wire for his cables--it was lighter and stronger than the iron wire cables used by his father on the Niagara bridge--and he had every one of the 5,180 wires galvanized as a safeguard against rust. The first wire was drawn across the East River in 1877, and for the next twenty-six months, from one end of the bridge to the other, the small traveling wheels--looking like bicycle wheels with the tires missing--spun back and forth on pulleys, crossing the East River 10,360 times, each time bringing with them a double strand of wire which, when wrapped, would form the four cables that would hold up the center span of 1,595 feet and its two side spans of 930 feet each. This technique of spinning wire, and the use of a cowbell attached to each wheel to warn the men of its approaches, is still used today; it was used, in a more modern form, even by O. H. Ammann in the cable-spinning phase of his Verrazano-Narrows Bridge in the 1960s.

The Brooklyn Bridge was opened on May 24, 1883. Washington Roebling and his wife watched the celebrations from their windows through field glasses. It was a great day in New York--business was suspended, homes were draped with bunting, church bells rang out, steamships whistled. There was the thunder of guns from the forts in the harbor and from the Navy ships docked near the bridge, and finally, in open carriages, the dignitaries arrived. President Chester A. Arthur, New York's Governor Grover Cleveland, and the mayors of every city within several miles of New York arrived at the bridge. Later that night there was a procession in Brooklyn that led to Roebling's home, and he was congratulated in person by President Arthur.

To this day the Brooklyn Bridge has remained the most famous in America, and, until the Williamsburg Bridge was completed over the East River between Brooklyn and Manhattan in 1903, it was the world's longest suspension. In the great bridge boom of the twentieth century nineteen other suspension spans would surpass it--but none would cast a longer shadow. It has been praised by poets, admired by aesthetes, and sought by the suicidal. Its tower over the tenement roofs of the Lower East Side so electrified a young neighborhood boy named David Steinman that he became determined to emulate the Roeblings, and later he would become one of the world's great bridge designers; he alone, until his death in 1960, would challenge Ammann's dominance.

David Steinman at the age of fourteen had secured a pass from New York's Commissioner of Bridges to climb around the catwalks of the Williamsburg Bridge, then under construction, and he talked to bridge builders, took notes, and dreamed of the bridges he would someday build. In 1906, after graduating from City College in New York with the highest honors, he continued his engineering studies at Columbia, where, in 1911, he received his doctor's degree for his thesis on long-span bridges and foundations. Later he became consulting engineer on the design and construction of the Florianopolis Bridge in Brazil, the Mount Hope Bridge in Rhode Island, the Grand Mere in Quebec, the Henry Hudson arch bridge in New York. It was Dr. David Steinman who was called upon to renovate the Brooklyn Bridge in 1948, and it was he who was selected over Ammann to build the Mackinac Bridge--although it was Ammann who emerged with the Verrazano-Narrows commission, the bridge that Steinman had dreamed of building.

The two men were never close friends, possibly because they were too close in other ways. Both had been assistants in their earlier days to the late Gustav Lindenthal, designer of the Hell Gate and the Queensboro bridges in New York, and the two men were inevitably compared. They shared ambition and vanity, and yet possessed dissimilar personalities. Steinman was a colorfully blunt product of New York, a man who relished publicity and controversy, and who wrote poetry and had published books. Ammann was a stiff, formal Swiss gentlemen, well-born and distant. But that they were to be lifelong competitors was inevitable, for the bridge business thrives on competition; it exists on every level. There is competition between steel corporations as they bid for each job, and there is competition between even the lowliest apprentices in the work gangs. All the gangs--the riveters, the steel connectors, the cable spinners--battle throughout the construction of every bridge to see who can do the most work, and later in bars there is competition to see who can drink the most, brag the most. But here, on the lower level, among the bridge workers, the rivalry is clear and open; on the higher level, among the engineers, it is more secret and subtle.

Some engineers quietly go through life envying one another, some quietly prey on others' failures. Every time there is a bridge disaster, engineers who are unaffiliated with its construction flock to the site of the bridge and try to determine the reason for the failure. Then, quietly, they return to their own plans, armed with the knowledge of the disaster, and patch up their own bridges, hoping to prevent the same thing. This is as it should be. But it does not belie the truth of the competition. When a bridge fails, the engineer who designed it is as good as dead. In the bridge business, on every level, there is an endless battle to stay alive--and no one has stayed alive longer than O.H. Ammann.

Ammann was among the engineers who, in 1907, investigated the collapse of a cantilever bridge over the St. Lawrence River near Quebec. Eighty-six workmen, many of them Indians, who were just learning the high-construction business then, fell with the bridge, and seventy-five drowned. The engineer whose career ended with this failure was Theodore Cooper, one of America's most noted engineers--the same man who had been so lucky years before when, after falling one hundred feet into the Mississippi River while working on James Eads' bridge, not only survived but went back to work the same day.

But now, in 1907, it was the opinion of most engineers that Theodore Cooper did not know enough about the stresses involved in the cantilever bridge. None of them did. There is no way to know enough about bridge failure until enough bridges have failed. "This bridge failed because it was not strong enough," one engineer, C.C. Schneider, quipped to the others. Then they all returned to their own bridges, or to their plans for bridges, to see if they too had made miscalculations.

One bridge that perhaps was saved in this manner was Gustav Lindenthal's Queensboro Bridge, which was then approaching completion over the East River in Manhattan. After a re-examination, it was concluded that the Queensboro was inadequate to safely carry its intended load. So the four rapid-transit tracks that had been planned for the upper deck were reduced by two. The loss of the two tracks was compensated by the construction of a subway tunnel a block away from the bridge--the BMT tunnel at Sixtieth Street under the East River, built at an additional cost of $4,000,000.

In November of 1940, when the Tacoma Narrows Bridge fell into the waters of Puget Sound in the state of Washington, O.H. Ammann was again one of the engineers called in to help determine the cause. The engineer who caught the blame in this case was L.S. Moisseiff, a man with a fine reputation throughout the United States.

Moisseiff had been involved in the design of the Manhattan Bridge in New York, and had been the consulting engineer of the Ambassador Bridge in Detroit and the Golden Gate in California, among many others, and nobody had questioned him when he planned a lean, two-lane bridge that would stretch 2,800 feet over the waters of Puget Sound. True, it was a startingly slim, fragile-looking bridge, but during this time there had been an aesthetic trend toward slimmer, sleeker, lovelier suspension bridges. This was the same trend that led David Steinman to paint his Mount Hope Bridge over Narragansett Bay a soft green color, and to have its cables strung with lights and approaches lined with evergreens and roses, costing an additional $70,000 for landscaping.

There was also a prewar trend toward economizing on the over-all cost of bridge construction, however, and one way to save money without spoiling the aesthetics--and supposedly without diminishing safety--was to shape the span and roadway floor with solid plate girders, not trusses that wind could easily pass through. And it was partially because of these solid girders that, on days when the wind beat hard against its solid mass of roadway, the Tacoma Narrows Bridge kicked up and down. But it never kicked too much, and the motorists, far from becoming alarmed, actually loved it, enjoyed riding over it. They knew that all bridges swayed a little in the wind--this bridge was just livelier, that was all, and they began calling it, affectionately, "Galloping Gertie."

Four months after it had opened--on November 7--with the wind between thirty-five and forty-two miles an hour, the bridge suddenly began to kick more than usual. Sometimes it would heave up and down as much as three feet. Bridge authorities decided to close the bridge to traffic; it was a wise decision, for later it began to twist wildly, rising on one side of its span, falling on the other, rising and falling sometimes as much as twenty-eight feet, tilting at a forty-five-degree angle in the wind. Finally, at 11 A.M., the main span ripped away from its suspenders and went crashing into Puget Sound.

The factors that led to the failure, the examining engineers deduced, were generally that the tall skinny bridge was too flexible and lacked the necessary stiffening girders; and also they spoke about a new factor that they had previously known very little about--"aerodynamic instability."

And soon, on other bridges, on bridges all over America and elsewhere, adjustments were made to compensate for the instability. The Golden Gate underwent alterations that cost more than $3,000,000. The very flexible Bronx-Whitestone Bridge in New York, which Leon Moisseiff had designed-- with O.H. Ammann directing the planning and construction--had holes punched into its plate girders and had trusses added. Several other bridges that formerly had been slim and frail now became sturdier with trusses, and twenty years later, when Ammann was creating the Verrazano-Narrows Bridge, the Tacoma lesson lived on. Though the lower second deck on the Verrazano-Narrows was not yet needed, because the anticipated traffic could easily be accommodated by the six-lane upper deck, Ammann made plans for the second deck to go on right away--something he hadn't done in 1930 with his George Washington Bridge. The six-lane lower deck of the Verrazano will probably be without an automobile passenger for the next ten years, but the big bridge will be more rigid from its opening day.

After the Tacoma incident, Moisseiff's talents were no longer in demand. He never tried to pass off any of the blame on other engineers or the financers; he accepted his decline quietly, though finding little solace in the fact that with his demise as an influential designer of bridges the world of engineering knowledge was expanded and bigger bridges were planned, bringing renown to others.

And so some engineers, like Leon Moisseiff and Theodore Cooper, go down with their bridges. Others, like Ammann and Steinman, remain high and mighty. But O.H. Ammann is not fooled by his fate.

One day, after he had completed his design on the Verrazano-Narrows Bridge, he mused aloud in his New York apartment, on the thirty-second floor of the Hotel Carlyle, that one reason he has experienced no tragedy with his bridges is that he has been blessed with good fortune.

"I have been lucky," he said, quietly.

"Lucky!" snapped his wife, who attributes his success solely to his superior mind.

"Lucky," he repeated, silencing her with his soft, hard tone of authority.

A High-Flying Fixer
Pat Walker

The Fresno Bee - April 4, 1984, Issue

In 1984 James van Hoften became the first civil engineer to work in space. Although the tasks he performed had little in common with those for which he had been prepared by his education as a hydraulic engineer, his six-day trip aboard the Challenger space shuttle has earned him a place in civil engineering history. Since the successful completion of his space mission he has spoken to student groups and to audiences at local and national meetings of the American Society of Civil Engineers about his space adventure.

The concept of civil engineers performing traditional civil engineering work in space--building structures, transportation networks, waste management systems, facilities for storing water and generating energy, and other types of civil engineering projects--has already progressed beyond the realm of science fiction, although the time when the concept will become reality appears to be in the relatively distant future. It is not inconceivable, however, that some civil engineers beginning their careers now will follow van Hoften into space. Like their civil engineering ancestors who have made it possible for man to occupy and work in areas of the Earth once considered uninhabitable, they will be solving the problems that have to be solved if man is to occupy and master a new environment.

When James van Hoften was a 5-year-old playing with the family cats outside his parent's Harvard Avenue home in Fresno, no one imagined he'd become the Mr. Fixit of the New Frontier.

Of course, that was in the late 1940s, and space flight was the thing of dreams. On Friday, it will become reality for astronaut-repairman Jim "Ox" van Hoften, 39.

Armed with wrenches and ratchets, he'll ride 300 miles high to make the first in-orbit service call, as he and the four other members of the space shuttle *Challenger*'s crew attempt to rescue and repair a solar satellite.

*   *   *

At 6 feet 4, native Fresnan van Hoften is America's tallest astronaut. He is said to be "as strong as an ox."

And his parents, Adriaan and Beverly van Hoften of Redwood City [CA], say their son "always gives 200 percent to whatever he does."

The van Hoftens will watch his fiery ascent into orbit aboard the space shuttle *Challenger* Friday from the Kennedy Space Center in Florida.

"I'll tell him to have fun . . . and bring me a star," said his mother, who plans to watch his untethered flight on National Aeronautics and Space Administration [NASA] monitors in Florida.

*   *   *

James van Hoften was born on June 11, 1944, three years to the day after his brother Scott. The family moved to Burlingame [CA] in 1950.

Now, van Hoften, who has three daughters, and his wife, Valerie, live in Houston.

He earned a bachelor of science degree in civil engineering at the University of California at Berkeley in 1966. He attended the University of Colorado and earned his master's degree in hydraulic engineering in 1968.

He went into the Navy in 1969 as a pilot. After his discharge in 1974, van Hoften earned a doctorate in fluid mechanics from Colorado State University.

Van Hoften was an assistant professor of civil engineering at the University of Houston when NASA eliminated height restrictions for its astronauts. He applied and was accepted into the program in 1978.

*   *   *

Van Hoften . . . will break out his tool kit on the third day of the six-day mission.

Floating to his position in the weightless workshop of the shuttle's cargo bay, the mission specialist will prepare his work station. Meanwhile, crew mate George "Pinky" Nelson, 33, will fire up a jet-powered backpack and fly away from the shuttle to capture the ailing satellite, a solar observatory known as the Solar Max Satellite--for Solar Maximum Mission.

The body of the Solar Max is 22 by 11 feet, but solar panels give it a "wingspan" of 100 feet. The craft, which weighs about 5,000 pounds and cost $100 million, was launched on Valentine's Day 1980 at a time of peak solar activity.

It was designed to send back information about how the sun affects Earth's atmosphere and environment. Scientists hoped it would yield clues to why sunspots and solar flares disrupt communications on Earth.

But nine months after launch, Solar Max failed.

Fortunately, it was designed for in-orbit repair. A small pin protrudes from the satellite to which an astronaut and the shuttle's robot arm can be attached.

Holding a "mating" device that resembles a white metal belly button on something the size of a small coffee table, Nelson will fire the thrusters on his backpack to match the satellite's rotation of one degree per second. He will dock to the pin and stop the satellite's spin by firing the thrusters again.

From his perch in the payload bay about 100 yards away, van Hoften will watch, ready to don his backpack and help in the event of trouble.

Stopping a satellite from spinning through space has never been done before, but *Challenger*'s commander, space veteran Bob Crippen, said, "I don't think there will be a significant problem, but it's not going to be a piece of cake."

Once Nelson has stabilized the satellite--expected to take less than 10 minutes--Crippen and mission pilot Dick Scobee will maneuver the shuttle within 10 yards of Solar Max.

Working from a set of complex controls on the shuttle's flight deck, mission specialist Terry Hart will position *Challenger*'s robot arm to pluck the satellite from space, then gently lower it onto a cradle in the cargo bay where it will be repaired.

The speed at which the shuttle travels--more than 17,000 miles per hour--and the absence of gravity make space repair work difficult.

"One of the biggest problems of working in zero gravity is that it is difficult to secure your body," van Hoften explained. "And unless you do, it is very difficult to perform simple and useful work."

When van Hoften begins the initial repairs, he will be anchored to a metal work station attached to the end of the robot arm; foot restraints will hold him in place.

Using tools with large, non-skid handles, van Hoften will remove a faulty attitude control box, one of three units that control power, command and position of the satellite.

The failure of that unit, which is about the size of a tissue box, left the Solar Max in a helpless tumble.

The most ambitious repairs will be made on the fifth day of the mission. Van Hoften and Nelson will replace the satellite's main electronics box in the coronagraph-polarimeter, a device designed to study the sun's outer atmosphere by creating artificial eclipses.

Nelson and van Hoften must work very slowly so the objects they are handling don't spin out of control and float away.

"It's going to be like working with boxing gloves on," van Hoften said. "It would be a much simpler procedure if performed on Earth."

To remove the electronics box, which is about the size of a briefcase, and install the new one, the astronauts will open a panel at the base of the satellite. They will cut through foil insulation and remove the 22 screws that secure a protective thermal blanket over the box.

After taping the thermal blanket and insulation out of the way, they'll install a hinge in the door they've created and remove the remaining screws to open the panel that covers the box.

Then they will unplug cables from the box and remove it. The replacement box will be installed, all connectors restored, the door closed and secured, and the protective insulation reattached.

Van Hoften said the repair is expected to take four or five hours.

To prepare for this mission, van Hoften and Nelson underwent intensive underwater training in the National Aeronautics and Space Administration's Weightless Environment Training Facility at the Johnson Space Center near Houston.

Wearing pressurized spacesuits, lead-weighted to give them neutral buoyancy, they were lowered onto a mockup of the cargo bay immersed in a 25-foot deep

103

pool of water. There they worked with models of the Solar Max and its faulty parts, testing tools to see which would best serve their needs.

"Working in the water gives the astronauts a feeling for what working on-orbit is really like," said astronaut Story Musgrave, who took the shuttle's first "space walk" a year ago. Musgrave said the space center offers training so close to the real thing that there are "no surprises."

Once the repairs are completed, ground crews from the Goddard Space Flight Center in Greenbelt, Md., who designed and built the solar satellite, will check its systems.

If the repair effort has been successful, Hart will use the robot arm to remove Solar Max from its cradle, point it toward the sun and release it back into orbit. For the next two years, it will provide the scientific community with valuable information about the center of the solar system.

When the shuttle lands at the Kennedy Space Center in Florida, van Hoften and his crew mates will return with experience that will help NASA in training other astronauts in the art of satellite repair and in developing more efficient techniques for in-orbit repair and maintenance.

"Our mission will extend man's use of space," van Hoften said.

"This is just a stepping stone for performing all kinds of tasks in all kinds of altitudes. The Solar Max mission will be a milestone for NASA in which man's role in space can be proven."

James van Hoften's trip into space turned out to be somewhat more eventful than anticipated when this article was written several days before lift-off. Astronaut Nelson's attempts to dock with the Solar Max satellite were unsuccessful, and Mission Specialist Hart was unable to maneuver the remote-controlled mechanical arm into position to grab the satellite which was now tumbling erratically. After 36 hours of anxiously observing efforts of ground controllers at Goddard Space Flight Center to stabilize the satellite, and when the satellite's solar-powered batteries were within five minutes of exhausting the energy they had drawn from the sun, the astronauts were relieved to see the Solar Max slowly begin to stabilize. By now, however, their own fuel for maneuvering was so depleted it seemed the whole mission might have to be aborted.

Solar Max was finally snared in a last-ditch effort which required the Challenger *crew to perform tasks somewhat different from those that had been scheduled if the mission had proceeded as planned. Van Hoften, using sheer physical strength, wrestled the satellite into position for the repairs once it had been captured. He and Nelson successfully completed the planned repairs in less than half the scheduled time. Together with other members of the crew they donned T-shirts identifying themselves as the "Ace Satellite Repair Co." and appeared in a tele-cast from the* Challenger *cabin to accept President Reagan's telephoned congratulations for their pioneering demonstrations of man's ability to work in space.*

Selected Additional Reading

The books listed here provide additional information on the history and heritage of the civil engineering profession. Although the list is not intended to be exhaustive it should serve as a good starting point for a reader who is interested in learning more about the historic aspects of the civil engineering profession. Most of these books focus on civil engineering in the United States, but a few books dealing with civil engineering history from a more general perspective have been included. Civil engineering history is not a particularly popular topic, so smaller libraries will not have copies of many of these books. University libraries and large public libraries are the most likely sources for these and other books dealing with the history and heritage of the civil engineering profession.

American Society of Civil Engineers Committee on History and Heritage. **The Civil Engineer: His Origins** (New York: American Society of Civil Engineers, 1970) 116 pgs.

Armstrong, Ellis (Ed.). **History of Public Works in the United States** (Chicago: American Public Works Assn., 1976) 736 pgs.

Billington, David P. **The Tower and the Bridge** (New York: Basic Books, Inc., 1983) 306 pgs.

Calhoun, Daniel H. **The American Civil Engineer: Origins and Conflicts** (Cambridge, MA: Harvard University Press, 1960) 295 pgs.

Condit, Carl W. **American Building Art: The Nineteenth Century** (New York: Oxford University Press, 1960) 371 pgs.

Condit, Carl W. **American Building Art: The Twentieth Century** (New York: Oxford University Press, 1961) 427 pgs.

DeCamp, L. Sprague. **The Ancient Engineers** (New York: Doubleday & Co., 1960) 408 pgs.

Finch, James K. **Engineering and Western Civilization** (New York: McGraw-Hill Publishing Co., 1951) 397 pgs.

Finch, James K. **The Story of Engineering** (New York: Doubleday & Co., 1960) 528 pgs.

Fleming, Thomas J. **West Point** (New York: William Morrow & Co., 1969) 402 pgs.

Gille, Bertrand. **Engineers of the Renaissance** (Cambridge, MA: The MIT Press, 1966) 256 pgs.

Hill, Forest G. **Roads, Rails and Waterways: The Army Engineers and Early Transportation** (Norman, OK: University of Oklahoma Press, 1957)

Hoy, Suellen and Robinson, Michael C. (Editors). **Public Works History in the United States** (Nashville, TN: American Association for State and Local History, 1982) 477 pgs. (An annotated bibliography)

Kirby R. S., Withington, S., Darling, Arthur B. and Kilgour, F. G. **Engineering in History** (New York: McGraw-Hill Publishing Co., 1956) 530 pgs.

Kirby, Richard S. and Laurson, Philip G. **The Early Years of Modern Civil Engineering** (New Haven, CT: Yale University Press, 1932) 324 pgs.

Layton, Edwin T. **The Revolt of the Engineers: Social Responsibility and the American Engineering Profession** (Cleveland, OH: Press of Case Western Reserve University, 1971) 286 pgs.

Merritt, Raymond H. **Engineering in American Society 1850-1875** (Lexington, KY: The University Press of Kentucky, 1969) 199 pgs.

Moses, Robert. **Public Works: A Dangerous Trade** (New York: McGraw-Hill Publishing Co., 1970) 952 pgs.

National Geographic Society. **Builders of the Ancient World: Marvels of Engineering** (Washington, D.C.: National Geographic Society, 1986) 200 pgs.

Pannell, J. P. G. **An Illustrated History of Civil Engineering** (New York: Frederick Ungar Publishing Co., 1964) 376 pgs.

Parsons, William B. **Engineers and Engineering in the Renaissance** (Cambridge, MA: The MIT Press, 1968) 661 pgs.

Sandstrom, Gosta E. **Man the Builder** (New York: McGraw-Hill Publishing Co., 1970) 280 pgs.

Stapleton, Darwin H. **The History of Civil Engineering Since 1600** (New York: Garland Publishing Inc., 1986) 231 pgs. (An annotated bibliography)

Straub, Hans. **A History of Civil Engineering** (London: Leonard Hill, Ltd., 1960) 258 pgs.

Stuart, Charles B. **Lives and Works of Civil Engineers in America** (New York: D. Van Nostrand, 1871) 343 pgs.

Wisely, William H. **The American Civil Engineer, 1852-1974** (New York: American Society of Civil Engineers, 1974) 464 pgs.

CIVIL ENGINEERS

The biographic sketches which make up this section have been chosen to portray the civil engineer in as many different dimensions as possible: as a leader of men; a creative genius; a technical expert; a sensitive and sympathetic human being; a wily schemer; a dedicated public servant; a literate and articulate observer of the human condition; a dreamer of great dreams; a promoter of grand schemes; a flamboyant publicity-seeker; and a quietly competent designer.

The civil engineers in these readings are not necessarily the most famous or the most successful American civil engineers. Some are virtually unknown, even within the civil engineering profession, but each person portrays some interesting aspect of what it has meant to be a civil engineer at various times during this nation's history. Although some of the portraits show civil engineers in roles that are seldom encountered in modern society, they are important to an understanding of how the profession has evolved--how we have come to the modern perception of the civil engineer as a problem solver engaged in a profession that has as its fundamental objective service to the needs of society.

A major part of the job of the civil engineer today is to help society reconcile the conflicts inherent in choosing from among a variety of alternatives. Where the civil engineer was once viewed primarily as the expert who was responsible for determining what needed to be done; the modern civil engineer is more likely to be viewed as the consultant or adviser who is responsible for describing to others the consequences of possible courses of actions, as the expert who recommends the best course of action for achieving a goal defined by others, as the impartial arbiter between competing and conflicting interests, or as the steward or manager of scarce natural resources.

The readings demonstrate that civil engineers have never been the tunnel-visioned technocrats that some critics would have us believe is both the heritage and inclination of the modern civil engineer. In fact, civil engineers have always had to be attuned to the social, political, economic and environmental consequences of their work. The degree to which these various dimensions of the civil engineer's work are reflected in their projects varies with the demands of society from time to time, but the successful civil engineer has always been able to incorporate these concerns into the planning and design of civil engineering projects.

Taken as a whole these readings reflect the diversity of the civil engineering profession and its inherent capability to accommodate a remarkable variety of personality types. It is true that a large majority of modern civil engineers are involved in work assignments that deal with the design and construction aspects of civil engineering--assignments that, in the normal course of things, tend to deal more with things than with people. However, there are many design and construction positions that provide opportunities for those who see interpersonal relationships as a primary strength--those who seek the challenges of exploring and understanding the ideas of other people rather than the nature and functions of new materials. In civil engineering planning and in operations and maintenance, the other two functional areas of the civil engineering profession, the ability to work well with others and the ability to communicate well, in both the spoken and the written word, are probably as important as the understanding of civil engineering concepts. Consequently, positions in these functional areas are likely to be attractive to civil engineers who are "people-oriented."

In addition to displaying the many different personality types that are found in the civil engineering profession, these readings, and those in the following section, also illustrate the variety of working conditions and work assignments encountered by civil engineers. From the smallest mid-American community to the largest metropolitan areas, from the consulting firm that limits its practice to local subdivision planning and layout to the international design and construction companies that work around the world, from positions where all of the work is accomplished at a desk or drawing board to positions where the challenges of dealing with the outdoor environment are as critical as the knowledge of engineering principles, the civil engineering profession encompasses every conceivable working condition.

Charles Ellet, Jr.
Civil Engineer, Author, Civil War Hero

Years of Apprenticeship

An excerpt from Chapter 2, **Charles Ellet, Jr.**
The Engineer as Individualist
Gene D. Lewis

*Like many other American civil engineers of the early nineteenth
century, Charles Ellet Jr. became an accomplished canal builder, bridge
builder, and railroad builder--an engineering jack-of-all-trades.
However, despite the fact that he was well-known among his
contemporaries for his engineering accomplishments and despite the fact
that he was a Civil War hero of sorts, he is not at all well known
today, even within the civil engineering profession. To some extent
lack of recognition may be due to the fact that his career was cut short
as a result of a Civil War injury which caused his death, but there
is considerable evidence that he was something of a professional
"loner"--a man who found himself at odds not only with those outside the
profession but also with his engineering peers who respected his
technical ability but were uncomfortable with his proclivity for
outspoken controversial opinions.*

*Much of what Charles Ellet thought and did is recorded in letters and
other writings preserved by his family. Historian Gene Lewis's
biography of Ellet is enriched by the availability of these documents.
Through this biography we are accorded a rare glimpse at not just the
works of this important civil engineer, but also his thoughts and
opinions as he himself expressed them.*

Members of Lincoln's cabinet, thoroughly alarmed and frightened, gathered at
the executive mansion, Sunday, March 9, 1862, and faced the prospect of a
Confederate flotilla anchoring in the Potomac near the White House.
Secretary of the Navy Gideon Welles found the President so excited that
"he could not deliberate," and Secretary of War Stanton "at times almost
frantic." Both the President and his war secretary went repeatedly to the
window and looked down the Potomac to see "if the *Merrimac* was not coming
to Washington."

Stanton was ready to direct naval affairs. What little confidence he had in the
Navy Department had vanished, and the next day he ordered the channel of
the Potomac filled with stone. A week later, he offered Cornelius Vanderbilt a
contract to destroy the *Merrimac*. And on March 14, he called for an interview
with Charles Ellet, Jr., a civil engineer, who for months had been attempting
to gain a hearing for his ideas on the steam ram.

In the next few weeks, following the famous encounter of the *Monitor* with
the *Merrimac*, Ellet was allowed to build and deploy a fleet of ironclads against
the Confederate Navy on the western rivers, including the Mississippi. It was
largely thanks to his efforts that the North regained the initiative in naval
warfare. But before he could gain recognition for his achievements he died on
June 6, 1862, as a result of a wound received while leading the successful naval
battle to capture Memphis--a victory for which others have unfairly taken the
credit. Ellet's uphill struggle to gain acceptance for his ideas on the naval ram,
his dramatic demonstration of their efficacy, and his ultimate failure to

receive the credit for them, are illustrative of a notable career filled with frustration and lack of recognition.

Ironically, Ellet's main right to fame is not based on his Civil War record, even though historians have chosen to emphasize it, but rests, in the first instance, on his recommendations for preventing floods and for improving navigation on western rivers--recommendations which have received their proper recognition only in the twentieth century. He was the first person to advocate a comprehensive plan of river improvement based on the utilization of upland reservoirs as a means of impounding surplus waters.

Second, his reputation is based on the fact that he introduced the principle of the suspension bridge into America, although, here again, others have received credit for the innovation. Before building the first permanent suspension bridge in the western hemisphere in 1842 across the Schuylkill River in Philadelphia, he advocated (unsuccessfully) the construction of a suspension bridge over the Potomac that would have proved of great assistance to the Union forces during the Civil War. He engineered and supervised the construction of the first bridge over the Niagara gorge in 1848, and the following year built a 1,010-foot suspension bridge, then the world's longest span, over the Ohio River at Wheeling.

In addition to Ellet's contributions to practical engineering, he often wrote in defense of his controversial ideas. During his lifetime he published at least forty-six different volumes, as well as a number of popular and technical articles. Adversaries and admirers alike recognized his penetrating and precise style. At the age of twenty-nine he published a scholarly book on transportation rate-making entitled **Essays on the Laws of Trade**. This was followed by a number of pioneer contributions to transportation economics. After 1860 his grasp of wartime problems was revealed in articles for American and English periodicals. One of the chief critics of McClellan's tactics, he won wide support and attention when, on numerous occasions, he presented his own plans for crushing the Confederacy. Thoroughly familiar with the geography of Virginia, he had been chief engineer during the construction of the James River and Kanawha Canal, the main internal improvement in Virginia prior to the Civil War. Later, as chief engineer for the Virginia Central Railroad, he became intimately acquainted with the entire transportation network upon which the Confederate Army in Virginia was to depend.

Ellet's interest thus ranged broadly from bridge-building, canal and railroad construction, and the economics of transportation to the improvement of western rivers, and to projects for defeating the Confederacy. Of the many individuals who contributed significantly to the advancement of the industrial revolution in America, few were responsible for more in both theory and practice than Ellet. Yet he has remained an obscure figure, whose name and career are known to but a few specialists.

Although he was a significant innovator, Ellet lacked those personality traits necessary to attract attention to his achievements. Possessed of a first-rate mind and a fund of intense energy, he was perhaps for this reason impatient and intolerant of the slower and more prosaic understanding of others. Often lacking the funds to develop his ideas, he would lash out bitterly in exasperation at those near to him, thus alienating the most promising source of support available to him. This personality trait was accentuated by the precarious state of his health. Although he was essentially a warm-hearted person, extremely solicitous for his family, illness often was responsible for Ellet's apparently harsh and arrogant behavior. Moreover, he shared the misfortune of many other inventors whose conceptions were brought into the world before it was prepared to accept them. Perhaps the most important consideration, however, in explaining Ellet's relative obscurity today is the emphasis hitherto given by historians to the glamorous personalities of politics, law, and the military, to the neglect of those concerned with technological advance.

Ellet's career is significant on at least three counts: most obvious is his personal responsibility for the introduction into America of the suspension bridge; his initial advocacy and demonstration on a significant scale here of the merits of the steam ram in naval warfare; and his projection of a comprehensive system of river improvement and flood control. It is noteworthy, moreover, that although the importance of his innovations did not achieve appropriate recognition in his lifetime, he did encompass them with sufficient flourish and drama that they have eventually redounded to his fame and to the benefit of mankind.

Second, there is here an illustration of the unresponsiveness of antebellum America to technological personalities and ideas, as mentioned above. This was further emphasized by the relative scarcity of capital at that time, which inhibited the application of novel and expensive technological innovations. This stands in contrast to the post-Civil War era when the greater affluence of the country greatly facilitated the application of the ideas advanced by technical men.

Finally, and more important, Ellet's career affords an ideal case study of the problems and conflicts associated with the beginnings of civil engineering as a profession in American. When Ellet came upon the scene in 1828, civil engineers, as an occupational group, were hardly recognized as belonging to a separate occupational category. When he died in 1862 the engineer--while his profession was not completely institutionalized--was generally beginning to think of himself as a distinct member of an occupation working mainly as a "salaried professional bound to whatever organization he might be working for at the time."

For his part, however, Ellet characteristically resisted the direction the emerging profession was taking. His broad interests, as indicated by his writings and activities, reveal him as more than an individual concerned only with the narrow technical aspects of engineering. At the same time that he was not the nonprofessional, jack-of-all-trades engineer of the early nineteenth century--those craftsmen possessed of some knowledge of most technical matters but expert in none--neither was he the narrow organization-type of professional engineer so characteristic of the late nineteenth century, and unfortunately of the twentieth century. Rather he was an engineer vitally interested in the implications of technology for society as a whole.

In another sense his career illustrates the frustration often resulting when individualism conflicts with bureaucracy, with group desires and professionalism. Although Ellet always sought professional standing and reputation, he preferred to achieve them as an individual, and not as an

111

organization engineer so typical of corporate America. Indeed his was the individualism of an earlier day, though his training and breadth of interests were unique.

*　　*　　*

. . . it was on the Susquehanna Branch Canals, in 1827 at the age of seventeen, that [Ellet] obtained his first position as an assistant engineer.

At this time, Ellet was already a part of an emerging profession, which had its origins in the building of the Erie Canal. Prior to the construction of this waterway, engineering work required on the early, small, essentially local internal improvements was performed by men with a nontechnical background. In 1815, on the eve of the great development in internal improvements, the thirty or forty men who were active or available as "something like engineers" had very few traits in common. Many were surveyors; others were contractors or builders. Landholders and speculators and even two lawyers and an architect were included in the small number of available consultants. Most of them did not confine themselves to their engineering tasks, and when they did perform technical jobs they almost always had some proprietary relationship to the project. But with the emergence of large-scale internal improvement projects, the proprietary engineer began to give way to the professional civil engineer.

The experience on the Erie is instructive, for the pattern it set in securing professional engineers was followed by the other large enterprises engaged in building public works. The alternatives were two in number: either the canal commissioners could seek trained engineers abroad, or they might hope to train their own engineers through on-the-job experience. In another generation, a third alternative would become available, namely, the hiring of engineers produced by certain colleges, primarily West Point. Two factors operated against securing foreign engineers: native pride and the prohibitive cost of hiring trained engineers from Europe. As a consequence the Erie Canal commissioners decided to upgrade their surveyors into engineers. Of all the engineers associated with the building of the Erie, none had any formal education for the job and only one went to Europe for training, and this only after the canal project was well under way.

For future training of engineers, a rather simple hierarchy developed within the Erie organization that provided the new recruits with both the necessary theoretical knowledge and experience. It evolved gradually, beginning with assistant engineer to principal engineer and then to chief engineer. Later, on other projects, the hierarchy included resident engineers whose duties usually included the operations on a specified section of the project. In addition, there were senior assistant engineers and junior assistant engineers.

On the Erie another pattern developed that was copied in other states on their public projects. This was a system whereby assistant engineers were recruited from the lower ranks of those working in the engineering field parties. Besides the axmen and other camp hands, there were those whose jobs more closely related to engineering: on the lowest level was the chainman; on a higher level, the rodman, who was employed in the leveling required in the surveying. There were some variations in the hierarchy from project to project. On occasion the assistant engineers, for example, performed the less

112

technical tasks of the surveying party, but there were instances when they performed duties that could well belong to a higher rank of engineer. The former circumstance appears to apply most nearly to Ellet's position in his fist job.

The Ellet papers do not reveal the way in which Charles secured his first employment, but several avenues were open to him. As a result of a scarcity of young men needed as engineers on transportation projects, employers were busy actively recruiting in neighborhoods along the route of the project as well as using other methods of personal contact and even advertising in newspapers. Perhaps the more important question is why Charles did not seek a job away from home earlier than he did. The most probable explanation . . . was the influence of a domineering father who wanted to keep his sons at home as long as he could, hoping to interest them in agriculture.

Employed as an assistant engineer, Charles was a member of a crew of fifteen men who, during the summer months of 1827, surveyed the North Branch of the Susquehanna River for a distance of 170 miles, from Northumberland to the New York state line. Two months after going to work, he reported that "my business is to traverse the shore, roads and islands to go ahead of the level and explore the country." His duties were varied, but the elements involved in surveying became second nature to him by the time the job was completed. The most difficult aspect of the work was the great amount of walking the job required. He mentioned that the 170 miles had to be surveyed twice and leveled on both sides, making a total of 340 miles, over which he walked four times.

In 1827, this section of Pennsylvania was relatively unsettled and the terrain along the river was covered with undergrowth, which had to be cut away before the surveying party could make progress. Even with a pioneer background, this rugged life, made more severe when cold weather came, brought many hours of difficulty to young Ellet. His performance, however, apparently satisfied his employers, for he confessed to his mother that he had every reason to believe his wages would be $2 a day, the highest the law allowed to any assistant.

*   *   *

The exact date when Charles left the survey crew has not survived. In August, 1827, he reported that the work would detain him until Christmas. The following May John Randel, Jr., the civil engineer in charge of the survey, recommended Ellet as one "employed by me the whole of the last season on the North branch of the Susquehanna . . . I found him to be a young gentleman of amiable manners, industrious habits; of strict integrity, sound discretion and good judgement; and now he has considerable experience in his profession: he is deserving of public and private confidence." . . .

After this Pennsylvania experience Charles was employed as an assistant "for nearly two years" on the Chesapeake and Ohio Canal. This waterway was an early attempt to connect the tidewater of the Atlantic Ocean with the Ohio River by means of a canal. With the exception of the James River and Kanawha Canal, it was the most ambitious attempt to link the two waterways, though it failed to achieve its goal of providing a short, cheap route to the

113

Ohio. At the age of eighteen, Charles was a member of the engineering crew responsible for getting the project under way.

On June 23, 1828, the company appointed Judge Benjamin Wright of New York as its chief engineer. A country lawyer, Wright, with James Geddes and Canvass White, had been largely responsible for the engineering work on the Erie Canal. He and Geddes had done land surveying but they were not professional engineers. No engineer connected with the Erie, except Canvass White, had ever seen an actual canal. It was White who had gone to England and studied her canal system after construction of the Erie had already begun.

As a consequence of his work on the Erie, Benjamin Wright became the best-known canal engineer in America. It is possible that he was a friend of the Ellets and that Charles's employment and rapid promotions were due to his family connections. More likely, however, they were a result of the young man's thoroughness and industry on the Susquehanna survey. In this period, a year's experience, such as Charles had acquired, was an important asset. The demand was great for engineers' services, especially if, like Ellet, they possessed initiative and dedication. At any rate Ellet traveled in the summer of 1828 to Maryland to begin his rather indefinite duties as a volunteer assistant on the Chesapeake and Ohio Canal. He arrived at his destination in time to witness the ceremony at which President John Quincy Adams turned the first spadeful of earth for the new project.

Following the ceremony the directors made plans to commence construction immediately. They divided the canal into three parts and then selected the engineers for each division. In making the engineering appointments, the board relied heavily on experienced canal engineers from the North or those of foreign origin. The rodmen were the principal exception to this policy. The directors accepted inexperienced applicants, often from neighboring towns, who were seeking a career in engineering. Some were appointed as apprentices and received their board and room, while others were only taken on trial and had to pay their own expenses. In this way the canal became a school in practical engineering, a common procedure in the period.

Charles did almost all the office work of his party, drew the maps and made the computations, and walked from ten to twenty miles a day surveying the route, all the while without having any designated position or salary. It is surprising that despite this strenuous schedule he also found time to study foreign languages, but report has it that he had a marked talent for them and possibly they provided a form of recreation in contrast to his more technical studies. His services soon brought him the award of the position of assistant engineer on the supposition that he was "at least twenty-two years of age" and had considerable experience in engineering.

After working on the canal for a year, Ellet appears to have been well pleased with his position. The job was not a lucrative one by present-day standards, but he was receiving $800 a year, an enormous sum at that time for a nineteen-year-old lad. Charles was satisfied that the construction of the canal was of high quality. "The entire work," he wrote, "is planned in the most effective, durable, and permanent manner; the canal is 60 feet wide at the water line, 6 feet in depth and 42 feet at bottom; the tow path is 12 feet wide and perhaps for a portion of the distance it will be paved on the inner slope with stone."

For the remaining two years that Ellet remained with the Chesapeake and Ohio Canal Company he struggled to resolve a decision as to his future. It involved a choice between whether he should proceed to Europe to secure further training as an engineer or seek work with the Illinois public improvements, an enterprise that had attracted the attention of the Ellet family for several years. Eventually he convinced himself and his family that the greater advantage lay in going abroad.

His father, however, was never completely reconciled to his European trip. He warned his son that there were many young men of talent "rising up in our country which will very likely supplant you, in your absence and after acquiring scientific knowledge you may be destitute of employment." On the other hand, his mother, while she cautioned him to avoid vice, especially the lures of females, possessed a great fund of optimism.

*   *   *

By February, 1830, Ellet was convinced that he must go abroad for further study. He wrote his brother John that he was waiting to receive more advice from his parents, "to see whether they still continue to think it preferable to follow the old and beaten track--the path to certain livelihood, and as certainly to but a medium station." Ellet argued that there were "not above 3 Engineers, who can be called men of science in the United States--having at the same time a knowledge of their immediate profession . . . therefore I ask whether there is any danger of being surpassed by those now engaged, whom I should leave at home?"

Charles concluded that the surest path to advancement in his profession was study of the engineering works of Europe. He anticipated other benefits of a less tangible and rewarding nature from his tour abroad. "I ask with confidence," he wrote, "will any one think to compare the pleasures, (the real pleasures) of a man of letters, poor though he may be, with those of an Engineer, reared on a canal bank, nursed in a lock pit, fed in a puddle ditch and oft-time bedded on a couch as hard as the wall he builds--and ignorant of all the refinements of life and mysteries of science?"

*   *   *

Graduate engineers were few, but their numbers were increasing rapidly. In 1830, when Charles left for France, there were only fifteen West Point graduates working as civilian engineers. Within a year or so, however, West Point equalled the New York canal system as a "supplier" of civil engineers and far exceeded it by the late thirties. Even so, one authority has concluded that "the American civil engineering profession of the generation around 1837 consisted mainly of men with no school training as engineers. By 1837, with the dying off of most of the miscellaneous group of engineers of 1816 and before, the experience-trained engineer was generally a man who had worked up within the engineer corps of an internal improvement project." In origin, the civil engineer was undoubtedly the creature of the organizations in which he worked. . . .

Ellet's pertinacity, his determination to go to France, and his desire to make good in his first major job left him little time to concentrate on those aspects of life with which many young men were concerned. His brother John, a year younger than Charles, asked him his advice on a wife. Ellet replied in his serious and characteristic manner: "You wish me to describe a beauty that would suit your taste--a matter that I have not settled for myself. It is beyond my abilities to say more than this--let neither the color of her eye, or the thickness of her lip--the span of the waist or the turn of the ankle, have any weight in influencing your choice, but look alone for refinements, education and modesty." Even in adolescence Ellet's correspondence contained little lightness and no humor.

When his parents were finally persuaded that Charles should go abroad, they cooperated to make the adventure successful. His father, contrary to his customary hostility, secured him letters of introduction, one to Lafayette, and the other to an unnamed American "gentleman of science" residing in Paris. Similarly, his mother obtained letters through the efforts of Charles's grandfather Israel, as well as $500 to defray his expenses.

Ellet's ship docked at Le Havre, from whence he made his way to Paris. . . . He found a third-story room in a hotel on the "literary side of the river" within two hundred yards of the Pont Neuf. After rising at five, his day was strenuous, with studying until he ate a breakfast of bread and milk at eight, and then resuming his studies until five in the afternoon. After dinner, he spent the evening reading and writing letters. He made an effort to limit his social calls to visiting Lafayette. He presented his letters of introduction to the general and described his first visit to his residence. ". . . The General himself entered and received me with a welcome so cordial and sincere, as to arouse all the gratitude an American citizen feels for his name, and his course." He conversed during the evening with Lafayette about "the state of America," but the "present disturbances in France occupied the chief point of conversation of the evening--in which the general has taken an active part, as well as his son--both of whom express their opinions very freely: and the latter made some remarks, upon intentions and probabilities that I did not expect to hear." Ellet felt he had never enjoyed such an agreeable evening.

His primary purpose in visiting Lafayette was to gain his assistance in obtaining admission to the Ecole des Ponts et Chaussees, [School of Bridges and Roads] an institution where lectures were delivered exclusively for engineers. All French engineers were required, after passing through the polytechnical school, to spend one year at the Ecole prior to entering upon the practice of their profession. With the additional aid of W. C. Rives, American Ambassador to France, he was admitted to the school at the end of July.

*   *   *

For seven months there is no account of his activities, but this tall lad, who had just attained his majority, can be pictured pursuing the course of study at the polytechnical school, observing all the engineering works in the vicinity of Paris, and becoming thoroughly familiar with the French language. In the spring of 1831, after four months of lectures, Ellet decided to tour southern France and Switzerland for the purpose of observing the public improvements. He provided an elaborate account of his trip in his letters to his sister Mary. They are valuable both for their description of France by an observant American traveler at the end of the first third of the nineteenth century and for the reactions of Ellet to the novel environment.

*   *   *

Ellet described only briefly the public works which he observed because he knew that his sister would not share this interest. He wrote of a dam at St. Feriol, the principal reservoir of the Canal du Midi, as "one of the most extraordinary works in existence." The dam, 128 feet high and 250 feet at the foundation, was thrown across a chasm between two spurs of the Montagne Noire, "forming a reservoir a mile in length, capable of containing nearly 7,000,000 cubic meters." He undoubtedly observed other reservoirs, although he mentioned none. His later concern with reservoirs as a means of improving the navigation of rivers in the United States stemmed from this experience in France.

His great interest in suspension bridges likewise originated while he was on this trip. He wrote of inspecting many such bridges on the Rhone, Loire, Garonne, and Seine. His letters are characterized by frequent references to these structures: "I made a few notes, observed a suspension bridge being constructed across the Loire and witnessed the manner in which the wire cables for these bridges are manufactured . . . At 5 o'clock, when within 4 miles of my place of destination, seeing a singular piece of canal work, I sprung ashore to take a drawing of it, and followed the barge to Adge . . . I arrived at Valence . . . in the morning of the 29th of March. Went to examine a fine suspended bridge across the Rhone."

Later Ellet indicated he had visited England, but he left no written account of his visit. Viewing the entire European trip, one can easily conclude that it was one of the most valuable experiences of his life. In addition to the ideas he gained on reservoirs and suspension bridges, he was able to indicate to prospective employers that he had studied abroad and observed the internal improvements of western Europe. This was particularly important at a time when an engineer acquired his training primarily through on-the-job experience. In the course of his European travels Ellet gained an advantage in the profession such as only a few other American engineers shared. Wright, his old mentor, who has been described recently as the "Father of American Engineering" because of his role in the building of the Erie Canal, took special note of Ellet's experiences. In a letter of recommendation written shortly after Ellet's return he indicated that his protege had examined all the public works, "and more particularly Railroads" in France and England. He concluded the letter by indicating that he believed Ellet to be qualified to manage a railroad.

What specifically Ellet learned in a technical sense by his European observations, especially as they relate to canals and railroads, is difficult to evaluate. With respect to the canals, which had been built extensively in Europe, he learned primarily what America should avoid. Fifteen years earlier, Canvass White, as Wright's chief assistant, had already returned to America with information concerning the design of the locks of the English canals. But what impressed Ellet most was the financial difficulties European canal projects experienced because of the lack of foresight and system with which they were built. There was, particularly in England, little planning to allow for new construction fitting into a perfect whole. Locks were never uniform even on the same canal, which limited all boats on the waterway to those that could be accommodated by the smallest locks.

Ellet learned more about European railroads than would be immediately applicable in America. Although the world's first railroad constructed for general transportation purposes began operation in England in 1825, there were, as late as 1830, the year Ellet first went to Europe, only seventy-three miles of railroad in the United States. By 1840 railroad exceeded canal mileage although the latter had more than doubled during the decade.

Within a short time Ellet was able to put his knowledge about railroads acquired in Europe to practical use on the surveys he conducted in New York state for both the Utica and Schenectady and the New York and Erie railroads. The reports on the projects indicated that he was an innovator in America with respect to locating railroads in areas that would necessitate traversing steep grades. He had studied closely the operation of the inclined plane and the stationary engines used to surmount steep gradients while in Europe.

Ellet derived, however, other advantages from his trip that may be more difficult to evaluate. His great desire to learn foreign languages and to teach them to his children was a result of this journey. And the emphasis he always placed on the study of history sprang, in part, from his early trip abroad. The most important consequence of the European experience was its effect on the development of his personality. A reserved but eager young man, accustomed to pioneer conditions in Pennsylvania, returned a matured, sophisticated, less provincial, but no less discriminating individual. Eager as he was to view European conditions, and despite the real pleasure he derived from sightseeing, he was suspicious by nature, and he had little use for many European customs, which he preferred to label as affectations. On balance, he considered the trip well worthwhile, sufficiently so to motivate him to return twice in later years, once accompanied by his family. It was a maturing experience, and in a very real sense the trip was a continuation, as well as a capstone, of his formal education.

Upon his return from France early in 1832, the Chesapeake and Ohio Canal Company offered him his old job at a salary of $1,000 a year with the opportunity for promotion the next year to superintendent of a residency. Ellet did not accept the offer. Perhaps he thought he could do better, but there is no further mention of an engineering project until October, when he became interested in a projected bridge across the Potomac. His refusal of the job, however, is illustrative of much in Ellet's career. The challenge afforded in building this particular work was far outweighed, he thought, by the difficulties it was experiencing, both in regard to adequate financial support and from the competition of the Baltimore and Ohio Railroad. For the moment, Ellet felt it would be more advantageous and challenging to introduce into America the concept of the suspension bridge. He was essentially an innovator; the bridge would be a new engineering feat while the canal was already a traditional means of transportation in America. Only when he was thwarted by lack of funds, or by the public's refusal to accept his ideas, would he return to the conventional in engineering. In addition, he always was more contented working by himself on essentially small projects, as most bridges were, than when engaged in large transportation enterprises. Unfortunately for Ellet the personnel of the engineering profession was fast coming to consist of men who worked within the constricting confines of a corporate organization. Although the role of the individual did not disappear, it was becoming ever more limited.

Plans for the construction of a bridge across the Potomac at Washington were circulated in an advertisement by the United States Treasury Department in August, 1832. For young Ellet it appeared that here was the opportunity to put the knowledge he had gained in France of suspension bridges into practice.

118

On October 6, he submitted his bridge design to Secretary of the Treasury Louis McLane. He proposed a bridge forty-six feet high, which would allow most boats to pass beneath it without lowering their funnels, and the span of the arches would be 600 feet, 572 feet between the piers. These distances would prevent ice from collecting, and allow steamboats to navigate the river with a minimum of difficulty. The stability of suspension bridges was perhaps their main attribute, and as Ellet explained: "The suspension bridge enables a light and weak structure to yield repeatedly to a heavy body passing over it, to acquire a new state of equilibrium, and return to its former situation as soon as the disturbing force is withdrawn. Whereas if the structure were of wood, and more particularly if it were of any of those systems, of which some peculiarity of the framing is intended to supply the place of curved ribs, after once yielding, or once bending it would never return." In Ellet's view, the most serious deficiency of wooden bridges was this tendency to change their position under stress. "There are few bridges," he wrote, "even among the most approved bridges in this country, that could be occupied without danger by a body of troops closely drawn up."

The plans Ellet submitted reached the committee too late for its consideration, and he was forced to write twice asking that his designs and plans be returned. In any event a suspension bridge was not constructed over the Potomac, but rather a contract was let for the repair of the old wooden bridge, at a cost of $113,126.

Even if the plans had arrived earlier, it is extremely doubtful that the Congress would have accepted them. Suspension bridges were a novelty in America, although they had already been erected throughout England and France. Ellet's innovations in the field of bridge-building would have to wait. It is important to note, however, that his advocacy of the suspension bridge in America was well ahead of those who later claimed primacy. Ellet was to be a prophet in more than one area of engineering endeavor.

With the rejection of his plans for a suspension bridge, Ellet turned to more conventional lines of work, and accepted a job making surveys for the Utica and Schenectady Railroad. This railroad was the second link in the plan to connect Albany with Buffalo by rail. The route, beginning at Schenectady, went in a northwesterly direction for seventy-eight miles, following the Mohawk River through Scotia, Amsterdam, St. Johnsville, and Manheim, to Utica. Charles was engaged in 1833 as an assistant engineer to locate the western division of the line.

In spite of the seemingly easy job of locating a railroad in the Mohawk Valley, it took over a year to complete the surveys on the Utica and Schenectady Railroad. The rocky "near-impasse" at Little Falls and the spring floods of the river presented difficulties for the surveyors. "But when it was finally located, it was well located," according to one writer, "and it has remained in that location from that day to this." Ellet left the survey in May, 1834, a year before it was completed, and two years before the first train traversed the line. The railroad was later incorporated into the New York Central system.

In May, 1834, Ellet accepted an appointment by Judge Wright to conduct the survey of the western end of the New York and Erie Railroad. His sections of the line extended for 260 miles from Binghamton to Lake Erie. When the original survey for the route was conducted in 1834, there was not a town with a population exceeding 3,000 between the Hudson and Lake Erie along the line of the proposed railroad. . . .

*　　*　　*

On June 22, 1834, Ellet began his survey for the New York and Erie Railroad at Tioga Point, New York, the same location where he had commenced his engineering career on the Susquehanna survey seven years before. One month later he gave his address to his sister as "52 miles west of civilization." He superintended two survey crews fifty miles apart, and in addition was obliged to explore an immense area of the country. His labor force, completely untrained, led him to comment after a difficult day: "A more stupid race of jackasses it has seldom fallen to the lot of an engineer to deal with."

Ellet wrote a forty-nine-page report dealing with the survey, his first publication. . . . It was published in February, 1835, four months before he began his work for the James River and Kanawha Canal Company.

As on former occasions, Ellet received the appointment of the canal project through the influence of Judge Wright, who was appointed chief engineer. The James River and Kanawha Canal Company, besides being the most ambitious plan for internal improvements in the South, was considered the most likely project to link the Atlantic coast to the Ohio River. Ellet's years of training were about to bear fruit. He was no longer to be looked upon as a young and promising engineer, but as an engineer of distinction in his own right.

The year 1835 marks not only Ellet's emergence as an engineer of note but the end of his formative years. As an adult he presented a striking appearance, standing slightly more than six feet two inches in height with sharp features and thick hair worn long. His large nose was frequently the subject of comment, if not of ridicule. His extreme thinness combined with his height often singled him out in a crowd. Similarly, his personality traits were sharply etched, for he possessed in extremes the qualities of independence, forthrightness, ambition, and determination. Less friendly observers interpreted these traits as evidence of inflexibility and priggishness, characteristics Ellet also possessed.

Ellet's personality was more like his father's than his mother's, although he never felt close to his father and frequently ignored his advice. Like his father he was often harsh and irritable, and could seldom bring himself to tolerate honest differences of opinion. Like his mother, he possessed a stern and unyielding sense of duty, but as in her case, it was softened in later years by a complete devotion to his wife and children.

In evaluating the influences in Ellet's life, his early practical engineering experiences and his trip abroad are significant, but his own innate brilliance can not be ignored. It was his intellectual capacities, evident equally to friend and foe, that made the difference between an average and an extraordinary engineer. These qualities surmounted an undistinguished family background and lack of formal education that might otherwise have proved to be deterrents to a significant career in technology. Ironically, it was his own emotional makeup, combined with poor health, that prevented him from seeing many of his ideas and projects brought to fruition.

120

"Crazy" Judah and the Big Four

Chapter 11, **Men To Match My Mountains**
Irving Stone

Theodore Judah had a vision which men without his education and experience could not see. Unlike many other engineers, he had the ability and perseverance to sell his vision to others who were necessary partners if the vision was to become reality. Unfortunately for Judah, his ethical standards eventually cost him the support of the less ethical men who financed his dream--a transcontinental railroad.

The civil engineer's Code of Ethics requires the civil engineer to put the well-being of society above his personal gain. In Judah's case, living up to that ethical standard cost him his dream and eventually his life.

The decade of 1859-1869 was forcibly pried open by that *rara avis*, an authentic genius, Theodore D. Judah, who was brought from New York in May of 1854 to build California's first railroad. Described in Oscar Lewis's **The Big Four** as "studious, industrious, resourceful, opinionated, humorless and extraordinarily competent," Judah was the fifth wheel without which the Big Four of Huntington, Stanford, Hopkins and Crocker* would never have rolled . . . not all the way across the United States on a transcontinental railroad.

Theodore Judah was born in Bridgeport, Connecticut, in March, 1826, son of an Episcopal clergyman. When the family moved to Troy, New York, Judah studied engineering at a local technical school, moving without a wasted day from his classroom to engineer on a railroad being built from Troy to Schenectady. Before he was twenty he had surveyed railroads in Massachusetts and Connecticut, built a section of the Erie Canal; at twenty-two he helped to plan and build the Niagara Gorge Railroad. He also built a cottage at the edge of the Niagara River for the pretty Anna Pierce, his bride. Judah told his wife admiringly: "You always have the right gaiters on."

She needed them. Her peripatetic husband moved her twenty times in the first six years of their marriage.

When Judah was twenty-seven and building part of the Erie Railroad system, he received a telegram from the governor of New York, Horatio Seymour, summoning him to a conference with C. L. Wilson, president of a group of Californians who wanted to build a railroad from Sacramento to the Sierra Nevada mines.

Governor Seymour recommended Judah as the most competent young railroad builder in the East. But Judah did not want the job. There were too many railroads to be built in the East; his brother, Charles, who had gone to

----

* Collis P. Huntington, Leland J. Stanford, Mark Hopkins and Charles Crocker were the four entrepreneurs who financed the development of railroads to California.

California in the gold rush and was now practicing law in San Francisco, was one of the few pioneers who found the Far West "harsh" and uninviting.

Wilson held an ace. He must have played it at the right moment, for on the third day after Judah had left his wife to meet with the governor, Anna received a telegram: "Be home tonight; we sail for California April second."

Wilson could only have asked Judah: Once you have built a railroad to the foothills of the Sierra Nevada, is not the next logical step to cross the Sierra Nevada? And with the mountains bridged, was not a transcontinental line inevitable? This dream was not new to Judah. His wife says: "He had always read, talked and studied the problem."

There was need for a railroad. The trip around Cape Horn was nineteen thousand miles. Across Panama the arduous journey took four weeks. The Overland Stage needed only seventeen days from Missouri to San Francisco, but could carry little freight. The heroic Pony Express riders made it in eight days, but they could carry only two saddlebags of mail. San Francisco, Virginia City, Salt Lake, Denver were split off from the rest of the nation by a geographic chasm. The rails . . . would not only serve as iron links connecting the East and West but would quicken the settling of the wastelands between. . . .

Could it be built? The central core of the Sierra Nevada was granite. A railroad would have to pierce that granite many times. The lowest summit at which a pass was forceable was seven thousand feet, where fifty feet of snow fell in winter. There was not merely one range, but two that had to be cut through to tame the upward grade to a rise of one hundred forty feet to the mile so that a locomotive could pull its cars. There were precipices of solid rock; gorges that would have to be bridged or filled with a million tons of dirt. Beyond, in Nevada and Utah, lay hundreds of miles of parched desert with no water for the men or animals; here it would be a matter of supply. And in Colorado there were the Rockies.

Judah set up an office in Sacramento, then went into the field to survey and draw his maps for a Sacramento Valley railroad. Never a man for waste motion, he completed his report in a few weeks, ordered iron rails from New York and, in February of 1855, set a hundred men to grading a roadbed and laying the first rails in California. With an eighteen-ton locomotive and two small flatcars that came around Cape Horn, Judah took a San Francisco delegation for a ride northward through the valley; and in February of 1856 there was an excursion and ball to celebrate the successful completion of the line to Folsom, in the Sierra Nevada foothills, shortening by a full day the wagon and stage journey between the two towns.

His job completed, Judah was free to give himself to what had now become his dominating passion: a railroad which would link the Atlantic and Pacific oceans. When he was hired to survey a possible wagon road across the Sierra Nevada to Gold Canyon in Nevada, he returned enthusiastic about a railroad pass he had located over the mountains.

He was too early. No one would back him. Unemployed, short, punchy Theodore Judah, with the broad flat face and unswervable eyes of a sphinx, used up his savings and devoted full time to talking and writing about a transcontinental railroad. He stood alone, a brilliant monomaniac, who worked his vision so hard that Sacramento people grew weary, turned away from him saying: "Here comes crazy Judah!"

Three years later, in September of 1859, in San Francisco's largest assembly hall, a meeting of the Pacific Railroad Convention provided the incubator where the highly improbable seed of a railroad to the Pacific was given the warmth of life. Theodore Judah had organized the meeting, masterminded it with charts, maps and reports, and probably paid all the expenses as well. In recognition of his one-man show the convention unanimously elected him their delegate to go to Washington, also at his own expense, to convince Congress that it should pass a bill appropriating money with which to start a transcontinental railroad to run through Colorado, Utah and Nevada to California.

Mrs. Judah again donned the proper gaiters. The Judahs set sail for New York. By a fortunate coincidence Congressman-elect John C. Burch was traveling on the same ship.

"No day passed on the voyage that we did not discuss the subject," reports Congressman Burch. "His knowledge was so thorough, his manners so gentle and insinuating, that few resisted his appeals."

Nor did Congress. He was given a room in the Capitol where he opened the Pacific Railroad Museum, assembled and dramatized railroad exhibits, educating congressmen, writing letters to the newspapers, publishing brochures, lecturing: a one-man lobby for a coast-to-coast railroad, liked and respected by everyone.

But it was the winter of 1860. Slavery was convulsing the nation and its lawmakers. The problem of whether the railroad should follow a northern or southern route created an impasse higher than the Colorado Rockies which had turned back John C. Fremont. And so, after a year of effort, Theodore Judah returned to California empty-handed.

Back at Sacramento he went again into the Sierra Nevada, determined to map a practical route across the mountains, and to establish an accurate cost survey. He stayed so long he was nearly trapped by the winter snows. He came down to Dutch Flat, focal point of his route through the foothills, assembled his drawings on the counter of Dr. Daniel W. Strong's drugstore, and then drew up the "Articles of Association of the Central Pacific Railroad of California." By his calculations it was one hundred fifteen railroad miles from Folsom across the crest of the Sierra Nevada and down to the Nevada line. Since California law decreed that $1000 of capital stock had to be subscribed for each railroad mile before work could begin, Judah would have to assemble $115,000 in stock pledges before he could begin a roadbed.

The "Articles of Association" were "passed from hand to hand around the village of Dutch Flat." Judah and his druggist friend Dr. Strong subscribed for more stock than either could afford, the little town and countryside subscribing the balance up to $45,000, a vote of confidence in Judah's engineering. Needing only another $70,000 to start work, Judah left for San Francisco and a meeting with the city's capitalists. He assured his wife, who spent most of her time waiting for him in strange hotel rooms, that their troubles were over.

He returned to the hotel, his lips trembling but his eyes snapping with anger, crying in the bitterness of his disappointment: "Anna, remember what I say to you tonight: not two years will go over the heads of these gentlemen . . . but they will give up all they hope to have from their enterprises to have what they put away tonight!"

He was as sage a prophet as he was a railroad engineer.

Judah and Anna packed their bags, took the morning boat for Sacramento. Sacramento would benefit enormously from such a railroad, yet for five years the town had given him little but indifference and hostility. Lionhearted, he walked the streets, collared people, got groups into conference where he could lay out his charts, demonstrate the idea of California and the East being connected by an inexpensive daily means of travel. When he considered he had worked up enough interest, he called a meeting at the St. Charles Hotel.

Few came, only curiosity seekers. He resumed his efforts to stir up interest, called a second meeting in a room above a hardware store owned by Collis P. Huntington and Mark Hopkins, local merchants. About a dozen men attended, among them Dr. Strong, a surveyor by the name of Leete who seems to have helped Judah plot the route, two Rayburn brothers who aspired to be railroad promoters, James Bailey, a local jeweler, Lucius Booth and Cornelius Cole, Judah's two hosts who had loaned him the room, Huntington and Hopkins, a lawyer turned wholesale grocer by the name of Leland Stanford, and a dry goods merchant, Charles Crocker.

"A druggist, a jeweler, a lawyer, the owner of a dry goods store, two hardware merchants: this hardly seemed promising timber to carry out the vast scheme Judah had envisaged."

Judah knew better than to try to sell such hardheaded businessmen a transcontinental railroad. They would not have risked a dollar on so visionary a scheme. Instead he presented the idea of backing a survey across the Sierra Nevada for a good and shorter wagon road, if not a railroad, to the Comstock Lode, which Sacramento was supplying with hundreds of thousands of dollars worth of foodstuffs and heavy mining equipment.* The men who could corner this traffic would make millions.

Since the investors had to put down only ten percent of the stock value, they would be risking less than $7000 among them. Huntington, Stanford, Hopkins and Crocker filled up the stock quota and voted themselves officers. Judah

----

    * The Comstock Lode was the unbelievably rich silver mining area in western Nevada, near Virginia City.

was content to remain chief engineer. In his hotel room he exulted to his wife: "If you want to see the first work done on the Pacific Railroad, look out your bedroom window this afternoon. And I am going to have those men pay for it."

"It's about time somebody else helped," replied Anna Judah wistfully.

That afternoon Judah began his survey line in the street below his wife's window. He then moved eastward to tackle the towering Sierra Nevada whose western flank rises sharply to seven thousand feet, torn and lacerated by swift-flowing mountain streams descending through jagged canyons. It was a formidable task, made to order for an engineering genius.

By the end of the summer Judah emerged with an almost precise railroad route. He also brought news from Nevada which interested his associates: a railroad would be able to bring out the low-grade ores now being discarded at the top of the mine shafts, worth a fortune to those who owned it.

The Big Four were shrewd and successful traders; they were also pugnacious, acquisitive and largely conscienceless. They quickly pushed out of the picture Judah's druggist friend from Dutch Flat, the surveyor who had helped Judah, the two promoter brothers, the jeweler. They would provide the organization to build the railroad!

The Judahs once again sailed for the East in the hopes of persuading Congress to finance the railroad. Judah reached Washington in October, 1861, three months after the Union defeat at Bull Run; Congress knew it was in for a long war. Judah revised his approach accordingly, rewriting the Pacific Railroad bill as a war measure urgently needed to keep California and Nevada, with their rich gold and silver deposits, in the Union. He then adroitly managed to have himself appointed secretary of the Senate Committee on Railroads and clerk of the House Subcommittee on Railroads. Even from this vantage point it took him a full year to maneuver the measure through both Houses. But what a bill it was: giving the builders, free and clear, ten alternate sections per mile of the public domain on either side of the line, and lending them millions of dollars of the public funds with which to build the railroad. Judah sent a message along the newly completed transcontinental telegraph: "We have drawn the elephant. Now let us see if we can harness him up."

When Theodore Judah returned to Sacramento he was no longer "Crazy" Judah, but a man of remarkable accomplishment who had talked Congress out of millions of dollars and millions of acres of land. This man was no visionary, he was a financial wizard!

Judah's troubles began when he wanted to build his railroad, and the Big Four wanted to build their fortunes. Because the government was lending the railroad from $16,000 a mile in flat country to $48,000 a mile in mountain country to build their road, the Big Four organized their own construction company and gave this construction company a contract at the highest possible price. If the railroad never ran, the Big Four could not lose, they would take their millions out of their construction company profits.

Judah fought the combine heroically. He obliged them to abandon their scheme. By his victory he was doomed.

When Huntington, Hopkins, Crocker and Stanford tried the next maneuver on their cost maps, moving the foothills into the flat of the Sacramento Valley because the government gave them $32,000 per mile to build in the foothills against $16,000 per mile in the valley, Judah again protested.

The Big Four wanted him out. They took away his powers as chief engineer. He was obviously not the man to build their railroad.

Defeated by a you-buy-us-out or we'll-buy-you-out ultimatum, Judah finally accepted their offer of $100,000 to resign and turn over his stock. In October 1863 he and his wife took ship once again for New York. Judah came down with yellow fever at Panama, and died in New York before his thirty-eighth birthday.

The Big Four set out to obliterate his name. From then on the Central Pacific would be theirs and theirs alone. Theodore D. Judah was one more man sacrificed to the concept of a transcontinental railroad.

----

* The Central Pacific geologists claimed that the Sierra Nevada foothills began in the absolutely level Sacramento River Valley, seven miles east of Sacramento, instead of where they actually begin--twenty-five miles further east. The federal government agreed and paid Central Pacific an additional $500,000. Wags said: "The Central Pacific is the strongest corporation in the world, because they could move a mountain range twenty-five miles."

Building the Union Pacific Railroad

An Excerpt from Chapter 13 of **Trails, Rails and War**
J.R. Perkins

The working and living conditions for civil engineers involved in building railroads during the middle third of the nineteenth century bear little resemblance to what most modern civil engineers encounter. Sometimes one had to be soldier, sheriff, politician and diplomat as well as engineer. Perhaps more than any other man, Grenville Dodge displayed the personal and professional characteristics needed for success during this era.

Dodge was a graduate of Alden Partridge's Norwich University civil engineering program (the first civilian school of engineering in the United States). By the time of the Civil War he had acquired considerable experience in railroad engineering, having moved from the eastern railroads to the Chicago and Rock Island Railroad near the western frontier. While working for the Rock Island he conducted surveys in the Platte valley and other parts of the western territories.

When the Civil War broke out he accepted a commission in the Union Army and had a distinguished career as a military leader, rising to the rank of Major General. He was successful as a troop commander and as a railroad engineer in charge of railroad repair for General Ulysses S. Grant. He was wounded twice, once in leading a charge during the battle of Pea Ridge (Arkansas) and once in Atlanta. After recuperating from his wounds he was assigned to duty in the West, under General William Tecumseh Sherman, where his knowledge of the area was sought for use in the Army's assignment to make the region safe for settlers.

The other main character in this story, Thomas C. Durant, was educated as a physician, taught in a medical school and engaged in a private practice of medicine, but while still a young man was attracted to the growing railroad industry as a source of adventure and spectacular financial rewards. He became one of the country's most visible promoters of the new mode of transportation. Moving west, he became involved with the Chicago and Rock Island Railroad and with the Missouri and Mississippi Railroad. Here he met Peter Dey and Grenville Dodge, both capable railroad engineers.

In 1863 Durant arranged a capital infusion for the struggling Union Pacific Railroad Company, managed to get himself elected vice president and proceeded to take over its operation. Durant instructed chief engineer Peter Dey, who was now chief engineer for the Union Pacific, to complete a survey of the Platte valley and beyond to the Salt Lake. Meanwhile, he was attempting to open negotiations with Dodge, hoping to persuade him to accept the position of chief engineer for the Union Pacific. Dodge, however, was fully occupied with his military duties at the time.

By the end of 1864 Durant's various promotions and schemes were enriching him and his friends but doing little to advance the completion of the railroad. Peter Dey, by all accounts an honest man, tried to expose Durant's shady deals but was unable to garner any support. In frustration, he resigned what he called "the best position in my profession this country has ever offered to any man." Once again, Durant began his pursuit of Grenville Dodge for the position of chief engineer of the Union Pacific. He was to get more than he was bargaining for.

Seven months after Dodge discovered the Lone Tree Pass over the Wyoming Black Hills, or on April 24, 1866, he met Durant at St. Joseph, Missouri, and held a final conference on the subject of becoming chief engineer of the Union Pacific. The meeting was hardly peaceful. Dodge's old boss, Peter Dey, the first chief engineer of the road, had severed all connection with its construction, being unable to get along with Durant. Dodge was perfectly conversant with all the issues between the two, and he knew Peter Dey to be as high-minded as Thomas Durant was crafty and bellicose.

But there was something else--something far more serious than a squabble over routes out of Omaha: Durant was under fire, both in his own company and in Congress, charged with having built up his personal fortune, through construction contracts, at the expense of the Union Pacific and the Federal government, joint agencies in the building of the road. . . . "I will become chief engineer only on condition that I be given absolute control in the field, Durant heard Dodge say in his deliberate manner. I've been in the army long enough to know the disastrous effects of divided commands. You are about to build a railroad through a country that has neither law nor order, and whoever heads the work as chief engineer must be backed up. There must be no divided interests; no independent heads out West, and no railroad masters in New York."

*　　*　　*

It was on May 6th that Dodge entered the chief engineer's office in the second story of a little brick building occupied by the United States Bank at Omaha. Disorganization was apparent, for there was no regular head to the company west of the Missouri River, and the engineering, the construction and the operating departments were all reporting separately to New York. And in New York City was a little group of railroad promoters who knew nothing of building across the plains and who, as a consequence, were quarreling among themselves.

The Union Pacific Railroad Company, at this hour, may be likened to that individual in the popular song who said, "I don't know where I'm going but I'm on my way." In other words, the final surveys for the road were never far ahead of its actual construction, and when Dodge assumed charge no one knew whether it would be built out the North Fork Platte [River] toward Fort Laramie; out the South Fork Platte [River] to Denver; or due west from where the Platte divides, out Lodge Pole Creek. And as to building across the mountains after traversing the plains, there was neither agreement nor understanding. Isolated engineering parties were roaming both the plains and the mountains, and had been for upward of twelve months but, decimated by the Indians and discouraged by receiving no pay, some of them had disbanded and others sat down to await developments.

One may keep well within the bounds of truth and say that General Dodge knew more of the possibilities of the country from the Missouri River to Salt Lake, from a railroad standpoint, than any other American engineer. He was the first engineer to be employed by a railroad company to make surveys out the Platte River Valley and on to Salt Lake. He had been sent by Henry Farnam of the Rock Island ten years before the Union Pacific drove its first spike at Omaha, and his surveys were something more than horseback reconnaissances, for he had used his instruments and the Indians had named him "Long Eye."

But nothing is more difficult in the history of the first transcontinental railroad than to determine the value of surveys and the place of surveyors. First, there were the buffalo trails, and no one knows how old they were before the Indians rode them; there were the emigrant routes superimposed on the Indian trails; there were the routes of the overland mails superimposed on both; and then came the builders of railroads. As Dodge said:

> There was never any very great question, from an engineering point of view, where the line, crossing Iowa and going west from the Missouri River, should be placed. The Lord had so constructed the country that any engineer who failed to take advantage of the great open road out the Platte valley, and then on to Salt Lake, would not have been fit to belong to the profession.

In thirty days Dodge completed his organization, and on a military basis. Isolated engineering parties, scattered from Fort Kearny to Salt Lake, were brought into coordination, provided with heavily armed escorts and ordered to swing into action again; construction parties, long idle, were stimulated to activity; materials began to move, and the thud of the sledges on the spikes told of the new beginning.

*   *   *

Three months after Dodge became chief engineer, Jack Casement* assembled a thousand men and one hundred teams out on the prairies of Nebraska, forty miles from Omaha, and told them what he expected. It was a mixed crowd of ex-Confederate and Federal soldiers, mule-skinners, Mexicans, New York Irish, bushwhackers, and ex-convicts from the older prisons of the East. Somewhere in California was another group pushing the Central Pacific eastward.

With a wild raw yell Casement's men swung into action and the track-laying of the Union Pacific increased to three miles a day within a month. The East heard of it and out came bankers, statesmen, magazine writers, and special correspondents. The Union Pacific Railroad, a dream, a theory in the opinion of many, was lifting to reality. And this is what they saw:

> Long lines of grading teams sinking scrapers into the soil of Nebraska where plows had never gone before; great wagon trains of ties rolling in from the Far West; a hundred bronzed men dropping the timbers in their place; a hundred others pulling iron rails from flat-cars and dumping them along the embankment; a dozen brawny Irishmen tugging the rails to their position; the falling of the sledges; the rhythmic bending of the

----

* Jack Casement and his brother Dan had been hired by Durant to be in charge of track laying for the Union Pacific. They had worked on railroads in the east beginning as common laborers while in their teens and working their way up to foremen and finally to contractors. Jack supervised operations in the field and Dan kept the accounts.

bolters; the laying of four rails to the minute; and the steady creeping, like a great brown worm, of the track to the west.

But this first group of easterners saw something else: they saw the beginning of the "moving town," of the "hell on wheels," for at each base, in increasing ratio, there assembled the strumpets, the gamblers, the liquor dispensers of a dozen states; and Paddies, troopers, Indians, and frontiersmen drank, sang, danced, gambled and fought.

The first moving town sprang up at Fort Kearny on the Platte River, the second at North Platte, and the third at Julesburg, which turned out to be the worst of the three. The rougher element figured that Julesburg was far enough west to be beyond the pale of law and order. A group of gamblers took possession of the town as soon as Casement and his crew began to work west of it, jumped the land that Dodge had set aside for shops and defied all creation. Dodge ordered Casement to return to Julesburg and restore order. Three weeks later the chief engineer had occasion to visit the town.

"Are the gamblers quiet and behaving?" he inquired of Jack Casement.

"You bet they are, General," Casement replied. "They're out there in the graveyard."

He had descended on the town with a hundred seasoned soldiers and wiped out the ringleaders.

It was in August, 1866, that the Indians first began their attack on the builders of the Union Pacific. The road had reached Plum Creek, two hundred miles west of Omaha, when a powerful band of Indians swept down on one of the freight crews, captured it, and held the train. The situation was acute from more angles than one. Dodge had no hesitancy in cleaning out troublesome whites, but the Indians, being wards of the government and the subject of much emotional oratory, presented another problem.

*   *   *

He was ten miles west at the end of the line when word came to him of the trouble. He ordered his private car--an arsenal on wheels--hooked on to the handiest engine and with twenty men raced back to the scene of the capture. Halting his train he deployed his men on either side of the track and opened fire on the Indians, who had set fire to the freight. They mounted their ponies and ran without making a show of resistance. It wasn't much of a fight, but it marked the beginning of twenty months of bitter warfare against the building of the first transcontinental railroad.

But more than Indian troubles now confronted the chief engineer: there were bitter disputes between the government directors and the Union Pacific officials over the route of the road through the mountains. Some favored one pass and some another, but the chief engineer had plans of his own. He headed straight for the lone tree that marked the defile he had discovered the year before when beset by Indians.

A thorough examination of both the eastern and the western slopes of the Wyoming Black Hills convinced Dodge that the road should be built through

130

the pass he had discovered and he instructed his engineers to make the exact location. His decision marked the beginning of his disputes with Durant over the location of the line from the eastern slopes of the Black Hills to Green River, far to the west--disputes that all but halted the building of the road.

*　　*　　*

On October 8th, Dodge and escort plunged into Boulder Canyon to hold a conference with one of his engineering parties when a heavy snowstorm swept down on them and caught them in a critical situation. The teams refused to face the snow that was turning to a stinging sleet. Dodge ordered the packs to be taken off the mules to let the animals shift for themselves, and after a hard struggle he succeeded in leading his men to the shelter of an old stamp mill near a mine. For three days they were snowed in and provision ran low, and Dodge saw enough of Boulder Pass to convince him that the main line of the Union Pacific could never be built any closer to Denver than a point a hundred miles to the north. While cooped up in the mountain storm he was elected to Congress, having been drafted by the soldier element of Iowa early in the spring to oppose John Kasson for the Republican nomination. "But I'd even forgot," Dodge said, "that it was election day."

*　　*　　*

Dodge went east, . . . presented his plans for building the Union Pacific through the mountains, secured the consent of the directors to construct over the Lone Tree divide, and then hurried back to the West. More difficulties than he had ever dreamed of were at hand, for Durant was beginning his struggle with the Ames brothers[*] for the control of the road; moreover, the Indians were fully organized to give their best blows.

*　　*　　*

Dodge wrote [General William Tecumseh] Sherman a long letter, told of his plans to reach Fort Sanders--288 miles west--in another twelve months, and frankly declared that he must have more troops and more authority.

Grant and Sherman moved at once and General Augur was given command of the Department of the Platte and told to cooperate with Dodge in the building of the Union Pacific Railroad. Augur came and asked Dodge what he wanted and Dodge replied:

> "I want strong military escorts with each party of engineers and I want detachments strung all along the line from Alkali to the Laramie River."

During the winter and the spring Dodge fought snows and floods. In March tracks were blocked, for no one had anticipated snow blockades. In April great

----

* Oakes and Oliver Ames were wealthy brothers who were involved in both politics and railroading. Eventually they acquired a financial interest large enough to dominate the Union Pacific Company.

floods swept through western Nebraska, tore out miles of track, bridges, and telegraph poles. The damage at the Loup Fork bridge alone exceeded fifty thousand dollars. And just at this inopportune time the company sent its representatives west to see what Dodge was doing. No visit could have been more inauspicious. Moreover, Durant and Dodge were quarreling. Dodge was frank with Oliver Ames, the president of the road, who came out.

"Durant is in the way," he told Ames.

Oliver Ames sized up the situation and sat down and wrote:

> "It shall be the duty of the chief engineer of the Union Pacific Railroad to take charge of all matter pertaining to the construction of the road."

* * *

The floods of nature and the disputes of human nature had been unable to halt the chief engineer of the Union Pacific, so the Indians took a hand. From the deep ravine of the Wyoming Black Hills they swept down to Lodge Pole Creek, pulled up the stakes that marked the line of the road, stole the teams and drove the workmen back upon the base; they struck another party on the Laramie Plains, cleaned it out and burned everything in sight; and they wound up by tackling one of the best protected engineering groups on the road; killed a soldier and a tie-hauler, and playfully burned the stage stations along a fifty-mile front.

Dodge, greatly troubled if not discouraged for the first time, wrote Sherman on May 20th: "I am beginning to have serious doubts of General Augur's ability to make a campaign into the Powder River country and at the same time give ample protection to the railroad, the mail routes and the telegraph." . . .

* * *

The last week in May three government commissioners[*], White, Simpson, and Frank P. Blair came west and went to the end of the track to examine the road. They had just completed their task and were standing talking with Dodge when more than one hundred Indians suddenly swept down a ravine and made a fierce assault upon the workmen at their lunch. A company of soldiers was less than a mile away, but they were unable to render any assistance, so swift was the rush of the red men. The whine of the bullets uncomfortably close to the government commissioners, the desperation of the attack, and the indifference of the Indians to their own fate convinced Blair and the others that the red men were now fighting for their country and not merely to steal a few head of cattle.

----

[*] In July, 1864 Congress had passed legislation which stipulated that three commissioners, appointed by the President of the United States, would be responsible for inspecting the work of the Union Pacific Railroad Company and reviewing its books.

Dodge left the commissioners standing on the hill, jerked his revolver, and hurried down to the tracks, yelling at the chief graders to grab their guns and go after the Indians, but the workmen were demoralized and sought the shelter of the freight cars and would not budge. He stormed about, upbraided the graders for allowing a band of savages to rush them without firing a shot in return, and then went back to the commissioners and tartly said:

> "We've got to clean the damn Indians out or give up building the Union Pacific Railroad. The government may take its choice."

He was despondent, ill, and had begun to feel that he stood alone in his efforts to complete the Union Pacific in the time allowed under the Act of 1862. But Blair told him that he had seen enough to be convinced that the road could not be pushed through the mountains without heavy reinforcements from the army and he promised to go back east and stir up things. He did so and three additional companies of cavalry were stationed along the line for scouting purposes, and to keep the Sioux and the Cheyennes pushed back into their own country.

* * *

Two weeks later Dodge and his party started on the long march to Salt Lake. . . . The escort toiled to the summit of what is now known as Sherman Pass, 8200 feet above sea level, and descended to Dale Creek, the most serious obstacle to the building of the Union Pacific, for this stream required a bridge 125 feet high and 1400 feet long.

Fort Sanders was reached in a few days, and there Dodge met General Gibbons, one of Grant's most trusted officers of the Civil War. Nothing could have been more fortunate in the building of the Union Pacific from this fort west to Green River than the presence of General Gibbons, who entered sympathetically into all of Dodge's problems. He was in conference with Gibbons when a rider, with foaming horse, dashed up to the post with the news that the Percy T. Brown engineering party, engaged in the difficult work of making locations across the Great Divide, had been severely beaten by the Sioux, and Brown had been killed.

Dodge bowed his head and groaned, for not only was Percy Brown a capable engineer but he was devoted to his chief and trusted above anyone who had ever operated between Rattlesnake Pass and Green River. Then Dodge heard the story. Brown and thirteen men, beset by a powerful band of three hundred Sioux Indians, fortified themselves on an elevation in the Great Basin and fought from noon until night, when their foes, for reasons unknown, withdrew. Brown, badly wounded, begged his men to leave him, but they refused, made a litter of their carbines, and carried him twelve miles to a stage station, where he died within an hour.

* * *

Dodge pushed on west to strengthen his badly disorganized engineering parties all along the line. At Rattlesnake Pass he discovered coal and later sank a mine, which was the first that ever supplied the Union Pacific along its own route. On reaching the North Platte he found it to be swollen from heavy mountain snows. He ordered it to be forded, but two of the young officers who attempted it were swept back to their starting point, and the remainder of his escort refused to budge.

Dodge jumped on his own horse, Rocky Mountain, and plunged into the stream, calling on his entire command to follow. It proved to be a desperate undertaking and three of his men had narrow escapes from drowning.

"If you are going to help me build the Union Pacific through this country you've got to learn to swim horses across more rivers than this one," the chief engineer said.

Near this crossing of the North Platte he and General Gibbons established Fort Steele, for just to the south was the Medicine Bow range, rendezvous of the Crows, the Sioux, and the Cheyennes, and no railroad could be built until they were held back.

*   *   *

They came upon the Percy Brown engineering party, strongly reinforced it, started it to work and then pushed toward the Great Divide. At their feet was a vast basin. Dodge lifted powerful glasses and looked in every direction.

"I see some teams down there. Must be white men--perhaps returning emigrants."

He plunged down into the basin, followed by his escort, and an hour later came upon one of his own engineering parties--the one headed by Charles Bates that had been ordered to survey from Green River back east to meet Brown. Bates and his men were in bad condition, without water, and with swollen tongues. But they were at work running a true line, for they were the kind of men who made the Union Pacific possible. . . .

*   *   *

Dodge's party reached Salt Lake the afternoon of the thirtieth [of August] and camped south of town. . . . On September 4th, Dodge led his command out of Salt Lake for the 700-mile trip back to Fort Sanders. . . .

On arriving at Fort Sanders, Dodge made preparations to go to Washington and take up his Congressional duties.

. . . But not for a day did he relinquish his hold on the railroad situation in the West. From his office in the Department of the Interior, especially placed at his disposal for his double task, he sent out his orders to the division engineers who, in turn, kept him posted on all developments. . . .

Near the close of the session of the Fortieth Congress word came from the West that Durant, in violation of every agreement made with Dodge when he became chief engineer, was changing locations, altering the line over the Wyoming Black Hills, and circulating rumors that the Union Pacific Company would make Laramie City, and not Cheyenne, the mountain division of the road. Moreover, the construction crews between Cheyenne and the new terminus town of Laramie City were threatened with extinction by a powerful gang of gamblers and whisky vendors operating in a half-dozen camps.

When Dodge received this news he drew down the top of his desk, packed his valise and caught the first train out of Washington for the West. On reaching Cheyenne, he found a muddled state of affairs. Scores had left the town, trekking across the mountains to the new base in Laramie City which, according to Durant, would soon have the shops of the Union Pacific Railroad. The citizens of Cheyenne--those who had invested heavily in town lots--called on Dodge and demanded an explanation.

"You can't prevent an exodus from an old base town to a new one. You fellows ought to know that," Dodge said. "But on the score of the removal of the shops, well, Durant lied. The shops will remain in Cheyenne; the branch line will be built from Denver, and this will be the division."

Dodge ordered his private car hitched to an engine, wired for the track to be cleared and raced west across the mountains toward Laramie City, then the end of the line. It squatted on the western slope of the Wyoming Black Hills and felt its importance, for it was, in May, 1868, the end of the freight and passenger division and the beginning of all the construction work west to Green River. The same crowds that had flocked from Fort Kearny to Julesburg and from Julesburg to Cheyenne had poured into Laramie City, and they had grown wilder as they progressed.

                    *   *   *

Dodge's visit to Laramie City was a dash of cold water.

"The shops will remain at Cheyenne," he said. "And if the gamblers and saloon keepers here don't let the railroad employees alone, I'll have General Gibbons send down a company of soldiers and we'll proclaim martial law. Take your choice."

Then he hunted up Thomas C. Durant, and the meeting was far from pleasant. Durant was making a final bid for power and authority, for he was at war with the Ames brothers who were pushing him steadily into the background. But Durant held as much stock as anyone, and other directors of the company feared him. He had used Dodge in pushing the Union Pacific across the mountains, but he now believed that he could get along without him; indeed, he felt that Dodge was in his way, and he planned to elevate Colonel Seymour, consulting engineer, to the position of chief engineer.

"Durant," Dodge said in his deliberate way, "you are now going to learn that the men working for the Union Pacific will take orders from me and not from you. If you interfere there will be trouble--trouble from the government, from the army, and from the men themselves."

Dodge turned abruptly and left Durant standing in the dusty Main Street of Laramie City, and the rails of the Union Pacific began to be laid faster than ever before. Thousands of emigrants rolled along the trails in covered wagons; the engineers, obeying Dodge to the letter, linked up their lines from rivers to mountains; the road crossed the North Platte and pushed into the desert; the town of Benton--the last wild terminus of the road-- was born; the grasshoppers came in great armies and ruined the crops of the settlers; flour jumped to eight dollars a hundred pounds; the Cheyennes struck savagely at a dozen points, killing and scalping graders and even the crews of freight trains;

135

but the chief engineer of the Union Pacific, dominant and in his full powers, raced from one end of the track to the other and drove unceasingly.

*   *   *

On July 23rd, while camped near Salt Lake, Dodge received a telegram from Sidney Dillon, a director of the Union Pacific, requesting him to return to Fort Sanders with all possible speed to meet Grant and Sherman. The dispatch also carried the information that Durant had, in Dodge's absence, secured larger powers from the company and would stand on that authority in the conference to be held with government commissioners and military heads en route to Fort Sanders. . . .

Dodge staged it back to Benton as fast as he could and then caught a train for Fort Sanders, where he met his old commanders. . . . Grant had made up his mind to take a hand in the affairs of the Union Pacific, for the troubles of the company had aroused the authorities at Washington to the seriousness of the situation.

It was on Sunday, July 26, 1868, that this notable group, augmented by Sidney Dillon, Thomas Durant, Colonel Seymour, and Jesse L. Williams, a government commissioner, met the chief engineer of the Union Pacific in a conference that had marked bearing on the building of the final 600 miles of the road.

Durant, believing that his new powers with the Union Pacific Company gave him undisputed authority, took the floor and boldly charged the chief engineer with having selected impossible routes, wasted money in useless experiments, ignored the sound judgment of his associates and failed to locate the line as far as Salt Lake.

"What about it, Dodge?" General Grant inquired, leaning back in a cane-bottom chair and smoking vigorously.

"Just this," Dodge began deliberately, "if Durant, or anybody connected with the Union Pacific, or anybody connected with the government changes my lines I'll quit the road."

There was a tense pause; Grant shifted his cigar, Sherman's seamy face was immobile, but the others were ill at ease. Durant's delicate fingers pulled at his Van Dyke beard; he glanced at Colonel Seymour, his henchman, but he said nothing. Grant finally broke the silence.

"The government expects this railroad to be finished," he said slowly. "The government expects the railroad company to meet its obligations. And the government expects General Dodge to remain with the road as its chief engineer until it is completed."

It was a dramatic moment; it was even a critical moment in the building of the first great transcontinental road. Durant looked at the man who would soon become president and doubtless did some quick thinking. Anyhow, whatever he thought, he turned to Dodge and said: "I withdraw my objections. We all want Dodge to stay with the road." . . .

136

"What Hath Man Wrought!"

An Excerpt from **The Builders of the Bridge**
David B. Steinman

David B. Steinman, himself a great bridge designer and builder, dedicated his biography of John and Washington Roebling, **The Builders of the Bridge,** *"To The Memory Of THE BUILDERS OF THE BRIDGE That Their Unconquerable Spirit May Live On As An Inspiration To THE BUILDERS OF TOMORROW." This selection, although primarily focusing on the life of Washington Roebling, conveys the sense of awe and inspiration he felt in their accomplishments.*

John Roebling, a twenty-five-year-old German citizen, came to the United States in 1831. He came seeking an opportunity to put to use his education as a civil engineer, to fulfill his dream of building bridges, and to found a community whose inhabitants would be free to pursue their ideals unhampered by the bureaucratic limitations that had restricted him in his native land. Part of his group migrated to the existing communal society at New Harmony, Indiana; another part migrated to a plantation in South Carolina; and John Roebling and his brother Carl, together with one other family, purchased land in western Pennsylvania and founded a community which they called Saxonburg. There, in 1837, John and his wife Johanna became parents of their first son, Washington.

John Roebling found opportunities to use his civil engineering knowledge, first as a surveyor, then as an inventor, and finally as a bridgebuilder, as he had dreamed he would. He founded a company to manufacture the wire rope needed to construct the suspension bridges he wanted to build. He located the factory in Trenton, New Jersey, to take advantage of improved transportation facilities, a greater supply of skilled labor, and closer proximity to related industries. In 1849 the Roebling family moved to Trenton.

In later life Washington Roebling invariably spoke with the highest regard of his father's genius and professional achievements. But in his reminiscent moods, he would also recall, without reserve and even with a certain apparent gusto, the unrestrained punishments which he had so often endured at the hands of an irascible parent. The severe castigations, the almost unremitting scoldings to which the boy was subjected during the growing years left their mark upon mind and character. Neither the rod nor the child was spared. Fortunately these punishments, though often unmerited, did not embitter, but were recalled in riper years with a philosophic calm mingled with an appreciative sense of their seriocomic aspect.

Of his mother, who was evidently devoted in every way to her children, he always spoke with reverent affection. And it was doubtless from her that he inherited the milder qualities of his personality--gentleness of heart, serenity of spirit, and patient resignation under adversity and affliction.

* * *

After the family moved to Trenton in the autumn of 1849, the youngster was enrolled in the Trenton Academy, which was one of the distinguished private schools of the time. The academy catalogue gave this significant warning: "If there is a necessity for corporal punishment it is promptly and judiciously administered." This phrasing must have caused a wry smile to appear on Washington's face, for "corporal punishment--*judiciously* administered" would be for him a new and rare experience. Here the youth secured the classical Latin education which the father had missed. And here the son of the immigrant had as his fellow students the sons of American aristocracy-- politicians, judges, senators, educators, and authors.

During the years of Washington's attendance at the academy his father was away from home much of the time, completing his suspended aqueducts at High Falls and Neversink, and then designing and building the Niagara bridge. In the bridge engineer's absences his faithful steward, Charles Swan, looked after affairs at the mill, aided by young Roebling in such time as he could spare from his studies. Even before he reached his teens, the boy had proved useful as an assistant to his father and had already demonstrated his aptitude in things mechanical.

Four years of study at the Trenton Academy found Washington prepared to enter the Rensselaer Polytechnic Institute at Troy, New York, then the foremost engineering school in America.

When young Roebling entered the institute, the students still wore a prescribed uniform; it consisted of a dark-green frock coat, with a black velvet collar, and a cap with a gold symbol on the band in front. He doubtless purchased an outfit upon his arrival at Troy.

There were no dormitories for students in those days. The printed catalogue supplied information and advice on living expenses:

> Members of the Institute find Board and Lodgings with respectable private families in the City. . . . The total living expenses, which include board, furnished lodgings, laundry, fires, lights and attendance, vary from about $150 to $300 for the scholastic year, according to the habits and inclinations of the student. . . . There is little necessity for sending much money during the student's life at the Institute; and the supply of any more money than what is sufficient is very apt to be *worse than useless* . . .

Washington was not subjected to the hazard of having "too free command of pocket money." The Roebling children had to account for every penny they expended, and even had to write home to Mr. Swan to send them a few postage stamps when needed.

Unfortunately the scholastic records of young Roebling's college career during his three years' attendance are not available--they were lost, together with his graduation thesis, in one of the fires suffered by the institute. . . .

* * *

To engineering graduates of today, and their teachers, a glimpse of the curriculum of a century ago, when engineering education was in its infancy, would be a fascinating and startling revelation. In the list of nearly a hundred subjects Washington Roebling had to master in three years of concentrated application were such diverse and advanced courses as Analytical Geometry of Three Dimensions, Differential and Integral Calculus, Calculus of Variations, Qualitative and Quantitative Analysis, Determinative Mineralogy, Higher Geodesy, Logical and Rhetorical Criticism, French Composition and Literature, Orthographic and Spherical Projections, Gnomonics, Stereotomy, Acoustics, Optics, Thermotics, Geology of Mining, Paleontology, Rational Mechanics of Solids and Fluids, Spherical Astronomy, Kinematics, Machine Design, Hydraulic Motors, Steam Engines, Stability of Structures, Engineering and Architectural Design and Construction, and Intellectual and Ethical Philosophy. It was a stiff schedule. Under such a curriculum the average college boy of today would be left reeling and staggering. In that earlier era, before colleges embarked upon mass production, engineering education was a real test and training, an intensive intellectual discipline and professional equipment for a most exacting lifework. Only the ablest and the most ambitious could stand the pace and survive the ordeal.

Washington A. Roebling was graduated in July, 1857, with the degree of Civil Engineer. For his Senior Thesis, always an important requirement for graduation, he chose as his subject, "Design for a Suspension Aqueduct."

Colonel Roebling has left on record his conviction that seventeen, at which age he entered the engineering school, was at least a year too soon for what he called "that terrible treadmill of forcing an avalanche of figures and facts into young brains not qualified to assimilate them as yet." He declared that the director of the institution was the hardest taskmaster that ever lived, and that "the boys were ground down and crammed with knowledge and mathematics that their poor young brains could not make use of. . . . I am still busy trying to forget the heterogeneous mass of unusable knowledge that I could only memorize, not really digest. When a class starts with 65 and only graduates twelve it is proof of the terrible grind." He added that "the few who graduated left the school as mental wrecks."

In all probability the Colonel in his memories exaggerated somewhat the strenuous conditions of his student life, and his delight in forceful verbal utterances carried him beyond the boundaries of sober fact. In any event he never appeared to suffer very much in later life from the evils which he depicted, for obviously he never became one of the "mental wrecks" which he accused the institute of producing in the educational grinding mill of his descriptive metaphor.

His class numbered only twelve at their graduation in 1857, but all twelve men subsequently gave evidence of the character of the training they had received. Four of them made names for themselves as bridge engineers. Charles Macdonald built the railroad cantilever bridge over the Hudson at Poughkeepsie and the Hawkesbury River Bridge in Australia. Francisco Trujillo laid out the modern city of Havana. Other classmates achieved success in related fields, especially in railroad engineering. Two became engineering teachers of eminence. Four of the twelve were Lieutenant Colonels in the Civil War, including young Roebling himself.

During his three years at Troy Washington established lasting friendships not only with his classmates but also with fellow students in the other classes entering or graduating while he was there. Among these were a number who achieved distinguished careers. Two attained the rank of Chief Engineer of the United States Navy. One became Engineer of Bridges of the Pennsylvania Railroad, and another, A. J. Cassatt, became President of that railroad. Another contemporary student was Theodore Cooper, whose brilliant career as one of America's leading bridge engineers was tragically terminated by the collapse of the ill-fated Quebec Bridge during its erection in 1907.

Five fellow alumni later worked loyally under Washington Roebling on the design and construction of the Brooklyn Bridge, notably Charles C. Martin as Principal Assistant Engineer and Francis Collingwood as Assistant Engineer. Lefferts L. Buck, who graduated after the Civil War, began his engineering career as a draftsman on this work and, following this apprenticeship, he became the builder of other bridges, including two spans over Niagara and the second bridge over the East River, the Williamsburg Bridge.

At an alumni reunion in 1881, two years before the completion of the Brooklyn Bridge, Francis Collingwood paid a tribute to his absent chief, Colonel Roebling, who "despite a nervous shock, from which he never recovered, has never lost his hold on the work"--and then the speaker gave due credit to the school that had trained them for the task: "The men who have come from the Institute to the bridge have come to stay; and this seems to be true on whatever enterprise they are employed; they seem to have a wonderful sticking power. . . . Higher praise can hardly be accorded the Institute than this; for it shows that her men are all equipped for their work, that their worth is everywhere recognized; and that, with experience, they are fitted for the highest engineering positions."

Despite his vivid criticism of the scholastic procedure of the time, the Colonel retained a grateful and loyal regard for his alma mater. In after years his benefactions to the Rensselaer Polytechnic Institute aggregated over $100,000; in addition he left a bequest of $50,000 to the institution in his will. His son, John A. Roebling, 2nd, and his brother Charles G. Roebling, also graduates, added another $100,000 to these testimonials of loyalty and affection.

About 1906, after a disastrous fire at the institute, Washington Roebling sent a check for $10,000. In thanking him, Director Palmer C. Ricketts explained the needs of the school and asked the Colonel for his advice. On receipt of this letter Roebling replied: "Dear Ricketts, What you evidently need is not advice, but money." Attached to this brief note was another check for $10,000.

*   *   *

Like the Brooklyn Bridge, his alma mater still bears visible testimony to the devotion of Colonel Roebling.

*   *   *

Washington Roebling's preparation for his lifework--home discipline, schooling, professional education, apprenticeship in the shop and in the field-- closely followed his father's concept, his father's plan. Even John Roebling's impatience and severity toward the growing boy were manifestations of an impelling urge to mold the youth to the exacting pattern of a foreshadowed destiny. As the bridgebuilder designed a bridge, and forged and hewed the members to fit, so he planned his son's career and obstinately endeavored to shape his mind and character to fit into a preconceived design.

Following his graduation at Troy, the young engineer returned to his home at Trenton and took up work in the mill. As the senior Roebling and the plant superintendent, Charles Swan, were away for a part of this time, Washington was left in charge of the works--a heavy responsibility for a youth of twenty. He wrote regularly to his father--addressing his as "Dear Sir"--giving him full reports on the business of the plant, and particularly on the matter of credits, which were then causing much concern. Collections were slow, and substantial amounts due were lost, for 1857 was the great panic year when widespread failures of banks and business enterprises played havoc with the nation's financial structure.

In that year John Roebling received a commission to build still another bridge at Pittsburgh--this one to cross the Allegheny at Sixth Street. . . .

*   *   *

In the spring of 1858 Washington joined his father at Pittsburgh to assist him in erecting the new suspension bridge over the Allegheny. At last John Roebling's dream had come true--father and son were working together on the building of a bridge. . . .

During the two years that followed, both father and son wrote many letters to Swan, affording a sketchy record of the progress of the work and some glimpses of life in this growing city at the head of the Ohio. John Roebling's letters were principally confined to business matters, including inquiries concerning affairs at home and orders for supplies needed for the bridge construction. The younger engineer's correspondence was more breezy and personal, revealing a permeating sense of humor and a wide range of interests. Happy on his first bridge job, he was in high spirits. He wrote informatively and entertainingly on community progress and local lawlessness, weather and sunspots, music and chess, Christian morality and Democratic politicians. Although he was working diligently and strenuously, his letters gave but little inkling of the fact.

*   *   *

For this contract the Roeblings, father and son, had to build three new granite piers in the river, in addition to the two massive anchorages on the banks. The tall piers were built with round ends, to deflect current, drift, and ice, and were neatly finished with copings of large slabs of granite.

*   *   *

On each granite pier two handsome iron towers were erected to carry the cable saddles. Each tower consisted of four cylindrical columns of cast iron, slightly inclined inward to form a neatly tapering structure. The spaces between these corner columns were filled with artistically effective latticed ironwork for bracing, with openings left above the sidewalks to form tapered portals. Above the cable saddles each tower was finished with ornamental spires, one at each corner and a higher one in the center. The resulting effect of this finial construction might have been tawdry if unskillfully handled; but

141

in Roebling's planning, with careful design and innate sense of artistic proportions, a pleasing composition was obtained, accentuating the graceful lines of the spans.

In June Washington reported progress on the bridge: "By the middle of this week the cast-iron towers will be up in place. . . . We put up two columns a day; they are awkward things to handle because they can't stand up alone. . . . About the 4th of July the masonry will be all finished--Thank the Lord."

In erection and adjustment operations the building of a suspension bridge demands the nicest calculations and the widest range of engineering ability. The labor involved in the construction is far greater than the casual observer would imagine. The heart of a suspension bridge is in the quality and strength of the cables. The making of the cables, stringing them wire by wire, requires so much time that the public is often impatient with the delay in the progress of the undertaking, only to be surprised by the rapidity with which the rest of the work is completed after the cables are finished.

The traveling wheel, hung from an endless moving rope, had to make the complete trip over the towers 4,200 times from shore to shore in the stringing of the two main cables; and then the spinning of the two outside cables, which were smaller, meant 1,400 additional trips from one anchorage to the other. In October Washington happily recorded completion of the stringing of the main cables: "Yesterday the last strand of the big cables was finished; tomorrow we commence wrapping."

Washington Roebling was enjoying the strenuous life of a bridgebuilder, comprising involved office computations alternating with active physical exertion out on the work. On anchorages and piers, on high towers and swaying catwalk, stringing the cables and hanging the ironwork, he was there with the bridgemen--planning, helping, watching, inspecting, and supervising-- with eyes and ears open to learn and to make improvements, and with a heavy burden of responsibility on his young shoulders.

One day, during the wrapping of the main cables and the hanging of the suspenders, the young bridgebuilder met with an accident. While he was out on a suspended "buggy," a falling wrench struck him on the head, knocking him unconscious. He described the experience in a postscript to Swan: "I was delayed in putting this letter into the post office by having had a hole knocked into my head accidentally, but I am happy to inform you that it is healed up again. I am all right, however. I experienced for the first time the delightful sensation of being knocked insensible."

In November Washington reported gratifying progress on the construction: "We have got along well on the bridge; the big cables are wrapped, all but about 50 feet at each end when we ran out of wire--You will have to make some more. The suspenders are all hung up in the main spans and in a day or two we hang on beams. Wrapping goes very well; no trouble after once getting started fairly."

*   *   *

The stringing of the two smaller cables was in progress when a violent February storm almost wrecked the footbridge. Young Roebling described the experience: "Yesterday we had a furious storm. All hands had to quit; they could not stand up, and I expected to see our footbridge go to the dickens, every moment; she jumped and pitched and reared like a wild horse; the new bridge shook a great deal sideways also because no planks are on yet to make her stiff in that direction. Yesterday morning we had 65 degrees heat and in the night it snowed; that explains the storm."

* * *

During the early spring the erection of the floor system was under way, and the work proceeded without a hitch. The spans of the new structure were stiffened with wrought-iron stiffening trusses extending the full length of the bridge. These were effectively proportioned and neatly designed. For additional stiffening, a system of inclined stays was installed. These wire ropes ran over the saddles and radiated downward from the tower tops for a distance of a hundred feet on each side of the tower. Where these diagonal stays crossed the vertical suspenders, they were lashed together with small wire.

* * *

. . . [Later] [John] Roebling wrote that he and Washington would soon be able to leave Pittsburgh: "I expect I & Washington can leave here on the 1st or 2nd July. . . . We are now finishing Tollhouses and the ornamental parts of the Work, which is a slow business. The Bridge will be beautiful, when entirely completed."

The thrill of the artist is revealed in the last sentence, "*The Bridge will be beautiful.*" The man had planned and wrought and, as he saw his work approaching completion, his heart leapt at the realization that his dream had taken shape as a thing of beauty. The builder's pride was justified when others--artists, engineers, and laymen--proclaimed this bridge to be one of the most graceful structures in the land.

In his next letter Swan sent word that he had been injured in an accident at the mill. Washington had to return to Trenton to take over the work at the plant, while his father remained at Pittsburgh to look after the final details which would complete the bridge.

* * *

When the Civil War descended upon the nation, few families escaped its wounds and its tragedies, its sufferings and its upheavals. The Roeblings felt its impact and its tension, and shared in the heartache and the heroism of the struggle.

Upon completing the Allegheny bridge at Pittsburgh, in the autumn of 1860, Washington Roebling, had returned to Trenton and had taken up his work in the mill.

During the next few months the national crisis was rapidly coming to a head. Lincoln was already feeling the weight of the impending tragedy. On February 21, while on the way to Washington for his inauguration, the President-elect stopped at Trenton and made a short speech. The Roeblings, father and son, heard the address, eloquent in its simple sincerity. Their earlier sympathies were strengthened; and from that day they were resolved, should the need arise, to support the President in his efforts to preserve the Union, even if civil war were the price that had to be paid. But each kept his thoughts to himself.

On April 12 came the startling news that Southern troops had shelled Fort Sumter, and three days later came President Lincoln's call for volunteers.

That evening an unspoken tension pervaded the Roebling household. Washington's mind was filled with thoughts of enlisting, but he hesitated to give voice to his half-formed resolve--he did not know how his parents would take it. But John Roebling was not the man to spare himself--nor even his eldest son, who was his right hand--in a cause in which he felt that great principles were involved. Suddenly, at the supper table, the father turned to the son and said to him, with characteristic harsh abruptness, "Don't you think you have stretched your legs under my mahogany about long enough?" According to the story that has come down in the family, Washington was eating a hot potato at the moment, and thus was unable to make an immediate reply, even if he would. As soon as he could recover himself, he dropped the half-consumed potato, rose from the table without a word and walked out of the house. The next morning he enlisted--never to return to his parental roof. For four years thereafter he neither saw his father nor communicated with him.

It was on April 16, 1861, the day after President Lincoln had issued his call for 75,000 volunteers, that Washington A. Roebling, then twenty-four, was enrolled as a private in Company A of the New Jersey State Militia. On the same day Company A, composed of the young men of Trenton, reported for duty in defense of the Union.

For two months young Roebling remained at Trenton with Company A, which was performing all the duties of a military garrison for defense, including the guarding of bridges, and was at the same time rendering important service in the work of arming and equipping seven regiments of troops for the field. But mere garrison duty in his home town did not satisfy the impatient youth. Looking for more active service, he secured a discharge from the New Jersey Militia and immediately enlisted again as a private in Company K, Ninth Regiment, of the New York State Militia. With that unit he soon saw action at the front, acquiring some artillery experience, and there he was not long in obtaining the promotions for which his natural abilities and his engineering training qualified him. In four months he was made a sergeant, and in another four months a second lieutenant.

*   *   *

In the spring of 1862 Lieutenant Roebling was transferred to the staff of General McDowell and assigned to various engineering duties, notably planning and supervising the construction of suspension bridges for military use. He had to prepare a number of standard designs that could be used for unknown crossings; and he had to assemble the material and equipment. Without guiding precedents, this kind of bridge designing called for special skill and resourcefulness, since the conditions were quite different from those of peacetime construction.

The young bridgebuilder prepared complete drawings and detailed instructions to facilitate erection of the suspension bridges by officers and men who knew very little about bridge construction. He sent the plans and descriptions to General [Montgomery] Meigs, who ordered five hundred copies printed. Lieutenant Roebling wrote home: "The written instructions are being printed very nicely. The drawings are being engraved in Philadelphia. I have received several proof sheets already for correction. The engraving is done very well--500 copies are to be struck off. I wish they were done; it would save me a heap of trouble in explaining to folks."

A few days later he went to Front Royal, Virginia, to erect a small suspension span across the Shenandoah, but the wire rope he had intended to use for it failed to arrive. Anxious about the rope, as he was charged with military responsibility for the material, he returned to Washington to see what had become of it. There he found that, during his absence, General Meigs had taken the rope and had sent it to Harpers Ferry to take the place of a ferry rope which had been ordered from Trenton. Everything was explained, and the young staff officer's anxiety was relieved.

During this brief visit to Washington, Lieutenant Roebling received orders to go at once to Fredericksburg to rebuild the road bridge across the Rappahannock, which had been swept away by a flood. Upon his arrival he at once prepared a plan for the proposed bridge. The length of the structure was over a thousand feet, divided into some fourteen short spans, so that he was able to get along with three wire ropes on each side to serve as the suspension cables. On account of the length of the bridge he did not have quite enough rope and had to order more from Trenton. Planking for the bridge floor was so scarce around Fredericksburg that he had to have it sent down from Alexandria.

*   *   *

In the meantime he was busy with the construction of the anchorages and the towers, to have everything ready for erecting the cables.

This was Washington Roebling's first suspension bridge--the first bridge independently designed and built by him. Time did not permit the spinning of wire cables, so he used prefabricated wire ropes instead. He was thrown completely on his own resources, acting as designer, computer, engineer, superintendent, foreman, and master mechanic. He had to break in green men, teach them to perform highly specialized operations such as connecting and anchoring wire ropes, and train them to work on the swinging cables.

He found some soldiers who could do rough carpentry work. Two of the old piers were gone, and he had to replace these with timber trestle construction. Then, on each pier, he erected a tower of bolted timber framing. On either shore he built an improvised anchorage of concrete and stone. He was then ready to erect the cables. With one end of a wire rope secured to the anchorage, the reel was loaded on a flatboat and towed across the river, allowing the rope, as it was paid out, to lie on the bottom of the stream; the other end was then secured to the anchorage on the opposite shore. Men climbed to the tops of the towers and, by means of attached manila ropes, lifted the cable to the iron saddles that had been provided on top of the timber

145

frames. Each of the wire ropes was erected in the same manner, three on each side of the bridge. The anchored ends were adjusted until all six ropes had identical sags in each of the fourteen spans.

With Lieutenant Roebling setting the example, the amateur riggers then went out along the cables in suspended slings, lashing the ropes together and attaching the suspenders, which were also made of wire rope. To the dangling ends of the suspenders timber crossbeams were secured. Upon these crossbeams the floor planking was laid, forming a continuous wagon roadway from shore to shore. Lighter wire ropes were strung to form the railings.

Although the bridge was not built under fire, there was no time to be lost. The soldiers enjoyed the novelty and excitement of the work. They had to do some swimming and diving in the course of operations, as well as some climbing and tightrope walking. There were few mishaps--an occasional tumble into the Rappahannock was all in the day's fun. Swinging out on the cables or walking on the narrow beams held no fears for men who had been dodging shells and bullets.

After the wire ropes arrived, it was only a matter of a week or two before the thousand-foot bridge was finished.

During the summer of 1862 Lieutenant Roebling served on the staff of General John Pope and took part in the campaign which ended in the Second Battle of Bull Run. He was also in the fighting at Antietam and South Mountain.

In October he was building another suspension bridge, this one over the Shenandoah River at Harpers Ferry. He was handicapped by the lack of mechanics and skilled workers. "The only men I can obtain," he wrote, "are prisoners, and a very lazy set they are."

*   *   *

The spans had to be longer than those at Fredericksburg. Some method of stiffening was required, since the bridge was to carry cavalry and possibly artillery. To provide the structure with framed trusses would involve too much delay. The bridgebuilder therefore decided to install a system of inclined rope stays to stiffen the spans. In addition he braced the structure against wind and lateral sway by adding rope stays in the horizontal plane under the floor. He thereby secured a bridge of unusual security and rigidity for this type of quick, light, temporary construction.

The erection operations were rendered more difficult by the raw, wintry weather. Cold rains alternating with chilling winds were not conducive to rapid progress, especially when the work was done by unwilling prisoners. The weather was so bad that the men could scarcely work half the time.

. . . Despite various delays and difficulties, the bridge was completed in December, less than two months after young Roebling had received the assignment. He wrote to Swan: "The bridge has turned out more solid and substantial than I at first anticipated; it is very stiff, even without a truss railing, and has been pretty severely tested by cavalry and by heavy winds."

146

The bridge at Harpers Ferry was later captured by the Confederate forces. To prevent its use by the pursuing Union Army, they completely destroyed the flooring system. While it was still in the hands of the enemy, Lieutenant Roebling was commissioned to reconnoiter it in order to ascertain the damage. When he found what was needed, he prepared a complete new flooring at a distance from the crossing; and when the Union forces reoccupied the position, he was able to restore the bridge without loss of time.

*   *   *

On the second day of the Battle of Gettysburg the young staff officer was on Little Round Top with General Warren, when the latter discovered the beginning of Hood's furious attack, which threatened to outflank and defeat the whole Union Army. Lieutenant Roebling helped with his own hands to drag up the first cannons, which did so much to save the critical situation.

*   *   *

From August, 1863, until March, 1864, Captain Roebling was attached to the general staff of the Second Army Corps, under the command of Major General Gouverneur K. Warren. Mutual respect and loyalty cemented this friendship. Washington met Emily Warren, his future wife while she was visiting her brother the General in camp.

*   *   *

Roebling accompanied General Warren when the latter was transferred to the command of the Fifth Army Corps; and in this assignment he served through the bloody campaign around Richmond--from the Battle of the Wilderness, through Spotsylvania, the Crater fight at Petersburg, and nine other battles to that of Hatcher's Run.

*   *   *

For his brave conduct during the campaign before Richmond, Major Roebling was breveted* Lieutenant Colonel, U. S. Volunteers.

In December, 1864, in icy weather, he took part in a raid into enemy territory to destroy the Weldon Railroad, which he effectively accomplished. This was Washington Roebling's last duty as a soldier. On that front the fighting was over.

At this time he received word that his father needed him at Cincinnati. On January 1, 1865, the young officer obtained his honorable discharge from the army, and two months later he was given the full brevet rank of Colonel, U. S. Volunteers, "for gallant meritorious service during the war."

Before leaving for Cincinnati, Colonel Roebling married Emily Warren, of Cold Springs, New York . . .

----

  * A brevet commission is an increase in rank granted in recognition of bravery under combat conditions.

A prevision of fate seemingly guided the soldier in his recognition of the life mate he needed. Mrs. Washington Roebling was a woman of strong character, rare intellect, and loyal devotion--a staff to lean upon in crisis and adversity.

Through four years of war the soldier-engineer had risen in rank from private to colonel. He had ridden captive balloons above Chancellorsville, and had built military suspension bridges in record time at Fredericksburg and Harpers Ferry. He had personally dragged the first cannon up Little Round Top to halt Hood's famous charge at Gettysburg. He had received three brevets for courageous conduct. And he had been pronounced "ablest and bravest" by his commanding general.

In the tense years of the Civil War Washington Roebling had matured. At the campfire and on the battlefield, the man had found himself, and had added years of understanding and fortitude to his stature. He had been tested in the fires of war and not found wanting. Toughened in fiber and strengthened in soul, he was now ready for the next great ordeal that awaited him.

*　　　*　　　*

The call for assistance from John Roebling to his son Washington came as a result of the resumption of work on the senior Roebling's suspension bridge across the Ohio River at Cincinnati. Work on the bridge had begun in 1856 but a series of adverse circumstances--inordinately severe winter weather; floods; the financial panic of 1857; and finally the outbreak of the Civil War--had limited the construction activity to almost nothing for the first seven years. Then, however, the ebb and flow of the war convinced Cincinnati authorities that completion of the bridge was important to the defense of the city, and work on the bridge resumed in 1863. Joining his father at Cincinnati, Washington Roebling was named assistant chief engineer. In less than two years the Roeblings completed the bridge, the largest suspension bridge in the world at that time.

Even before the Cincinnati bridge was completed John Roebling had begun work on a proposal for what was to become the Brooklyn Bridge. In 1865, with work barely underway in Cincinnati, he submitted a set of plans for his proposed bridge over the East River. In May 1867, less than six months after completion of work at Cincinnati, he was hired as chief engineer for the construction of the Brooklyn Bridge. For two years he worked to complete detailed plans for the bridge. Then, in July, 1867, while working on a survey to determine the final location of the Brooklyn tower, his foot was crushed when a ferryboat struck the pier on which he was standing. Despite the amputation of his foot, he died of tetanus less than three weeks later.

With its inspiration gone, the Brooklyn Bridge seemed impossible to build, for John Roebling had been more than the designer. His had been the vision, the conception, the knowledge of the whole and of every part, the anticipation of every problem of execution. A new guiding spirit was needed to carry the great work forward.

By destiny or prevision the master builder had prepared a deputy who was indeed a part of himself--his own son, Colonel Washington A. Roebling. This young but able engineer had been trained and prepared for this task. He had been given the best engineering education of the time; he had received his

148

practical training under the greatest bridgebuilder of his day--his own father; and his own judgement, initiative, and courage had been tried and tested on the battlefield.

For years he had shared his father's professional confidence and labors. They had worked together on the construction of the Pittsburgh-Allegheny bridge and the great Cincinnati span, and the disciple had become thoroughly schooled in the master's methods. And in the Civil War he had devised and built his own suspension bridges under conditions testing both resourcefulness and fearlessness.

When the elder Roebling began planning the Brooklyn Bridge, the son took an immediate interest in the great adventure. He helped his father in the preliminary studies; and then he journeyed to Europe, visiting all the important engineering and metallurgical works and studying caisson work and the fabrication of the new structural material, *steel*. With the mass of data and information he had collected, he returned to America to serve as his father's aide for twelve months before the latter's death. He was thoroughly familiar with the plans for the bridge, having helped to create them.

And, most important of all, Washington Roebling had the endowments of character and personality that the grave responsibility demanded. He had inherited not only his father's true engineering instincts but also his concentrated devotion and unconquerable tenacity of purpose.

To this son the great engineer on his deathbed turned over the completion of the task. The following month--August, 1869--at a meeting of the directors Colonel Washington A. Roebling was officially appointed to succeed his father as Chief Engineer of the work.

Courageously and loyally the son assumed the gigantic responsibility of carrying forward the construction of this, his father's greatest dream. To this consecrated task the young engineer gave all he had--his health, his strength, his career.

He resigned as president of the wire company at Trenton to devote all his time and energy to the bridge, and his two brothers, Ferdinand and Charles, took charge of the wire-rope plant. From that day until the job was ended, Washington Roebling lived only for the Brooklyn Bridge.

*　　*　　*

During the days and nights that the work was going on under the bed of the East River, the young chief engineer was continually in the caisson, personally directing the efforts of his men and setting an example of devotion and courage. From him flowed the energy and impulse that animated the entire force. Mindful always that any slip, no matter how trivial, at this stage of the work, might prove disastrous, he actually spent more hours in the working chamber than anyone else. Night and day he worked with the men under the crushing air pressure, until it wore out his strength.

One afternoon in the early summer of 1872 Colonel Roebling had to be carried out of the caisson, nearly insensible, a victim of the dread caisson disease. All night his death was hourly expected.

149

In a few days he rallied and, mustering all his reserve of strength, he attempted to return to work. But he again collapsed and had to be brought home.

At the age of thirty-five his days of physical exertion were ended. He remained painfully paralyzed--doomed to lifelong suffering--with progressive blindness and deafness setting in, vocal cords affected, and every nerve and muscle tortured with pain.

Thenceforth this man, who had been full of life and hope and daring at the inception of the work, was a crippled invalid, confined to his home. Just as his brilliance as an engineer was at its height, and just as his masterwork was passing its crisis, he was struck down.

The chief engineer was helpless, except for his fighting spirit and his active brain. From the hour of his breakdown in the caisson there was not a day when his injured body and nerves were not racked with pain.

But the work had to go forward, and, from his sickroom, the stricken man continued to direct it in every detail. Although his nervous system was wrecked, his mind was not affected; indeed, his intellect appeared to be quickened. Realizing how imcomplete his plans and instructions for completing the bridge were, and fearful that he might not live to finish the work himself, he fought his pain and feverishly spent his time in writing and drawing. His plans, his ideas, his specialized knowledge, could not be allowed to die; they must be recorded and made available to his assistants. The papers and sketches he prepared contained minute and exact directions for finishing the towers, for building the anchorages, for stringing the cables, and for suspending the spans. Every difficult or specialized erection operation he anticipated and worked out, with calculations, diagrams, and specific instructions.

In this manner the work continued; and it can safely be said that no great project was ever conducted, before or since, by a man who had to work under so staggering a handicap. From an upper bay window of his home on Columbia Heights he anxiously followed the work through field glasses, while he prayed for strength and struggled for time.

Through the long winter he painfully wrote out memoranda and instructions covering future stages of the work--page after page of detailed notes. Writing so much in his weakened condition further impaired his eyesight. More and more he came to depend upon his wife to act as his amanuensis and to write out his notes and instructions.

His nerves were shattered. He partly recovered the use of his voice, but he was too weak to carry on a long conversation; and some days he could not talk at all. He had become so morbidly sensitive that the mere sound of a strange human voice was unbearable. Not one of the assistant engineers could consult with him; and yet the most intricate technical problems, to which he alone held the key, had to be worked out for the erection operations.

His wife was as rare and strong as the man she tended. She met the emergency by becoming his capable aide. Instead of crying hysterically over her husband's misfortune and sinking down nerveless and helpless into the privileged softness of her sex, she set herself at once to the task of acquiring the knowledge necessary to become a real helpmeet to him in the critical situation. Under his guidance she studied higher mathematics, the calculation of catenary curves, strength of materials, stress analysis, bridge specifications, and the intricacies of cable construction; and she applied her new education to the problems of the bridge. She grasped her husband's ideas and she learned to speak the language of the engineers. She made daily visits to the bridge to inspect the work for the Colonel and to carry his instructions to the staff. She became his co-worker and his principal assistant--his inspector, messenger, ambassador, and spokesman--his sole contact with the outside world.

But in developing scientific and technical skills to meet the emergency, she did not sacrifice her womanly qualities. She remained the comforting companion and tender nurse. As Thomas Kinsella, one of the bridge trustees and editor of the Brooklyn *Eagle*, remarked at the time: "The most abstruse study has not interfered with Mrs. Roebling's ministrations at her husband's bedside. If he is restored to health, it will be largely through her patient and intelligent attendance upon him, and Colonel Roebling will be indebted to his noble wife even as the people are."

In the summer of 1873, on his physician's advice, the Colonel went abroad--to Wiesbaden--for several months in a vain quest to restore his shattered health, but on his return he suffered another attack of acute prostration.

During the following years the Chief Engineer continued to superintend construction from his sickroom. An artist of the day has left us a drawing, published in a popular magazine, depicting the paralyzed bridgebuilder seated at his window, gazing wistfully at the unfinished structure in the distance; his field glasses rest on the sill; and on a table near by is his beloved violin, its strings now mute--for never again will the invalid's fingers be able to produce the music he loved.

The Colonel's continued planning and direction of the work, despite constant pain and a nervous collapse which would have cut off an ordinary man from mental as well as physical activity, was an outstanding triumph of consecration and will power. His own self-sacrifice inspired the most devoted loyalty in the corps of engineers about him, and they directed all their energies to carrying out his wishes. Not the slightest advantage was ever taken of his illness to deviate from the letter or spirit of his plans; and Colonel Roebling, physically helpless, away from the work, always remained complete master of the enterprise. Like a wounded general who directs a battle from a hilltop, Washington Roebling issued his orders and superintended the completion of the bridge.

Through field glasses, he watched the building of the towers and later the stringing of the cables. Though he was absent in body, it was still his spirit that prevailed and his brain that directed the work. The bridge was his--by right of inheritance, by his own labors, and by his own sacrifice.

Physical hardships, dangers, and afflictions were not the only ordeals that had to be faced by the Chief Engineer in the execution of his task. He was also harassed and beset, through the long years of painful toil, by financial and political complications arising from conditions beyond his control. Political corruption, civic jealousies, lack of funds, injunction proceedings, and dishonest contractors were among the tribulations.

151

The whispering campaign of criticism and doubt kept gaining in volume and vehemence. The enemies of the project were now "hitting below the belt"--the physical disability of the Chief Engineer was made a focal argument in the attacks.

Toward the end of June, 1882, the trustees passed a resolution requesting Colonel Roebling to be present at the next board meeting. He replied in writing:

> I am not well enough to attend the meetings of the Board, as I can talk for only a few moments at a time, and cannot listen to conversation if it is continued very long. My physicians hope that living out of doors and away from the noise of a city may lessen the irritation of the nerves of my face and head. I am now able to be out of my room occasionally. I did not telegraph you before the last meeting that I was sick and could not come, because everyone knows I am sick, and they must be as tired as I am of hearing my health discussed in the newspapers.
>
> I believe there is not a day that I do not do some sort of work for the Bridge. . . . My assistants do the work assigned them with perfect confidence that they can always refer to me for any advice or assistance they need. The work to be done this summer is very plain routine. There is nothing to be done that has not been done before, and the interests of the Bridge do not suffer in any way from my not being there. If the Edgemoor Iron Company will furnish the steel as fast as we need it, the work will go on all right and without delay.
>
> It is very important that authority be given me at your next meeting to advertise for the iron for the elevated termini, the spruce planks for the floors, the steel rails for the tracks on the Bridge, and the paint. Specifications for all these things are now ready, and they should be obtained as soon as possible, so that they may be ready when we want to use them.
>
> I shall be most highly honored to be present at meetings of the Board as soon as I am well enough to be of any use there.

The underground campaign went on, and soon it took the form of an attempt to displace Colonel Roebling as chief engineer of the bridge. Through twelve years of trials and tribulations he had carried the project forward successfully; and now, after all his work and sacrifice, it was proposed to take the work away from him only nine months before its completion. At a meeting of the trustees in August, Seth Low, then mayor of Brooklyn introduced the following resolution:

> Whereas, The Chief Engineer of the bridge, Mr. W. A. Roebling, has been for many years and still is an invalid; and

Whereas, In the judgment of the board the absence of the Chief Engineer from the post of active supervision is necessarily in many ways a source of delay, therefore,

Resolved, That this board does hereby appoint Mr. Roebling Consulting Engineer, and Mr. C. C. Martin, the present First-Assistant Engineer, to be the Chief Engineer of the New York and Brooklyn Bridge; and

Resolved, In so doing the board desires to bear most cordial testimony to the services hitherto rendered by Mr. Roebling, and to express its regret at the necessity of making the change at this time.

After some discussion it was decided to lay the matter upon the table until the next regular meeting, to be held in September.

As soon as the crippled engineer heard of this proposal to take his bridge away from him, he decided to fight. He was physically incapacitated, but he had kept his fighting spirit. He started to dictate a statement to his wife. After many starts he found what he wanted to say. Without asking for sympathy, he told what he was doing on the bridge and why he should not be displaced. It was an appeal to the engineering profession to prevent an injustice.

The American Society of Civil Engineers was meeting at the time. Mrs. Roebling went to the meeting and secured permission to read her husband's statement. It was the first time that august body had been addressed by a woman. The drama of the situation, the simple eloquence of the appeal, and the impressive presentation produced an immense sensation. The backing of the engineering profession was secured, and the confidence of the public was won.

*   *   *

May 24, 1883.

It was glorious spring day. Dawn revealed the twin cities resplendent with waving banners and streaming colors. From windows and roofs, from the forests of shipping along the wharves and from the vessels riding the sparkling waters of the bay, flags were floating proudly in the breeze; while high above all, from the massive, time-defying towers of the Bridge, the Stars and Stripes signaled to the world--from the gateway of the continent--the arrival of the long-awaited day.

The Bridge was completed. The dream of years was at last a reality. Once more the unconquerable spirit of Man had triumphed. Of faith and courage, a masterpiece had been wrought.

A spirit of tense excitement was in the air, a sense of high significance, a feeling of joyous celebration.

Almost before the sun was up, the thoroughfares of both cities had put on a festival appearance. Shops and offices were closed and all business was suspended. The streets and avenues were a maze of bright and stirring colors.

153

Tens of thousands of men, women, and children, in their Sunday best, surged through the flag-draped streets. Bands played. Street hawkers did a lively trade in souvenirs of all descriptions. Incoming trains continued to bring thousands of eager visitors to witness the celebration.

The crowds began to flow in the direction of the great span. Young and old, on foot, in carriages, stages, and horsecars, moved toward the center of the day's festivities, and there sought vantage points to see the show.

The scene along the river front was unique and beautiful, with hundreds of vessels, including five ships of the North Atlantic Squadron at anchor below the bridge, decorated from stem to stern. In the bright spring sunshine the rippling reflection from every ferryboat, warship, and tug, and from the hundreds of pleasure craft, made the waters of the river run with colors, like a dye vat.

And above this sparkling scene was the great vaulting span--the embodiment of a city's dream in granite and steel.

*   *   *

This was the moment for which the crowds had been waiting. As the officials set foot on the span high above the river, the historic event was proclaimed by the thunder of guns. The heavy cannon at Castle William on Governors Island thundered out across the harbor, the artillery at the Navy Yard in Brooklyn boomed out over the river, and the battery of guns on the summit of Fort Greene took up the reverberating salute. Boats set their horns and steam sirens shrieked in chorus. And a million men and women cheered.

*   *   *

As the official procession moved out in triumphal progress along the span, the air was filled with a tremendous flood of joyous sound from ships and docks and factories and from the streets below--the clanging of bells, the roaring of a thousand steam whistles, and the frantic cheers of the crowds--while sounding from afar, clear and distinct above the clamorous din, the silver chimes of Trinity rang out over the river.

*   *   *

During these ceremonies a lonely man--paralyzed, crippled, and racked with pain--sat at an upper window of his home on Columbia Heights, viewing the scene from afar. He saw the crowds surrounding the bridge approach, and the host of more privileged citizens pouring into the reserved space. Through his field glasses he saw the distinguished procession coming over the bridge--the President of the United States, the Governor of the State, and all the other notables, together with the glittering military escort. His gaze was riveted on the scene. The breeze carried the cheers of the crowd and the strains of the band, but he could not hear the glowing tributes of the orators.

His throat was choked, and he could hardly keep back the tears; for this was the great moment of his life. Through long and weary years he had been enduring pain and fighting on for this consummation. This day gave meaning to his life, and to his father's before him. For this was his father's Bridge--

154

and his Bridge. The father had dreamed the dream, the son had wrought the dream. Both had lived, and battled, and sacrificed, that this Bridge might be built, and both had given their last supreme effort to this achievement.

Beside the crippled man at the window stood his wife, who had been his ministering angel through the years of pain and struggle. She had been his eyes, his hands, his feet--recording his notes, carrying his instructions to the job, and bringing back reports of its progress. When others had tried to displace him just as the work was drawing to completion, she had eloquently presented his plea to stop the injustice. Brave and loyal and tender, she had given her strength to carry her husband through his soul-straining ordeal.

If the paralyzed engineer watching the celebration through his field glasses could have heard the words of the speakers, his throbbing heart would have beat faster. For there, at the scene of dedication, standing before the vast assembly of citizens, an honored civic leader whom they had selected as their spokesman--the Honorable Abram S. Hewitt--was eloquently recording Humanity's indebtedness to the Builders of the Bridge:

> "When we turn to the graceful structure at whose portal we stand, and when the airy outline of its curves of beauty, pendent between massive towers suggestive of art alone, is contrasted with the over-reaching vault of heaven above and the evermoving flood of waters beneath, we are irresistibly moved to exclaim, 'What hath *man* wrought!'
>
> "Man hath, indeed, wrought far more than strikes the eye in this daring undertaking, by the general judgment of engineers, without a rival among the wonders of human skill. . . .
>
> "But the Bridge is more than an embodiment of the scientific knowledge of physical laws. It is equally a monument to the moral qualities of the human soul. It could never have been built by mere knowledge and scientific skill alone. It required, in addition, the infinite patience and unwearied courage by which great results are achieved. It demanded the endurance of heat, and cold, and physical distress. Its constructors have had to face death in its most repulsive form. Death, indeed, was the fate of its great projector, and dread disease the heritage of the greater engineer who has brought it to completion. The faith of the saint and the courage of the hero have been combined in the conception, the design and the execution of this work.
>
> "Let us, then, record the names of the engineers who have thus made humanity itself their debtor for a successful achievement, not the result of accident or chance, but the fruit of design, and of consecration to the public weal. They are: *John A. Roebling*, who conceived the project and formulated the plan of the Bridge; and *Washington A. Roebling*, who, inheriting his father's genius, and more than his father's knowledge and skill, has directed this great work from its inception to its completion. . . ,
>
> "During all these years of trial, and false report, a great soul lay in the shadow of death, praying only to stay long enough for the

155

completion of the work to which he had devoted his life. I say a great soul, for in the springtime of youth, with friends and fortune at his command, he braved death and sacrificed his health to the duties which had devolved upon him, as the inheritor of his father's fame, and the executor of his father's plans. . . .

"But the record would not be complete without reference to the unnamed men by whose unflinching courage, in the depths of the caissons, and upon the suspended wires, the work was carried on amid storms, and accidents, and dangers sufficient to appall the stoutest heart.

"One name, which may find no place in the official records, cannot be passed over here in silence. With this Bridge will ever be coupled the thought of one, through the subtle alembic of whose brain, and by whose facile fingers, communication was maintained between the directing power of its construction and the obedient agencies of its execution. It is thus an everlasting monument to the self-sacrificing devotion of woman. The name of Mrs. Emily Warren Roebling will thus be inseparably associated with all that is admirable in human nature and with all that is wonderful in the constructive world of art."

One after the other the following speakers added their words of acknowledgement and praise. From one who had been intimately identified with the bold enterprise from its inception, the Honorable William C. Kingsley, the hushed audience heard the story of the planning and the building, the heroism and the sacrifice:

"With one name, this Bridge will always be associated--that of *Roebling*. At the outset of this enterprise we were fortunate to secure the services of John A. Roebling. . . . While testing and perfecting his surveys . . . the man who designed this Bridge lost his life in its service. He was succeeded by his son, Colonel Washington A. Roebling. . . . Down in the earth, and under the bed of the East River . . . within the caissons . . . always on hand at the head of his men to direct their efforts and to guard against any mishap or mistake which might have proved disastrous . . . Colonel Roebling contracted the mysterious caisson disease. . . . For many long and weary years this man, who entered our service young and full of life, and hope, and daring, has been an invalid, confined to his home. He has never seen this structure as it now stands, save from a distance. . . . But every step of its progress he has directed. . . . Colonel Roebling may never walk across this Bridge, as so many of his fellowmen have done today, but while this structure stands all who use it will be his debtors."

The Master Builder
Elting E. Morison

American Heritage of Invention & Technology - Fall, 1986 Issue

Elting Morison's biographic sketch of his great-uncle, George Morison, calls to mind a time when engineering was learned "on the job;" when intelligent and industrious men learned by doing, working in concert with others who had previously learned the same way. That day, of course, no longer exists, but it behooves us to reflect on the extent to which an engineering education can be (and should be, many would argue) based on experience and observation rather than on academic problems and examinations.

My great-uncle, George S. Morison, one of America's foremost bridge builders, died July 1, 1903, exactly (as he undoubtedly would have said) six years, five months, fourteen days, and six hours before I was born. What follows begins with some incidental intelligence that has nothing to do with his work; these, listed in no order of relative importance, are just some of the things I know about him:

He had, like Zeno, a conviction that time was a solid. If he made an appointment to confer with a person at 3:15 p.m., or as he always put it, at 15:15 hours, that was when they met. Those who arrived earlier waited; those who came at any time after 15:15 never conferred at all.

He read the *Anabasis* in Greek, the *Aeneid* in Latin, and the dime novels of Archibald Clavering Gunter in English.

He had a substitute in the Civil War.

He invariably referred to Mexico as Pjacko.

He thought that people who were good with animals, particularly horses, were popular with their fellows and loose in their morals. When he himself drove a horse, he brought it to a full stop by saying, "Whoa, cow"; and at least once while trying to turn a Concord buggy around, he turned it over in front of White's Machine Shop.

He was rude to waiters.

One Sunday morning he walked out of church after telling the minister, who was explaining to the congregation why he thought silver should be coined at a ratio of 16 to 1, that he should never try to deal with a subject he obviously didn't understand.

Of his neighbor Edward MacDowell, student of Liszt, composer of "To a Wild Rose" and the well-regarded Second Piano Concerto in D Minor, he said he was "a man with whom I had absolutely nothing in common."

A bachelor, he built a house in the years from 1893 to 1897 that had, by one way of counting, fifty-seven rooms, so that he would have a suitable place to eat Thanksgiving dinner and to watch the sun set over Mount Monadnock.

Of those who have written about him, one spoke of his ability to enforce a decision taken "with a tenacity and ruthlessness that bore down all opposition. . . ." Another called him "a bulwark." And a third said: "Force was the striking impression. When he entered a room, power came with him." They were all trying to explain the source of his remarkable works--he did in fact put a satisfying dent in oblivion by the things he made. But he bore down on the opposition of time in quite another and less obvious way. That ability to fill a room with power turned out to be sufficient to project the force of his character through three generations of his family.

In March 1902 George S. Morison appeared before the Senate Canal Committee. He explained at length why he believed that the best way to join the waters of the Atlantic and Pacific lay through the Isthmus of Panama. The only real difficulty was posed, he said, by the Culebra Cut. "It is a piece of work that reminds me of what a teacher said to me when I was in Exeter over forty years ago, that if he had five minutes in which to solve a problem he would spend three deciding the best way to do it." Because the Culebra Cut was a big problem, more time would be required. It would take two years to figure out what to do and how to do it.

There were many times when he was put in mind of his old teacher and quoted him on problem solving for the benefit of others. It was, said one associate, "one of the principal rules" of his life. He sought beforehand to take everything into account, analyze the evidence, determine the "best possible solution," and then reach the "inflexible, intractable decision." That, in fact, is the way he decided to become a civil engineer.

It took some time to do so. Born in New Bedford, Massachusetts, in 1842, the son of a Unitarian minister, he was educated, like his father before him, at Phillips Exeter Academy and Harvard College. From there he went South as the government superintendent of plantations on Saint Helena Island. The object was to bring some order out of the chaos produced by the Civil War among the resident whites and freedmen. After a year he returned to enter the Harvard Law School, in 1864, where he won the Bowdoin prize for the best dissertation. In 1866 he joined the great New York law firm of Evarts, Southmayd & Choate.

"Exactly one month later" he confronted the problem of what to do with himself--practice law, study the principles that lay beneath the practice and teach them at some university, or go west as a civil engineer. He set May 1 of the following year, seven months later, as the date to decide the matter. On that day he informed the firm of his intention to leave the law, and five months thereafter he went out to Kansas City, Missouri, to build a bridge with Octave Chanute. I have the distinct impression that he was turned in this direction by some work he did while in the law firm, on the bankruptcy of a small Western railroad. I cannot verify this by the documents now available, but it has the support of a fairly reliable memory, and it suggests a link in the causality he always sought. When he started work "calculating the cubical contents of stone for the masonry piers," the "four years of doubt, vacillation and search" which had "formed the introduction to my life" were ended.

He could not have landed in a better place at a better time with a better man. The Missouri was a wild and willful river often disturbed by heavy floods and destructive ice jams. It constantly filled up old channels and cut out new ones. No serious bridge had ever been built across it, and the received judgment was that if a bridge were built, it could never be maintained. For someone who knew no engineering, it was a great place to begin.

There was also Octave Chanute, who had never built a big bridge before. But he had worked for a dozen years in various capacities constructing small Western railroads, and he had learned a lot on the job. At a time when there was really no other way to learn, Chanute was, at thirty-four, near the top of his class. He was, as all his later career indicates, an "acute and accurate observer," an "inventive engineer," a "truly scientific spirit," and, withal, a man possessing the "Gallic power of clear and forceful expression." When in middle age he turned his attention to "aereal navigation," his experimental glider flights greatly expanded the knowledge of the field. To the success of the Wright brothers he contributed both useful principles and actual designs.

What it meant to start on such a job with such a man was made clear in a journal written in Kansas City on Thanksgiving Day 1867. After laying out his daily work and study schedule from 0800 to 2130 hours "with not more than one evening a week being excepted," Morison went on to plot the move into the future. He was "ambitious, very ambitious." What he had set his sights on was not a financial fortune but "a good and useful life." With that as his purpose he would, when the Kansas City bridge was finished, "cross the Atlantic and devote a year to the study of French and German, and the acquirement of scientific knowledge; it being my wish to make the profession of engineering a truly liberal profession and through it to rise to science and philosophy, raising it with me rather than to prostitute it to mere money making. . . ." Not many of those who at the time were calculating the cubic contents of stones would have put it quite that way, and even now it must appear a very large and liberating definition of the possibilities in the field.

Given such attitudes and such a personal program, it is probably not surprising that he rose rapidly to the position of associate engineer on this first job and that, as soon as the bridge was finished, he went to work on a book that described the solutions to the problems encountered in the building of it. What followed--in a rare departure from his program--was not France and Germany but a six-year internship of steadily increasing responsibility in the design and construction of small, short Western railroads with names like Leavenworth, Lawrence & Galveston or Detroit, Eel River & Illinois. Near the end of this period Chanute called him back to serve as his principal engineer on the Erie Canal.

On May 6, 1875, the bridge at Portageville, New York, said to be the largest wooden trestle in the world, was consumed by fire. Morison was put to work drawing up the design and specifications for an iron structure that would replace it. On May 10, four days later, the first building contract was let, and he assumed the direction of the construction. Eighty-two days after that the bridge was open for rail traffic. It was 818 feet two inches between abutments and it gave him, at age thirty-three, an "international prominence."

For the next seventeen years he devoted most of his time and thought to building railroad bridges in the West. He built these bridges across the

Missouri, Mississippi, Ohio, Snake, Columbia, and Willamette rivers. They all had certain common characteristics. Their specifications filled the requirements of the particular situations to a T, and in the building those specifications were satisfied precisely. As at Plattsmouth, Nebraska, where the "total deflection of the main span under the test load of 800,000 pounds was exactly" as previously calculated, so with all the others. They were also on the grand scale. At Memphis, Tennessee, the main span was 790 feet, which made it the longest truss in the country. At Cairo, Illinois, the metalwork was 10,560 feet--two miles--in length, the longest steel bridge in the world. And they were all structures in which the function was obviously made to determine the form, in studied austerity.

It was said that in this period he compiled a record that was "unrivaled in the history of bridge construction." Whatever the truth of this evaluation, it is certain he acquired a reputation that made him sought after for many different kinds of services. He joined the boards of four railroads. For fifteen years he provided Baring Brothers of London with comprehensive analyses of the physical condition, financial structure, and managerial competence of American railroad companies. He played a large part in the study that led to the reconstruction of the Erie Canal. President Cleveland put him on one commission that selected San Pedro as the deepwater port for Los Angeles and on another that started the action that produced, nearly forty years later, the George Washington Bridge across the Hudson.

Then in 1899 he was appointed by President McKinley to the Isthmian Canal Commission. For the next two and a half years he devoted himself to an exhaustive examination of the political difficulties and technical factors, past and present, that were involved in the great enterprise. Twice he went to Europe; once he made a four-month exploration of the isthmus itself; and he attended all the fifty-one meetings of the commission in Washington. In November 1901 the members signed a report that, reflecting a powerful combination of historical, political, and technical pressures, recommended Nicaragua as the site for the canal. Appended to this document was the dissenting opinion of a minority of one. It recommended, with much careful explanation, the choice of the Isthmus of Panama as the preferred site; and it was signed by George S. Morison.

There followed weeks of argument within the commission, debates in Congress, discussion in the press, and earnest consideration in the White House. In January 1902 the commission rendered a supplementary report that unanimously concluded that the "feasible route for an Isthmian Canal to be under the control, management and ownership of the United States is that known as the Panama Route."

In such a tangle of historical, political, international, and technical considerations and in such a concert of dominant personalities, it is hard to determine final causes. David McCullough, who has made the most recent and careful investigation of the situation, concludes as follows: "If one traces back through the chain of events. . . . and if it is remembered that Morison. . . made no effort to glorify his contributions, at the time or later, then Morison emerges a bit like the butler at the end of the mystery--as the ever-present, frequently unobtrusive, highly instrumental figure around whom the entire plot turned." It is an image he would, beyond much doubt, never have chosen, but it makes a point he would never have made for himself.

160

Such, briefly, was the nature of his principal works. Before trying to establish a more coherent explanation of the man himself, it may be useful to say something about the man among his fellows. Was his record indeed "unrivaled," should he be called "the leading bridge engineer in America, perhaps in the world," did he deserve the title of Pontifex Maximus bestowed on him at one college commencement? That is a very doubtful kind of exercise that leads to no useful conclusion. What is far more to the point is that he was a contributing member of a remarkable company, some of whom held his achievements in a good deal higher respect than his person. And what is interesting is not what set the members apart but what they all had and did in common.

There were, of course, some distinguishing temperamental differences. John Roebling played the flute and allowed a caller to be five minutes late before canceling the appointment; Octave Chanute made witty remarks; James Eads interrupted a stunning career for four years because he preferred the "happy environment of his family"; Charles Latrobe liked to go about in society and worked in watercolors; and so on. What really matters is the shared experience of those who practiced civil engineering in the last half of the last century and the effect of that experience on themselves and those around them.

They came up, for the most part, the hard way. Leaving college and, more often, high school, they started out on the ground floor, measuring stone, surveying lines, calculating stresses. They did these things more often than not on a new railroad, which for them, like the Erie Canal for the preceding generation, was the only available institute of technology. Here they learned from men who knew a little more than they did because they had been a little longer on the job. Frequently they followed these instructors into the engineering division of one of the larger, more stable roads in the East. And from there, after a time, they usually struck out on their own as "consulting engineers, which meant they were ready to deal with whatever propositions came to them.

Wherever they went, whatever they did, they found the subject matter was always changing. Larger loads, longer spans, deeper excavations, new materials, novel procedures. In such conditions the name of the game was figuring out sensible new departures from what had been tried and true for centuries. And if the figuring wasn't right, the cost of going wrong could be measured out and the source of difficulty explicitly defined. When, for instance, the bridge Amasa Stone had built at Ashtabula, Ohio, went down one stormy night, it took a train of passengers with it. And after a jury found that the bridge had been an experiment "which ought never to have been tried," Amasa Stone, "as exacting of himself as he had been of many others," took his own life.

Those who started on the ground floor and worked their way through to the top of such a calling were often said to be bold, self-reliant, independent, secure, powerful, daring, resolute, and, sometimes, arrogant and overbearing. At this distance it may be seen that their most continuing collective contribution was not the things they built but their way of going at things. They gave a significant push to the developing new method of solving certain kinds of problems that occur in life.

Over and over they demonstrated that the ingenious solution that worked was reached through accurate observation, exact knowledge of the strength of

161

materials, precise calculation, due respect for the laws and forces of nature, and the resourceful ordering of evidence obtained by the unclouded intelligence. They could be daring when the findings from the hard data-- subjected to the logical process--supported the bold conclusion, and they were resolute because, within their scheme of things, they could prove they were right. Faith might well have its uses, but they had found a surer way to remove a mountain. This method, increasingly refined, has put us wherever it is that we are today.

On this subject he had ideas which in his closing years he put down in a small book. It demonstrates the extent to which he had fulfilled his early intention to rise through his profession to philosophy, and it still speaks to our condition. Our ability to manufacture power in unlimited quantities, begun with Watt and the condenser, had opened up what Morison called a new epoch for mankind. Carried to its logical conclusions, it would in time give men the capacity to create all the essential conditions for their living and to determine their own fate. He foresaw a future when "material developments will come to a gradual pause," when "an immense population will live comfortably and happily, and the qualities which make the good citizen and the contented man will be more in demand than those which make leaders in periods as we are familiar with."

But he also believed that the new epoch, before it reached this possible end, would "destroy many of the conditions which give most interest to this history of the past, and many of the traditions which people hold most dear." Among other things it would "destroy ignorance, as the entire world will be educated, and one of the greatest dangers must come from this very source, when the number of half-educated people is greatest, when the world is full of people who do not know enough to recognize their limitations. . . ."

How do we assemble the bits and pieces of Morison's personality and character in a more intelligible mosaic? If the design is supposed to fulfill a familiar expectation, this is a hard question. Remember that until he built the great brick house, at the age of fifty-five, he had no place to call his own, and during the remaining years of his life his accumulated occupancy of that house came to little more than forty-nine days. Though he had apartments in Chicago and New York, he didn't use them much, and then only for bed and sometimes break-fast. For the most part he stayed in hotels and sleeping cars, and ate in clubs and restaurants. Considered as a social being, he seems a programmed nomad.

There are some family letters, but for the most part they have to do with the arrangement for a proposed visit or the details of some small errand he wished a member to perform. There is also the daily diary he kept throughout his life. In the entries are faithfully reported temperatures, rainfalls, and the number of minutes the train he was riding on was behind schedule.

In such conditions one must respect the dead air spaces, accept the fact that what you see is all you're going to get, and recognize that he planned it that way. If you look back to the journal entry for Thanksgiving Day 1867, you will find his program for a good and useful life. What he did with himself from those first calculations of cubic quantities to his closing consideration of engineering as the source of a new epoch satisfied the terms of that program-- not less, not more, but exactly.

The Rise and Fall of Michael O'Shaughnessy

Excerpts from **The Gate**
John Van Der Zee

The municipal engineer must be a person of many talents. With responsibilities for roads, sewers, water supply, waste disposal and a variety of other types of projects and with those responsibilities including planning, design, construction, and operation and maintenance of all these types of projects, the municipal engineer's need for diversity of knowledge probably exceeds that of almost all other career choices within the civil engineering profession.

Michael O'Shaughnessy, city engineer for the City of San Francisco after the Great Fire, was such a man. However, even great men can become the victims of their successes. O'Shaughnessy's story, created here by excerpts from the periphery of John Van Der Zee's fascinating account of the design and construction of the Golden Gate Bridge, reveals some of the pitfalls encountered by those engineers who fail to recognize the need for balance between professional opinion and sensitivity to public desires. Good engineers are sometimes undone because their professional success causes them to lose sight of the fact that the public, in the end, is the client, and without the support of the public there can be no professional success.

At three o'clock in the afternoon of February 20, 1915, President Woodrow Wilson pressed a gold key in the East Room of the White House; an instant later, across the continent in San Francisco, an elaborate "Fountain of Energy" leaped to life. . . . The Panama-Pacific International Exposition was officially opened.

The exposition was a celebration of progress incarnate: the completion of the Panama Canal, the mythical linking of the oceans, made real by the greatest feat of engineering the world had yet seen, an ebulliently American enterprise at which the French had tried, and failed. It was engineering, above all, that people had come from all over the world to San Francisco to honor, and 257 different engineering societies were scheduled to meet at the exposition. For engineers generally, particularly sensitive now to their emergence from the level of artisans and craftsmen into the ranks of licensed and better-paid professionals, the exposition had an entire second level of significance: it was being held in a city that had been all but destroyed only nine years before [*] and had now been restored, largely through the work of engineers, a fact that *Engineering News*, the bible of the civil engineering and construction industry, considered worth celebrating with a double-sized exposition issue featuring the "Notable Engineering Works of the Engineering Department of San Francisco."

There was something in the reconstructed San Francisco with which an engineer of almost any specialty could identify: a new sewer system; a fire-fighting network that utilized both fresh and salt water reserves; a completely new municipally owned transit system that would eventually include tunnels, forty-eight miles of track, and "iron monster" railway cars built in the

----

[*] By the earthquake and fire of 1906.

Municipal Railway's own shops; and, most significantly, the most sweeping and elaborate water supply system ever devised for an American city, the Hetch Hetchy system, which involved dams, lakes, and more than two hundred miles of aqueducts and conduits and which would require more than twenty years and over a $100 million to complete.

Especially poignant to the engineers was the fact that this enormous reconstruction was almost entirely under the direction of one man, a knock-about, practical civil and construction engineer not so different from themselves. And that, in San Francisco, he seemed to be the living embodiment of the engineering ideal of professional technical leadership showing the way to political and moral progress.

Born in County Limerick, Ireland, on May 28, 1864, Michael Maurice O'Shaughnessy was educated at the Queen's colleges of Cork and Galway, constituent colleges of the Royal University of Ireland, from which he received his bachelor's degree in engineering in 1884. The following year O'Shaughnessy arrived, "with brogue and parchment," in San Francisco, where he took a job in the engineering department of the Southern Pacific Railroad.

A strapping young man with a piercing, steady gaze, a thinning head of hair, a thick, trimmed mustache, and a taste for elaborately formal dress almost always including a boutonniere, O'Shaughnessy, as assistant engineer of the railroad and, later, civil engineer for townsites, rapidly learned the ins and outs of California construction and its politics. Bright, articulate, vigorous, and capable, O'Shaughnessy established himself rapidly in his adopted city. He married a young local woman and in 1893, while still in his twenties, was named chief engineer of San Francisco's first notable fair, the Midwinter International Exposition, designed to promote California's produce and her climate.

An ambitious young man who was gaining recognition in an expanding profession, O'Shaughnessy now went, for a time, wherever opportunity took him, although he was almost always either an independent consultant or else running the whole show: chief engineer of a copper company in Shasta County; construction and hydraulic engineer on sugar plantations in Hawaii. Returning to California in 1905, O'Shaughnessy became chief engineer of the Southern California Mountain Water Company at San Diego, for whom, between 1907 and 1912, he supervised the building of the Morena Dam, the largest rock-filled dam built to that time in the United States.

As O'Shaughnessy's reputation and career expanded, San Francisco, the city where he'd first lived and worked in America and his wife's hometown, endured what must still rank as the most severe natural convulsion, and among the worst politically, suffered by any major American metropolis.

Perhaps the best way to gauge the devastation of the San Francisco earthquake and fire of 1906 is to examine the photographs of the burned and shaken city. In the extent and depth of the damage--the heaps of rubble, the scarred, standing hunks of wall, the terrifying emptiness--the pictures are like a preview of the bombed-out European and Japanese cities of World War II. Nothing like the damage inflicted by nature on San Francisco was to exist for the world's cities for another thirty-five years, until the advent of long-range multiengine aircraft and the blockbuster bomb.

In addition to this physical devastation, San Francisco, in the years immediately preceding and following the earthquake, was probably the worst-governed city in America.

Under the administration of a stooge mayor, Eugene Schmitz, an ex-bandleader and local head of the musicians union, and a wily political boss, Abe Ruef, San Francisco in the years after 1902 was like a municipality bearing a giant "For Sale" sign. Prostitution was open and rampant, and the dens and deadfalls of the Barbary Coast for years maintained an average of a murder a night. More serious, although less apparent, was the infiltration of the city's public utilities by graft-paying private interests. Patrick Calhoun, organizer of the United Railroads, the city's combined streetcar system, took advantage of the postearthquake confusion to string the city with the overhead electric power lines that had formerly been widely opposed, then sealed the deal by bribing members of the city's Board of Supervisors to pass an ordinance approving the unsightly lines. The city's water service, supplied by the private Spring Valley Water Company, which resisted public outcries for improvements, was so poor that in the Richmond District there was often insufficient pressure to produce more than a trickle out of householders' taps. The water company kept the entire Board of Supervisors on retainer.

Public indignation, and an effective reform movement led by former mayor James D. Phelan, industrialist Rudolph Spreckels, and Fremont Older, editor of the San Francisco *Call*, had led ultimately to the ouster and trial of both Schmitz and Ruef. As long as the city's water and transit utilities remained in the hands of the Spring Valley Water Company and the United Railroads, however, those hands could always line the official pocket and clutch the civic throat. By 1912, encouraged by the reform spirit of the Progressive movement, and particularly the California Progressives, led by Hiram Johnson, now governor of California and a man who had prosecuted Ruef, the people of San Francisco had voted to build their own municipal transit system, to run in competition with what was now the Market Street and United Railroads, and had decided to build a new water system or to acquire the Spring Valley Water Company, or both.

What was required to keep these new systems too from falling into the hands of the grafters was a super-engineer, a man who could oversee the physical reconstruction of the city while remaining free of the pressures and plundering that had brought about its political decline.

Mayor James D. Rolph, who took office in 1912, was convinced he knew the man to handle both tasks.

"In the later part of August, 1912," Michael O'Shaughnessy recalled years later, "Mayor Rolph wired me at San Diego . . . asking if I would be available for the position of City Engineer of San Francisco."

The task was enormous: the earthquake and fire had destroyed 2,300 acres of the city and left 100,000 people homeless; the entire public utilities system would have to be reconstructed. Also, O'Shaughnessy had been exposed to San Francisco politics before, an experience he remembered as "discouraging." And the salary, $15,000, was half the total of his income for the previous year.

Yet there was the challenge of undertaking the reconstruction of an American city--the chance to do it all as it really should be done, and the assurance from the mayor that, as city engineer, O'Shaughnessy would be politically autonomous. There was also something else, O'Shaughnessy admitted: "My wife being a native of the City, influenced my decision and favorable consideration of the Mayor's proposal."

Rolph, a shrewd campaigner who was to serve four terms as mayor of San Francisco, then one as a governor of California, had the successful politician's instinct for pinpointing exactly what his audience most wants to hear.

"You must look on the City as your best girl," Rolph told O'Shaughnessy, "and treat her well. Do what you think is best for her interests."

When O'Shaughnessy expressed "my strong objections to political interference by elected officials," Rolph capped his sales pitch by offering the engineer carte blanche: "Chief, you are in the saddle. You're it. You're in charge."

Riding the wave of a new, reform administration, responding to a public mandate to "get things done, (we don't care how)," O'Shaughnessy plunged into the problems of San Francisco with the bold, personal style that was to fill, and eventually overflow, the boundaries of his job.

"He has not waited for the reports of his subordinates," a San Francisco paper reported approvingly, not long after O'Shaughnessy took office, "nothing of the kind for this virile man. He investigates first personally, visits the spot where a correction is needed, makes a hasty, though no less accurate survey; arrives at his conclusions; tells what should be done . . . and the blueprints follow later."

The engineer as a doer, the thinker as man of action, the theoretician combined with the practical man, all in the interest of public service. This was the O'Shaughnessy that O'Shaughnessy himself wished to be. And that, during these years at least, he was--to the administration, to the press, and to the overwhelming majority of the citizenry of San Francisco.

O'Shaughnessy delivered results. The city of vacant lots, newly built homes, and rising skeletal buildings bustled with the activity of his department's construction crews: track laying for the streetcars that were being welded together in the Muni's shops; tunnels under the Stockton Street hill and beneath Twin Peaks; pipelines and pumping stations for an auxiliary fire protection system; a sewer outfall at Baker's Beach--all working against the deadline of the 1915 fair. And, most impressively, Hetch Hetchy.

It was the development of the city's independent water supply that forced O'Shaughnessy's vision beyond the existing physical limits of San Francisco. Isolated on the tip of a peninsula, the city was both cut off by water--and in need of it.

To satisfy the city's thirst for fresh household water, San Francisco voters had agreed to reach some 245 miles into the Sierra Nevada Mountains and acquire rights to a valley called Hetch Hetchy, a neighbor of Yosemite Valley, and regarded by many people who had seen it, among them John Muir, as Yosemite's equal in beauty. In an act politically inconceivable today, and one opposed by Muir and the Sierra Club in a battle that dragged on for years and ultimately reached the Supreme Court, Hetch Hetchy was logged, dammed, flooded, and turned into the reservoir for the water supply of San Francisco. The project, which included the building of a railroad, tunnels, and access roads and the laying of a wilderness pipeline, cost eighty-nine lives and was not completed until 1934. It provided San Francisco, a city today of less than 750,000, with a water supply for 4 million people and is considered the model for urban water systems everywhere. That its design, construction and completion represent the organizational ability and will of a single man can best be testified to by the fact that the dam at the mouth of the valley, the most significant structure of the project, was named at its completion, in 1923, after Michael O'Shaughnessy.

By the time of the 1915 exposition, O'Shaughnessy was a man of independent political power in San Francisco. Backed by the unquestioning support of a popular mayor, not answerable to the electorate himself, he wrote signed articles in the San Francisco newspapers on the state of the city and often advised readers how to vote on local bond issues. He served as a consultant on urban utilities to other cities and was retained in an advisory capacity by the boards of local municipalities' water districts. His reputation, expanding with the prestige of his profession, grew mightily on the national and local level. In 1919 Rolph, in a speech at Hetch Hetchy, compared his city engineer to Hezekiah, the Old Testament king who "stoppered the watercourse of Gihon and brought it straight down . . . to the city of David."

A man of abundant physical and mental energies, O'Shaughnessy seemed not to be burdened by the breadth and depth of his responsibilities, but instead to be stretched by them. His imagination, he found, was like a muscle, which grew stronger with use, and the more local projects he took on and completed, the more O'Shaughnessy found his range of ideas expanding beyond San Francisco to the entire bay region. This city, this farthest west that he had come to, represented not only his own best possibilities but the country's. Here, where everything had begun over again, the future could be not only planned but engineered.

On Sundays, the tireless O'Shaughnessy took the ferry from San Francisco to the village of Sausalito in Marin County, where, accompanied by his two dogs, an Aberdeen Scotch terrier and a bulldog, he hiked through the territory he called Hobbylogland, the thickly wooded slopes and canyons of Mill Valley, often continuing up over the Coast Range into the cathedral redwood groves of Muir Woods.

"A chunky lump of a man," a reporter for the San Francisco *Call* described him, "with a chest broad and deep, bushy eyebrows behind which blue eyes twinkled, and he walked with the gait of a man accustomed to getting there." Led by his two leashed dogs, a cigar usually clutched in his mouth, O'Shaughnessy customarily devoted his day off to a hike of between twelve and fifteen miles.

His weekend hikes in the Marin woods stimulated O'Shaughnessy's active expansionist imagination. Here, around him, was water, plenty of it, tumbling down the fern-lined Marin creeks; and here also, more importantly, was room: space for the homesites that O'Shaughnessy could already see would be severely limited by the fixed dimensions of San Francisco. Here, in the most beautiful--and the least developed--of the counties bordering the bay, was the repository of the engineer-and-planner's hopes. O'Shaughnessy bought rental property in Marin County. And on his way back from his weekly hikes, riding on the ferry past the Golden Gate, or more memorably, waiting in his car in the long line that backed up from Sausalito almost every Sunday evening, he pondered the possibilities of engineering San Francisco's next great extension.

The 1915 exposition, while not planned itself by O'Shaughnessy, was nevertheless an impressive demonstration of his works. Visitors rode to the fairgrounds on streetcars built in O'Shaughnessy's Municipal Railway shops, on tracks laid by his crews, on streets paved by his gangs, with asphalt from his city asphalt plant. Indeed, the very arteries of the fair, from the water that first whooshed through the Fountain of Energy at Woodrow Wilson's touch to the sewage system that drained the fair's wastes, were appendages of O'Shaughnessy's reconstructed San Francisco. "The City That Knows How," President William Howard Taft had dubbed San Francisco at the groundbreaking ceremony for the fair in Golden Gate Park in 1910; but it was one man, above all, who'd overseen the job, in the city's greatest burst of civic construction, before or since. The shattered, corrupt San Francisco of 1906 had been replaced by another city, white and shining, full of hope, and as committed to progress as the imposing yet temporary domes and palaces of the fair.

O'Shaughnessy . . . was an outsider, an immigrant Irishman whose speech was still seasoned with a brogue, yet an outsider who had made his way directly to the control room of local political power. As city engineer of a reconstructed and still expanding city , he was the operator, the man who did the doing when something had to be done and who, always thinking expansively, was feeling confined by the physical limits of the city whose future was now linked with his own. ... O'Shaughnessy had built or overseen the building of railways, tunnels, and sewer and firefighting systems and was building an immense water supply complex. Yet the one thing he was not, and could not convince his constituents he might become was a bridge builder. . . .

It had become a kind of game with Michael O'Shaughnessy. He would meet a bridge builder--and in the course of his job as city engineer of San Francisco he was to meet just about every construction man of any real significance in America--and, as a sort of intellectual puzzle, he would pose the problem of bridging the Golden Gate.

That some sort of bridge system would eventually be built at San Francisco seemed inevitable: the city was the largest American metropolis still served primarily by ferry boats. Choked off at the tip of a peninsula, San Francisco faced a future of increasing congestion and economic strangulation. In 1920, local civic pride suffered its worst shock since the earthquake when the federal census showed that Los Angeles had replaced San Francisco as the largest city in California. It was a gap that the southern metropolis, with its unlimited access to adjacent space for homes and businesses, and its hunger for incorporation, was to widen throughout the next decade. Los Angeles had become the fastest growing city in America, while growth rates in San Francisco for both population and industry had fallen below the national average. San Francisco's expansion had fallen behind that of the city's own suburbs, a fact that seemed to suggest even more strongly the need for a permanent link between the city and its communities around the bay. The city's future clearly depended on San Francisco's collective determination and resources and its willingness to commit them to the building of bridges.

The distance at the Golden Gate--a little over a mile--was the shortest of any point on the bay. A bridge here would link San Francisco with the resource-rich and underpopulated counties of all of Northern California. There was a natural symmetry to the site, two facing points of land, separated by a narrow strait; just looking at it, the Golden Gate seemed to present one of the most obvious places in the world to build a bridge. It also represented a variety of obstacles, the like of which had never existed in a single construction site before.

Geologically, the Golden Gate is a gap in a mountain range through which pours the outflow of seven different rivers. This gorge, carved out over the millennia by the actions of the rivers, also fronts directly on the Pacific Ocean, producing a complex of tides and currents not duplicated anywhere in the world. At the center of the channel, the Gate is 335 feet deep. Any structure built here would have to be anchored in such a way that its base could absorb and withstand the force of both tidal currents and ocean waves while at the same time its superstructure would have to be firm enough and flexible enough to resist the battering of gale winds. Added to these hazards was the fact that the site was within twelve miles of a major earthquake fault that, within the last two decades, had caused an epic catastrophe.

These natural obstacles were compounded by other difficulties, mechanical, social, and political. The Golden Gate was the lone entrance to one of the world's great harbors, and no bridge had ever been built at a harbor entrance before. Not only would the bridge have to be anchored deep enough to survive tides and ocean waves, it would have to be tall enough for the largest ships to pass beneath its roadbed at high tide. There were naval bases inside the bay and army installations on either side of the Gate; any structure built here would require the approval of the departments of the Army, Navy and War. How were the functions of a great seaport to be carried on while a construction project, which would undoubtedly take years, cluttered its mouth? In the event of war, couldn't a single well-placed bomb send a bridge roadway here crashing down to block the entire port? There were also aesthetic considerations. The Golden Gate represented one of the earth's most dramatic meetingplaces of land and water. Was a man-made structure, erected here, altering the landscape forever, a wise choice? Was expansion by this means really in San Francisco's best interests, or would it help to destroy the fragile uniqueness of the city and make expansion alone its strongest characteristic, as in Los Angeles?

169

The men to whom O'Shaughnessy posed his problem were practical men, builders familiar with the realities of subcontracts and bids. Most of them did not need to consider beyond the initial questions of time, design and the cost of materials and money. Bridge building, after all, followed a certain logic, one theorem leading to another, with the solution to a particular problem usually being an extension or an adaptation of some other solution. The larger and more complicated the problem was, the larger and more complicated, usually, was the solution. A bridge could conceivably be built anywhere, provided someone was determined enough to build it and, more critically, to pay for it.

At the time of his initial collaboration with O'Shaughnessy . . . Joseph Strauss was forty-five years old, affluent, and unfulfilled. His patented . . . bridge design, easily adapted to streets, highways, and rail systems, had made him independent and kept his firm busy on jobs all over the United States and in many places abroad; these works, however effective, were merely functional devices, pieces of other systems that would never be remembered in and of themselves. The great and lasting work, the indisputable achievement that Strauss had always longed for, had, as yet, eluded him. . . .

Strauss's most ambitious and industrious endeavors, however, always carried a desire for acceptance, which was somehow always denied him. Although there were hundreds of Strauss bridges and numerous Strauss patents, there was no Strauss paper read before a meeting of the American Society of Civil Engineers, no publication of Strauss engineering theory bearing any sort of official imprint, no body of work by Strauss that was the subject of study by aspiring student engineers. He was the promoter, the salesman, at best, the tinkerer and gadgeteer, the entrepreneur, but never the poet-engineer, the John Roebling he so clearly longed to be. . . .

How accurate a measure Michael O'Shaughnessy had taken of Joseph Strauss we can only guess. As one of the nation's leading city engineers, O'Shaughnessy surely knew that Strauss had never supervised the building of any bridge on a scale comparable to that required at the Golden Gate; he was essentially a builder of highway bridges, the lip-to-lip movable structures dictated by America's vast river system. He was not a profound or original thinker, and his success as a builder was based on easily reproduced adaptations of a single idea. Still he was a bridge builder, and he had an obvious yearning, typical in men of a certain age, for an achievement that would catch the eye of posterity. . . .

O'Shaughnessy was convinced that San Francisco's future depended on expansion to the north, and of all the builders to whom he had proposed the idea of a bridge at the Golden Gate, Strauss was the only one willing to give it a serious try. . . . Beginning in 1919,. . . Strauss and O'Shaughnessy undertook the first serious examination of the problems of building a bridge over the Golden Gate.

In March 1923, Strauss testified before the [California] Assembly's Committee on Roads and Highways on formation of the Bridge District[*] "The engineering

170

plans are fully solved," he announced to the legislators. "It is up to California to go ahead."

Presented with this authoritative endorsement, backed by what seemed to be an endless prairie of grass-roots support, the legislators yielded to the challenge of inevitable progress. The Coombs Bill, enabling the formation of a tax district for the purpose of building a bridge at the Golden Gate, was signed into law on May 25, 1923.

The whirlwind acceptance of the bridge proposal could only have aroused profoundly conflicting feelings in Michael O'Shaughnessy. While the enthusiastic response to an idea that he had been nurturing for more than four years was gratifying, as was the mention of his name in the press in association with the bridge, Strauss's leap to the forefront of the project must have caused him a certain alarm. The idea of a bridge, which O'Shaughnessy had initiated, had been seized by the terrier Strauss, who had run away with it. . . . This was not at all what O'Shaughnessy, who normally insisted on controlling all his jobs, had in mind. The city engineer had underestimated Strauss's promotional ability and his political skills, with the result that O'Shaughnessy's professional pride, already of a sensitivity that, in a few years, would lead him to threaten resignation rather than accept criticism of his performance as city engineer, had been wounded. Strauss and O'Shaughnessy remained cordial and continued, for the present, to cooperate on the bridge project, but it was a wound that would continue to fester over the next few years.

O'Shaughnessy had a number of increasingly critical rivals on the board,[**] men acutely sensitive to the city engineer's penchant for mighty projects and his ability to use these works to assure and expand his own authority; as the decade wore on there would be growing sentiment among board members to bring O'Shaughnessy to heel. The situation put a certain political edge on every issue that involved O'Shaughnessy, and assured that any proposal with which the city engineer was associated would be subject to intense scrutiny by at least some members of the board.

O'Shaughnessy's career, a mountain range of peaks so far, was now showing its inevitable valleys. In 1920, the city engineer had arranged to make a triumphant return visit to his mother's home in Limerick, Ireland. On the eve of his departure, he was presented, on behalf of members of the Board of Supervisors and his friends at City Hall, with a diamond stickpin in recognition of his services to the city. Even more gratifying was the ceremony in May 1923, christening the 344-foot-high dam at the mouth of Hetch Hetchy

----

[*] The Golden Gate Bridge and Highway District, a local tax district which would be formed for the purpose of building a bridge.

[**] The San Francisco Board of Supervisors, the city's governing body.

Valley O'Shaughnessy Dam, an unusual act of hagiography for a live and sitting city engineer. O'Shaughnessy's prestige would never be as great, or his authority as unquestioned, again. Already there had been rumblings of resentment over the city engineer's exemption from accountability to the usual public bodies. There were cutting references to him in the press and among the city supervisors as "The King." O'Shaughnessy's reputation and prestige had attracted a certain critical attention in themselves: was it really appropriate for an appointed city official to possess unquestioned power?

Among the critical eyes that O'Shaughnessy's string of successes had caught were those of William Randolph Hearst, who maintained a certain watch over San Francisco politics from the bridge of his flagship San Francisco *Examiner*. Hearst had now decided to start firing a few shells in the city engineer's direction.

In a page-long bold-type "Open Letter To Our City Engineer" on December 10, 1923, the *Examiner* which described itself as "always your friend in the past," had taken O'Shaughnessy to task for failing to enforce the Raker Act of 1913, which had required that the electric power generated by the Hetch Hetchy system be distributed from public, not private, bodies. Instead, the letter charged, "There are indications that you and your assistants obstinately oppose municipal electricity, and openly espouse turning over the city's power output to Pacific Gas and Electric Company." Although professing to admire O'Shaughnessy's engineering skills, the *Examiner* suggested he leave matters of public polity to elected representatives and the mayor's advisory committee on Hetch Hetchy. The city engineer, warned *Examiner* editorialists, was faced with two choices: "Either to forget your opinions on matters of polity and finance; or to acknowledge frankly that your differences of viewpoint are too great to warrant your remaining in office."

O'Shaughnessy was defended by the other San Francisco papers, who accused the *Examiner* of trying to turn the city engineer into a Hearst stooge, but the issue was a live one, one that remained unresolved and has been periodically revived over more than sixty years. More immediately, the letter demonstrated that the city engineer, long regarded as a local fixture beyond politics, was no longer immune to public criticism. O'Shaughnessy's unchallenged honeymoon with his wife's hometown was over. By November 1924, a measure had been introduced before the Board of Supervisors that would "shear" O'Shaughnessy of his "carte blanche" to spend city bond issue money on Hetch Hetchy. O'Shaughnessy would continue to defend his hard-won autonomy and prestige vigorously. He would also grow increasingly touchy about what he considered attempts to diminish it.

When, in September 1926, a pair of San Francisco supervisors, Warren Shannon and James McSheehy, accused the city engineer of low-balling (deliberately underestimating) bids on a portion of the Hetch Hetchy water line and considering only one bid for the building of the dam that had been named after him, O'Shaughnessy was so stung that he confided to Mayor Rolph that he was contemplating resigning. To O'Shaughnessy's dismay, the mayor announced to the press that the city engineer so resented the statements made by the supervisors that his resignation was "probable." Rolph said that he was "very much afraid that we shall lose O'Shaughnessy" and that "it is well-known that he has several outside offers carrying a much larger salary."

For O'Shaughnessy, any advantage he might have gained by having his importance to the city endorsed by its mayor was offset by the public admission that the supervisors had got his goat. "I thought it was a private conversation between the Mayor and myself," O'Shaughnessy responded. He admitted that he resented the continuing attacks and that they had become "almost unbearable." Yet he now announced, "I will never quit under fire."

In part, O'Shaughnessy was now paying the price of the extraordinarily favorable press he had received earlier in his career. The making and unmaking of public personalities has always been a reliable stock in trade for certain elements of the press, and it had a particularly strong history in San Francisco . . . The making and unmaking of celebrities had news value, and any individual involved had to develop a thick skin and a strong survival instinct. If this were the way the game was played, O'Shaughnessy concluded, then he would learn to play it with the best of them.

When, in October 1927, Mayor Rolph vetoed an attempt by the supervisors to appoint an outside engineer to make a street railway transportation survey, O'Shaughnessy made a public statement that he had not been consulted in the choice of engineer and that his own proposal to conduct the survey himself at a cost less than the outside engineer's fee had been submitted to the supervisors, who had tabled it. This growing instinct to go public in matters that he would previously have dealt with privately or remained aloof from, O'Shaughnessy now extended to the city's Board of Public Works. In an interview with an engineering trade magazine, O'Shaughnessy described this civic watchdog group as being composed of "a green grocer, a boilermaker and a retired military officer," men so technically unsophisticated that they were political weathervanes. The temperature of invective between O'Shaughnessy and his enemies would continue rising.

In 1929, however, O'Shaughnessy was still a man of enormous power and prestige in San Francisco. Mayor Rolph, his patron, still reigned at City Hall. The Hetch Hetchy project, still uncompleted and requiring periodic transfusions of capital in the form of new bond issues, still represented the city engineer's great vision made manifest. A new road, linking the city's Glen Park District with the thoroughfare around Twin Peaks, was designated to be named O'Shaughnessy Boulevard. Even his severest critics among the Hearst press and on the Board of Supervisors did not dispute the value of his services to the city. It was his power and autonomy they feared, and they let slip no opportunity that offered the possibility of curtailing them.

In the matter of the bridge, O'Shaughnessy, with his heightened political sensitivity, could feel himself being pushed out of the picture entirely. . . .

For the most part, all these recent political jolts represented the normal ups and downs of political life that O'Shaughnessy had learned to endure with a certain amount of equanimity; but beyond them were certain foreseeable long-term changes that O'Shaughnessy must have found profoundly disturbing. If the bridge project went forward without him, it would represent the first major work undertaken in San Francisco since 1912 that did not bear his stamp. In a sense, it would mark the limits of his professional life: he'd been good enough to go this far, but no farther. It suggested that all his earlier efforts had somehow been flawed and that later reconsideration might find them wanting. There was also the fact that O'Shaughnessy's works, no matter how respected among civil engineers, were, to the people of San Francisco, all but invisible. Everyone in San Francisco might one day use the water from Hetch Hetchy, but hardly anyone ever went there, stood on the dam, traveled the railroad, marveled at the size of the culverts and tunnels and pipelines and the mighty effort that had been required to build them. It was the same with the city's streets, sewers, tunnels, rail lines and streetcars. People used all these things, but they did not marvel at them. With a great bridge it was different. The works of it were there, for all to see, an enabling power that, like a giant sculpture, incorporated into its vision even the people beholding it. Almost everyone who crossed a great bridge must be aware, if only for an instant, of his dependence upon its engineer. O'Shaughnessy had earned the right to be part of such a work, and now it was about to be denied him.

What must have been most infuriating of all was the fact that O'Shaughnessy, the engineer who in many ways had originated the bridge project, was being excluded from it, while Joseph Strauss, to whom he had first suggested bridging the Gate, was not only still part of the proposal but among the engineers being considered to be put in charge.

In fact, Strauss had paid a stiff price to get the chief engineer's job. According to Richard Welch,* who, as the only man associated with the bridge longer than Strauss, probably made him the offer, the terms of Strauss's contract were "the lowest ever written for a bridge job of such magnitude." For a fee of 4 percent, as compared to the standard 7 percent suggested by the American Society of Civil Engineers, Strauss's firm would be expected to absorb all engineering expenses, and there would be no royalties beyond the fixed construction percentage. On a total construction estimate of $30 million, for example, Strauss would have only $1.2 million to cover salaries, plans, travel expenses, testing and consulting fees. Strauss, who had campaigned long and hard for the job, had to swallow his financial as well as his professional pride to get it. . . .

For Michael O'Shaughnessy, the selection of Strauss, in so many ways his own creature, as chief engineer was the final, unendurable affront of his ten-year association with the bridge. Almost immediately following Strauss's appointment, O'Shaughnessy . . . was quoted as saying that he was opposed to building the bridge because in his opinion, at least for the record, its costs would be in excess of $100 million.

In an era of dwindling civic income, with city officials struggling to come up with money enough to provide adequate unemployment relief, O'Shaughnessy had been forced to go to the electorate again and again with appeals for more

----

* Welch was a member of the San Francisco Board of Supervisors and a member of the Board of Directors of the Bridge District.

bond issue money to complete the great water project at Hetch Hetchy. The *Examiner*, riding him hard now, had tagged him "More Millions," and repeatedly accused him of selling out the city's private power resources to the Pacific Gas and Electric Company. Mayor Rolph, O'Shaughnessy's patron, was about to embark on a campaign for the California governorship, leaving his city engineer exposed. Now, for the first time, serious investigations of the "Czar of City Hall" were either threatened or impending. O'Shaughnessy responded to these imputations about his integrity with outbursts of professional pique. In the spring of 1931, when an independent investigation by an outside engineer, retired Army Captain John A. Little, was critical of O'Shaughnessy's recent management of the Hetch Hetchy project, the touchy city engineer responded by filing a formal complaint, charging "gross incompetency" against Little with State Department of Professional and Vocational Standards. The charge backfired, instead producing among the San Francisco Board of Supervisors, O'Shaughnessy's old pursuing pack of hounds, demands for the city engineer's ouster and public questioning of the legality of his job. O'Shaughnessy, it seemed, wearing his detachable wing collar and inevitable boutonniere, was still living in the era of "Sunny Jim" Rolph when that sun, in fact, had set.

Michael O'Shaughnessy, once the avatar of Progress in San Francisco, had become the symbol of diehard resistance to change. His diehard opposition to the bridge had linked him with the most resented of entrenched local interests, and his outspoken fears of extravagant costs, made groundless by the cold reality of construction bids, had brought into question not only his political but also his technical and managerial competence. With Rolph's departure from the mayor's office, the politically untouchable city engineer had become fair game, and he was now under fire, it seemed, from every quarter. There were charges in the Board of Supervisors of cost overruns on the Hetch Hetchy project, conflict of interest, and the use of O'Shaughnessy's staff for outside work. There were also personal incidents that added fuel to the various flickering feuds O'Shaughnessy had sparked over the years, combined with an overall public desire, intensified by the Depression, to share and dramatize bad luck by seeing the once-mighty brought low.

In December 1931, the investment property that O'Shaughnessy had bought in Marin County, a Mill Valley cafe that he had leased to a pair of Italian proprietors, was raided by federal agents as a speakeasy; there was a possibility that O'Shaughnessy might be held responsible for the fine. His dispute with Captain Little, the consulting engineer hired by the supervisors to check on the Hetch Hetchy pipeline figures, had been aggravated into an ugly row by O'Shaughnessy's contesting the engineer's competence and, in a letter to the state engineering board, his reference to Supervisor James R. McSheehy as a "political buccaneer and agitator." The letter, and the remark, were made public, and the Board of Supervisors passed a resolution censuring O'Shaughnessy.

Beneath these surface ripples was a deeper tide of public change. People were tired of O'Shaughnessy and his ceaseless demands for money to complete his distant, unseen Hetch Hetchy project. They had grown weary of this combination engineer-official who always seemed so sure he alone knew what was best for San Francisco. The times had passed him by. The city had rebuilt itself thoroughly from the earthquake. The new bridges, one to the East Bay and one to the north, promised the dawning of a new era of predominantly regional and national, as opposed to local interest. O'Shaughnessy, even though he had prophesied and helped bring out this new age, was too strongly associated with the old. By giving himself totally to his city, he had engineered himself out of a job.

In 1932 San Francisco adopted a new city charter, eliminating the post of city engineer as occupied by O'Shaughnessy. In its place, a chief administrative officer system was introduced, with Alfred J. Cleary, formerly an O'Shaughnessy assistant, as the nonelective CAO. O'Shaughnessy was retained as a consulting engineer, with his duties confined to reduced responsibilities on the Hetch Hetchy job, which was finally winding to its conclusion.

The great water project, twenty-two years in the building, was at last nearing completion. In the fall of 1934, the first Hetch Hetchy water was scheduled to be delivered to San Francisco, and a great civic celebration was planned, with Secretary of the Interior Harold Ickes making the dedication speech, and featuring O'Shaughnessy as an honored guest.

He was still an imposing figure in San Francisco, representing as he did the spirit that had rebuilt the city after the earthquake, and his presence was requested at most important civic gatherings. Izetta Cone[*] remembers him from a dinner dance the American Society of Civil Engineers staged aboard an American President liner in San Francisco Bay, the great mover and shaker of the past, still distinguished in black tie and tuxedo. "I remember him and Mrs. O'Shaughnessy dancing together. They both must have been in their seventies. They were *enthusiastic* dancers."

The Hetch Hetchy completion ceremonies were scheduled for October 28, 1934. On October 12, O'Shaughnessy attended a meeting of the Federation of City Employees at the Dreamland Arena in San Francisco. Feeling ill, O'Shaughnessy, who had been suffering recurring heart attacks over the past two months, left early and went home. Early the next morning, his wife called their five children to their father's bedside.

"God bless you all," O'Shaughnessy told his family. And he subsided into the last sleep of his vigorous and embattled life.

----

[*] Wife of Russell Cone, resident engineer for the construction of the Golden Gate Bridge.

The Forgotten Engineer:
John Stevens and the Panama Canal
Virginia Fairweather

Civil Engineering Magazine - February, 1975 Issue

This brief biographic sketch of John Stevens describes the classic civil engineer of the late nineteenth and early twentieth century: independent, resourceful, dedicated and technically competent.

In the latter half of the twentieth century the requirements for dedication and competence are as important as ever, but the American civil engineer tends to be less independent and resourceful than his professional ancestors. Larger engineering organizations (consulting firms, construction companies and government agencies) with their inevitable bureaucracies have restricted the independence of individual engineers. Improvements in communication of technical knowledge and the widespread adoption of various technical standards and codes have diminished the need for individual resourcefulness.

Despite these changes, the civil engineering profession remains a profession which provides great opportunities for personal growth and creativity. In the place of individual resourcefulness in problem-solving the modern civil engineer enjoys a working environment in which advances in technical knowledge are continuously revealing new problems, providing new insights into old problems and producing new problem-solving techniques to be learned and employed. Furthermore, increased attention to social and environmental aspects of civil engineering projects has provided opportunities for civil engineers to expand the scope of their studies and evaluations to incorporate these types of considerations.

John Frank Stevens was the personification of American faith in progress and technology, of individuality combined with Yankee know-how at the turn of this century. The Maine-born and self-taught engineer had a career that began on the frontiers of the American northwest and took him to remotest Russia at an age when most men would be retired. The capstone of that career, to the minds of most observers, should have been his appointment in 1905 as chief engineer for the Panama Canal, a spectacular achievement that captured the imagination of the nation and the world. Yet by his enigmatic departure less than two years later, Stevens became the forgotten man when plaudits for the accomplishment were bestowed. Why did he leave so abruptly? Perhaps some clues may be found in what we can determine of the man's personality.

Fame, money and power do not seem to have been motivating forces in this engineer's life. Stevens' personal lexicon was studded with Horatio Alger-like maxims. "Work and study" was his slogan and the secret of 90% of all achievement, he said. "The man who continuously strives to make himself more capable makes it difficult for opportunity to ignore him." "The most interesting task in the world is the one that forces you to reach beyond what you had thought your limit." This kind of work ethic credo combined with a self-acknowledged wanderlust served Stevens well in his profession.

Born in Maine in 1853, he started work for the railroads as a rodman in Minneapolis, moved to Texas and elsewhere when opportunity beckoned. According to contemporaries, young Stevens worked hard days and studied far into the night. His diligence paid off; by 1905 he was probably the top railroad engineer in the country. His major accomplishments were made during his association with James Hill, owner of the Great Northern Railway. Under Hill, Stevens discovered two passes (one of which bears his name) that saved the railway miles and money, built railroad tunnels and bridges all through the northwest. All in all, he was probably responsible for laying more miles of track than any other single individual. In legend, Stevens defied weather, wolves and Indians in his efforts to extend the railroad. In fact, he did seem to defy topography. One of his more striking engineering efforts is his route

John Stevens

through the Cascades, tunneling through mountains and bridging abysses to bring coal from Canada and iron ore from the Mesabi Range.

Stevens worked as a reconnaissance and location engineer, a task that he was to look back on in later years as his most exciting. The western railroad extension presented a kind of challenge hard to recapture now with current concern over the environment, second thoughts about the effects of technology and latter-day consciousness over the plight of the American Indian. In those days, railroad engineering was heroic, a part of the fulfillment of America's destiny.

In 1903, he had been hired away from Hill to the Chicago Rock Island and Pacific Railway and was a vice-president when he was approached (at Hill's suggestion) about Panama.

President Teddy Roosevelt had identified his administration with this mammoth and seemingly doomed undertaking. He was furious when Stevens' predecessor John F. Wallace quit. (An interesting sidelight--both Stevens and Wallace were later to become ASCE [American Society of Civil Engineers] presidents.) Stevens was persuaded to take over the chief engineer's job against his real wishes, according to his biographers.

He made conditions to his taking the job; he was to have a free hand with no interference from politicians and an understanding that he would leave when

178

he had either proved the project a failure or assured its success. Roosevelt, equally blunt, told Stevens that things at Panama were in "a devil of a mess."

So they were. According to accounts of the time, work was at a standstill, yellow fever and malaria were rampant. Boredom, discouragement and dirt competed to lower the morale of those who avoided disease. Housing was deplorable; decent food hard to come by and expensive when available. Eggs sold for $1.50 a dozen, partly because local merchants saw they could capitalize on the captive consumers.

The French, under Ferdinand de Lesseps had failed twenty years earlier and, to many, it looked as though the Americans would also fail. Others thought the United States' claim to the zone was questionable. To Roosevelt and the Republicans in power, the completion of the canal became of overriding importance.

In his brief eighteen months tenure, Stevens seemed to do just that. How did he accomplish so much where others had failed? The first thing Stevens did, paradoxically, was to stop work, if the little that was being accomplished could be called that. He took immediate steps to improve the lot of the workers and to halt disease. His oft-quoted "There are three diseases on the Isthmus, yellow fever, malaria and cold feet. And the worse of these is cold feet," reveals his stoic nature and his approach to the problems that greeted him in Panama. He was able to instill morale into the workers both by the sheer force and example of his character and by improving their working conditions.

The first he did by donning boots and going out into the fields with the men in all kinds of weather, through jungles and swamps. He disdained the offer of a mansion such as the Governor of the Canal Zone was having built for himself and instead lived with his family in a corrugated roofed house on the Culebra Cut where he could see the work progress. For his men, he had decent living quarters built and set up a commissary where workers could eat fresh refrigerated food at cost. He built hospitals, schools, churches, paved streets, put in a sewage disposal plant, a telephone system and organized band concerts and a baseball league.

A second and crucial step taken by Stevens was to place full confidence -- backed by money--in Dr. William C. Gorgas. Gorgas had rid Havana, Cuba, of yellow fever but many people including Commissioner Shonts* thought him a crank. With Stevens' blessing, Gorgas fumigated every building in the Zone, sprayed stagnant pools, built drainage pools and took other measures to rid the area of the mosquitoes that carried malaria and yellow fever. In five months there were no cases of yellow fever at all. Stevens permitted Gorgas to spend $90,000 for window screening alone during the campaign against the diseases that had so shadowed the project before his arrival. The extent to which the disease had haunted the men may be imagined by the story that Stevens' predecessor had arrived with coffins for himself and his wife.

The immediate task that confronted Stevens once his workers were provided

----

* Theodore Shonts, commissioner of the Isthmian Canal Commission created by Congress to oversee the construction of the Canal.

for was the Culebra Cut, where the workers would have to scrape through 9 miles of clay and shale mountain. This unstable material had defied all previous attempts to get through; every time a cloudburst occurred, landslides would undo whatever had been accomplished.

Stevens' approach was that of the seasoned railroad man. He regarded the Cut as a gigantic railroad pass and employed steam shovels along strips of benches the way he had seen on the Mesabi Range of Minnesota. He double-tracked the Panama railroad and set up a highly efficient schedule. All earth removal equipment was to be operating every moment; all excavated material removed immediately so there would be no accumulations to cause landslides. The schedule worked, but success was undoubtedly also due to the fact that Stevens worked a 14-hour day out in the field with the men throughout the venture. He encouraged rivalry between steam shovel crews resulting in productivity unthinkable with the dispirited workers who had greeted him on his arrival six months before.

Finally, Stevens was instrumental in winning approval for the lock-type rather than the sea-level canal, a decision that surely was a key factor in the success of the project. The sea-level canal had been the popular choice, but there were two obstacles.

One was that the Pacific side of the Isthmus had a maximum tide of 20 ft., while on the Atlantic side, tides were only 20 in. at most. This difference would have required some locks anyway. The turbulent Chagres River in the center of the Isthmus was the other problem. The Chagres was a trickle in dry weather, a torrent after rain. When Stevens saw how time-consuming and expensive it would be to divert the river, he was converted from a sea-level to a lock canal man and managed to convince others that he was right. President Roosevelt and Secretary Taft came over to this opinion and, in 1906, Stevens testified before the Senate Committee on the matter. He argued that the great depth necessary for a sea-level canal would make it too expensive and worse, susceptible to landslides. Further, it would more than likely have to be so narrow as to impede the crossing of ships travelling in opposite directions.

Stevens' solution to the problem posed by the unstable Chagres was to resurrect a 30-year-old plan made by a French engineer, Adolphe de Lepinay. This was to bridge the highest part of the Isthmus with a man-made lake formed by damming the Chagres at Gatun. The lake and the river would control and supply each other while acting as the central link in the canal system. Eventually the official decision was made for the lock-lake canal, a decision according to many historians that guaranteed the ultimate success of the project.

Six months later, Stevens wrote to President Roosevelt, evidently to the latter's great surprise, suggesting that he wished to leave Panama. Roosevelt chose to interpret the letter as a resignation, apparently indignant at what he felt was a slight by one of his favorites. He went further; he never again gave Stevens public credit for his accomplishments at Panama.

While Stevens' name is remembered at the Canal Zone, his contribution has been largely obscured here, much due to Roosevelt's silence. One mural at the Museum of Natural History in New York accords Stevens his rightful place among the men who made the Panama Canal possible. Stevens and Teddy

180

Roosevelt stand side by side flanked by Dr. Gorgas and Goethals, the man who took over after Stevens.

What did the famous letter say? It mentioned money, saying that it cost Stevens $100,000 a year to remain in Panama compared to what he could be earning back in the U. S. He spoke of red tape, of being "continually subject to attack by a lot of people, and they are not all in private life, that I would not wipe my boots on in the United States." He suggested that a military man should replace him, one who could be ordered to cut through the red tape and proceed with construction. The latter suggestion was taken; Colonel (later General) Goethals was an army engineer.

Stevens' action has been interpreted in countless ways. Among the conjectures: that he was overworked, afraid of disease, bored once the creative work was done, even that he had uncovered a scandal within the Republican party. Stevens' only comment, and that many years later, was: "I resigned for purely personal reasons, which were in no way, directly or indirectly related to the building of the canal, or with anyone connected with it in any matter."

The mystery of the resignation pursued Stevens to the end of his life. He outlived Gorgas, Taft, Roosevelt and Goethals and could have redeemed himself and gained his place in history. But he never did. It is possible that this self-made man who made his name in the unsophisticated American west simply could not endure the bureaucracy of the Civilian Commission, the necessity to convince the American government to "do it his way." One of his conditions to taking on the job, after all, had been to leave when he either proved the project a failure or had ensured its success.

Most agree that he had done the latter, including his successor, General Goethals, who later said "the canal was laid out by one of the best engineers God ever put on the face of the globe--John F. Stevens. I was merely the hired man carrying out Mr. Stevens' ideas." In any case, Stevens did not seem to care about public acclaim and on his deathbed, spoke of Roosevelt as one of the two men who had most influenced his life (the other was James Hill of the Great Northern Railroad).

What did Stevens do after his abrupt departure from the drama of Panama? He returned to his first love, the railroads, holding executive positions with several before setting up his own railroad consulting practice in 1911. But he was not to remain in private life. Events transpired thousands of miles away that would bring Stevens back into government service when most men would be retiring.

In 1917, Stevens was asked to go to Russia to aid the reorganization of the Trans-Siberian railways. The breakdown of this system was a U. S. problem because Russia was our ally in World War I and the railroad was vital in transporting troops and supplies across the continent to Europe. The confusion was aggravated by the fact that a revolution was taking place in Russia.

Stevens agreed to go, believing he would be taking part in a two to three months mission. Instead he spent six years in Russia, Manchuria and Japan under circumstances in many ways more frustrating than those at Panama, first as a member of a Technical Advisory Commission, then after the Armistice, as head of the Inter-Allied Technical Board.

This Board was comprised of representatives of eight nations, each of whom had troops remaining in Siberia dependent on the disrupted railroad to get home. Throughout his stay, Stevens was beset by difficulties. The overseeing nations struggled among themselves; a few tried to usurp control of portions of the system. Outmoded equipment and methods and severe financial problems defied progress--some member nations never even paid for use of the railroads. Cossack brigands raided and terrorized workers; the path of the revolution with its attendant violence also endangered men and impeded the route. The Russian government itself appealed to the Board for assistance in times of famine and epidemic. In those days, and in that place, the railroad was the only means of communication and transport for thousands of miles and millions of people.

At first Stevens had the assistance of the Russian Railway Service Corps, a group of expert railroad engineers recruited from the U. S. at his own suggestion. After 1919, he carried on with one or two assistants. Throughout this turbulent period, Stevens presided over the multinational Board and kept the railroad operative, "the only disinterested person in the whole bewildering situation," in the opinion of one observer.

Stevens' occasional pleas to come home from Russia were not greeted as his famous letter from Panama was; instead he was repeatedly persuaded to stay. And stay he did, until the last Japanese soldier was transported home in October of 1922. For this task, less spectacular than Panama, but extraordinarily arduous, Stevens was decorated by his own country, and by France, China, Japan and Czechoslovakia.

Stevens continued to practice his profession and was elected president of ASCE in 1927. When he died at 90, he had practiced engineering for well over half a century. He had had a hand in transforming the planet, linking oceans and extending civilization, spanning continents with his railways, bridges and tunnels. In his own mind his most important service to his country had been to convince Roosevelt and the Congress to build a lock canal at Panama. It is indeed unfortunate that Roosevelt, his most important if not most accurate historian, failed to credit this achievement to John Stevens.

The Man at the Helm

Excerpts from Chapters 18 and 19, **The Path Between the Seas**
David McCullough

Path Between the Seas, *David McCullough's powerful book about the development of the Panama Canal may be one of the best-selling books ever written about a civil engineering project. The success of the book was probably due to a combination of factors: the book was well written, it dealt with interesting subject matter and its publication coincided with the national debate concerning the United States relinquishing its rights to the ownership and operation of the canal.*

In this selection from the book we learn a little about the character and philosophy of Colonel George Goethals, the man President Theodore Roosevelt selected to be chief engineer for the Panama Canal after John Stevens resigned from that position. The reasons for Mr. Stevens' resignation have never been fully explained, although many believe that he was simply exhausted from the demands of the position (which he had filled admirably).

Colonel Goethals' brought to the job a concern for the employees, an understanding of what needed to be done, and a willingness to do whatever was necessary for successful completion of the task. He was the right man at the right place with the right knowledge.

With the appointment of George Washington Goethals, [President Theodore] Roosevelt's worries over the work at Panama came to an end. The canal would now be the "one-man proposition" John Stevens had called for, only the one man was to be an entirely different sort from Stevens.

At forty-eight Goethals was the same age as Roosevelt and of similar ancestry. His Flemish father and mother had arrived in New York with the great wave of immigration in 1848. The second of three children, he had been born in Brooklyn on June 29, 1858, and later, when he was eleven, moved with his family to a house on Avenue D in Manhattan, a block from the East River. But his family had been poor and struggling and unlike Roosevelt he had had to make his way "exclusively by his own exertions." Starting at age fourteen he had worked his way through City College in New York, then went on to West Point, where he was elected president of his class and finished second in his class in 1880, the same year Roosevelt was graduated Phi Beta Kappa from Harvard.

Goethals' career in the Corps of Engineers had been exemplary. In the Department of the Columbia [River] in 1884, William Tecumseh Sherman had singled him out as the finest young officer in his command and predicted a "brilliant future." He had worked on "improvements" in the Ohio River valley (1884-1885); as an instructor of civil and military engineering at West Point (1885-1889); on improvements on the Cumberland and Tennessee rivers and particularly on the Muscle Shoals Canal (1889-1894), where he designed and built a lock with the record lift of twenty-six feet; as assistant to the Chief of Engineers (1894-1899); and harbor works from Block Island to Nantucket (1900-1903). In 1903, the year of Elihu Root's reorganization of the Army, he had been picked to serve on the new General Staff, a corps of forty-four officers who were relieved of all duties in order to assist the new Chief of Staff.

*   *   *

He was a model officer, but a soldier like many in the Corps of Engineers who had never fought in a war, never fired a shot except on a rifle range, and who seems in fact to have had little affection for conventional "soldiering." Once on a parade ground in Panama, while watching some troops pass in review on a broiling-hot drill field, he would mutter to a civilian companion, "What a hell of a life."

Cool in manner, capable, very correct, he was a man of natural dignity and rigorously high, demanding standards. He had had no experience with notoriety, nor apparently any craving for it. And it would be hard to imagine him losing himself in **Huckleberry Finn** or anything other than his work. Asked years later how "the Colonel" had amused himself, a member of the family would respond, "He did not amuse himself."

A reporter wrote that "above everything he looks alert and fit." Six feet tall, he was in fine physical trim. The salient features were his intent, violet-blue eyes--"rather savage eyes," Alice Roosevelt Longworth would recall--and his close-cut, silvery hair, which he parted in the middle and washed daily. If a bit stiff socially, he was never pompous, largely because he was almost incapable of talking about himself. To pretty young women he could be especially gracious, in a rather fatherly fashion, and they considered him extremely attractive.

George W. Goethals

He was also a chain smoker and he detested fat people--with the one exception of William Howard Taft. Secretary [of War] Taft, Goethals was once heard to remark, was the only *clean* fat man he had ever known.

On the night that he was first summoned to the White House, Goethals and his wife had been entertaining an old friend, Colonel Gustav Fieberger, head of the engineering department at West Point, at their home on S Street. A messenger arrived with a note from William Loeb, Roosevelt's secretary, asking if Goethals would be free to come by the first thing in the morning. Goethals had immediately telephoned Loeb, who told him not to wait until morning but to come over that night at twenty minutes after ten. So Goethals had excused himself from his guest, changed into dress uniform, and left the house having no idea whatever as to why he was being sent for. Nor had he ever met Theodore Roosevelt.

"He entered at once upon the subject of the Canal," Goethals would recall. The canal commission was again to be reorganized and for the final time. Goethals was to be both chairman and chief engineer. Jackson Smith[*] and Dr. Gorgas[**] were to be members of the commission, along with four new men: a former senator from Kentucky named Joseph C. S. Blackburn, Rear Admiral Harry Harwood Rouseau, and Major David Du Bose Gaillard and Major

William Sibert, both of the Corps of Engineers. Gaillard was the only one on the list with whom Goethals was personally acquainted--Gaillard, too, had been a member of the first General Staff--but he knew Sibert and Rousseau by reputation and agreed to their appointments.

The critical decision, however, concerned Goethals. "He [Roosevelt] expressed regret that the law required the work to be placed in charge of a commission or executive body of seven men," Goethals remembered, "but . . . his various efforts to work under the law . . . were so unsuccessful that he resolved to assume powers which the law did not give him but which it did not forbid him to exercise."

So while all members of the commission were to be on the Isthmus henceforth, Goethals was to wield supreme authority, an authority that would be backed by another new executive order the following year. Goethals was to be a virtual dictator--"Czar of the Zone"--responsible only to the Secretary of War and the President. In the words of his biographer, Goethals at once became one of the world's absolute despots, who "could command the removal of a mountain from the landscape, or of a man from his dominions, or of a salt-cellar from that man's table."

----

* Mr. Smith had been brought to Panama by John Stevens, Colonel Goethals' predecessor as chief engineer. He was in charge of Labor and Quarters {housing} in the canal zone.

** Dr. William Gorgas, an Army doctor and an expert on tropical disease, had been named to the Commission because tropical diseases had played a significant role in the failure of previous efforts to construct a canal across the isthmus.

A common misconception later was that the canal was built by the Army, that it was the creation of the Corps of Engineers. It was not. Goethals and the other engineering officers were detached from the Army to serve in Panama. They did not report to the Chief of the Corps of Engineers; they, like the civilian engineers, reported to the canal commission--which was Goethals--and Goethals reported to Taft, exactly as Stevens had according to the previous reorganization.

The critical difference now was that an Army man could not and would not quit. For a West Point graduate to abandon his appointed task in the face of adversity or personal discomfort was all but inconceivable.

In the next several days, Blackburn, Rousseau, Gaillard, and Sibert appeared at the White House one by one to meet with the President and Goethals in the President's office. The same scene was repeated in each instance. Having introduced Goethals, Roosevelt would ask the man to be seated, then would inform him that he was to be appointed to the commission. "It will be a position of ample remuneration and much honor," Roosevelt said. "In appointing you I have only one qualification to make. Colonel Goethals here is to be chairman. He is to have complete authority. If at any time you do not agree with his policies, do not bother to tell me about it--your disagreement with him will constitute your resignation."

Goethals' salary, Roosevelt had decided, would be $15,000 a year, which was substantially more than he had been earning, but only half what Stevens had been paid.

A week or so after his new assignment had been announced in the papers, Goethals wrote in reply to the congratulations of a friend, "It's a case of just plain straight duty. I am ordered down--there was no alternative."

To a whole generation of Americans it was Theodore Roosevelt who built the Panama Canal. It was quite simply his personal creation. Yet the Panama Canal was built under three American Presidents, not one--Roosevelt, Taft, and Wilson--and in fact, of the three, it was really Taft who gave the project the most time and personal attention. Taft made five trips to Panama as Secretary of War and he went twice again during the time he was President. It was Taft who fired Wallace* and hired John Stevens, Taft who first spotted Goethals. When Taft replaced Roosevelt in the White House in 1909, the canal was only about half finished.

None of this made much difference, however. Nor ought there ever be any question as to the legitimacy of the Roosevelt stamp on the canal. His own emphatic position was that it would never have been built but for him and it

----

* John F. Wallace, the first American chief engineer for the Panama Canal. Wallace's resignation from the position was demanded by Taft when Wallace came to Washington, after less than a year and a half as chief engineer, to request that his salary be increased to match an offer he had received from a company in the United States.

was a position no one tried to dispute. To Goethals, "The real builder of the Panama Canal was Theodore Roosevelt." It could not have been more Roosevelt's triumph, Goethals wrote, "if he had personally lifted every shovelful of earth in its construction . . . ."

*   *   *

Even [Roosevelt's] Panama visit, however brief, he achieved at a stroke something that had never been done before: he made the canal a popular success.

And finally, he had entrusted command of the work to one extremely well-chosen man. "I believe in a strong executive," he once wrote to a correspondent, "I believe in power . . . ."

*   *   *

For the man who now bore the burden of responsibility for all that occurred, the initial hurdle had been primarily personal and as difficult as anything in his experience.

Goethals' reception upon arrival had been pointedly cool. Plainly, neither he nor the Army was wanted by the rank and file of Americans on the job and everyone seemed eager to make a special point of Stevens' tremendous popularity. Thousands of signatures had been gathered for a petition urging Stevens to withdraw his resignation and stay. No one, it seemed, had anything but the strongest praise for him and all he had done. Never in his career, Goethals remarked, had he seen so much affection displayed for one man.

Stevens and Dr. Gorgas were at the pier the morning Goethals and Major Gaillard landed. No real reception had been arranged; nothing had even been done about a place for Goethals or Gaillard to stay. Stevens still occupied the official residence of the chief engineer, a new six-bedroom house at Culebra that was to be Goethals' once Stevens departed, but since Stevens "didn't seem inclined to take us into his house" (as Goethals wrote to his son George), the two officers had moved in with Gorgas at Ancon, where there was little privacy, not even a desk at which Goethals could work. His letters to his family those first weeks were written on his lap as he sat in a straight-backed chair in one of the bedrooms.

To add to the spirit of the gloom, the <u>Star & Herald</u>[*] openly deplored the prospect of military rule. Probably no workers would have to wear uniforms, the paper presumed, but neither should anyone be surprised if he had to answer roll call in the morning or salute his new superiors.

That the railroad men around Stevens had scant regard for Army engineers seemed also abundantly plain to Goethals. "Army engineers, as a rule, were said to be, from their very training dictatorial and many of them martinets," he would write, "and it was predicted that if they . . . were placed in charge of actual construction the canal project was doomed to failure." The Army men

----

[*] The American-owned Panamanian newspaper

187

had only technical training, it was said; they had never "made a success as executive heads of great enterprises."

His own private estimate of the state of the work was entirely favorable. The difference between what he saw now and what he had seen in 1905, during the visit with Taft, was extraordinary. As he wrote to his son, "Mr. Stevens has done an amount of work for which he will never get any credit, or, if he gets any, will not get enough. . . ."

Several days passed before he was granted a more or less official welcome--a Saturday-night "smoker" given as much to entertain a party of visiting congressmen. John Stevens declined to attend and Goethals, at the head table, sat listening without expression as the toastmaster extolled Stevens at length and made several cutting remarks about the military. It was an evening he would never forget. With each mention of Stevens' name there was a resounding cheer, while the few obligatory references to Stevens' successor were met with silence. Goethals was furious at what he regarded as "slurs" on the Army, but kept still until it was his turn. He had come to the affair not in uniform but in a white civilian suit. In fact, he had brought no uniforms to the Isthmus and never in the years to come would he be seen in one.

He was, he told the assembled guests, as appreciative as they of the work Stevens had accomplished and he had no intention of instigating a military regimen. "I am no longer a commander in the United States Army. I now consider that I am commanding the Army of Panama, and that the enemy we are going to combat is the Culebra Cut and the locks and dams at both ends of the Canal, and any man here on the work who does his duty will never have any cause to complain of militarism."

He took over from Stevens officially at midnight, March 31, 1907, and a week later Stevens sailed for home. One of the largest crowds ever seen on the Isthmus jammed the pier at Cristobal to see him off, everyone cheering, waving, and singing "Auld Lang Syne." Stevens was noticeably amazed and touched by the outpouring of affection. This time it was Goethal's turn not to attend.

Having none of Stevens' colorful mannerisms or easy way with people, Goethals impressed many at first as abrupt and arbitrary, a cold fish. The word "goethals" in Flemish, it was soon being said, meant "stiff neck."

He hated to have his picture taken. He found the visiting congressmen rude, tiresome, terribly time-consuming. Callers were "an awful nuisance." It was expected that he appear at every dance and social function at the Tivoli or the Culebra Club. He would "brace up" and go "out of a sense of duty" and spend the evening sitting on a porch listening to the music, waiting only for the time when he could politely withdraw.

Stevens' former secretary, having agreed to stay and help with the transition, suddenly resigned. William Bierd, the railroad boss, made a surprise announcement that he was retiring because of his health, but then Goethals learned that Bierd was taking a job with Stevens on the New Haven Railroad.

Frank Maltby* decided no civilian engineer had a future any longer at Panama and so he too quit. Then the steam-shovel engineers, sensing the time was at last right for a show of strength, threatened to strike unless their demands were met. Goethals refused and they walked off the job. It was the first serious strike since the work had begun. Of sixty-eight shovels, only thirteen were still in operation. He recruited new crews.

Even the newly arrived Major Sibert was proving "cantankerous and hard to hold" in meetings. Mrs. Sibert, Goethals learned, was "disgusted" with the Panama weather.

From surviving letters written to his son George, then in his senior year at the Military Academy, it is apparent that he was also extremely lonely. Mrs. Goethals was still in Washington "doing society at a great rate"; another, younger son, Thomas, was at Harvard. He felt very out of touch, he wrote; there was not time even to read the paper. His sole source of amusement was the French butler, Benoit, who still spoke practically no English but went with the official residence at Culebra, Goethals being his seventh chief engineer.

The day began at first light. At 6:30, with Benoit standing stiffly in attendance, "the Colonel" had his breakfast--one peeled native orange stuck on the end of a fork, two eggs, bacon, one cup of coffee. By seven he had walked down to Culebra Station to catch either the No. 2, northbound, at 7:10, or the No. 3, southbound, at 7:19. The morning was spent inspecting the line. He carried a black umbrella and customarily wore white. Invariably he looked spotless; invariably he was smoking a cigarette.

Back at the house again, immediately upon finishing a light lunch, he would rest for half an hour, then walk to his large, square corner office on the first floor at the Administration Building. There he would receive people until dinner at seven. In the evening, unless otherwise engaged, he would return to the office to concentrate on his paper work until about ten.

To most observers he seemed wholly oblivious of his surroundings, intent only on his work. One employee, relaxing on his own porch one particularly beautiful moonlit evening, witnessed the following scene:

> "There were only a few lights here and there in the Administration Building. One by one they went out, all except that in the old man's office. It was getting on toward ten when his window went dark . . . .A full moon, as big as a dining-room table, was hanging down about a foot and a half above the flagstaff--a gorgeous night. The old man came out and walked across the grass to his house. He didn't stop to look up at the moon; he just pegged along, his head a little forward, still thinking. And he hadn't been in his own house ten minutes before all the lights were out there. He'd turned in getting ready to catch that early train. . . ."

----

* Maltby had come to Panama when John Wallace was chief engineer and had stayed through John Stevens' term as chief engineer. He wrote of his experiences in an article "In at the Start at Panama" in Civil Engineering magazine {1945}.

To his elder son, Goethals wrote that he was better off occupied, since there was nothing else to do. He confessed to working so hard that he would often end the day in a kind of daze. He was not the "clean-desk" man Stevens had been. His "IN" and "OUT" baskets were always jammed. Papers were piled wherever there was room on his desk--correspondence, folded maps, specifications, plans, half a dozen black notebooks, reports in heavy dark-blue bindings. The bit of clear desk surface he managed to maintain directly in front of him was soon peppered with cigarette burns.

He liked things on paper. If during his morning excursions along the line a department head or engineer urged some new approach or improvement, the inevitable response was "Write it down."

It was not in him to court popularity. He wanted loyalty first, not to him but to the work, that above all. He abhorred waste and inefficiency and he was determined to weed out incompetents. Nor was there ever to be any doubt as to his own authority. "What the Colonel said he meant," a steam-shovel engineer remembered. "What he asked for he got. It didn't take us long to find that out." Requests or directives from his office were not to be regarded as subjects for discussion. When the head of the Commissary Department, a popular and influential figure, informed Goethals that he would resign if Goethals persisted in certain changes in the purchasing procedure, Goethals at once informed him that his resignation was accepted and refused to listen when he came to retract the threat. "It will help bring the outfit into line," Goethals noted privately. "I can stand it if they can." He put Lieutenant Wood* in as a replacement. ". . . I just put it up to him to make good . . ." he wrote.

"Executive ability," he observed on another occasion, "is nothing more or less than letting the other fellow do the work for you." But to some he gave every appearance of wishing only to dominate everything himself. Marie Gorgas, in particular, found him "grim, self-sufficing." He was much too abrupt for her liking. "His conversation and his manners, like his acts, had no finesse and no spirit of accommodation." She grew to dislike him heartily. Even Robert Wood, who admired his "iron will and terrific energy," found him "stern and unbending--you might say a typical Prussian. . . .I was his assistant for seven years," Wood recalled long afterward, "and I might say that everything in my life since has seemed comparatively easy."

But if the manner was occasionally severe, the standards demanding, he was invariably fair and gave to the job a dignity it had not had before. "I never knew him to be small about anything," recalled an electrical engineer named Richard Whitehead, who joined the force that same summer of 1907. Goethals knew how to pick men. He knew how to instill determination, to get people to want to measure up. He was not loved, not then or later, but he was impressive. And by late summer he had "the outfit in line."

----

* Lieutenant Robert E. Wood was one of the first Army officers sent to Panama (in 1904). He eventually became Chief Quartermaster of the Canal Zone, and still later he became a corporate leader of Sears, Roebuck and Company.

190

"Another week of observation has confirmed my view ... that the discontent and uneasiness which followed the departure of Stevens have nearly passed away ..." wrote Joseph Bucklin Bishop to Theodore Roosevelt in mid-August. Undersized and grouchy-looking, with a little, pointed gray beard and a shiny bald head, Bishop was another new addition. He had been transferred from the Washington office on Roosevelt's orders and was to be at Goethals' side from then on, as secretary of the commission, ghost writer, policy advisor, alter ego. And not incidentally he was to feed confidential reports to the White House on how things were going.

Goethals, reported Bishop, was "worn and tired and says that he has had a veritable 'hell of a time,' but I believe he has won out. When I told him so, he said, 'Well, I don't know.'"

Mrs. Goethals had arrived and departed meantime. So his marriage, characterized years afterward by members of the family as "difficult," became still another topic for local speculation as the lights in his office burned on into the night.

*   *   *

To give employees opportunity to air their grievances, Goethals next established his own court of appeal. Every Sunday morning, from about 7:30 until noon, he was at his desk to receive any and all who had what they believed to be a serious complaint or problem. He saw them personally, individually, on the basis of first come, first served, irrespective of rank, nationality or color. By late 1907 there were thirty-two thousand people on the payroll, about eight thousand more than when he took over. By 1910 there would be nearly forty thousand. Yet once a week, beginning in the fall of 1907, any of these people--employees or dependents--could "see the Colonel" and speak their minds.

The scene was unique in the American experience, unique and memorable in the eyes of all who saw it. Jules Jusserand, the French ambassador, likened it to the court of justice held by Saint Louis beneath the oak at Vincennes. "One sees the Colonel at his best in these Sunday morning hours," wrote a reporter who had been greatly frustrated by what seemed a congenital inability on Goethals' part to talk about himself. "You see the immensely varied nature of the things and issues which are his concern. Engineering in the technical sense seems almost the least of them."

Some advance screening was done. Bishop saw the English-speaking workers, while the Italians, Spaniards, and other Europeans were seen by a multilingual interpreter, Giuseppe Garibaldi, grandson of the Italian liberator. And often these preliminary interviews were enough to resolve the problem-- the mere process of free expression gave the needed relief--but if not, Goethals' door stood open.

On an average Sunday he saw perhaps a hundred people and very few appear to have gone away thinking they had been denied justice. They came to the front of the tall, barnlike Administration Building, entered a broad hallway hung with maps and blueprints and there waited their turn. Their complaints included everything from the serious to the trivial: harsh treatment by a foreman, misunderstandings about pay, failure to get a promotion, dislike of

191

the food or quarters, insufficient furniture. He listened to appeals for special privileges and financial dispensation. One request was for the transfer of a particular steam-shovel engineer to a different division where a particular baseball team needed a pitcher. (The request was granted.) He was given constructive ideas regarding the work and was made party to the private quarrels between husbands and wives or families in adjoining apartments. By all accounts he was a patient listener.

Many complaints could be settled at once with a simple yes or no or by a brief note sent down the line. A serious situation of any complexity was promptly investigated. "He was a combination of father confessor and Day of Judgment," wrote Bishop. The vast majority who came before him were almost excessively respectful. Rarely would anyone challenge his authority and then to no avail. "If you decide against me, Colonel, I shall appeal," one man declared. "To whom?" Goethals asked.

Some of the remaining officials from the Stevens regime had expressed vehement disapproval when these Sunday sessions were first announced. Jackson Smith, of the Labor Department, had been especially exercised, since his own policy in past years had been to tell anyone who had a complaint to feel free to leave on the next ship. And this, apart from Smith's own rude manner, had been considered a perfectly appropriate policy. Stevens had been in full accord. The new approach was in fact wholly unorthodox by the standards of the day. In labor relations Goethals was way in advance of his time, and nothing that he did had so discernible an effect on the morale of the workers or their regard for him: "they were treated like human beings, not like brutes," Bishop recalled, "and they responded by giving the best service within their power."

In Goethals' own estimate, expressed privately many years afterward, it was thus that he won "control of the force," and control of the force was "the big, attractive thing of the job."

When another delegation of congressmen, members of the House Appropriations Committee, arrived in November, they were impressed as much by Goethals as by the strides being made, a point of special satisfaction at the White House. "I was present at all the hearings . . ." Bishop wrote to Roosevelt. "Not only did he [Goethals] show that he knew his business thoroughly, had absolute grasp of the work as a whole, but that he had at his tongue's end more knowledge of details than any of his immediate subordinates."

Before leaving for Washington the chairman of the delegation, Congressman James A. Tawney, told Goethals privately not to worry about appropriations-- he could count on whatever he wanted. The committee reported the situation in Panama to be in "excellent shape." And as time went on, Goethals' standing on Capitol Hill was to be a factor of the greatest importance. Money sufficient to do the job correctly was never to become an issue.

"There is only one man who should be heard at Washington on the Canal, and that is Goethals," Bishop stressed to Roosevelt. "He has absolute knowledge, perfect manners, and can talk . . . He says I am the man who should be spokesman rather than he, but don't let him persuade you into such a belief. He is the man at the helm . . ."

"Mr. Strauss Gave Me Some Pencils..."

An Excerpt from **The Gate**
John Van Der Zee

*John Van Der Zee's book about the planning, design and construction of the
Golden Gate Bridge provides interesting descriptions of the circumstances
and people involved in bringing the bridge into existence. He believes that some
of the key people involved in the bridge's design and construction have never
received the credit they deserved for the work they performed on the bridge.*

*Joseph Strauss, who was Chief Engineer for the bridge, studied civil engineering
at the University of Cincinnati and was awarded a degree in 1892. He went to
work for a Chicago consulting engineering firm that specialized in bridge
construction and, in 1902, formed his own bridge design and construction
company, eventually acquiring a considerable reputation as a bridge builder.
His company specialized in building relatively small bridges of a single type,
and he had no experience in the design and construction of suspension bridges.
He became interested in the Golden Gate project during the 1920s and spent
years promoting the idea of a bridge there. However, after voters finally approved the bridge concept, the design he submitted (an adaptation of the types of
bridges his firm had been building) was criticized as being ugly and impractical.*

*Strauss's firm eventually won the contract to design and build the bridge, but
the contract required that he share with a three-man Board of Consulting
Engineers the responsibility for major design decisions. All three members of
the Board were more knowledgeable about large bridges than Strauss. Leon
Moisseiff and Othmar H. Ammann were world-renowned bridge engineers and
Charles Derleth Jr., the third member of the Board, (who was also the Dean of
Engineering at the University of California in Berkeley) had been a consultant
on large bridge projects in California. Strauss hired Charles A. Ellis, a
civil engineering professor from the University of Illinois and an expert in
structural design, to be responsible for the actual design which would be
carried out by Strauss Engineering Company.*

*This selection describes the work carried out by Ellis and the roles played by
Strauss and the Board of Consulting Engineers during the design stage. Ellis
was eventually discharged by Strauss (in the middle of the Depression). He
was unemployed for several years before finding employment as a structural
engineering professor at Purdue University where he remained until his
retirement.*

With its mile-wide stretch of moving, swelling water, backed by the sudden
upthrust of the Marin hills, its intermittent rivers of fog, and the broad sweep
of the Pacific opening out on one side, and the bay on the other, the Golden
Gate at San Francisco is one of the world's great natural stage settings. It
lends an aura of drama to anything happening there. This in part was why, on
December 9 [1929], the day the test drillings for the water foundations began, more
than a thousand spectators appeared at either shore of the strait.

Out in the strait some one thousand yards from the San Francisco shore, a
large dredge was anchored in place; powerful motors were ready to be started,

great shafts about to be sunk beneath the waters of the strait. The first steps were being taken toward the mastery of the Gate itself.

On the shore at Fort Point, the San Francisco Municipal Band was playing, while on both the Marin and San Francisco sides of the bridge a full program of speakers made addresses that were carried by radio throughout California and up to the Oregon border. Joseph Strauss, as chief engineer, was the principal speaker at both sides of the strait.

Strauss, returned from the East, had plunged into his pet project with renewed vigor: the stimulus of travel always seemed to restore him. It seemed he was involved in everything: chairing meetings, making speeches, meeting the press, conferring with experts. Strauss seemed to be everywhere, but then, abruptly, he was gone. His presence was intermittent. A few days in one place and he was off, promoting another project or overseeing one in progress. It seems that, as much as winning the chief engineer's job had meant to Strauss, his real interest was more in promoting the bridge than in building it. Indeed, throughout this period, the man who was there on the job every day was Ellis. He was at the test site, not just on dedication day but every day. He oversaw the borings as they were made, making sure the contractors adhered to the guidelines established by the surveyors. He met not only with Strauss when Strauss was in town but with MacDonald[*] and Lawson[**] and Captain Savage, the army engineer, and the concrete man who wanted to talk about aggregates. He was the man on the line, the man people came to each day with questions, expecting to leave with answers. He was also, presumably in his free time, supposed to be thinking about the mathematical formulas involved in developing the design for the bridge.

Ellis's relationship to Strauss had become like one common on ocean liners, where, because of the press of social functions, there are actually two captains. One, the ceremonial captain, hosts different passengers at the captain's table at dinner, takes people on tours of the ship, and poses for photographs. The other, the working captain, oversees the day-to-day running of the ship and the performance of the crew. Charles Ellis had become very much the working captain on the Golden Gate Bridge, and if he felt at all neglected by the lack of official notice, it was something he was too busy to devote much thought to; more than that, it was fundamentally contrary to his nature to complain. Besides, he was enjoying himself enormously.

One day, while the borings were still going on, Ellis was meeting at Fort Point with Professor Andrew Lawson, the Berkeley geologist who would be analyzing the samples from the test borings, when the question of earthquakes arose. Professor Lawson, who, with his snow-white hair, walrus mustache, and horn-rimmed spectacles, looked like a cartoon absent-minded professor, had made a more thorough study of earthquake questions than any other man in America, according to Ellis. Now, as the two men watched the drillers at work on the rise behind the fort where the giant cables of the

----

[*] Alan MacDonald, general manager of the Golden Gate Bridge District.

[**] Andrew Lawson, the geology consultant for the bridge project.

bridge were to be anchored, Lawson asked Ellis what would happen if one of the anchorages should instantly slip six inches.

Ellis, a man who was at home with questions of structural engineering and who had been devoting concentrated study to the behavior of cables, replied that as far as the vibrations of the bridge were concerned, a slip of six inches instantly would be the same as if the anchorage slipped twice as much gradually.

They were two professors talking now, illustrating points of theory by using simple anecdotes, working preferably with the material at hand, as one would in a classroom. "Supposing I have a fish," Ellis improvised in explanation, inspired no doubt by the fishermen who invariably line the seawall near the fort, "and I lay my fish on a scale pan, letting go of it gradually as the pan sinks. When I have completely released it, and the pan is carrying the fish, the recorded weight is, say, five pounds. Now, if instead of letting go of the fish gradually, I hold it down very close, practically touching the pan, and release it instantly, the scale sinks twice as far and registers momentarily ten pounds. The scale vibrates back higher than its original position and down again, back and forth, until it comes to rest again at five pounds." The question, in other words, becomes "what would be the effect if one of the anchorages slipped a foot gradually." The answer, said Ellis, "is that nothing would happen--the practical effect would be a lengthening of the cable by one foot"--something that could happen anyway with a simple rise in air temperature.

Lawson, clearly enjoying this interplay of theorizing and pedagogy, next asked Ellis what would happen if one of the main piers dropped instantly six inches. The answer again, said Ellis, "is that nothing would happen that would seriously affect the bridge. It would simply increase the length of the cable slightly. The structure is a limber structure and is susceptible to considerable variations in deflections."

This was for both men like faculty club conversation at its best, an interchange of disciplines between peers, yet more invigorating than that because it was real, because they stood twelve miles from a major earthquake fault, discussing its possible effect on a structure their ideas were to help create.

Professor Lawson proposed a more difficult question: What would happen if one of the piers were to be moved sideways, to slide transversely? According to studies of earthquake situations in Japan and San Francisco, scientists had arrived at the conclusion that for structures built on any kind of soil and made up of heterogeneous material, such as steel, terra cotta, brick, and concrete, the effect of a severe earthquake shock is equivalent to a force equal to 10 percent of the acceleration of gravity. What would happen to the tower above the pier in such an event?

Ellis replied that this bridge would be built of homogeneous material and on rock, and therefore 5 percent would give an equivalent force of a severe earthquake. Say the total dead load* at the top of one tower made up of two

----

    * Dead load is the term used to describe the load imposed by the weight of the structure itself. The load induced by the traffic the bridge would carry is called the live load.

posts was 100 million pounds. Applying a transverse force of 5 million pounds at the top of the tower would not exceed the working stresses[*] at nearly all points in the tower. Double the transverse force to 10 million pounds and the working stresses might be exceeded, but they would not exceed or even equal the yield point[**] of the material.

Ellis, who was to refer to this conversation later in the year, in an address before the National Academy of Scientists, summarized his feelings regarding the bridge and earthquakes to the scientists as this:

> "If I knew that there was to be an earthquake in San Francisco tomorrow and I couldn't get into an airplane and had to remain in the city, I think I should get a piece of clothesline about 1,000 or 2,000 feet long, and a hammock, and I would string it from the tops of two of the tallest redwoods I could find, get into the hammock and feel reasonably safe. If this bridge were built at that time, I would hie me to the center of it, and while watching the sun sink into China across the Pacific, I would feel content with the thought that in case of an earthquake, I had chosen the safest spot in which to be."

On February 3, 1930, Joseph Strauss proudly announced to the press that a rock formation so hard that it had worn away more than three carats of diamond drills had been discovered at Lime Point, the proposed Marin County pierhead of the Golden Gate Bridge. The rock in this leg extending out from the shoreline was so hard that it had kept an expert diamond setter working continuously. Its discovery fully vindicated earlier reports made by the bridge geologists and, said Strauss, "clears the way for work to commence within the next two weeks on final plans and specifications."

The Bridge District directors were elated. The findings of the drillers had refuted the testimony of engineers employed by opponents of the bridge, who had repeatedly asserted that no suitable foundations for so heavy a structure existed on either side of the Gate.

Privately, the conclusions drawn from the test drillings were somewhat less enthusiastic, particularly those of Professor Lawson, the consulting geologist. The rock on the Marin shore, Lawson informed Strauss, consisting partly of sandstone and partly of greenstone, was indeed strong enough and amply sufficient to support the bridge pier. The south pier, to be set in the strait some one thousand feet off the San Francisco shore, would be founded on serpentine, which is not uniform in strength. Because of the effects of water, the rock had suffered internal shearing, so it now "consists of an aggregate of ellipsoids or spheroids of strong rock embedded in a matrix of sheared rock of very small strength."

----

* Working stress is the stress the bridge is designed to withstand.

** Yield point is the term used to describe the stress at which a material actually begins to fail.

While the drilling was going on, Ellis had conducted a pressure test at Fort Point on a typical stretch of the same kind of serpentine. He had found and demonstrated to Lawson that the rock could sustain a load of 32 tons per square foot, or 450 pounds per square inch. Since the supporting strength of the same kind of rock below the pier, confined by concrete, would be in excess of this, Lawson concluded that the foundation rock would be adequate, providing "the excavation for the foundation be not less than 25 feet below the lowest point of the rock surface."

It was the south anchorage of the bridge, on the hill behind Fort Point, that represented the greatest problem to Lawson. Here the rock was made up of "serpentine in a rather badly sheared condition with feeble coherence and very low tensile strength." Because of the structural weaknesses in the rock, the south anchorage of the bridge, binding its two great supporting cables to the San Francisco shore, would, in Professor Lawson's view, "have to be designed to depend upon dead load (the weight of the concrete base anchoring the cable) rather than upon the tensile strength of the rock, or the frictional resistance of the latter to the pull of the anchor."

Professor Lawson also had some chillingly fatalistic conclusions about earthquakes, pointing out that any earthquake so violent that it would destroy the bridge would also destroy San Francisco. Yet, "tho it faces possible destruction, San Francisco does not stop growing and that growth necessarily involves the erection of large and expensive structures."

More surely destructive, Lawson pointed out, would be the fact that the life of a steel bridge near salt water is limited and that, from an economic view, the fact would have to be faced that the bridge would have to be replaced once or twice a century owing to deterioration by rust.

Lawson's conclusions, while favorable, did not provide an airtight argument in support of the bridge. People who wished to take issue with conclusions could conceivably find openings here, and with a bond issue election on the horizon, Strauss and the directors could be sure that the opposition would exploit any weakness.

Within the Engineering Board, the emphasis had shifted from making theoretical decisions to designing the structure itself. Nine years had been spent fine-tuning and deliberating Strauss's original design for the bridge. Now a new design would have to be prepared, from scratch, in a matter of months. In their deliberations, extending through several sessions and amplified by correspondence, Strauss, Ellis, and the consulting engineers had arrived at certain overall conclusions about the type and style of the bridge. These guidelines, reached amid the deepening shadow of the Great Depression, show an increasing obsession with economy, and many of them were dictated by necessity.

It was to be an all-suspension bridge, since that was the most economical kind to build and, according to Moisseiff's recent discoveries, also the most appropriate. There would be two piers, one in the strait, 1,100 feet off the San Francisco shore, and the other at the Marin shoreline. Again, this positioning was justified by its saving of a million dollars in construction costs. For the sake of symmetry, it was decided to make the two side spans--the parts of the bridge stretching from shore to tower--the same length: that of the stretch required between the San Francisco tower and the shore at Fort Point. Since the topography on the two sides of the strait differed, there would be no savings in using straight backstays in the side spans. Therefore, Strauss was able to recommend the more aesthetically satisfying curved backstays.

In the interests of lightness and economy, the rail lines proposed for the original roadbed would be dropped. This was justified by the conclusion that the day of the trolley as the most widely accepted means of mass transit was ending and that rail vehicles everywhere would be replaced by cars and buses. It may have been one of the few instances when the engineers' projections would later be proved wrong, although the jury on rail mass transit is certainly still out.

The roadbed would be made of concrete slabs, seven inches thick, weighing eighty-five pounds per square foot. If, at some later date, a slab of sufficient strength could be designed of lighter aggregate, the concrete could be replaced and the live-load capacity of the bridge increased. The floor beams would be designed to carry six 24-ton trucks abreast, with the heavy axle loads directly over the floor beam. There would be wide, ample pedestrian sidewalks.

"All these decisions," Charles Ellis recalled a few months later, "were the result of deliberations of the board through several sessions. All the major questions requiring experience and sound judgment having been settled, there was little left to do except to design the structure. At this point, Mr. Strauss gave me some pencils and a pad of paper and told me to go to work."

Charles Ellis's dry sense of humor was definitely at work when he made his statement, in a speech to a group of scientists, about the simplicity of designing the bridge. It was a simple problem, but only once you understood it and had it thoroughly defined. This was something new. Moisseiff understood it, and Ellis, and that was about all. Derleth, who admitted it, did not understand it; nor did Strauss, who never would have admitted anything of the kind. Ellis, working now mostly in Chicago with the Strauss staff, and in consultations with Moisseiff in New York, was now the man completely on the line.

An engineer, any engineer, is a man always working within the state of the art--the sum of the achieved knowledge and accrued experience of his time. Ellis was now going to jump across it--go beyond the state of the art. Assumptions had been made about the way certain things must be done. Now Ellis was challenging those assumptions.

According to Ellis, any young engineer who has mastered the fundamental theories of moments, shears, and equilibrium or coplanar forces should have little difficulty in computing the stresses in a plate girder or simple truss. With a few years of experience, he can design a girder or truss himself. "He will find that it is but a step further to the analysis and design of a cantilever bridge." If he advances to the study of an arch, the engineer meets one or two interesting new problems, but nothing for which his earlier work has not prepared him. When he comes to the stress analysis of a suspension span, however, the engineer "will find himself in a totally undiscovered country."

In undergraduate courses in bridge analysis, applications of fundamental principles are almost always based on the beam, plate girder, and simple truss types of bridges. The reason for this is that an engineering student, in order to develop suspension bridge theory, must of necessity have had more mathematics than is given in a four-year undergraduate program. "Of course, the young engineer can take the formulas as given in treatises on suspension bridges, but his success will be very doubtful, because any person who uses a formula which he cannot develop is on dangerous ground."

Even an engineer who has had sufficient math to master suspension bridge theory and understand it thoroughly will find, on his first application of it to a real problem, "some very mysterious things."

To begin with, a suspension bridge does not behave like any other type of bridge. It behaves like a clothesline: it sways with the wind, expands in the sunshine, and contracts when it rains or grows cold. The suspension bridge engineer is being asked to design a clothesline capable of supporting six 24-ton trucks, traveling abreast.

In every aspect of its design, a suspension bridge challenges the traditional assumptions of bridge design, but in ways that require an understanding and appreciation, an informed view, of the principles being challenged. Some things considered essential become negligible. Other things, considered negligible, become crucial.

As to the wind stresses, according to Ellis, there had been no theory published anywhere that was even approximately correct for the analysis of them. At least there hadn't been until Moisseiff's breakthrough idea of balancing the horizontal deflections in the cable of a suspension bridge with those in the stiffening truss. "Now it takes a genius to discover a new idea," Ellis conceded, "but it doesn't take very many brains to make slight improvements in it."

Ellis's calculations had to take into account not only the shape and structure of the bridge and the forces of heat, cold, and winds, but had to calculate, in advance, the amount of stress to be borne by each of the bridge's hundreds of suspender ropes. He was not merely entering unexplored territory; he was mapping it.

In his computations for the bridge towers, Ellis was also able to incorporate recent findings that further improved the bridge's lightness, flexibility, and economy. "In considering any type of bridge," Ellis explained, "the real test is not whether the structure can be designed, nor indeed whether it can be fabricated in the bridge fabricating plants. The real test is whether the bridge, after it is fabricated, can be shipped piecemeal to the site and there be put together.

At this time, and for this type of bridge, the state of the art was the Philadelphia-Camden Bridge towers, which were made of steel cells. Ellis proposed for the Golden Gate Bridge towers that were similar in structural form, although very much larger. These could be made from either carbon or silicon steel, depending on the anticipated dead and live load. Ellis based his calculations on a live load of 4,000 pounds per linear foot, a figure reckoned on an extreme estimate of the bridge being packed full of the heaviest automobiles imaginable--sixteen-cylinder Cadillacs, seven-passenger Packards and Buicks--bumper to bumper, and jamming all six lanes. Under these conditions, Ellis calculated the total live load would be slightly more than 3,000 pounds per linear foot. Of course, in practice, there would probably be four or five lighter Fords, Chevies, or Plymouths to every seven-passenger stretch sedan on the bridge, and the distance between cars, even in heavy traffic, would probably be ten or fifteen feet, so that in reality it would be difficult to imagine a load greater than 1,000 or 1,500 pounds per linear foot. Nevertheless, the towers were designed for a load over all spans of 4,000 pounds per linear foot.

In addition to the live and dead loads on the towers, allowances had to be made for bending, both lengthwise and sideways. When the center span of the bridge and the far side span were fully loaded and the temperature was lowest--in this climate, say, 30 degrees--the top of the tower adjoining the unloaded side span would bend 23 1/3 inches toward the channel. When the near side span was loaded, with no load on the center and far side spans and the temperature was the highest, at, say, 100 degrees, the top of the tower would bend 6 2/3 inches toward land. The total deflection, then, would be 30 inches. Ellis suggested that this be equalized when the tower was completed. "One way of doing this will be to attach lines to the top of the tower and literally pull it shoreward nine inches and hold it there while the cables are being placed on the tower." There would then be a uniform lengthwise deflection at the top of the tower of 15 inches.

The towers would also have to bend sideways, because of wind load. The bridge had already been designed for a wind load of thirty pounds per square foot of exposed surface on the floor system, including trusses and cables. And for fifty pounds per square foot on the exposed surface of the towers. The maximum actual velocity--as opposed to indicated velocity--of wind ever recorded by the weather bureau at San Francisco was, according to Ellis, fifty miles per hour, equivalent to a wind pressure of ten pounds per square foot. A wind pressure of thirty pounds per square foot is officially classified as a hurricane, strong enough to overturn empty boxcars. "We have used this wind load, " Ellis commented dryly, "in designing the towers and trusses. This wind load causes transverse moments in the towers and also stresses due to the eccentric loading." Under a hurricane wind pressure of thirty pounds per square foot, the towers would deflect transversely, move sideways, about five inches. One can almost hear Ellis add in: big deal.

Ellis then considered what the stresses would be if all four of these conditions were combined at the same time. If, under some bizarre combination of circumstances, the whole bridge could be loaded with traffic to the weight of 4,000 pounds per linear foot but with only the center span and one of the side spans so loaded; and if, in addition, the temperature was either down to 32 or up to 110; and if this whole dead and live load were subjected to a wind strong enough to tip over a line of boxcars a mile long--if these four unlikely and even contradictory conditions should occur simultaneously, what would happen then?

"In an ordinary design," explained Ellis, "where dead and live loads are combined with wind loads, it is customary to increase the allowed unit stresses by at least twenty-five percent. This is done in all standard specifications. In this design, however, we have assumed that all these four extreme conditions may happen simultaneously. The structure has been designed so that in no part of the towers will the allowed unit stresses be equalled or exceeded, even should all four of these conditions happen at the same instant."

200

On March 1, Ellis had sent copies of his preliminary design specifications to the members of the Engineering Board for their comments. The engineers had approved of these early designs and figures, with certain specific changes. Strauss now directed Ellis to devote all his time to the detail computations. "Before I return West," the chief engineer wrote the three consultants, "sufficient progress should have been made on the general design to enable us to jointly determine a fairly definite schedule."

While Ellis, in Chicago, labored with the thousands of calculations for the suspension ropes, highway deck, floor beams, highway track and sidewalk "stringers," or longitudinal members, stiffening trusses, cables, and towers, Moisseiff, in New York, worked on calculations of his own. The two men checked figures with each other frequently. It was an enormous and complicated job . . .

Ellis was being forced to accomplish an enormous amount of complicated and important thought in a very short time. "The pressure was constantly being brought to bear to rush the work," he confided later to Ammann and Moisseiff, "and give a definite date of completion. I always explained"--presumably to Strauss, his boss and the logical applier of the pressure--"as clearly as I could that it was impossible to make any estimate which could be relied upon; that this was not just another structure to be designed similar to many others in which the computations were more or less a mechanical process, but one that required considerable original thought."

Strauss was growing impatient with these delays. Undoubtedly he was feeling considerable heat from the Bridge District directors, who were now promoting an invisible bridge, one for which they had no agreed-upon, presentable design. He was also under considerable financial pressure. "Although I have made many sacrifices," he wrote to Moisseiff, who had complained about slow pay, "in order to insure success at the bond election, and the cost has become a considerable burden, I have not asked and do not now ask anyone else to make any sacrifices." As an employer, Strauss was sensitive to being imposed upon or taken advantage of. Impatient with theory, he had begun to feel that Ellis was magnifying his role by fussing unnecessarily over details. Hadn't Strauss put his design staff at Ellis's disposal? Couldn't Ellis delegate some of these detailed computations?

Ellis had indeed assigned some of the computations to a Strauss staff engineer, Charles Clarahan, who had been a student of Ellis's at the University of Illinois. He was, in Ellis's view, "an able chap, but this was, to the best of my knowledge, his first introduction to a problem of this nature." Clarahan, working on the computations for the bridge towers, had run into difficulty. He came to Ellis for advice. Ellis, under time pressure and busy with the bridge trusses, had intended to tackle the towers later. He told Clarahan that it would be best if, for the present, they continued to work independently, as Strauss had instructed. Clarahan spent some three months on his tower computations. When they were finished, he presented his results to Ellis. "You or I," Ellis wrote to Moisseiff, "would not have to look at them ten seconds to know absolutely that they are very much in error." Ellis did not mention this to Clarahan.

*   *   *

There continued to be irritating interruptions and delays. Professor Lawson, the geologist, was recommending that the pier base go down twenty-five feet below the rock bottom surface; Strauss and the Engineering Board were trying to persuade him to settle for ten. There was a question of earthquake insurance and the desirability of an additional check of all computations on the bridge, to be conducted by Moisseiff and his staff. Nevertheless, Strauss assured Ammann, "the work at Chicago is progressing at full speed," and Strauss was diligently at work on his engineer's report, which was due, along with the completed plans and estimates, "on or about the 20th of June."

At this point, the time constraints worked both for and against the engineers designing the bridge, particularly Ellis and Moisseiff. While pressured with deadlines, they were also able to employ, to their advantage, the blank paper syndrome. Most people, confronted with a creative or design problem, are physically incapable of putting down, on paper, a solution to it. The very emptiness of the space becomes intimidating. Yet many of these same individuals, once an idea has been conceived and committed to paper by someone else, suddenly find release. Examining a preliminary drawing, often with pen or pencil already in hand, the consciousness that was blocked before now finds itself flooded with minor improvements that can be made, ways the bare-bones idea can be adorned, corrected, embellished, redirected, saved. They are suddenly eloquent.

The more time there is to do this, the more the formerly blocked consciousness may be stroked. But because of the pressures--financial, political, procedural--surrounding the design of the Golden Gate Bridge, the time and trouble required for this customary ego stroking were denied. Individuals who would have found the bridge design regrettably flawed or incomplete and in need of their personal improvement and embellishment were forced, because of the constraints of time, to let the design pass relatively unscathed. In short, they were desperate enough to buy something good.

There is another aspect to this, and it was definitely operating in the design of the bridge. Excellence stimulates excellence. The fact that an associate has done--in some cases, gotten away with--something exceptional can encourage others around him to try the same. The standard of the acceptable, the normal, is raised a notch, and the dimensions of the possible expand. The wall had been pushed back.. Moisseiff's theory had done this for Ellis; now Ellis had the opportunity and the obligation to meet the standards of dedication and imagination that had been applied to the new theory of long-span suspension bridge design. The side must not be let down. The leap, once made, must be sustained. There is no turning back or reining in without plummeting to failure.

The most striking feature of the new bridge would undoubtedly be its towers, the tallest bridge towers ever built, rising in one of the world's most dramatic settings. Here was an opportunity and a challenge to move beyond traditional engineering design into something that could become a fulfillment of, rather than an intrusion on, its environment.

Bridge towers, even those used in suspension bridges, had traditionally been massive structures, great bulking works of steel or stone or combinations of both. In a typical design, two posts, rising from a pier, would be joined by a great arch over the roadway, or by a webwork of giant crossbeams, or again, by a combination of the two. In order to preserve the feeling of massiveness and

202

solidity, tower details were often designed to be larger at the top than at the base to preserve the look of uniformity against the eye's adjustments for distance and perspective. Although Moisseiff's and Lienhard's theories had already changed the nature of suspension bridge cables, suspenders, and trusses, tower design tended to lag behind. The towers of the George Washington Bridge, for example, were still the traditional Erector-Set crosshatch of exposed steel beams.

At the same time, changes in metallurgy, extending the allowable narrowness and flexibility of suspension bridges, were suggesting new, lighter, less cluttered possibilities for bridge towers.

Ellis, the author of a textbook on the theory of frame structures, was keenly aware of these developments in his field. He was working with the man most responsible for them. Now was his chance to advance the state of the art of tower design. In the Ambassador Bridge, the giant cross-members had been extended below the roadbed; now, why not confine them there entirely? The Golden Gate had to have the highest over-water clearance of any bridge ever built: there would be room for two huge cross-members, with cross-beams, between the pier and the roadbed. With such a strong base, the towers above the roadbed, already the tallest of any bridge, could be made to seem even lighter, longer, and more graceful. If the giant cross-members were dropped from the towers above the roadbed, they could be replaced by the smaller rows of cross-beams at intervals between the towers and covered by aluminum alloy housings, which would greatly reduce corrosion and remove the last trace of Erector-Set mechanics from the superstructure of the bridge. Where the giant cross-members between the towers had been, at Philadelphia and at Detroit, there would now be great open rectangles, huge windows or portals framing the intense blue sky, the scudding clouds, and the frequent swirling fogs of the Golden Gate.

The idea of the portals and the use of cellular construction inspired another innovation, this one the suggestion of Joseph Strauss. "Referring to the towers," Strauss wrote to Ammann in March, "I have in mind the stepped-off type of architecture."

The practice of stepping off, or indenting structures as they rose in order to accentuate the feeling of soaring height, was a practice increasingly in vogue in other kinds of architecture, particularly the Chrysler Building and the proposed Empire State Building in New York.

At the Strauss offices in Chicago, Ellis, in addition to calculating the dimensions of the materials for the bridge, was also computing estimates of their costs. Using figures from the George Washington Bridge job, which Ellis felt were too high, he had to make estimates for everything from the per-yard cost of excavation to the wire wrapping used on the cables. These figures also had to be checked by Moisseiff, and the added time this took further delayed the completion of the final plans and estimates, which in turn further irritated Strauss. The June 20 due date came and went; so did July 20. "I have not yet received from Mr. Ellis the final total cost," Strauss wrote Ammann on July 28, "nor can this be done until the Consultants check estimate is completed." He was determined to submit his engineers' report at the next meeting of the board in August, even if it meant that the amount of the total cost had to be left out. The possibility of completing the bridge plans in time to hold a special election on the bond issue had vanished; now they would have to struggle to make the statewide election in November.

While the engineering and design plans were in progress, an eager watch was kept on them by the Bridge District Board of Directors, There was first of all Derleth, who, through MacDonald, was keeping the directors informed of developments within the Engineering Board. In addition, there was the situation, common among directorships, in which a few members lead the way, with the rest of the board generally falling into line. Richard J. Welch in particular, with his moral authority as the earliest elected advocate of the project and his political clout as a congressman, seems to have been the official liaison between the directors and their engineers at this point. There was a certain grass-roots appropriateness to this: Welch had originally entered San Francisco politics from a job in an ironworks.

On July 9, Welch assured the directors at their monthly meeting that design work was proceeding rapidly and that prospects were bright for wrapping up final details "preliminary to actual construction work on the bridge structure." The directors were understandably anxious: it was now only four months to the next general election, for which they expected to propose a bond issue for a bridge whose design they did not yet have in hand. Welch may not have relieved these anxieties by reminding the directors that the project had now assumed a nationwide importance, or by predicting that the electors of the district would vote the bonds by an overwhelming majority. There still had to be a bridge.

Thus it was with understandable relief that on August 21 a call was issued for a special meeting of the board to be held on August 27 in the chambers of the Board of Supervisors at the San Francisco City Hall. Its purpose was (1) to receive, consider, file, and adopt the report of the chief engineer, and (2) to call a special election, submitting to the voters of the district a proposition to issue $35 million in bonds.

On the appointed day, it was Joseph Strauss, accompanied by Derleth who presented the engineer's report to the board. They were also joined by MacDonald, the general manager, and Harlan, the Bridge District's legal counsel. Neither Ellis nor Moisseiff was present.

The report, together with architectural studies and the fact-finding investigation conducted by the Board of Directors, runs to some 285 pages and includes everything from a prefatory eulogy by the San Francisco poet George Sterling, who had committed suicide in 1926, to engineering drawings and specifications, to cost breakdowns, to Lawson's geologist's report. Capitalizing on the nine years of promotion, frustration, negotiation, and imagination that had preceded his report, Strauss had approached the question of the bridge from every conceivable angle. Each of the elements or factions anticipated as being in opposition to the bridge is addressed with a full armory of factual weaponry: the antidevelopment interests, the shipping interests, the ferry boat interests, the taxpayer, military, political, aesthetic, earthquake, finance, travel, traffic, commerce and future concerns of an entire metropolitan area are touched upon in this document. Clear, direct, long on information, mercifully short on grandiloquence, yet with a firm sense of the emotions involved and a sure grasp of the momentousness of what is being proposed, the report is an example of Strauss's generalship at its most effective. Here he had marshaled his promoter's positive vigor, his salesman's sense of the territory, his Napoleonic ego, and his ability to recognize and attract to his cause people of genuine ability, all in behalf of the great work he had yearned to be part of since his boyhood days in Cincinnati.

There is, however, another aspect to this report, and it sheds a certain light on why engineers of solid credentials and national reputations had such misgivings about working with Strauss.

In volume I, section I, article 8, which describes the engineering personnel proposed for the bridge, Strauss is listed as possessing the "degree of C.E., University of Cincinnati, 1892." While it might be claimed that this referred only to Strauss's liberal arts B.A. in commerce and engineering, it is obviously intended to show comparable academic status to the "degree in C.E." listing for O. H. Ammann, Charles Derleth, Jr., and Leon Moisseiff, all three civil engineers of national distinction.

Additionally, Strauss, in the same article, is described as consultant to the New York Port Authority on the Bayonne Arch, which was to be completed in 1931, and the Hudson River (George Washington) bridge, which was to open in 1933. According to John Hughes, publicist for the NYPA, Strauss was not involved with the George Washington Bridge at all, had no bearing on it, and is not listed among its engineers. And while there are no records of consulting engineers on the Bayonne Arch, the NYPA's staff expert on the bridges, whose experience goes back more than 30 years, "has never heard of Strauss being connected with that or any of our other projects."

Since both bridges were directed by Ammann and, in part, designed by Moisseiff, perhaps Strauss felt he could claim consulting engineer credit for having offered a few opinions and suggestions. Or he may have assigned Ellis, his employee, to work with the other two civil engineers and, through the Strauss firm, assumed credit for Ellis's work. What is clear is that Strauss, especially sensitive on the grounds of professional credentials and attributed work, was more than capable of ethical misdemeanors and sharp practice.

There is, in this report, which was written by Strauss, no indication of Ellis's role in conceiving the overall design for the Golden Gate Bridge. Instead Ellis is described, along with Clifford Paine, another Strauss engineer, as one of Strauss's two chief assistants on the project. There is no special thanks, no individual credit, no division of labor as to who was responsible for what. The bridge design is represented as being the work of the Strauss Engineering Corporation: in other words, Joseph Strauss.

What had been achieved, up until now, in advancing the bridge project had been due almost entirely to the promoter in Strauss. The speeches, the newspaper articles and editorials, the crafting of a government framework with which to manage and finance the project, the crucial approvals by county, state, and federal governments--all this had been accomplished with precious little engineering and critical analysis. It had come largely on the authority of Strauss's reputation as an engineer and on the basis of hastily drawn estimates, erroneous ground surveys, and a paper sketch of an ugly structure of mixed parentage. Now Strauss's dream of a great bridge was close to being funded. The desired state of assured financing would require a jaundiced appraisal of the technical foundations of a project that took it beyond the limits of Strauss's ability as an engineer. What was required now was the exacting and inspired thought of an engineer who comprehended both the theory and technology necessary to make the dream come true, who could make possible a bridge within the feasible price range Strauss had recklessly promised. In Charles Ellis, Strauss had found the man who had delivered an engineering design that had surpassed Strauss's most fervent hopes. Yet just as it was not in Strauss himself to conceive such a design, it was not in him to delegate or even share recognition for it. In assuming full credit for what was actually Ellis's work, Strauss the promoter, now fully ascendant, had overruled Strauss the engineer.

Moisseiff, who understood the significance of Ellis's work more than anyone else, fully appreciated the irony and potential unfairness of this situation. On August 19 he wrote two versions of a letter expressing his verification of the fundamental data on which the design of the main structure was based.

The first version was addressed to William Filmer, president of the Bridge District. It begins:

> "Dear Sir:
>
> In accordance with the request of your Board of Directors, I beg to submit to you herewith, the results of the independent check made by me of the design and preliminary plans and specifications of the Golden Gate Bridge.
>
> I and the engineering staff under my direction have carefully examined and checked the design of the Golden Gate Bridge as prepared by Joseph B. Strauss, Chief Engineer. I have also checked the estimated quantities of this design as well as the estimated cost.
>
> For this purpose, Mr. Strauss has submitted to me forty-one drawings together with photostats of his computations and estimates."

Moisseiff goes on to summarize his verification process, and then concludes,

> "I take this opportunity to express my gratification that the work done by your Chief Engineer and the staff under his direction show careful consideration of the problems involved and has produced a good, workable design of the bridge."
>
> > Respectfully submitted,
> > Leon S. Moisseiff
> > Consulting Engineer

The second version of the letter was written to Strauss, Ammann, and Derleth, his fellow engineers. It began like this:

> "Gentlemen:
>
> In accordance with our understanding at the meeting of the Board of Consulting Engineers, in Chicago, on June 11th, and 12th,

1930, I and the engineering staff under my direction have
carefully examined and checked the design of the Golden Gate
Bridge as prepared by the Strauss Engineering Corporation under
the direction of Mr. Charles A. Ellis. I have also checked the
estimated quantities of this design as well as the estimated cost.

For this purpose, Mr. Ellis has submitted to me forty-one
drawings together with photostats of his computations and
estimates."

This version concludes:

"I take this opportunity to express my gratification that
the work done by Mr. Ellis and under his direction with the staff
of the Strauss Engineering Corporation has shown careful
consideration of the problems involved and has produced a good,
workable design for the bridge."

The first version is what was required according to the rules of accepted
business practice. The second version was Moisseiff's attempt to set the
record straight. As a consultant, in private practice, there was a practical set
of rules Moisseiff had to live by; as a civil engineer, a professional dedicated to
the primacy of fact, there was another set of ethics he had to live with; by
writing two versions of this letter, Moisseiff had attempted to honor both.

The directors, thrilled and relieved at being given such a tangible, promotable,
detailed presentation of their proposed bridge, resolved to produce 2,500
copies of the engineer's report for distribution. They also resolved that, it being
necessary for the Golden Gate Bridge and Highway District to incur a bonded
indebtedness, a special election would be called, for the purpose of submitting
the proposition of issuing bonds to the electors, to be held the same day as the
statewide election, on November 4, 1930.

Now, at last, Strauss's army had its flag, the image of a bridge unlike any
other in the world. From above, the perspective adopted in the Maynard
Dixon painting, it was a narrow ribbon, elongated as though stretched in a tug-
of-war between two mountainous and muscular points of land. Viewed head
on from the roadway, the towers, with their ladderlike rise, carried the eye
and the spirit upward; in profile, the towers shrank to slender ellipses, and the
great cables and long suspenders turned the bridge into a giant harp, hung in
the western sky. There was, in this bridge, no bad side, no unflattering
perspective, no camouflaged ugly hodgepodge of mechanistic works. The most
powerful emotional argument against the bridge--its looks--now worked in its
behalf.

The proponents of the bridge--Strauss, MacDonald, the directors, Derleth, and
others--seem to have seized immediately upon its metaphoric power: this
bridge, as designed, was a statement of faith in the future, radiating
confidence amid the bleakest of American economic times. The country, this
bridge declared, was not at the end of things economically any more than it
was geographically, but instead at the beginning of something new,
unobstructed by the past, with renewed aspirations to excellence.

*   *   *

Meanwhile, as the promotional effort in behalf of the bridge was intensifying, work continued on the details of the bridge design itself. Proposals must now be turned into specifications, overall formulas broken down into detailed computations, plans expanded into blueprints. This work was now being done in the Strauss offices in Chicago, under the direction of Ellis.

Derleth, the seasoned veteran of years of classroom discussion and debate, became the engineering point-man for the probridge forces, dispatched to address groups where questions might be raised requiring specific, quotable answers. . . . Then, in September, Derleth was presented with an opportunity to raise both appreciation of the bridge's design and the discussion concerning its practicality.

On September 18, the first conference of the National Academy of Sciences ever held on the Pacific Coast was to be convened on the University of California campus at Berkeley. Almost one hundred of the most prominent scientists in America would be attending the conference, which would include discussions covering scientific subjects ranging from anthropology to molecular physics. It occurred to Dr. Robert G. Sproul, president of the university, host for the occasion, and an enthusiastic supporter of the bridge, that the conference represented a unique opportunity to present, describe, and explicate the "insoluble" engineering problems that had been overcome in the design of a bridge at the Golden Gate, a site dramatically visible from the California campus. Sproul proposed the idea to Derleth, who responded enthusiastically: not only would such a gathering provide an ideal forum for presenting the engineering and design story behind the bridge, Derleth could arrange for the address to be made by the man most qualified to tell it, the design engineer, Professor Charles Anton Ellis.

On the opening morning of the conference, the scientists gathered at the Greek Revival Hearst Memorial Mining Building on the university campus. After a welcoming address by Sproul and a response by Dr. Thomas Hunt Morgan, president of the academy, Derleth spoke briefly, introducing Ellis.

Ellis's speech ran the better part of an hour. Its tone was modest and understated; Ellis's delivery was droll, wry, anecdotal. Like a practiced university lecturer, he led his class step by step through the thickets of theory, pausing to point out reassuringly familiar landmarks along the way. The design of a suspension bridge was no great mystery--once you knew what you were doing, Ellis assured the scientists in his own dry way. Then he proceeded to lead them through a ten-minute explanation of Moisseiff's theory of wind stresses and his application of it to this bridge.

This was Ellis's subject; the idea he had been incubating all these months now had been hatched and was seen to be not only perfectly formed but with the possibility of becoming aesthetically beautiful. His pride in what he had done, his absorption and application, in an era before computers, of a phenomenal amount of theory and mathematics in an incredibly short time, his role in applying all he had learned about frame structures to the design of a work of engineering that also promised to be a work of art, could not be entirely concealed. Although in his talk Ellis deferred repeatedly to Strauss, referring to him, with respect, as "Mr. Strauss," treating him always as the initiator and overseeing eye of the project, when it came time to talk of the design itself, the author in Charles Ellis stepped forward: "at this point, Mr. Strauss gave me some pencils and a pad of paper and told me to go to work."

To Joseph Strauss, a man with an ego as touchy as an inflamed corn, the news that a man who worked for him, an underling, was to address a gathering of the most prominent scientists in America on the subject of his bridge must have been felt with the excruciating pain of an elephant's tread. All the old rejections--on his shortness, his Jewishness, his lack of peer or academic recognition as an engineer--must have welled up in him. Under such circumstances, before such an audience, to such a personality as Strauss's, Ellis's statement about taking a pad of paper and going to work could have appeared to be an attempt to claim sole credit for conceiving a bridge already a step beyond any ever designed before. It was also uncomfortably close to the truth. The design of this bridge had entered realms of theory and mathematics in which Strauss was entirely unfamiliar; it had replaced the earlier, cherished, and patented design of his own; in order to be part of it he had accepted the shared authority of a consulting board and a reduced rate of pay. Now he was faced with the possibility of being denied conceptual credit for the bridge at all. It was something that Strauss, a dedicated scrapper, would have to do something about.

Feelings of this nature, rooted so deeply in the personalities involved, tend to put every aspect of a relationship on edge. Every word, each gesture becomes part of a pattern, an attitude, a scheme of some sort that reduces the usual complex mix of emotions that people inspire in each other to a simple, final, and usually unfavorable judgement.

On October 16, before another important audience, there was another blurted admission of who, exactly, had done what, and this time Strauss was present. The Commonwealth Club of California, which conducted the most influential speakers' forum in San Francisco, had commissioned an independent check of the plans for the bridge. The conclusions, viewed from the point of traditional engineering practice, were critical of the bridge's design in several important aspects, and the chief engineer was invited to respond. A luncheon crowd including prominent civic and business figures heard Strauss open with a few digs at the motives of the outside experts responsible for the report, then settled back to listen to Derleth respond to the specifics.

Speaking impromptu, his words taken down by a stenographer, Derleth addressed himself to . . . issues raised by the report: . . . the questions of overall cost, and the threat of earthquakes. "You know," said Derleth, with an edged amiability,

>"all projects of this kind seem to arouse groups of antagonists. I call them my friends, the objectors. Oh, I met so many of them and they made good friends for me when we built, against terrific obstructions, the Carquinez Bridge. And we expect the same obstruction here. And we must do missionary work, and I, as a professor, must help to instruct these men that their fears should be alleviated."

*   *   *

Derleth turned to the matter of earthquakes. The panel investigating the plans had cited Professor Lawson's geology report on the lack of tensile strength in the rock at the San Francisco anchorage and suggested that the bridge, oscillating in an earthquake, might rip that anchorage out. The gentlemen did not know, said Derleth, "that we have made two designs for the anchorage, one of which, in accordance with Professor Lawson's recommendations, is a gravity anchorage, which makes it entirely unnecessary to consider the sealing strengths in these terrifically unfavorable rocks."

If, because of an earthquake, the bridge were to fall, Derleth advised his audience, "you won't care. There will be nothing else left."

On the matter of costs, the engineering critics had pointed to the lightweight stiffening trusses proposed for the bridge and suggested that, in comparison to those used in other bridges, they had been designed to be too light, in the interests of lowering the estimate. They were approaching the "mysterious things" associated with suspension bridges that were so perplexing in terms of conventional bridge engineering. Here, Derleth, not all that comfortable with the details of the bridge's design himself, felt obliged to present outside credentials.

The original design for the bridge, he explained, "was made in Chicago, under Mr. Strauss's direction and ably assisted by a man named Charles A. Ellis, who stands high as a structural bridge engineer." Ellis, he explained, had been a professor of engineering first at Michigan and then at the University of Illinois, "where they have some of the best-known experts in the world. From that he went to Mr. Strauss's Engineering Corporation."

"Mr. Ellis," Derleth declared, as Strauss and the audience listened, "was in charge of that first design. And then those results were taken to New York. There are engineers here who will corroborate what I am about to tell you, that Mr. Leon Moisseiff is recognized in Europe and America as the greatest living analyst of stress and computation in long span bridges. And he is the one who has made the advances possible in long span work, so that now we can build suspension bridges of long spans at costs which are far below what anyone dreamed of when they used the old methods."

The new lighter trusses specified for the bridge by Ellis were in accordance with this new theory and could only be considered insufficient by people measuring the bridge by materials used in other bridges built according to antiquated thinking, Derleth said. Viewed in this way, it was only natural for "some of my friends here, who have collaborated in this report," to look at the design for the Golden Gate Bridge and conclude, "Here is a mistake, here is an error. Somebody made a blunder."

At last, responsibility for the bridge's design had been revealed. Strauss, seated at the speakers' table, was publicly endorsing it. The public admission of his own ignorance of the theory and computation behind the design of his most famous work must have made him cringe. He would devote a consider-able part of his energies for the rest of his life to the cosmetic task of papering it over.

"Never My Belly to a Desk"

An Excerpt from **Hoover Dam - An American Adventure**
Joseph E. Stevens

*If there is a stereotype of the modern civil engineer it is probably the image of
the hard-hatted construction boss, on the job in the field, supervising men,
directing the use of construction equipment, studying blueprints, and generally
overseeing a beehive of activity. While there are civil engineers who, in fact, are
employed in such positions, their number is probably not now as large as it once
was. Much of the construction superintendent's work is now done in the office,
and much of the responsibility and authority that was once concentrated in the
hands of a single individual is now distributed to a number of persons, each
with a somewhat smaller area of expertise and authority. Nevertheless, the
construction field is still the place to be employed for the civil engineer who
seeks the challenge of day-to-day involvement with bringing projects into
existence.*

*This selection provides a biographic sketch of a man who, perhaps more than
any other American civil engineer, created the image that became the stereotype.
Frank Crowe was the foremost construction engineer in the country, if not the
world, at the time of the Depression. There may have been civil engineers
smarter than Frank Crowe, or civil engineers who worked harder than Frank
Crowe, and there may even have been civil engineers who achieved more than
Frank Crowe. However, few civil engineers, before or since, have found
themselves at center stage, with the attention of a whole nation focused on the
performance of their engineering duties, and none have ever delivered any better
than Frank Crowe.*

Within the construction fraternity, a group not noted for paying deference to
anyone or anything, [Frank T.] Crowe commanded respect bordering on awe.
He had gained his reputation through hard work, dedication to his craft, and
consistent success bringing difficult projects in on schedule and under budget.
His drive, ingenuity, organizational flair, and physical toughness epitomized
the can-do spirit held in highest esteem by field engineers, and his efficiency,
punctuality, and instinctive frugality endeared him to his employers, who
profited handsomely from his efforts. Crowe's loyalty to the men he
commanded on the job won him a large and dedicated following of construction
workers who went from project to project with him and affectionately called
him "the old man" even though he was only in his forties. "The best
construction man I've ever known," Frank Weymouth, a brilliant engineer and
builder in his own right, said of Crowe. "Nothing stumps him. He finds the
way out of every difficulty. And he is not conceited."

Crowe's father, a woolen mill operator who came to the United States from
England in 1869, had hoped his son would enter the ministry, but
mathematics and science appealed more to the young man than did religion,
and he entered the University of Maine in 1901 to study civil engineering.
During his junior year he attended a lecture about the work of the U. S.
Reclamation Service and was so taken with the romantic vision of a
magnificent but still wild West that, when the lecture was over, he asked
the speaker for a job. The lecturer, none other than Frank Weymouth, obliged
him, and that summer Crowe worked on a Reclamation Service crew

211

surveying the drainage basin of the Yellowstone River near Glendive, Montana. Smitten by open western landscapes and the rugged outdoor lifestyle of the field engineer, Crowe finished his degree requirements in 1905, and without waiting to attend commencement exercises or receive his diploma, he returned to the Reclamation Service and the Mountain West.

During the next twenty years the engineer from Maine worked on a variety of projects in Idaho, Wyoming, and Montana, earning rapid promotion and recognition as the government's best construction man. It was at the Reclamation Service's most glamorous and technically challenging work--the construction of high concrete dams--that he excelled, demonstrating both his engineering skills and his managerial and leadership abilities. These jobs appealed to Crowe not just because of their size but because they put him on the cutting edge of his profession; dam building was being revolutionized by dramatic advances in concrete technology and by introduction of motorized earth-moving equipment and pneumatic excavating tools. In 1911, while working as assistant superintendent of construction at Arrowrock Dam, he pioneered the development of two forward-looking mechanical methods that would help make possible a generation of super dams: a pipe grid for transporting cement pneumatically and an overhead cableway system for delivering large quantities of concrete rapidly to any point at a construction site. He was constantly devising new techniques for increasing speed and efficiency and always was on the lookout for equipment that might prove useful. His wife, Linnie, complained that the most interesting thing he could find on his honeymoon in New York City was a large automatic dump truck that unloaded eight tons of coal into a small chute in a matter of minutes without spilling a single lump on the sidewalk.

In 1924, after completing Tieton Dam, near Yakima, Washington, he was appointed general superintendent of construction for the Bureau of Reclamation, making him responsible for construction projects in seventeen western states. He looked forward to new field challenges, but a year after he was promoted a policy change upset his plans: construction work previously handled by government forces now was to be performed by private contractors. Overnight, Crowe's status changed from hands-on builder to paper-shuffling administrator. For a man whose motto was "Never My Belly to a Desk" and who made it an unbreakable rule never to write a letter more than one page long, this was an intolerable state of affairs. His experience in the field and his dealings with other government engineers, such as Arthur Powell Davis, Raymond Walter, and Frank Weymouth, had convinced him that a new generation of dams, far bigger than any then standing, would rise during the next decade. He passionately believed that he was the man to build them, directing construction at the site rather than from an office hundreds of miles away. He was incapable of becoming a desk-bound bureaucrat; the odor of dynamite and concrete was in his nostrils, the music of rock drills sang in his ears, and in 1925 he resigned from the bureau and joined the Morrison-Knudsen Company, which was teaming up with Utah Construction to bid on government dam jobs.

As superintendent of construction for the Utah-Morrison-Knudsen combine, Crowe was completely in his element: in the river bottoms building bigger and better dams, working shoulder to shoulder with men whose drive and ambition matched his own. Guernsey, on the North Platte River in Wyoming, was finished in 1927; Coombe, on California's Bear River, was completed in

1929; and Deadwood, on the Deadwood River in Idaho, was brought in in 1930. With each project Crowe better integrated the construction techniques that allowed him to build bigger faster, and groomed the cadre of workmen that traveled with him from Wyoming to California to Idaho.

The profits earned and the experience gained on these jobs were satisfying, but Crowe knew that Guernsey, Coombe, and Deadwood were merely warmups for the biggest dam of all. He had followed the Boulder and Black Canyon investigations closely, had talked at length about the proposed dam with Arthur Davis, Frank Weymouth, and others, had even made a rough cost estimate for Davis in 1919 and worked with Raymond Walter and Jack Savage on the preliminary designs in the bureau of Reclamation's Denver office in 1924. His interest in Hoover Dam was more than proprietary: it was a burning passion. "I was wild to build this dam," he said years later. "I had spent my life in the river bottoms, and [Hoover] meant a wonderful climax-- the biggest dam ever built by anyone anywhere."

Stevens' book continues with the description of the construction of Hoover Dam, which, of course, is the story of Frank Crowe's greatest hour. He was the centerpin about which the consortium of companies which ultimately became known as Six Companies revolved. His cost estimates were the estimates that won the contract for Six Companies. His leadership as general superintendent of the job for Six Companies was instrumental in the successful completion of the project despite numerous unforeseen problems. Not only did he produce solutions to the technical problems that materialized in the construction of one of the world's largest dams, he also demonstrated a considerable talent for coping with a variety of logistical and managerial problems. Some of those achievements are recounted in selections elsewhere in this book.

In his summary of the significance of Hoover Dam Stevens describes briefly the career of Frank Crowe after the completion of his masterwork.

For the principal author of Six Companies' success in Black Canyon, completion of Hoover Dam was a bittersweet triumph. Frank Crowe had performed brilliantly in an extraordinarily demanding job, rising to every challenge and proving beyond a shadow of a doubt that he was the finest field engineer in the world. In the process he had become wealthy: besides an $18,000-a-year salary, he had earned a bonus of 2.5 percent of profits--between $260,000 and $350,000--a princely sum in the middle of the Depression and enough on which to retire comfortably if he so chose. He was also in a position to capitalize on his new celebrity status by becoming a professor of engineering or a highly paid consultant.

But neither retirement or consulting held any appeal for the man the workers called Hurry Up. *Las Vegas Evening Review-Journal* writer Al Cahlen interviewed Crowe on March 3, 1936, two days after the government's acceptance of the dam and the conclusion of the general superintendent's responsibilities. He expected to find the engineer in a happy, relaxed mood, savoring his accomplishments and looking forward to a well-deserved vacation. Instead he found a man anxious and depressed, beset by doubts and misgivings and desperate to get back to work as soon as possible. "I feel like hell," Crowe told Cahlen. "I'm looking for a job and want to go right on building dams as long as I live. Somehow I can't imagine myself in a big city skyscraper acting as a consulting engineer. . . . I'm going to keep my feet in the dirt someway-- that's where I'm happiest. . . . I've got to find a dam to build somewhere."

Finding work was not Frank Crowe's real concern. He already had a new dam to build for Six Companies--Parker, 155 miles downstream from Black Canyon-- and when that was finished, he could have his pick of the big New Deal construction projects planned for the Columbia River Valley and the Central Valley of California. The real problem was more fundamental: how to fill a gaping void in his life. Hoover Dam had been his obsession for ten years; in his mind it had loomed as the ultimate challenge of his career, the focal point of all his plans and ambitions. From the day in 1925 when he left the Bureau of Reclamation and joined Morrison-Knudsen, all his efforts had been concentrated on preparing himself for the task of directing its construction. Now he had realized his lifelong ambition, but he was only fifty-four and the demons that had driven him from construction site to construction site until he finally reached Black Canyon and the pinnacle of his profession would not let him rest. Frank Crowe stood alone at the apex of the Great Pyramid of the American Desert, unwilling to rest on his laurels, looking desperately for a new job worthy of his talents, realizing that there would never be another challenge that could measure up to Hoover Dam.

To his credit, Crowe did not let the knowledge that Hoover Dam was his crowning achievement affect his performance on subsequent projects. He brought Parker Dam in ahead of schedule in 1938, directed construction of two small dams, Copper Basin and Gene Wash, that were part of the Colorado Aqueduct system, then turned his attention to the plans for Shasta Dam, the key structure in Northern California's huge Central Valley Project. Shasta would rival Hoover in size, and Pacific Constructors, the contracting organization that had won the right to build it, asked Crowe to be the general superintendent. He accepted and spent the next six years working on the big dam, guiding the project to a successful and timely conclusion in spite of manpower and material shortages brought on by World War II.

When Shasta Dam was finished in 1944, Frank Crowe was sixty-two years old and had been engaged in stressful, physically demanding work for almost four decades. During that period he had had virtually no time off, not even weekends, and the continuous strain had taken its toll. In 1945, when the War Department asked him to organize and direct all the reconstruction work in the U.S. Zone of Occupation in Germany, he was eager to accept, but his doctor ordered him to forego the assignment because of his failing health. Reluctantly, Crowe turned the job down and retired to a cattle ranch he had bought in Redding, California, not far from the Shasta Dam site. He died there on February 26, 1946.

Obituaries described Crowe's long and distinguished career in heavy construction and lauded his ability as a field engineer and builder. Eulogists in Las Vegas and Boulder City spoke of his devotion to family, friends, and proteges, of his fairmindedness, of his sense of humor, and of his absolute integrity. But in the end Frank Crowe was remembered for one thing: he was the man who built Hoover Dam. Those closest to him knew that he had never hoped for anything more.

"Who the Hell Does He Think He Is?"

Excerpts from **Big Red**
John Haase

*Novels in which the main character is a civil engineer are extremely rare.
Historical novels about civil engineering projects are even more rare.* **Big Red,**
*by John Haase, is a historical novel which tells the story of the construction of
Boulder Dam. The main character, Frank Crowe, who is the project engineer
for the construction of the dam, was a real person with the kind of experience
indicated in the novel. As is usually the case in historical novels, many of the
events described in* **Big Red** *actually happened.*

*Young engineers usually do not visualize themselves engaged in solving "people
problems" and "money problems" such as those described in this selection. We
like to believe that an engineering career consists of solving the kinds of
technical problems encountered in our engineering courses. For most
engineers, however, the ability to deal with problems other than technical
problems is an essential skill. Observe Frank Crowe as a living, breathing,
laughing, cursing human being concerned not just with how to get a dam built,
but also with his own mortality and his future economic security.*

This selection includes most of the first five chapters of **Big Red.** *It describes in
abbreviated form the planning and other groundwork that precedes most major
projects. The time is June, 1927. Almost five years have elapsed since Secretary
of Commerce Herbert Hoover negotiated the Colorado River Compact with
representatives of the seven states whose boundaries include portions of the
Colorado River (Big Red) watershed (Arizona, California, Colorado, Nevada,
New Mexico, Utah and Wyoming). The compact allocated the water of the
Colorado River Basin among the seven states (and to Mexico) and provided for
the construction of a dam which would make it feasible to store excess water in
years of high flow. However, the compact also required ratification by all seven
states, and Arizona was balking.*

Dawn had always been Frank Crowe's favorite time of the day. He loved the
grays, the blues, the reflection of the mist as the land once more emerged
from darkness. More philosopher than he was willing to admit, he liked the
precision of nature, the perfect renewal of another day. Perhaps it was the
memories of Vermont, even the end of night, which he did not admit he
feared; yet the speculation occurred to him.

He straightened his legs and his torso as he arose from the rattan-covered
couch of the washroom of the railway car. He could have afforded a sleeper,
his wife had urged him this indulgence, but he had spent too many nights on
trains working for the Bureau of Reclamation, pocketing the difference in the
allowance for a sleeper against the fare on coach. He was used to fixing his
eyes on the washbowl in the corner, the metal basket beneath it holding used
towels, the familiar sound of the rattle of the door to the toilet, its hinges
mercilessly beaten by the endless, poorly maintained span of track bed.

He stood, washed his face with cold water, combed his sandy hair, and reached
for his slouch hat, good felt, shapeless after twenty years of wear. He was

forty-four years old, stood six feet four inches, lean and fibrous, a physique that many men would feel was frail.

He was not a physical man in the sense of brute strength or bulging muscles, yet he felt trim, knew that his body could take heat, cold, dampness, rain, that he could sleep on a hard bench, as he had, and dismiss the discomfort with one stretch. He had suffered much exposure to the elements, but it affected him little; there was always a greater purpose which eclipsed the physical discomfort.

He wore denim pants and a beige shirt, knee-high boots, a shapeless jacket, gray flannel, the remnant of some obligatory suit he had needed for some "damn-foolish" event in his long career of supervising engineer.

Crowe stepped out into the aisle and to the rear of the car. The porter had hardly placed the step on the platform when Crowe lighted on it and walked briskly toward the station and entered it. He took out his watch; it was a good Hamilton with a second hand ticking gracefully. He opened it and checked it with the wall clock in the station. Seven-ten. He closed the cover, felt the click of the tiny lock and headed for Mary's diner.

Anyone who had ever worked with Crowe, and thousands who had worked under him, knew that this watch was his master. Punctuality and utilization of time were a fetish with this man. He had been kidded about it, cursed for it, but it was as much of a trademark of his character as his felt hat or his denim work clothes. The latter came from L. L. Bean in Maine, and he liked the sparseness of their cut, like the precision of his watch and perhaps the precision of the dawn. He did not speculate about this as he put distance between himself and the station. He heard the sound of the locomotive as the train pulled out of town. He could not resist turning to watch the engine belch smoke. It was a sight he had beheld since he was a young boy, and he laughed at his constant fascination with the iron horse.

It was seven-twenty in the morning when he arrived in front of Mary's diner, where he was to meet Jack Williams and Tom Wanke. It was ninety-five degrees already, this June day of 1927, and Frank measured the town of Las Vegas briefly.

It was a desert town, raped by mining, short on good soil, depending now almost solely on the railroad shops of the Union Pacific for its income. Heat and despondency showed their effects. He could see no sign of fresh paint, or a seedling tree, or a crisply lettered sign. Main Street, with half its doors shuttered, mirrored the despair of the dying town's inhabitants. All was yellow and gray, dust and tin cans resting by the side of the road.

The evaporative cooler had not been invented yet; people sat on rickety porches, under the faint shade of cottonwood trees, waiting for the day that the desert would swallow them up. "I couldn't raise fifty men here who'd be worth a damn. This merciless heat has seared the will of most of them," Crowe thought to himself.

He entered the diner. Williams and Wanke were seated in the only booth. Crowe was revolted by the odor of the place. It mirrored the decay of the town. Mary's kids sat dolefully on the bar stools playing with a stack of packages of chewing gum, which Mary hoped would be sold when customers paid their bills.

Frank sat down quietly with Williams and Wanke. They shook hands easily and by agreement discussed nothing about their mission.

Jack Williams was portly, of Irish descent. His clothes resembled Crowe's: boots, denims. He sat next to a wooden box, two feet long, a foot high. He wore a leather string tie around his open shirt collar.

"All dressed up, Jack, I see," Crowe commented lightly.

"Yeah, well, Tom and I spent the night here, and thought we'd let the girls know they're dealing with gentlemen."

A waitress appeared and Crowe ordered coffee.

Wanke was almost as tall as Crowe. A studious-looking man, bespectacled, lean, wearing Levi's and a faded sport shirt. "We thought of letting you wait a couple of hours," he said. "Just see how you'd react."

"If you guys think I've mellowed since I quit the Bureau, forget it." He pulled out his watch and opened it. Both Williams and Wanke laughed.

"Okay," Crowe said, "enough bullshit. Let's get on the road."

They walked to the counter and Mary took their money.

Crowe looked at the woman. Perhaps fifty, perhaps twenty-eight, surviving on corn flakes and lemonade, he figured. The features were almost rachitic.

"You wouldn't have a taxi in town," Crowe asked.

Mary laughed. "You're kiddin'. Ain't nobody here's got cab fare. 'Sides, where's there to go?"

"I understand. Anybody's car for hire?"

"I reckon," Mary said. "Go over to Ned's Garage. Might be cheaper to buy one than rent one. This town is full of abandoned cars."

"Thanks very much."

The three men strode purposefully to Ned's Garage, the owner's face so dark from oil it was difficult to guess his age.

"We need a truck for a day. Something that'll hold up in the heat and on some back roads."

"You guys miners?"

217

"I'm looking for a car not a job," Crowe said sternly. "If you got one I'll pay you cash money."

"The only one I can trust is my own. It's a Ford pickup. Runs good. New tires. Five bucks a day."

Crowe reached into his pocket and brought out a worn leather coin purse. He knew Williams and Wanke were watching him as he carefully reached for a five dollar bill. He snapped it to make certain there were not two bills, and handed it to the garage owner.

"I need a deposit," he said.

"What for?" Crowe asked.

"I ain't giving up that car for five bucks."

"How much do you want?"

"Twenty bucks."

Williams reached into his pocket, took out a twenty dollar bill and handed it to the garage man.

"Give me a receipt," he said sternly, "and fill that thing up with gas."

"Yes sir." He fumbled through a maze of bills, road maps, battery cables, looking for a receipt book and a pencil. It was evident that he was much more proficient with a grease gun and a wrench than with paper and pencil.

The three men piled into the cab of the pickup. Crowe looked at his watch. It was eight o'clock as they headed toward Black Canyon. The Nehi thermometer in front of the Cottonwood Cabins was frozen at 105.

*   *   *

It took a scant five minutes to get out of town as they left the pavement and hit the dirt road leading to the river.

"Have a good trip, Frank?" Williams asked.

"Like any other." . . . "How about you boys?"

"No big deal. We stayed at the Colonial, had a few beers at Charley's Saloon. The locals stare you down, get a few drinks into them, and accuse you of being another damned-fool miner."

"We let it go at that." Wanke said.

Crowe nodded. There was silence, the heat intense in the cab of the truck. Finally Crowe broke that silence, looking ahead, the desert almost liquid with sunshine.

218

"Jackson Lake, Wyoming, Tieton in Washington, Arrowrock Dam in Idaho, Flathead in Montana, Deadwood. . ." His voice trailed off. In one sense he had chronicled twenty years of dam building, twenty years of brilliant engineering, and twenty years of camaraderie with Williams and Wanke. Williams was the best tunnel man in the country, Wanke the greatest concrete engineer, and Crowe by reputation the ultimate supervising construction engineer.

It was a bond between three men so deep and secure it needed little reminiscing to shore it up. They were, all three of them, basically silent men, tough, organized, weather-hardened, private.

"They say," Crowe continued, "a man has five dams in him in a lifetime."

"I've heard that too," Wanke said.

"This will be our sixth," Crowe said, swerving to miss a boulder.

"As my water boy said at Deadwood, 'I don't cotton to that bullshit,' when I asked him about voodoo," Williams said.

"I agree," Crowe said. "How about you, Tom?"

"I don't go to church or see a palmist."

Frank Crowe's single most outstanding quality was meticulousness. His Yankee shrewdness and a solid engineering education at the University of Vermont were combined with twenty years of service with the Bureau of Reclamation. Like a surgeon he had mastered the art of making decisions, vital decisions on the spot, and preparing for these decisions before the emergency arose.

This was Crowe's strength. Time and time again he had proven that the most carefully chosen set of plans, the most detailed researches of stresses, the most stringent estimates of weather were inadequate before reality. Each step he took, though honestly and thoughtfully prepared, would need a counter step and the counter step would need a counter step.

Project after project had proven that on-the-spot decisions by Crowe, although often seeming foolish to other engineers, to theoreticians at the drawing board, were nevertheless right. Dead right. "Six months on the job, ten crises down the road," Walker Young once said, "and Crowe's got them eating out of the palm of his hand." And, though Crowe had worked with Williams and Wanke in the wilds of Montana, the heat of Idaho, the floods, the rainstorms of Washington, he could predict their reaction at the first sight of Black Canyon. He remembered his own, and he would never forget it.

Outside of the sheer canyon walls and the rushing river, which were familiar sights to dam builders, Crowe knew what would hit these men: the nothingness, the vast, vast, unrelenting nothingness.

The Nevada desert, not unlike the moon, is the most inhuman stretch of landscape on earth. In the summer, for days and months, the sun rises full-blown from ragged craters in the east, describes its arc and descends unspent behind the craters in the west. The heat is so intense, so merciless, that everything under it grows painful to the touch.

A shovel left unattended resembles a bar of steel emerging from the furnace.

There is no life except for reptiles or vultures, whose black wings swooping over the lower landscape seem like an endless reminder that this is the territory of the dead.

But there is winter. Winds, icy winds whose prodigality is limitless because of the endless playing field, black clouds from which thunderstorms strike furiously, rains so devastating as to almost preclude breathing, frost, floods, all the inclement qualities of weather seem to grandstand in the desert of Nevada.

Frank Crowe knew this desert. He knew Black Canyon. He had traversed it with Herbert Hoover years ago, when Hoover was chairman of the Colorado River Commission; he had surveyed it with Walker Young, with Elwood Mead; and he had camped by the river and lived at the canyon's edge, with his binoculars in his briefcase, his slide rule, his graph paper and fountain pen, his canteen of water, and his poncho.

He knew not only the fury of the weather but also the subtlety of the desert. He knew spring when mesquite bushes sprout with green, yucca raises white flowers, and myriad wildflowers prevail despite the seeming lack of water or the sparsity of nutrients in the soil. It was this annual resurgence, the flowers and the foliage, which gave Crowe courage. He knew the desert was alive, but at this moment, at nine-thirty in the morning, temperature at 115, he could predict the reaction of Wanke and Williams.

The three men stood at the rim of Black Canyon looking at the river eight hundred feet below them, but Crowe prudently would not let them linger long. There was much work to be done this day and endless reflection would only depress his friends.

He sat on the ground, opened his briefcase, and took out his slide rule. The other two men joined him in a tight circle.

"As you guys know," Crowe said quietly, "I'm not a speechmaker. Perhaps they'll say someday that this was the most crucial day in the building of this dam."

He used his slide rule like a baton, punctuating his objective. "I've seen the plans at the Bureau for this dam. They're good. Damned good. Best I've ever seen, and believe me, the best you've ever seen. But, as you know, plans are on paper. Drawn by a lot of kids wet behind their ears, kids who've got a diploma in civil engineering but never owned a pair of boots or slept in a pup tent."

He stopped. "Sure, there's Walker Young, and he's great, and Herbert Hoover, who can't be beat."

"What's troubling you, Frank?" Williams asked.

There was a pause.

"I'm going to build this dam. I am. Not Walker Young, not Herbert Hoover, not the U.S. government or anyone else. I'm going to build that dam. I am, and you are, Tom, and you are, Jack." He looked at each of them for a long time. "I've been over every set of plans, I have read all the data, watched all the projections . . . ."

"Where's the hitch?" Wanke asked.

"To build it I need to know two things. I need two answers, and you can get them for me today. You, Jack, have got to get four tunnels through this rock in less than a year. Four tunnels fifty-six feet in diameter, fully lined, as pretty as the Taj Mahal."

"Fifty-six feet," Williams shouted.

"Fifty-six feet," Crowe reiterated. He looked down at the Colorado. "The locals call that river Big Red, and it's the toughest, meanest river in the world. It can rise seventeen feet in four hours, gouging out tons of earth and making that tame river down there look like flowing red lava. It tore out the Grand Canyon, it built up a riverhead of silt a hundred feet above sea level. It will make every other river we have dammed look like a spring creek."

"Fifty-six feet," Williams repeated, obsessed with the enormity of his job.

"Well, we've all done this before." Crowe looked at them and they nodded. "I want you, Jack, to go to the end of Black Canyon and blast test holes. Now. I don't have to tell you what to do.

"And you, Tom, you take the truck and find a gravel bed. The best gravel, the closest, and the greatest quantity. The Bureau claims it's five miles north of the Arizona border. Check it out.

"And, while you guys are cavorting in the beautiful desert, I'm going to make my last set of determinates. With all their plans, the Bureau has made one fatal error."

"What's that?"

"They want to bring the dam in at 560 feet. My figure, and I have figured it over and over, is 726 feet."

"One hundred sixty feet will add another year," Wanke said.

For the first time Crowe smiled. "Not the way I figure it. Get off your ass. I'll tell you all about it at dinner. Right now it's academic."

After twenty years of working with Crowe, both Williams and Wanke knew the conversation was over. Wanke jumped into the truck and said to Williams, "I'll take you as close as I can to where you're going."

It was deathly quiet when the sound of the pickup faded in the distance. Crowe set up his tripod and transit and slung a pair of Zeiss binoculars around

221

his neck. Carefully he wiped the lenses of the binoculars, adjusted the sights, and slowly he moved along the canyon wall across the river until he found the faded red markers he had staked out years ago. He spread out his graph paper and ignored the heat and the buzzards overhead.

He worked methodically, plotting, shooting, writing. Occasionally he heard the blasts of Williams's dynamite far down the canyon; once he sighted Wanke's truck. The sun was almost set when both men returned to Crowe's perch above the canyon.

"Got everything you need?" he asked.

They both nodded their heads.

"What do you think?"

"I'll get through the canyons," Williams said. "All fifty-six feet of them."

"I'll give you all the concrete you need," Wanke added. He threw a handful of pebbles at Crowe's feet. Crowe picked up several, inspected the lowly rocks like an aging jeweler examining a diamond.

"You got plenty of this?"

"Enough for five dams."

Crowe nodded. "I'll buy dinner. I know a good place."

Williams drove the pickup and Crowe slept on the way back to Vegas.

*   *   *

Frank Crowe asked for a private booth and he, Williams, and Wanke sat heavily on fragile woven-cane chairs. Crowe ordered a bottle of red wine. Dominick brought an ample plate of antipasto--salami, olives, peppers, garbanzo beans--and soon the three men settled down to a heavy, pleasant meal of pasta, meatballs, Italian sausage. The talk was easy, discussions of deaths and births, promotions and firings, other projects, other days. Neither Williams nor Wanke pressed Crowe, knowing he would divulge his plans sooner or later.

Another bottle of wine arrived.

Finally Frank began to talk about those plans.

"As you know, I quit the Bureau two years ago and joined Utah Construction Company."

"We heard about it," Jack Williams said. "We wondered too."

"Twenty years I'd spent," Crowe said. "Built five dams, just like you, and what do I have to show for it? A house worth maybe forty-eight hundred dollars, a beat-up Chevy, arthritis in my right shoulder, and the prospect of getting three hundred eighty a month for the rest of my life."

222

Williams nodded his head in agreement. "It's gone up to four ten now, Frank."

"Very generous. Finally it came to me that every son of a bitch that graduates from engineering school has three choices: he either works for the Bureau, or for a big company, or he buys a truck and a cement mixer and is in business for himself.

"Look at Barnes, Edward Chapman, Henderson Andrews, they're no better engineers than any of us, and where are they? On Nob Hill, in Salt Lake, getting fat. Why?"

"You're right, Frank. I've thought about it too, but maybe that's a decision we should have made fifteen years ago."

"Maybe it was, Tom, but I made the decision at forty-two, joined Utah, and even now make three times what the government ever paid me."

"You're worth it, Frank."

"Sure I'm worth it. What is it they say? Frank Crowe, the best dam builder in the world. Jack Williams, the best tunnel man. Tom Wanke, the best concrete engineer."

"What do you suggest?"

"Join me at Utah. I'll double your salary. Soon they'll start bidding on the Boulder Dam. Utah is going to be one of the bidders."

"How much are they willing to put up?"

"A million bucks."

"Wow."

"Wow is right. Shaughnessy who owns Utah isn't half the engineer we are, but he's got a million bucks. Earned it fair and square, I'll say that."

"It's not that easy, Frank. I could retire soon; so could Jack."

"I know. I know. Three more years and you're home free. Twenty-five years, five dams, maybe six, and you haven't got enough dough to take your wife out once a week."

"You got enough authority at Utah to hire us?"

"Shaughnessy paid for my trip to Vegas to meet you guys. He's even buying this dinner."

Crowe noticed the engineers looking at each other. Things were working fast. These men, sitting across from him, were equally methodical men. Their lives were orderly. Offers like this were unnerving.

Wanke took the initiative. "Look, Frank, I know what's going through Jack's mind, it's the same as me. You're betting on a couple of horse races. One is

223

whether the dam will be built, and two, whether Utah will get the low bid. Goddamned Arizona will never ratify that pact*."

"You're right," Frank said, "but you've forgotten a third factor."

"What's that?"

"Whether Hoover will make President."

"I never considered that," Williams said.

"I have," Crowe answered. "I went to Palo Alto and talked to Hoover. I went right to his house. Perhaps the greatest day in my life. He was humble, honest, brilliant. He told me he would run, he would win, and he'd build the Boulder Dam."

"Did you ask him about Arizona? It takes seven states to ratify that pact."

"It does now. I asked Hoover about it. He said, 'Frank, when I'm President, it will take six states out of seven. It's as simple as that. I can do it with the stroke of a pen.'

"I'll never forget that statement. Hoover laughed. Said yes, just like that. 'Bet you never learned that at the University of Vermont.'

"'No sir,' I said.

"'Well, Frank, you learned all there is to engineering and I hope to hell you'll get to build that dam.'"

"I'll say one thing," Williams countered, "you sure been getting around."

"That's not the only place I've been. Where do you think I've spent most of my days these last two years?"

"Where?"

"At the Bureau. Right next to Walker Young's drawing board."

"What about the others? The others who will want to bid on the dam?"

"The Bureau's offices in Denver are open. It's a government office. They're just as welcome there as I am."

"I'm not so sure about that."

"Well, let's say I'm on home ground," Crowe adjusted.

"One thing, Frank," Wanke added seriously. "When we were at the dam site, you mentioned something about one hundred sixty extra feet."

----

* The Colorado River Compact

"Right. Five hundred sixty feet of dam can't hold twenty million acre feet of water behind it. It'll wear out the spillways in two years."

"What does Walker Young think about that?"

"We've argued and argued. I've worn out three Keuffel and Esser thirty-dollar slide rules but he's stubborn."

"Maybe he's right."

Crowe looked at Wanke quizzically. "Maybe he is."

"You've got to bid on his plans."

"That's right. His plans aren't finalized yet. He listens to me, and if I can't make him see the light, one person can."

"Who's that?"

"Hoover. Walker idolizes the man."

"Okay, okay," Wanke said. "It'll still take one extra year."

"It won't."

"Why?"

"Because I plan to bring that dam in two years early."

"How?"

"I'll work twenty-four hours a day instead of eight."

"Jesus," Williams gasped.

"What difference does it make to you, Jack? You'll be underground day or night."

"It doesn't make any difference to me, but what will the government say, the unions, the Wobblies*."

"Three shifts a day means a payroll three times as large. What politician is going to argue against employment?"

"It's ingenious," Wanke said.

----

* "Wobblies" is a slang name for the Industrial Workers of the World, an organized labor movement which sought to overturn the capitalist system and set up a socialist government. Strikes, slow-ups and sabotage were common tactics employed by the Wobblies.

225

"It's secret," Crowe retorted. "Up to this minute, the only man who knew about this was me. Now you both know."

"That's no problem," Williams said.

"There's one more thing," Crowe continued. "The way I've been watching the plans, the computations I've made, I know I can bring in the low bid, I know I can come back in two years ahead of schedule, and I know we'll be in the black."

"I trust your figures, Frank."

"Thanks, Tom, but I also know this will be my last dam, and your last dam. Before I take this job, I'm going to demand ten percent of the profits."

"That's pretty steep, isn't it, Frank?" Wanke asked.

"Very steep. But I intend to give each of you twenty percent of my share."

He noticed both men trying to accept the reality of it all.

"You're better with figures than I am, Frank. What kind of profit do you project?"

"About $3 million."

"So you're talking about $300,000 for yourself."

"I am projecting $180,000 for myself and $60,000 for you each." He watched his friends trying to cope with those figures. "And I'll make one more projection. You'd have to live to the age of 190 to get that kind of money in pensions from the Bureau."

"When did you figure that out?"

"On the train coming here."

There was silence in the room. All the three men could hear was the dishwasher working. Frank drank his brandy in one swallow.

Finally, Frank Crowe said, "I never deal with anybody after booze. Let's all get a good night's sleep and take another run up to the site tomorrow. You guys talk it over. I'll see you at seven at Mary's."

"You wouldn't want to make that seven-fifteen?" Williams suggested, laughing.

"Seven o'clock," Crowe said, also laughing.

"I wonder," said Wanke, "what time Henry Barnes gets up?"

*   *   *

Once more they had traced their way to the dam site, standing on a promontory later to be known as Observation Point. Familiarity with the site did not bring affection. The utter desolation of the area reminded Wanke of a deserted battlefield, one in which the battle was really about to begin.

Yet, here in this wasteland, nature somehow managed to survive, thin-leafed aspens were replaced by water-filled cacti, animals had skin which not only concealed but protected, and the river, far below them, had a life which awed them all.

"Well, what did you boys decide?" Crowe asked.

"I've got some questions," Tom Wanke said. "So does Jack."

"I bet." Crowe knew they had spent most of the night talking.

Wanke with binoculars around his neck directed Frank to raise his own. He spoke carefully. "I want you to raise your glasses and follow this canyon wall east until we come to a ridge about five miles upstream."

"I'm with you."

"Now cross the river with me--straight across."

"All right."

"You see a level area there?" Wanke asked.

"Yes."

"Just beyond it is the gravel wash."

"That's what the Bureau figured."

"Well, they were right. I'm going to need a bridge across here, Frank, a good trestle railroad bridge that can take two tracks with a maximum load of forty gondolas filled with concrete, not counting the engines."

Crowe retraced the river in his glasses. "That'll have to be a hell of a bridge."

"You bet."

"You'll get it. What else?"

"I want some of that new high-carbon steel for the trackage. The lime has a way of eating steel. You remember Deadwood."

"I do indeed."

"Is that bridge coming out of our pockets or the government's?" Wanke asked.

"The government's."

"No wonder you're not worried." All three men laughed.

"What else?" asked Crowe, feeling the men were with him.

"That's all for now."

"And you, Jack."

Jack Williams sat on the ground and spread out a rough drawing. Crowe hunched beside him. "I figure roughly I've got to blast one and a half million yards of rock for those four tunnels. I'll blast a twelve-foot hole first starting at each end and then increase to fifty-six feet."

"What's this rig?" Crowe wondered.

"This will be known as the 'Jack Williams Special.' It's a rig three stories high. It will be on tracks. There will be six stations on the first level, six on the second. I'll have twelve men drilling twenty-foot holes at once, stick in the dynamite, pull back and BLOW. Then the muckers can haul away the rock."

Crowe studied the drawing carefully. "It's ingenious."

Wanke had joined the circle now and was equally impressed with the invention.

"What else?" Crowe asked.

"One thing," Williams said carefully. "What about the inspectors, . . .

"I'll handle them," Crowe said.

"How?"

"I'll tell you something, Jack. Walker Young is older than I am, so is Herbert Hoover. They're from our school of construction. Walker Young wants that dam built just as much as I do. He wants a cabinet post."

"Good man," Wanke said, "Walker Young."

"Right," agreed Williams.

"And Hoover?"

"Hoover, Tom," said Crowe carefully, "Hoover wants that dam for two reasons. The first one is because he's an engineer. He knows that building that dam will save the Imperial Valley, create jobs, produce energy. The second reason is because Hoover, like the rest of us, is human. He knows this thing will bear his name. Hoover Dam."

"I'll be goddamned," Wanke said. "You've got the Jack Williams Special, Hoover will get the dam, what glory do I get?"

"We'll name the shithouse after you," Williams said.

228

"Yeah." Crowe laughed. "The Tom Wanke Federal Relief Facility."

It was at this point of levity that Crowe asked the question he knew he did not have to ask. "Well, are you guys in?"

"We're in."

Crowe and Williams and Wanke, three men with buzzards over their heads and a rampaging river below them, shook hands solemnly. There would be no papers drawn; there would be no legal eagles involved. Heavy-construction men, like firemen or sailors, men pitted against the elements, would shake hands and that was their word.

"I've got a couple of questions for you boys," Crowe said.

"Shoot."

"I've got to bring in power from California. Two hundred miles from here." He pointed west. "I'm worried about these California boys. I've led 'em before. They're soft."

"Get Higgins or Jones. I don't know any tougher foremen."

"Good idea. Maybe I can get them both."

Wanke was almost speaking to himself. "You're going to light up that whole dam site."

"Yup," Crowe said. "Every inch of it."

"That's going to be quite a sight."

"I figure it would be. Might even invite Edison to look at it."

"Fine," Williams said, "and you'll have to wear a suit."

"Yeah, I forgot about that. Scratch that idea, I'll have enough trouble with all those lard-ass senators and congressman taking their jaunts."

"A little concrete on their blue suits will discourage that soon enough."

"One question," Jack Williams asked. "Do you hate this river, Frank?"

Crowe thought for a long time. "Right now," he said, "I respect it. Before we're done I might hate it."

Williams kept his eyes on Crowe and said nothing.

They drove back to Vegas in silence, returned the truck, and ate a final meal at Poppa Joe's.

"When can I expect you in Salt Lake City?" Frank asked.

"I figure a week," Williams said. "I'll give notice tomorrow."

Crowe laughed. "Construction stiffs. God damn. We're all construction stiffs. No roots. A week is all it takes to move your gear."

The men shook hands; Crowe boarded the train and headed for the rest room with its six-foot bench. Mercifully it was empty and he took off his coat and hat, made a pillow out of the coat, and waited for the train to gain momentum.

Twenty years and five dams, that's how long he'd known Tom and Jack. Twenty years in heat, snow, floods, in forests, disaster, fifteen hours, eighteen hours, side by side, seven days a week across the western slopes of the United States. Twenty years of work, hard work, . . . picking up the weekly check at the Bureau office, shopping for a new pair of boots, bitching about the bums in the crew. And still they were impersonal years. Sure, Tom was married, so was Jack, they had kids, they lived in houses, but above all, it was the job, the mountain, the river, the gorge, the jackhammers and the compacters and the miles and miles of graph paper which held them together. If there was a death, they stopped and had a funeral and if there was a wedding they stopped and got drunk, but these were short beats in a long concerto.

* * *

Herbert Hoover was right. He ran for President of the United States and he won, as he had told Frank Crowe he would in the garden of his home in Palo Alto, and on December 21, 1928, the Boulder Canyon Project Act was passed in Congress and signed by Calvin Coolidge. Arizona had held out to the end, fighting began over the public and private financing, but the act was passed.

Its vital section read:

> For flood control, improving navigation, and for delivery
> of water for irrigation and domestic purposes, the Secretary
> of the Interior is authorized to construct, operate and main-
> tain a dam and incidental work at Black or Boulder Canyon
> sufficient to store not less than 20,000,000 acre feet of water
> for the Imperial and Coachella valleys in California. He is
> also authorized to construct, or to cause to be constructed,
> at or near the dam, a power plant and incidental structures. . .

> If the Colorado River compact is not ratified by all of the
> States within six months, until it is ratified by six of them,
> including California. . .

Frank Crowe in his office at Utah Construction slammed his fist on his desk when he read this section in the paper, remembered what Hoover had told him: "You didn't learn that in engineering school." He ran into his boss's office, and excitedly waved the document in front of him. Shaughnessy had never seen him so emotional or even recalled his ever shoving into the office without knocking.

The Boulder Canyon Act had projected the $165 million estimated cost on the basis of $70 million for the dam and reservoir, $38.2 million for construction of a 1.8 million-horsepower hydroelectric plant, $38 million for the All-American

Canal; and interest on the whole sum during the construction period was figured at $17.7 million.

The government projected amortization of the project over a period of fifty years, and before anything could begin it was necessary to negotiate the sale of the projected power to finance the project.

There were twenty-seven applicants for the energy. The City of Los Angeles, the Southern California Edison Company, and the Metropolitan Water District, being the most affluent bidders, got the lion's share. The states of Arizona and Nevada each were to receive 18 percent of the available energy.

When these figures were released Arizona blew its legislative stack once more, claiming that California was raping not only their water but also their power. In desperation the state chartered a paddle-wheel steamer, filled it with armed National Guardsmen, and moored on the the Parker Dam Site to "defend" it from California invaders. "The Arizona Navy" as it got to be called, stayed for months and effectively won the battle. They left only after Los Angeles agreed to allow Arizona a share of its water.

On June 25, 1929, President Hoover issued a proclamation noting that the Boulder Canyon Project Act was in effect.

Order 436 instructed Dr. Elwood Mead, Commissioner of Reclamation, as follows: "You are directed to commence construction on Boulder Dam."

Williams and Wanke were in Crowe's office when he returned from his emotional visit to Shaughnessy. Crowe's office was large and meticulously neat. Along one wall were scores of cubbyholes for maps, carefully labeled and obviously delicately handled.

Crowe's desk was old and bare except for a green blotter, an inkwell and pen and a beer mug holding twelve highly sharpened pencils, black, red, green, and blue. There were a telephone and an unused ashtray. Along the last wall was a large drafting table, a map spread over its surface, and triangles and T-squares laid out as meticulously as a surgeon's instruments.

Williams and Wanke brought in two straight-back oak chairs and sat down facing Crowe, who stood. He detested chairs, feeling that it was better for a man to stand when he was working.

"Well, Frank," Wanke said, "so far you've called it one hundred percent."

"I listened to Hoover and I believed him. He's an engineer."

"Well, when do we get to work?" Williams wondered.

Crowe walked around the room. He was agitated, unsure of himself, a detestable quality. "I was just as excited as you guys when I burst into Shaughnessy's office with the act in my hand."

"What's the problem?"

"The problem is simple, Tom. Shaughnessy told me that, for a company to bid, the government is going to require a five-million-dollar-bond."

"Five million dollars," Wanke shouted. "There isn't a contractor in the West who's got that kind of money."

"Exactly."

"So some Eastern bastard is going to grab the ball and run with it, is that it?" Williams said angrily.

"Maybe," Crowe answered.

"What, maybe? We're through. Finished. Those sons of bitches!"

"The old man said we were not through."

"How come?"

"We'll have to join forces, maybe with Barnes, maybe with Chapman, maybe with both of them, and maybe with more contractors to come up with the bond . . ."

"Sure," Williams said, "they're all thinking the same thing, but what if they decide they don't want our million bucks?"

"I'll tell you what Shaughnessy said. He said, 'Sit down, Frank. This is the way I have it figured. I've got a million, I'm not ashamed of it, it isn't chickenfeed, that's that, all that Utah Construction has for that bond. Barnes has a million, Chapman has a million, Mackenzie-Stein has a million, I've been on the phone all morning. . . . But I've got one thing that all of them want.' 'What is that?' I asked. 'Frank Crowe,' he said."

It bothered Crowe to make that speech. All his New England reticence revolted against this display of braggadocio, but he had brought these men in, he had terminated their safe career at the Bureau, he had to give them hope.

"He's right," said Wanke. "The old man is right, but what about our deal? I figured you could swing it with Shaughnessy, but what about the others."

Frank Crowe picked up his slide rule from his drafting table and sat down on his desk. Somehow, he felt better with this instrument of precision in his hand; he enjoyed the interaction of the highly polished ash, tongue and groove, as he enjoyed the click of his watch case when it shut, the bubble of a level when it centered.

"My deal and your deal are the same. Whether I'm up against Shaughnessy or Chapman or Barnes or Stein or all of them. I'm desperate, Tom. I'm shooting craps. I'll get it my way or I'll walk out of this office and build sidewalks in Salt Lake."

"I'll get a wheelbarrow and help you," Williams said.

"I'll carry the water," Wanke followed.

232

"Thanks, boys," Crowe said, standing. The meeting was finished.

Crowe watched them leave and closed the door behind them. He was still playing with his slide rule. . . .

The days drew long before the opening of the bids on March 4, 1931. Merritt and Scott were already building railroads and highways to the dam site, Bureau engineers were all over the area checking and rechecking their calculations, and Crowe, never a desk man, sat caged in his office, waiting, waiting for the money to materialize for the bond. He had persuaded the Bureau to raise the dam to 726 feet, but all that seemed gratuitous now.

He rarely saw Shaughnessy, though he knew that this man was doing what he knew best--wheeling and dealing, figuring as Crowe figured elevations and stresses.

Finally in November, Shaughnessy, a short energetic man, called Frank to his office and handed him a list. "There it is," he said. "Five million bucks."

Frank Crowe read carefully and aloud. "Utah Construction Company, one million dollars. J. F. Shea of Los Angeles, half a million dollars. Williams will be happy with them; they're good on tunnels and sewers. Mackenzie-Stein, San Francisco, one million dollars. What in the hell did they ever build?"

"The Mark Hopkins Hotel," Shaughnessy said.

"That makes them dam builders?"

"They have a million dollars."

"Henderson-Andrews," Crowe continued, "Edward Chapman, and H. A. Barnes." He put down the list. "That's five million dollars."

"To the penny," Shaughnessy said.

"This Stein," Crowe asked. "He's a Jew?"

Shaughnessy, irritated, answered, "Yes, Stein is Jewish, but he doesn't worry me."

"Who does?" Crowe asked.

"Between us," Shaughnessy said, "Ed Chapman. I used to build gutters next to him right here in Salt Lake City. We used to find a lot of our shovels in his trucks. At any rate, that's it, Frank. That's three months of work on this goddamned telephone, and the only thing everyone agrees on is that you should be general superintendent, so get to Denver and wait for the Bureau's plans. They should be ready by December 1; then we can get on with the bid."

Crowe stood up and started to leave. "What kind of a place is the Mark Hopkins Hotel anyway?"

"It's the most beautiful hotel in the world and you'll never have enough money in your jeans to stay there overnight."

233

"We'll see about that," Crowe said, smiling.

All the principals of the Big Six Companies met at the Brown Palace in Denver on February 28, and while they represented the cream of builders in the West, some with $30 million of construction on their drawing boards, they teamed up in twos to fill the six bedrooms they had rented. They were all construction men and a buck was still a buck. This was a business meeting, and it came as no further surprise that the council meeting the next morning would be held in one of their rooms, the men sitting on unmade beds, windowsills, dressers, their backs against the wall. It was an atmosphere which made Crowe comfortable.

There were introductions, coffee was passed, and finally Frank Crowe, pointer in hand, had everyone gather around the model of the dam. He was aware that around him stood some of the best engineering talent of the West, and he took pains to be as brief and precise as he could.

He took the job step by step. Tunneling, cofferdams, unwatering the site, cherry picking, pouring the dam, grouting, finishing, housing, safety, labor, materials, wages, food, legalities, hazards. The presentation was so masterful it prompted not one question until the end. Mackenzie asked it.

"What is your bid, sir?"

"The cost," Frank Crowe said evenly, "will be $36 million. The profit you have to decide."

"Twenty-five percent," Old Man Donner said, "in order to cover the bond and pay off the backers. That leaves $3 million net profit."

"Then the bid will be $48 million."

"Hold on, Crowe, hold on," Chapman said nervously.

"Mr. Chapman?" Crowe looked at him.

"My boys have been in the field too, and listening good. The last I heard is that Arundel is going to bid $56 million. You're $8 million under."

"That's right, Mr. Chapman."

"That's one hell of a discrepancy."

Crowe paused for a minute to let the room settle down. "Arundel, to the best of my knowledge, and every other bidder is planning to work eight hours a day. I plan to work around the clock."

"Three shifts?" Mackenzie asked.

"Three shifts," Crowe answered.

"What about the Bureau," Stein asked, "the unions? How do you get around this?"

"There is nothing in the plans that states we cannot work twenty-four hours a day. The Bureau specifies that no man can work more than eight hours a day. No man will."

"What about the swing shift and the graveyard, won't they demand more money?" Barnes asked.

"I talked with Las Vegas yesterday, Mr. Barnes. There are two thousand stiffs sitting around the lawn of the Union Pacific in the heat, surviving on doughnuts and coffee. I'll have no labor shit on this dam, I promise you."

The room was filled with skepticism. Crowe seized the opportunity. "If my projections are accurate, and God willing, I'll bring this dam in two years early. I'll earn back your bond the first year. The profit will be three million dollars. I have figured and built five dams in my lifetime and come in within ten thousand dollars of my projection each time."

Crowe knew that money talked in this room and today he too could talk money.

Stein spoke next. "Mr. Crowe," he said quietly, "I have an engineering question, not a money question."

There was nervous laughter in the room.

"Do you feel confident that you can pour concrete continuously twenty-four hours a day, year in, year out, without causing too much heat?"

Crowe forgot all the misgivings he had about the man. He thought of the Mark Hopkins Hotel.

"That is a good question, Mr. Stein. Perhaps I did not make it clear earlier that I intend to pour in very definite patterns. Each grid is ringed with ice water, each grid is individually monitored. . . ."

"That's never been done before, has it, Mr. Crowe?"

"No, but before Edison we used candles."

"That answers my question, Mr. Crowe," Stein said respectfully.

Old Man Donner of Shea had known Crowe and admired his skills for years. "Frank," he said, "you know it's no secret that everyone here expects you to be supervising engineer. What is it that you want from us?"

Crowe had expected this question for years and knew the answer cold, but now at the point of delivery he could only think of the hills of Vermont, the sparsity of the winter landscape. He had never been taught to ask. But he did.

"I want a salary of twenty-five thousand a year, I want the largest house in Boulder City for my home. I want a number of men on my staff, and ten percent of the profits."

It was the last statement that jarred the room as Crowe had known it would. He stood and headed for the door. "I'll go find myself a beer. You gentlemen can make your decision."

Once Frank Crowe had closed the door everyone started talking at once. Chapman asked Shaughnessy whether he had known of Crowe's demand and Shaughnessy said no.

"The nerve," Barnes shouted. "Ten percent of the profits! Who the hell does he think he is?"

The talk was loud, gesticulations were wild, and finally Stein of San Francisco, the quiet-voiced gentlemen brought the room to order. "There are two questions, as I see it. The major one is whether we can get away with working around the clock. Let's take a poll. Remember that very little opposition will come from the government. Employment is the key word of the dam."

Everyone raised his hand.

"If we can get away with it, as you feel, then my second question is, do you think we can come in at a profit?"

Slowly, but steadily, everyone raised his hand.

"Then it is my conclusion that no one will work harder than Frank Crowe to earn us that profit, and his share to boot."

"I'll second that," Old Man Donner said.

But Barnes was not pacified.

"I'm not going to have some son of a bitch tell me what kind of house he wants and how much I should pay him. I'm putting up a million bucks. What's he gambling?"

Shaughnessy was angry now. Visibly. "You put up a million bucks and Frank Crowe put twenty years and five dams on the line. You can believe me or not, I knew nothing of his demands, they're his, not mine, but Frank Crowe works for me and he either gets what he asked for or I'll pull out my stake. That's final."

"Shall we vote, gentlemen?" Roger Stein said, still quietly.

"All in favor of the proposals of Frank Crowe say aye."

They all said aye.

Shaughnessy then spoke: "I request that the conversations which took place in his room remain in this room." Everyone nodded.

"I hope, gentlemen," Stein said, "we win the bid, build the dam, and celebrate as my guests at the Mark Hopkins."

Now Shaughnessy recalled Crowe's smile. God damn, he thought, he might afford a night there yet.

The Engineer-Soldier at His Best

An Excerpt from **Leif Sverdrup**
Gregory M. Franzwa and William J. Ely

The relationship between military engineering and civil engineering is closer than might be apparent to the casual observer. Most civil engineers are aware that civil engineering came into being when engineers involved in projects which had nothing to do with military activities wished to distinguish their work from that of engineers involved in military work. Until that time, about the middle of the eighteenth century, no distinction was made between civil and military engineers.

In the United States the civil engineering profession has always been closely allied with military engineering. From the earliest days of the nation's history when the Military Academy at West Point was producing the nation's only academically-trained native engineers to the present day when the U. S. Army Corps of Engineers retains a major capability for developing and managing the nation's water resources, the bond between civil and military engineers has been quite close.

There are a variety of reasons for the close relationship. The opportunity for military engineers to be involved in civil projects during peacetime provides hands-on engineering experience that is invaluable in maintaining readiness for military engineering tasks. Also, the civil engineers who work closely with the military engineers involved in civil works projects provide a pool of trained and experienced engineering manpower that could be mobilized rapidly in the event of a national emergency.

This selection illustrates the latter point. Leif Sverdrup, known to his friends as Jack, came to the United States from Norway as a young man. He obtained his civil engineering degree from the University of Minnesota and went to work as a highway and bridge engineer. After acquiring experience as an employee of the Missouri State Highway Department he decided to open his own consulting firm. He invited his structural engineering professor from Minnesota, John Parcel, to join him. Parcel agreed to work with Sverdrup on a part-time basis (while retaining his position at Minnesota), and the firm of Sverdrup & Parcel was born.

With offices in St. Louis, Sverdrup used his contacts from his days with the highway department to get the young business off the ground. By the time 1941 rolled around, his firm had weathered the rocky Depression days and was beginning to build a reputation as one of the nation's most capable consulting engineering firms. Sverdrup had become familiar with a number of Corps of Engineers officers through his work on various bridges over the Missouri River, where the Corps jurisdiction for maintaining navigation clearances brought them into contact with one another. As the nation began to mobilize for World War II, it was not surprising that the Corps of Engineers officers who had become familiar with Sverdrup's work in Missouri began to call on his firm for engineering assignments associated with the mobilization.

Jack Sverdrup was a happy man. His firm was busy and expanding its horizons into new fields of opportunity. But there was an uneasiness throughout the world, and Sverdrup could not have guessed that the stage was being set for his greatest adventure.

*   *   *

There was no question in the mind of any observer about the inevitability of war between Japan and America. The question was, when? A few felt that the Japanese would strike as early as December 1941, but many experts, including MacArthur, felt that the attack would be held back until the end of the monsoon season, in April 1942.

MacArthur regarded the time as heaven sent, and planned to use every minute of it in building a viable force to defend the Philippine Islands from the invaders from the north.

*   *   *

. . . A War Department directive of October 4, 1941, ordered Lt. General Walter C. Short, commanding general of the Hawaiian Department, to begin construction of a new route . . . for "movement of heavy bombardment aircraft between Hawaii and the Philippines."

On October 13, Short ordered construction at Christmas Island, Canton Island, Fiji, and New Caledonia. He ordered airfields built in Australia at Darwin and Townsville (or Rockhampton), and also at Fort Stotsenburg in the Philippines. Each had to have a runway at least 5,000 feet long, capable of taking heavy bombardment aircraft by January 15, 1942.

A $5 million fund was made available, with another $6 million in reserve if needed. Short delegated the construction job to Wyman[*] and as far as Wolfe[**] was concerned, all hell broke loose.

"Look," [Wyman] told Wolfe, "this [airfield construction] wasn't too hot in August but its damned hot now, and you and Jack are the logical people to do it. I'm giving you fair warning. I'm contacting General Reybold, and I'm going to do it this very day.

Eugene Reybold, a major general, had been appointed Chief of Engineers on October 1, 1941. He had attended a meeting of the Mississippi Valley Association (MVA) in St. Louis as Jack Sverdrup's guest on Monday and Tuesday, October 27 and 28, 1941.

Sverdrup attended the final session of the MVA meeting, and he and Reybold took a cab to Union Station and headed south to inspect work in Oklahoma. Sverdrup had packed his bag for three nights--he was expected back in St. Louis on Friday . . . Jack Sverdrup kissed his wife [Molly] goodbye that

----

* Major Ted Wyman, an old friend of Sverdrup's, was District Engineer for the Corps of Engineers Hawaii District.

** De Wolfe was a Sverdrup & Parcel employee stationed in Honolulu as project engineer for projects underway in the South Pacific.

Tuesday morning in October. He wouldn't see her again until the following April.

It was probably late Wednesday night, October 29, or early the following day that Wyman caught up with Reybold. The chief of engineers laid it out for Sverdrup very simply after talking with Wyman. "Look, Jack," he said, "you gotta do it."

*   *   *

Sverdrup caught a train from Tulsa on October 30 and arrived in Omaha that night, where he took a night flight to San Francisco. Dawn had broken when the plane touched down near the shores of San Francisco Bay.

This was no time for a Matson liner. There was something far more exciting awaiting Jack Sverdrup. He was about to become the first passenger of Pan Am's new Pacific Clipper flight to Fiji.

The airline had inaugurated its "China Clipper" service to the Orient in 1936, and had pioneered a new South Pacific route to New Zealand in September 1940. The aircraft scheduled to make the maiden voyage to Fiji was Boeing's B-314 flying boat, an enormous four-engine plane with two decks and a triple fin and rudder. . . .

Sverdrup boarded the plane at 11 a.m. November 5, at Pan Am's Treasure Island base in San Francisco Bay, and flew the 411 miles to Los Angeles Harbor. There were 13 other passengers aboard, but Sverdrup held Ticket No. 1, according to Pan Am's *New Horizons* magazine for December 1941.

"Engineer Sverdrup," the article said, "of course was discreetly silent on the object of his mission."

At 4:29 that same day, the Pacific Clipper rocked off her bubble, rode up onto her step and planed up out of the harbor. She headed nearly due west, chasing the setting sun at 140 knots. The plane was 2,100 miles from Honolulu.

"The trip was an uneventful one," Sverdrup wrote. "The weather was ideal and we traveled at 7-10,000 feet. Arose at 5 in the morning and saw the most beautiful sunrise above the clouds."

Wolfe, Wyman and Robinson[*] spent the day with Sverdrup, going over the job, getting passports, and talking of the old days. Before the day was over Sverdrup had signed a contract to cover planning and design for all work to be accomplished in the Fiji Islands, New Caledonia, New Hebrides and the Solomons--somewhat different from the order laid down by Short.

*   *   *

----

[*] Major Bernard Robinson was Wyman's assistant. He served as Operations officer for the Hawaii District.

Instead of submitting preliminary drawings to Honolulu for approval, Sverdrup asked permission to proceed with construction immediately, drawing plans as the work progressed. Sverdrup felt Wyman should get permission to pay more than the allowable architect-engineer (A-E) fee, due to the fact that the A-E would be undertaking several contractor functions also. Sverdrup said that Wolfe would be the chief engineer for all projects.

Wyman wanted action; he got it. He was also going to have to pay for it, but he suspected that anyway. Sverdrup got all he asked for and, in the final analysis, so did Wyman.

Early the next week Sverdrup learned that New Zealand could provide 3,000 native laborers and 1,200 skilled and semi-skilled workmen, all of whom would be housed in tents during the course of construction.

Sverdrup then took a flying boat to New Zealand, landing at Wellington, where he held an evening conference with the cabinet of the prime minister. The meeting didn't break up until 1 a.m.

On November 22, 1941, Walter Nash, minister of finance and deputy prime minister, announced the New Zealand government's approval of the project. Permission was granted for the United States to utilize New Zealand men and machines to expand the Nandi airfield, in the Fijis, with the United States paying all the bills. Things were beginning to take shape.

*   *   *

The evening of January 3, 1942, Sverdrup was visited by General George H. Brett, in command of U.S. forces in Australia . . . "General Brett offered to have me commissioned a B.G. [Brigadier General] and wanted me to take over all work in connection with Darwin--and is that a honey! Told him I appreciated the B.G. offer but that it was better to get a regular officer for the job and let us handle the work as engineers."

Brett, . . . Wolfe, Sverdrup and several others drove to the base of the Royal Australian Air Force outside of Melbourne--"arrived in one hell of a bad sandstorm--as good as any dust bowl storm there ever was," Sverdrup wrote. There they spent the evening or at least part of it. They all arose at 3 in the morning, took off before dawn and landed at Darwin, in the center of Australia's torrid north coast, "just 10 hours and 2,140 miles closer to the equator. The trip was uneventful, except that the emergency escape hatch opened, and General Brett . . . lost his briefcase with all the secret plans, but got them back eventually. Fell off just as we took off."

It was during this trip that Brett realized what kind of man Sverdrup was. The harbor at Darwin was choked with American cargo ships diverted from the Philippines by the Japanese onslaught, and the local stevedores were on strike. Sverdrup knew the Darwin harbor had a 22-foot tidal range, and he knew the Americans wanted the hundreds of 55-gallon drums of high-octane gasoline on those ships.

He came up with the idea of throwing the drums over the side at high tide. When the tide went out, the gasoline was on the beach--high and dry.

Sverdrup spent four days in Darwin. "I'll give anyone the place who wants it. My hours have been from 6 a.m. to 12 midnight, and I am sweating like I used to when I worked in the garden. The problem here is one of unloading ships and transportation--and it sure is some problem too."

He was in Darwin to supervise the unloading of the first of the supply ships carrying materials and equipment for the airfield construction. Here is his entry for January 11:

> My God, I forgot it is my birthday until I wrote the date down!
> And have I celebrated it! First I did my washing: socks,
> underwear and handkerchiefs. No laundry here. At noon we had
> a little excitement. A submarine came into the harbor and fired a
> torpedo at the *Houston*--but missed, and the next thing a
> subchaser had the sub. Three men came out alive, and the rest
> were killed. A hospital ship with seriously wounded is due in
> from Manila tomorrow.

On January 14 Sverdrup flew to Sydney and began a dizzying travel schedule between Melbourne, Sydney and Brisbane, involving nine plane or train trips between January 16 and February 7. In late January 1942 he signed a fixed-fee contract to provide architectural and engineering services for the United States Armed Forces in Australia. . . .

*　　*　　*

On March 31, shortly after his arrival in Australia, Pat Casey* . . . had become aware of the Army contract Sverdrup & Parcel had negotiated for work in Australia and . . . had started looking for his old friend, Jack Sverdrup. . . . [On] April 15 . . . Casey offered Sverdrup a job as head of his operations division, with the rank of colonel.

The next day, . . . Casey helped Sverdrup to the first of the planes which would take him to Honolulu and the States. Then he prepared a wire to the War Department in Washington, requesting that Sverdrup be appointed a colonel and assigned to MacArthur's staff. It was protocol to clear the wire through MacArthur's chief of staff, Major General Richard K. Sutherland. Sutherland, in an obstinate mood that morning, looked at the wire and said, "Why, I can't approve that."

Casey was in no mood for such foolishness, and although he was outranked, he said, "Then I'll just have to go in and see General MacArthur," and stalked past the suprised general.

"After my presentation," Casey said, "Mac took one look at the wire and initialed it 'OK--Mac', so you see, Sverdrup's [rank] was on the direct order of General MacArthur. It was one of MacArthur's best moves and he knew it within a very few weeks."

----

* General Hugh "Pat" Casey was chief engineer for General MacArthur's forces in the South Pacific.

Sverdrup spent April 17 in a futile attempt to get to Auckland. One of the engines of the flying boat started throwing oil just before reaching the point of no return, and the plane limped back to Sydney. Sverdrup knew that Wolfe, too, was itching to take a commission as a lieutenant colonel. As soon as he landed, he wrote to Wolfe telling him that he absolutely had to finish the . . . job [of supervising the design and construction of the airfields], and only then could he join the Corps of Engineers.

Sverdrup arrived in Auckland on April 18 and spent the evening in conference with New Zealand officials. Then he hitched rides on military aircraft to Hawaii, arriving at Hickam Field on Tuesday, April 21, 1942. He spent the next five days in Honolulu, winding up all the Pacific Ocean contracts with Colonel Albert K. B. Lyman, the native-born Hawaiian who had replaced Wyman as district engineer. On his last day there, April 25, Sverdrup wrote a long letter to Wolfe, explaining that Sverdrup & Parcel not only had the job of selecting the islands [for the airfields], which Sverdrup had already done, but now would serve as designers and builders, . . . with Wolfe in complete charge. The three-page letter detailed the job with unusual thoroughness. Wolfe didn't receive it for weeks, and to this day he has never seen a copy of the actual contract, although it is on microfilm in the Sverdrup Corporation headquarters in St. Louis.

There was no question whatsoever in Jack Sverdrup's mind about his immediate future. He had promised Casey he would wire his decision regarding the acceptance of a commission. He advised him he was now flying to Washington, D.C. to accept the colonelcy.

Sverdrup left Pearl Harbor on the now-resumed Clipper service on Sunday afternoon, April 26, and arrived in San Francisco the next morning. Molly and Jackie [Sverdrup's son] were waiting for him at Lambert Field in St. Louis when he arrived there at 9:30 Tuesday morning. He had been away from home six months to the day, almost to the minute. Next time it would be much longer.

*   *   *

Jack and Molly Sverdrup left St. Louis by train at noon on May 7, 1942, and arrived in Washington, D.C. the next morning. Sometime during that day he took the oath and became, by Congressional decree, "an officer and a gentleman" in the Army of the United States of America. He went to a uniform store and picked out the uniform recommended for his tour of duty as a colonel in the Corps of Engineers, and, as is customary, asked his wife to pin on his eagles. Then they left for St. Louis.

An officer friend of Sverdrup's happened to be on the same train. "He whispered to Jack that I had pinned the eagles on backwards," Molly said. "The heads were turned the wrong way. He quickly pinned them on properly."

Another trip to Washington was necessary, and after his return Sverdrup had only a few hours left before his departure for the Southwest Pacific. He, his wife and son, . . . and a very capable attorney, whom Sverdrup had known for some time, were in the old Lambert Field administration building at about 1 o'clock Thursday afternoon, May 14, 1942. The lawyer was W. Edwin Moser, senior partner in a large law firm in downtown St. Louis and a man who had impressed Sverdrup with his skill. He placed his briefcase on a radiator in the building lobby, drew out a sheaf of papers, and had Sverdrup sign at the bottom of each page. It was a will.

The plane to Kansas City was already at the ramp, and it was time for farewells. . . . There was warmth, a handshake and a big hug for young Jackie, a lingering farewell kiss for Mollie. The tall colonel, his blond hair already flecked with gray, climbed up the portable stairway into the plane. The door closed quickly, and the aircraft rolled to the end of the runway and took off. On the ground, the party stood silently and watched the plane until it disappeared over the western horizon. Sverdrup was bound for the Pacific, war, and three years of separation from his family.

*   *   *

In those dark months of mid-1942, there were thoughts in GHQ [General Headquarters] of an offensive against the Japanese, but only because a good offense is the best defense. . . .

Soon after his arrival in Australia, MacArthur had determined that the defense of Australia should take place in New Guinea. He planned to use Port Moresby as a major air and staging base and to develop other bases in New Guinea as his capabilities increased. The enemy already had a small force in the Lae-Salamaua area to the west and could be expected to make other landings in New Guinea from their huge base at Rabaul, New Britain, in the Bismarcks.

*   *   *

Earlier, MacArthur had decided to establish his headquarters in Brisbane, halfway up the east coast of Australia. At 2 p.m. July 21, the MacArthur entourage boarded a special train at the Spencer Street station in Melbourne for the 1,000-mile trip to the north. When the train reached Sydney, the midpoint of the journey, a courier brought discouraging news to MacArthur--at that very moment the Japanese were landing in great force at Buna [in New Guinea].

*   *   *

. . . There must be logistical support for troops which had to throw the Japanese from New Guinea. There must be food, clothing, medicine, transportation and ammunition. MacArthur ordered Casey to engineer a new road over the Owen Stanley [Mountains] to Buna, to facilitate that supply. Both Casey and Sverdrup protested to MacArthur that such a road would require more than the entire engineer force in the Southwest Pacific to build. They also argued that roads run in two directions, and this one might weaken the defensive barrier of the mountain range. In Casey's final presentation of detailed plans, he proved the adverse impact of the transfer of engineer effort from other vital operations and the loss of the major defensive obstacle opposing the enemy advance. General MacArthur said, "Pat, your logic is quite sound. We won't build that road."

243

Jack Sverdrup

[The] routes over the mountains . . . were virtually unknown to the Allied forces since the mapping of New Guinea was as primitive as the country itself. The only way to determine the condition of the trails was to hike them. . . .

Casey and Sverdrup left Moresby on September 16 and arrived at Abau the morning of the 18th. Their job was to determine whether the harbor could be used as a supply base . . . and to learn whether it would be possible to send either jeeps or mule trains over the [mountains]. Casey and Sverdrup flipped a coin to determine which of them would lead the reconnaissance over the mountains. Sverdrup won. Or lost.

*   *   *

Sverdrup left to start over the Abau Track. Casey reconnoitered the Abau area for several days to determine the potential of the harbor for port development. Sverdrup had with him a U.S. Army lieutenant with a military intelligence background, an Australian army warrant officer experienced in handling New Guinea natives, a medical corps sergeant from the U.S. Army, 10 native policemen and 26 bearers. They made 13 fairly easy miles on the 18th but ran up against the Owen Stanley foothills the next morning. That night Sverdrup sent a letter back to Casey via one of the natives: "Some of the hills we climbed are over 45 degrees--would like to see a mule climb that."

*   *   *

Sverdrup's notes for the fourth day, September 21, include these passages:

> Started off on the trail at 8 a.m., heading up the river to the Stanley range. Walking along, and mostly in, the river for about 30 minutes. Most crossings knee deep. Hills on both sides very steep, no chance for a road in this location. . . . 10:00 a.m., on top of intermediate ridge, elevation 2200. Main ridge to northwest is 800 feet higher. Clay-like soil though fairly light. On up--45% [grade] and full of small rocks--slippery--must be hell when it rains. Loamy clay. 11:55, elevation 3380 and still going up. . . . Track mostly wide and easy and could in good part be used by mules as is--if you could get them in. Soil mostly clay

244

with roots across track. Probably plenty slippery when wet.
Kept on about the same all the way and arrived at Aimari at 5:20 p.m.
Aimari lies in a kettle but the mountains are not so high but
what it could be used for a small air dropping ground--though all
the clearing has been made right out of the jungle and is mainly
used for gardens, 14 huts plus church, guest house, and carrier's
house. . . .

On September 22, Sverdrup continued . . .

> Hope my shoes hold out. Had two pairs when I started--one is
> gone completely and the other not so hot. When you have to
> cross streams all of the time by wading across, your shoes soon
> rot away. . . . Came into a little village about 4 this afternoon,
> and stopped for the night. It is pretty high up so the river in
> front of it was plenty cold. Had a fine swim and then dinner. . . .

*   *   *

Sverdrup felt that he was "on a wild goose chase" about this time, and he
wrote that it would be impossible "to build a motor transport road in here and
probably not a mule track either." When he reached Jaure on September 25,
he felt it would even be impossible for troops on foot to advance at a
reasonable speed over the Abau Track.

*   *   *

Sverdrup's report on the ruggedness of the Abau Track made it clear that the
Kapa Kapa Trail was the better of the two. For his feat, Sverdrup was
awarded the Silver Star, the first of a chestful of decorations he would win in
World War II. The commendation read as follows: "For gallantry in action
near Abau, New Guinea, from September 18, 1942 to October 6, 1942. Colonel
Sverdrup led a reconnaissance party into enemy-occupied territory, far in
advance of friendly troops, and thereby secured information of great value to
the command."

*   *   *

During a broadcast after the war Sverdrup described his report to MacArthur
after his return from Abau:

> I flew to Brisbane and reported [to General Casey] that it was
> impossible to build a road [over the mountains] in time. We
> didn't have the men; we didn't have the equipment; we didn't
> have the materials. I recommended instead that we organize a
> battalion of natives to go in there with hand tools, with which they
> could mow down the heavy grass, known as Kunai grass, on the
> flats--level it off a little bit with shovels and our C-47s could land
> there.
>
> He took me up to the chief of staff [Sutherland] who in turn took
> me to General MacArthur. After a while he said: "We will do it,
> and you are in charge." Just as I left the room he put his arm
> around my shoulder and in his inimitable voice, said, "Remember,

245

my boys--food and ammunition. Time is of the essence--of the essence."

I got the supplies that day and flew out that night. I had been two nights on the plane, two weeks before that walking. At 4:30 that [first] afternoon, I felt that I just couldn't walk any further. I was about to give orders to start a camp when I literally could feel that arm around my shoulder and I could hear him say, "Time is of the essence--of the essence."

*   *   *

. . . one missionary, Cecil Abel, proved to be a valuable ally. Abel owned a plantation near Abau and knew well the land north of the divide. He was flown to Port Moresby early in October and testified that a strip could be developed easily at a place called Farasi, in the upper valley of the Musa River. That was within easy striking distance of Buna.

Sverdrup asked Abel to fly to Wanigela, on the north coast, to recruit about 150 native laborers and walk them to Fasari. In the meantime, Sverdrup bought the trade goods to pay them--tobacco, calico, scout knives, and garden seeds. He was back at Abau again, and now he had with him one U.S. Army captain, a lieutenant, three sergeants, and 185 native carriers. . . .

*   *   *

Sverdrup . . . boarded a launch to travel up the Babuguina River to the landing at Abel's mission. Here are some excerpts from his field notes:

Oct 15 -- Left Abau at 10 a.m. Fixed all cargo so carriers had approximately 40-lb. loads. Had to leave about 25 loads as only 185 carriers available.

Oct. 16 -- Campsite to Ororu--arrived 6:30 p.m. Crossed up to 2,500 feet--some very bad hills. Relocation could be made, but would take long time. Track not good.

Oct. 17 -- Ororu to Silimidi. This was a real walk, two days in one. Arrived at Silimidi at 12:30 a.m. Should have arrived at 10:00 p.m. but was stranded in mountains without light. Very tough going and very bad trail--up to 4,000 feet.

Oct. 18 -- Silimidi to Safia. Arrived at 4:30; easy all the way. Crossed Musa River on raft two miles from Safia. Went out to drome and found that Cecil Abel had made fine start on strip. All we had to do was cut out stumps, widen in spots, smooth out and make turnaround.

Sverdrup . . . started the natives to work cutting out the stumps from the strip, and that afternoon a DC-3 touched down at what would henceforth be known as Abel's Landing.

*   *   *

Sverdrup's field notes continue:

Oct. 23 -- Left Kakasa at 8:00 a.m. Advance guard of commando
unit waiting here. As Japs from Kokoda and Fall River have
filtered through here we arranged matters a little differently.
Went ahead of a column with Thompson sub-machine gun myself--
having a 100-foot space behind me--then some carriers--then Abel
with other Thompson--then the remainder of carriers and our
lone armed policeman bringing up the rear. Kakasa [to] Korala
(3 hrs. 10 minutes) flat most of the way. Track not in good
condition due to lack of maintenance. Troops would have to walk
in single file. Several bridges (single logs) to cross. Left
Korala at 1 p.m.--arrived Kinjaki Barige at 5:30. Rain from 3:00
p.m. on; very heavy at times. Mosquitoes God awful!

Oct. 24 -- Went out at 6:00 to survey proposed strip. Found it
covered with elephant grass 8-12 feet tall. Ground looks a little
bumpy . . . Started cutting a line through at 7:30 a.m.--hotter
than the hinges of hell. At 9:30 had 400 yards cut through, with a
fairly large patch at beginning. Placed T there even though I am
not too sure of getting a strip suitable for large transports. At
9:45 a B-25 came over, circled and then came roaring down the
cleared line with bomb doors open and then it rained rice and
corned beef. Holy Mother of God--came within 3 feet of being
killed by 4 cases of corned beef lashed together. If the Johnnie
who dropped that could do as well with bombs the war would
soon be over. As it was, he scattered my natives from hell to
breakfast. Took an hour to get them all rounded up again. . . .
Very heavy rain at 2 p.m.; had to stop work. Went out at 3 p.m.
to walk over strip. Tramped down 300 yards of elephant grass to
get clear view of entire strip. . . .

*　　　*　　　*

Oct. 30 -- Work started at 6:30 a.m. Two gangs clearing
approaches and trees on runway; one gang putting in side and
cross ditches to drain northwest end of strip. Will put french
drains in across field, using coral to fill these trenches. [A french
drain is a perforated pipe in a trench, surrounded with coarse
rock.] Getting coral from floor of church, only place available.
The "brethren" ought to be pleased to know what an important
part their church has played against the Japs. 6:00 p.m., through
for the day after 12 hours of damned hard work. Call in for 5:00 a.m.

Oct. 31 -- Inspected drome at daylight--okay for landing. Will use
the boys to cut a few more trees along the approaches, polish up
the runway and then finish up drainage system. Here is what
the "Papuan Aviation Battalion" did in a little over two days
(average of about 200 natives per day):
     -- Cut 470,000 square feet of grass
     -- Cut 960 trees, 3" - 24" in diameter
     -- Dug 1,860 feet of side drains, 2' wide and 2' deep
     -- Dug 930 feet of cross drains, 1' by 1'

   -- Gathered 930 cubic feet of coral on beach and in church
     and carried one mile for cross drains
   -- Smoothed out entire surface of drome with pick and
     shovel
Somebody may tie it--but they will never beat it. It would have
taken Company C of the 114th Engineers at least two weeks if it
hadn't killed them.

First plane landed at 2:55 p.m. and took off at 3:25, and I was on
it for flight back to Port Moresby.

     *   *   *

Sverdrup stopped off at Port Moresby during his return flight from New
Guinea to Brisbane. There he found many of his engineers down with
malaria, dengue, dysentery and assorted skin maladies. The overworked
machines were in equally bad condition.

Those same engineers had a big job to do. Before the end of 1942 they, and
Sverdrup's "Papuan Aviation Battalion," would carve 15 airstrips in this area of
New Guinea to back up the Allied ground forces. Later, seven of these would
be enlarged to accommodate B-17s. Sverdrup didn't wait for them all to be
started, much less finished, before he took off to the highlands far to the
northwest of Buna to start several more.

He had an uncanny ability to get production from the native population of
New Guinea. In George Kenney's postwar book, **General Kenney Reports** the
author wrote:

> An engineer colonel by the name of Jack Sverdrup, a tall blond
> reincarnation of Leif Ericsson, was in charge of finding and
> preparing landing fields between Wanigela and Buna. He had
> already located and prepared six strips along that route and this
> had immensely facilitated the airborne movement. For some
> reason or other, Sverdrup worked miracles with the natives, who
> seemed to work harder and longer hours for him than anyone else.

     *   *   *

 . . . during the last week in December [Sverdrup] was at the front near
Buna. There he found a portion of an engineer company building a road
through a swamp where the water was sometimes as much as four feet deep.
He found the rest of the unit unloading ships at Oro Bay for the Australians.
Sverdrup blew his stack and brought the unit back together on the
construction task. He angrily reported to Casey that the diversion of engineer
troops by unauthorized commands "had resulted in the loss of valuable time
which will adversely affect construction of the airstrips."

On December 29, Sverdrup went to Buna Village to inspect a ramshackle
plank-floored bridge leading from adjacent Musita Island across Entrance
Creek to the beseiged Buna Mission, where the Japanese were holed up in
virtually impregnable bunkers. The night before Sverdrup arrived, one
engineer had been shot and two others stranded on the wrong side of the

Road Construction near Port Moresby

creek when the enemy discovered a midnight attempt to replace some planks in the floor of the bridge.

Since the Japanese were now alerted to American plans to cross the creek via the damaged bridge, Sverdrup felt that a second attempt to repair it would be suicidal at that time. Writing in **The Corps of Engineers: The War Against Japan**, Karl C. Dod, military historian, said that Sverdrup noted that a spit of land extending eastward from Buna Village came close to the one jutting into the mouth of the creek from the opposite bank.

"On returning to the infantry command post, he suggested that assault troops ford the narrows and make a flank attack on the pillboxes."

That night two companies did so. They crossed but were unable to destroy the bunkers. On New Year's Day, [a] force attacked across the coconut plantation from the southeast, and a detachment again waded the narrows in a flank assault. While the enemy was thus diverted, the engineers finally got the bridge over Entrance Creek repaired. The infantry then surged across, and resistance finally ended on the afternoon of January 2, 1943.

*   *   *

It didn't take MacArthur long to recognize Sverdrup's contribution to the combat operations. In accordance with MacArthur's recommendations, Sverdrup was awarded the Distinguished Service Medal, through War Department General Orders dated March 11, 1943. The citation read:

> For exceptionally meritorious service to the Government in a position of great responsibility in Papua, New Guinea, during the period July 23, 1942 to January 23, 1943. During the Papuan campaign, Colonel Sverdrup personally executed numerous

reconnaissance missions in New Guinea, over difficult mountains and through swamp and jungle, far forward of the areas occupied by our troops, in order to secure vital information needed for engineering operations. Utilizing native labor, which he recruited and trained, equipped only with hand tools, he constructed with great rapidity a series of air fields urgently needed for the transport by air of troops and supplies to distant and otherwise inaccessible areas. His success in completion of these essential advance airfields, accomplished under severe physical hardship and at great personal risk, made possible the effective coordination of land and air forces and contributed materially to the success of the Papuan campaign.

Sverdrup continued his service on MacArthur's staff throughout the remainder of the war. He was eventually designated as MacArthur's chief of construction and promoted, first to brigadier general and finally to major general. Sverdrup was awarded the Distinguished Service Cross, the nation's second-highest award for bravery in action, for his heroism during the invasion of Luzon. He also received the Purple Heart for wounds received in action. When American forces recaptured Manila from the Japanese, MacArthur placed Sverdrup in charge of reconstruction of the Philippine Islands. This assignment involved construction and reconstruction of roads and bridges, harbor and port facilities, electrical and water supply systems and housing and hospitals. It was a responsibility that required the full range of civil engineering expertise, much of which would have been beyond the scope of knowledge of an engineer who had been educated solely to meet the needs of military engineering requirements.

Bridge construction under combat conditions in the South Pacific

Gordon Maskew Fair
Abel Wolman

National Academy of Engineering - **Memorial Tributes**

Abel Wolman, himself a distinguished sanitary engineer, writes movingly of his friend and colleague Gordon Fair in this memorial tribute. Professor Fair is one of a relatively small number of American civil engineers who became internationally recognized for accomplishments as a civil engineering educator. He was a graduate of both the Massachusetts Institute of Technology and Harvard University, and he received honorary degrees from universities in Europe and South America as well as the United States.

Professor Fair taught sanitary engineering at Harvard for almost fifty years. He served as dean of the engineering faculty for three years. The sanitary engineering textbooks he authored or co-authored were among the most widely used civil engineering textbooks produced in the twentieth century.

Unlike most well-known civil engineers, Gordon Fair was world-renowned for the ideas he shared rather than the projects he designed or constructed. Dr. Wolman's memorial underscores the importance of engineers who contribute to society primarily through the quality of their thoughts rather than their deeds.

Following a successful engineering career, Gordon Maskew Fair, who was born on July 27, 1894, in Burghersdorp, Union of South Africa, died on February 11, 1970, in Cambridge, Massachusetts.

Philosophers have frequently pointed out that ideas have a greater impact upon society than do material consequences of ideas. To engineers the beautiful bridge, the soaring office building, or the graceful dam offer visible evidence of the translation of ideas into the service of man.

Professionals choose many routes to attain their major purposes in life. Whether consciously or not, Gordon Fair obviously chose to affect his fellow man through the route of ideas--as teacher, writer, investigator and mentor. That he chose well, his long and preeminent career gives ample testimony.

Gordon Fair brought to his life's work an unusual intellectual capacity, deeply sharpened by extensive and broad education in the great institutions of learning of his day. He included in his armamentarium a competence in foreign languages, not usually the hallmark of many engineers. The statement that evidence of his accomplishments was not to be found in monumental structures, or that he did not work with steel or concrete, is only half true. Most of what he taught, wrote, and preached did in fact find its way into structures throughout the world, through the more subtle route of the minds of men.

One could quantify, at least, his direct impact on man via a count of his hundreds of students. More difficult is the estimate of his impact upon thousands of students and practitioners throughout the globe. His textbooks, perhaps the most valuable yet available, mirrored the intellect that he possessed to an extraordinary degree. As a matter of fact, few were as well endowed as he with such lucidity of reasoning, precision of language, and accuracy of recording. He demanded of his students an equally high level of performance--sometimes impatiently, perhaps even harshly. Such is the habit of those more broadly endowed than many of their fellows.

If one were patient, however, one could soon discover that, while his demands were high, a strong thread of humor, good sense and even gentleness pervaded his life. Those of his friends who had the good fortune to sit and fish with him by the hour attest to these deep-seated softening qualities in an otherwise deceptively austere exterior. While he demanded high quality in the pursuits of his students, he asked no more than he persistently required of himself. The hallmarks of the man were orderliness of conception, honesty of diagnosis, sharpness of investigation, and clarity of exposition. And all his works stand as permanent monuments to these extraordinary virtues.

He was no "ivory tower" academe. He gave much of himself throughout his career to the needs of man throughout the world. He traveled widely to lend his competent aid in alleviating the lot of men, women, and children in almost every part of the disease-ridden and hungry universe.

One of his most fruitful contributions was to the Rockefeller Foundation, which he served as a Member of the Board of Scientific Directors--incidentally, the first engineer to be so honored. One of his colleagues in that activity describes him well in these terms: "Whether it be in the swamps of Sardinia, in the jungles of Brazil, in the lecture rooms of the Ecole Polytechnique in Paris or in the laboratories of the London School of Hygiene, the presence of Gordon Fair inspired all those with whom he came in contact."

He served long and contributed heavily to the peace-time and war-time activities of the United States and international agencies, notably, the League of Nations and the World Health Organization. His years of uninterrupted contributions to myriads of advisory committees of the National Research Council, in the National Academy of Engineering, on the Army Epidemiological Board, and in the earliest efforts of the Agency for International Development in Central and South America are legion. The number and variety of these services are astonishing in the lifetime of one man, no matter how genetically well endowed he happened to be. It is compulsory that even his friends review anew the list of his commitments enumerated in this memoir.

Gordon Fair was no "sitting member" of these groups. As he participated in these sessions, he was simultaneously busily engaged in the laboratory and library, producing new materials, new interpretations, and new guides and criteria for engineering action for the betterment of that environment--recently discovered by more naive crusaders. Gordon Fair antedated them by a mere half a century.

The outcomes of these wartime efforts, among many others, is that *vade mecum* of every global traveler, "globaline," still one of the excellent bactericides and amoebicides. It is well to remember this warborne asset to humanity that bears the hallmark of Gordon Fair's devotion to preventive action.

One of his perceptive admirers, Ed Cleary, properly noted, at the memorial exercises at Harvard University, that "he chose engineering as the fulcrum and teaching as the lever for moving the minds of men to cope with scientific and technologic change." He had an abiding faith in man's capacity to control his environmental fate with wisdom and logic. He needed no formal lesson in his own conception of engineering, that the engineer had a preeminent responsibility to society. He lived that way!

It may well be said of Gordon Fair what Nicholas Murray Butler said years ago of another great engineer, William Barclay Parsons, one-time distinguished member of the Army Corps of Engineers: "Parsons conceived of the engineer as an instrument of civilization." In any such Hall of Fame, Gordon Fair would qualify.

What of the man himself, that private self so often concealed behind the public facade? Those friends, long close to him at Cambridge, had years to view him more intimately. They saw him raising his voice in song. They even claim he had a fine tenor voice! Like all true Izaak Waltons he angled patiently and not too successfully. As Master of Dunster House, he presided for years over "the quick and the slow," fairly, judiciously, with reason and, most of the time, good temper. Good minds deserve *some* explosive moments.

Again, one of his close friends, Edward S. Mason, described him well as "preeminently a man of the age of reason, a classic rather than a romantic, a man with whom one could discuss any subject with the assurance he would come away with a balanced view."

The parading environmental activists of the coming decade will sorely miss the sense of equilibrium that Gordon Fair brought to the discussions of our ever-pressing ills. Although he recognized the ills, he also emphasized repeatedly the possibilities of solutions. These he did not feel would come from "the ravings of scaremongers or even by the practice of confrontation, as favored by the young, but through the careful scientific study that needs to precede action."

In his family life, as in his profession, Gordon Fair was fortunate. His wife, Esther, gentle and understanding, was devoted to him. He was proud of his sons, Gordon and Lansing, and they of him. The generation gap was not visible!

It's Been a Richly Rewarding Life

Margaret S. Petersen

In a professional career already spanning more than 40 years Margaret Petersen has worked as a planner, designer, researcher, and teacher in the field of water resources engineering. Now an Associate Professor of Civil Engineering at the University of Arizona, she worked as a hydraulic engineer for the U. S. Army Corps of Engineers for more than 30 years. In this personal memoir she describes the influences that led her to a career in civil engineering at a time when there were very few women in the profession. She describes the environments she encountered as a female engineering student and then as an engineer in those years when it was truly a rarity to see women in an engineering classroom or workplace.

Ms. Petersen's story, however, is not primarily a story of triumph in the face of overwhelming adversity, although there were adversities enough to dissuade anyone with only a lukewarm commitment to an engineering career. It is instead a story of personal and professional challenges, of supportive friends and co-workers, of dedication and industry, and of professional activities and achievements. Her story illustrates the importance of being prepared for unplanned and even unforeseeable choices that one makes over the course of a professional career. She recounts the feeling of satisfaction so many civil engineers express when seeing the effects of projects in which they have been involved during their careers. And finally she describes the joy of teaching, of passing on to another generation of young engineers those lessons learned in a lifetime of service to society.

How did I, a child who grew up in a working-class family in a working-class neighborhood during the "Great Depression," end up as a professional engineer in the 1940s and a university professor in the 1980s--when most of my schoolmates married early and raised families?

Many things influence each of us, and certainly no one thing or driving ambition led me to this point. Largely, I think, it was a series of circumstances and opportunities and great good fortune in the choices I made, coupled with a lot of hard work. There is much serendipity in all of our lives, and probably more in mine than any of us has a right to expect. I am indebted to countless people who helped me along the way--my family, friends, teachers, and mentors. We all need encouragement and assistance from time to time, and they were there to help me when I needed them.

For a woman (we were called "girls" in those days) to obtain a Bachelor of Science degree in civil engineering in 1947 wasn't as strange as it might appear to today's students, although it is true that there were few of us.

I was fortunate to have a mother who drilled my two younger brothers and myself in the philosophy that anyone could do anything if one wanted to enough and was willing to work to accomplish it. She also believed that education was priceless.

I was nine when the stock market crashed in 1929, so the early experiences that influenced me most in later life were etched in the depression years of

255

the 1930s. I was raised in Rock Island, Illinois, an industrial town of about 42,000 people at that time, located on the Mississippi River. I learned about such things as belts and pulleys, blacksmithing, coal-fired steam locomotives, dikes in rivers, and carburetors and batteries at an early age while tagging along after my father and playing with my brothers. My passion for color and design and for studying came mainly from my mother.

When we were small, we spent many Sunday afternoons visiting a great aunt who lived on a farm on a sweeping point bar of the Mississippi River. There I was exposed to the changing levels of the river, the flooding of farmland, and the sediment left on crops as floodwaters receded. My Aunt Minnie was a widow with six sons. They lived in a concrete block house with concrete floors and a kitchen on the second floor as well as on the first. The house was sparely furnished, and when the river rose--as it did almost every spring--they moved everything upstairs and lived on the second floor for weeks or months until the water dropped. The house had a wide shaded concrete porch on the north side, facing the river, with about three steps leading down to a boardwalk that extended out to near the water's edge when the river was low. Aunt Minnie once showed us a picture of the house taken twenty years earlier. In it there were fifteen steps leading up to the porch. Over those twenty years the river had left deposits that completed filled the area around the house, reducing the fifteen steps to three. It was my first real introduction to fluvial hydraulics.

Aunt Minnie was a wonderful, self-reliant independent woman, and my great uncle's second wife. He was a river man who had divorced his first wife (by mutual consent) because she had insisted on doilies on all the furniture and other refinements that were anathema to him. Minnie was determined to be a no-frills wife. To walk with her through her garden of vegetables and flowers and listen to her talk about the river was a never-ending treat. Of course, she was influencing my later life, but in a subtle way that I never realized until recent years.

Two of my teachers in junior high school also made major imprints on my life. Miss Cook was an English teacher who drilled us on grammar and the function of words and phrases in sentences, all the while trying to instill in us a love for literature and the ability to think for ourselves. Miss Birch was the other, always in flat-heeled shoes, skirts and blouses, and wearing a sweater, in contrast to Miss Cook who was prematurely white-haired and the picture of stately elegance. Miss Birch taught a seventh-grade general science course. Every class session was an adventure, every page of the thick textbook was stimulating to me, with ideas and facts I'd never heard or thought of before. It was in her class that I was introduced to centigrade and Fahrenheit scales, variations of climate, the different forms of clouds, and other wonderful things. Neither of their classes was easy; they both demanded a lot from their students, but their dedication to and love for teaching made them inspiring. Although the years of the early 1930s were harsh in many ways, it was also a wonderful time to be a child. There was time for everything; life was not as complex as it is today.

My brothers and I were all born with congenital cataracts, and mine developed rapidly so that I was blind in one eye and had little vision in the other as a child. We lived next door to my grandfather and two young aunts. As a toddler, I loved to spend time with them, but I routinely walked off the edge of

256

their porch and ended up in a heap five steps below on the concrete walk. They were concerned that I was mentally deficient or so uncoordinated that I couldn't learn to walk down stairs. It was years before anyone realized I couldn't see the edge of the porch or the steps.

Finding a specialist willing to operate on a child in the early thirties was not easy, but by the time I was 13 and finishing seventh grade it was obvious that to continue with math when numbers, fractions, and especially exponents appeared as a smudgey blur would be almost impossible. The surgery finally was done in Chicago in 1933 and 1934, and it changed my life. I know I was a trial to my early teachers. In the primary grades, I couldn't see the flash cards and was slow at learning to read; by the time I reached sixth grade I couldn't see their extensive notes on the blackboard, and so on. But they were all kind and caring, and I often think of their patience when I'm working with some of my students today.

We had family friends with an only daughter about ten years older than me, Norma Young. My first remembrance of Norma is of her taking me to a movie one warm summer afternoon when I was about nine and she was a classics student at Goucher College. Norma's mother died when I was a senior in high school, and she and her father became a part of our extended family. She attended graduate school at Iowa and received a doctorate in classical languages. She then taught in men's physical education at Iowa because Dr. McCloy preferred to have intelligent people on his staff rather than those of lesser intelligence who happened to be in the field. She taught there until her untimely death from cancer in 1963, actively participating in national and international professional societies and coauthoring books on tests and measurements in physical education and kinesiology. Through her I met many stimulating and fascinating people when I was an undergraduate at Iowa. I never thought of Norma as a role model, but looking back now I realize that she was. She was a fun-loving young woman who played pranks with us younger children. She became an independent woman who did what interested her, although in a non-traditional woman's role, and she did it with competence, grace and ease. Her students loved her. In thinking of Norma, I sometimes remember a quotation Miss Cook used, "Nor knowest thou what argument thy life to another's life has meant."

In high school I was interested in math and science and in my brother's mechanical drawing projects, but I never considered engineering as a career--I had no understanding then of what engineers do. I was an avid reader of everything, especially adventure stories in far away, exotic places, and Richard Halliburton was one of my early heroes. I had a sense then that life was a lot more interesting and a lot more fun for boys than for girls.

After high school, I attended Augustana College in Rock Island--a small liberal arts school at that time--for a few semesters, but couldn't afford to continue, since part-time jobs were hard to find at the end of the depression. My ambition at the time was to be an executive secretary. I found a place in an architect's office just prior to World War II. I was intrigued by the work they did, especially by the detailed drawings. I spent hours reading all the back issues of architectural and engineering magazines I could find. Meanwhile I continued with college classes at night in mathematics and drawing, and by 1942 I had acquired the background necessary to become a draftsman for the U. S. Army Corps of Engineers in its Rock Island office. That led me eventually to an awareness of what engineers do, and to the obvious conclusion that being an engineer would be a lot more challenging (and a lot more fun) than being a draftsman.

Prior to the beginning of World War II the Panama Canal Company had started construction of a third set of locks across the Isthmus of Panama, and from time to time people from the Rock Island District of the Corps were loaned to the Special Engineering Division of the Panama Canal Company where work on the Third Locks Project was underway. By 1942 German submarine activity in the Caribbean had escalated to the point that much of the construction materials and equipment shipped from the States for the Third Locks Project was being sunk. Consequently, the decision was made to finish the contract drawings for the new locks, but to defer further construction until the war ended. I was one of the draftsmen sent to Panama to assist in that work.

There were four young women and six men in our group. Canal Company apartments were assigned to the two older girls, and the two of us who were younger by a year or two were assigned to share their apartments. Irene Miller and I were assigned quarters together, and that was the beginning of a long-lasting friendship and professional association that lasted until her death in 1979. Although we lived in the Canal Zone and worked six and sometimes seven days a week, I spent as much time as possible in Panama itself, especially painting watercolors and photographing architectural subjects. Painting took me into the side streets and neighborhoods in downtown Panama City, and with my limited Spanish I made friends with many women and children who sat and talked with me as I painted. It was my initiation into another culture, and the experiences still color my interest in lesser developed areas of the world.

Irene and I found we shared many interests, from mathematics to arts and crafts; we had the same values; and we liked the same people. By the time we returned to the States from Panama we had both become interested in engineering, and with money saved while in Panama we decided to go back to school and study engineering. While working in Panama we had been given the choice of accepting either the standard 25 percent pay differential for working overseas or a per diem living allowance of $3.00 per day. For us the living allowance was greater than 25 percent of our pay, so we chose that option. Saving most of that allowance is what enabled us to return to college.

We applied to the University of Iowa because it was easily accessible from Rock Island by train and bus. Dean Francis Dawson, of the College of Engineering, did not initially receive our applications with much enthusiasm, but his secretary, Mary Sheedy (who had been dissuaded from entering medical school when she was younger), prevailed upon him to "at least meet with the girls." That was in 1944, and it did not seem daring to us to want to study civil engineering. It was, after all, the era of Eleanor Roosevelt who had encouraged women to become involved in all aspects of American life, and it was during World War II when American women first went to work in industry in large numbers.

Dean Dawson's initial reluctance changed to warm support once he was convinced of our serious interest in engineering, and we were admitted. Again we were lucky. The Corps of Engineers at that time had a suboffice of the St.

Paul District in the University's Hydraulic Laboratory where models were being tested of new locks and dams for the Ohio River and tributaries. As undergraduates Irene and I worked half-time for the Corps in the laboratory, an experience which changed our interest from bridge design, which seemed most attractive at the time we initially enrolled, to hydraulics.

I remember those undergraduate years in Iowa City with great affection. Working half-time and carrying a heavy class load to try to finish before our savings ran out left us with more money than time, a feeling not experienced by too many undergraduates. The faculty was enthusiastic, encouraging, and unfailingly kind and helpful. We never felt discriminated against because we were girls. Eventually we were to become the second and third girls to receive an engineering degree from the University of Iowa, being preceded by one real pioneer, a young woman who graduated in about 1914!

Iowa City was a small university town in those days. The Iowa River ran through town and through the heart of the campus. A small power dam near the Hydraulics Laboratory created a small pond in what was normally a gently flowing clear stream, but could become a raging torrent during periods of flooding in the spring. We walked across a bridge near the laboratory every day on our way to class or to work, observed the changing face of the river, and occasionally heard the roar of the water pounding snagged trees against the dam crest during flood periods. It was another fascinating lesson in fluvial hydraulics.

At that time, no women earned enough money at anything to live alone except in a rented room; usually four shared an apartment to make ends meet financially. So, after finishing our undergraduate work in 1947, Irene and I looked for jobs in the same town so we could live together. We soon found that most employers in those days had no interest in employing women as engineers. A paper manufacturer offered us positions in its technical library; but besides that our mailbox was filled with rejections. Finally, the Waterways Experiment Station (WES) of the Corps of Engineers in Vicksburg, Mississippi, offered us positions as junior engineers.

It didn't matter that we couldn't even find Vicksburg on any map we had readily available, we eagerly accepted. Again we were fortunate.

WES offered opportunities to become involved in professional engineering in ways not normally available to junior engineers. The foremost hydraulicians of that era were consultants on our projects: Boris Bakmeteff, Morrough O'Brien, Hans Einstein, Emory Lane, Art Ippen, Hunter Rouse and Lorenz Straub. We were introduced to local activities of the American Society of Civil Engineers (ASCE) and to its technical programs and publications, and we had the opportunity to meet and work with people from offices of the Corps throughout the Mississippi River valley as a result of our work with the Mississippi Basin Model (MBM), one of the largest physical models of a river basin in the world at that time.

The Mississippi Basin Model was one of the tools used in developing design and operating criteria for projects in the Mississippi River basin. It was also used occasionally in "real time" flood forecasting and floodfighting operations. During a major flood in the upper reaches of the Missouri River in the spring of 1952, engineers in the Corps District and Division offices in that area worked around the clock. Irene was is charge of work on the Missouri River section of the model at that time. The model was in operation around the clock, with almost all of us involved in one way or another. Every day, as new data on precipitation and streamflow was reported from gages in the upper Missouri basin, the expected flood hydrograph at Sioux City was modified to reflect the flood wave working its way downstream through the Dakotas. The model was operated with these flows to forecast future flood heights in the densely populated Omaha-Council Bluffs area. Engineers were concerned about whether the existing levees and floodwalls in that constricted reach were high enough to prevent flooding in those urban areas. The large storage reservoirs now in operation on the Missouri River in North and South Dakota had not yet been completed and there was no way to modify the flood wave traveling downstream. However, the model results could be used to determine whether it would be necessary to evacuate certain areas as well as where to concentrate activities such as sandbagging levees to provide additional temporary protection.

As a result of our work at WES Irene and I came to the realization that we needed more formal education to better understand the hydraulic problems we were encountering on a routine basis. We decided to go back to Iowa for masters degrees in hydraulics and engineering mechanics. One of the members of the Iowa faculty at that time was Ake Alin, a Swedish-educated engineer retired from the Corps, who had worked on Denison Dam on the Red River in Texas and on the large dams in the upper Missouri River basin. Ake and his wife lived in Omaha, and he came to Iowa City by train every third week for three afternoons of intense classes in dam design. He gave the students voluminous notes which I have retained and referred to throughout my career with the Corps. Those same notes served as a guide when I organized my first course in flow through hydraulic structures at the University of Arizona in 1981.

Irene and I finished our masters degrees in 1953, early in the Eisenhower administration when Federal water resources programs were being significantly curtailed from what they had been during the Truman years. The cutbacks sharply reduced the number of job opportunities for individuals with the kinds of experience and education we had acquired. But once again we were lucky. The Corps of Engineers' Missouri River Division office in Omaha, probably because there were engineers there who had become familiar with our work when we were at WES, agreed to hire both of us. Later we learned that their personnel office had initially refused to extend an offer of employment on the basis that "no woman is worth that much money," but eventually the insistence of the technical experts in the Engineering Division caused them to reverse their position.

In Omaha I worked on hydraulic structures, sediment problems, channel stabilization, and bank protection. I was involved in a number of model studies, with frequent visits to the models in Vicksburg. It was a stimulating job, and many of the problems we faced on the Missouri River then were problems for which there were no existing engineering design solutions. We were developing solutions based on theories that had not yet been tested on the scale in which we were working, studying the likely outcome of those solutions in physical models at WES, modifying the solutions to reflect needed changes indicated by the model study results, and building the structures and facilities needed to implement the solutions. For amusement

Irene and I often drove up to South Dakota on weekends to watch construction of Gavins Point or Fort Randall Dam, and our work often involved field trips to those dams and to others under construction farther north.

Most of the Corps people we worked with in Omaha were workaholics, I suppose. Probably over all the years with the Corps, only a few of the engineers with whom I worked were not. The questions and problems were so pervasive; there was so little in the literature; and the Iowa Hydraulics Conferences every three years were almost the only opportunities for discussions with colleagues in other Corps offices and in university research laboratories. There were not many of us across the country working in hydraulic engineering at that time and, largely through the Iowa Conferences, we all knew each other. I never felt discriminated against in that group of people because I was a woman. The Corps needed all the well-trained hydraulic engineers they could hire, and not many universities offered programs in hydraulics in those days.

The civil works program of the Corps was and is the nation's primary vehicle for addressing the planning, engineering and management problems inherent in developing and protecting its water resources. Even today the Corps civil works mission is not well understood by the general public. Working on the civil works program is not, as people often think, like being in the Army except that the general atmosphere reflects dedication to understanding the problems one is trying to solve and a commitment to getting the job done. The official Corps motto, "Essayons" (Let us try), was seldom voiced, but it characterized the spirit of the professional staff in all the offices in which I worked. However, the Corps' civil works program was not only engaged in developing and managing the nation's water resources, it was also serving as a educational program for developing the nation's water resources engineering profession. Most of the senior engineers with the Corps at that time were self-taught hydrologists and hydraulic engineers. They had learned through their work in the Corps an aspect of civil engineering which was only then beginning to be well enough understood to be formally incorporated into the civil engineering curriculum.

In 1955 Irene and I left Omaha to accept positions in the Little Rock District of the Corps where we would be involved in the planning and design of the Arkansas River Navigation Project. Once again we were fortunate to be in the right place at the right time with the right knowledge. The problems associated with building a series of navigation locks and dams on the Arkansas River, a river carrying one of the largest natural sediment loads in the world, were formidable. Our work on the Missouri River and our experience in the laboratories at Iowa and at the Waterways Experiment Station had provided the knowledge necessary to begin finding solutions to those problems. We walked the banks of the river looking at erosion; we traveled the river in small boats belonging to the District's streamgaging crew to inspect areas not accessible by road; we flew up and down the river frequently in small airplanes to obtain a perspective that could not be obtained on the ground; and we examined aerial photographs taken after almost every flood of any consequence. Out of all of these observations, our experience and education, and the experience, education and efforts of the engineers and consultants we worked with in the Hydraulics Branch of the Little Rock District, emerged a plan for dealing with the sedimentation and bank stabilization problems that had to be solved in order for the navigation project to be feasible. I was in Little Rock from 1955 to 1964, and in those nine years came to know every bend of the Arkansas River, every sand bar, rock bank, gravel bar, and so forth. Working on the Arkansas was probably the most fun I have had in my life! To go back now and see how the river has been changed by the navigation project and how the changes have improved the lives of the people living along it is to experience a warm satisfaction.

By 1964 most of the basic criteria had been established for the navigation locks and dams in Arkansas, and Irene Miller transferred to the Sacramento District office of the Corps. I went back to the Waterways Experiment Station in Vicksburg to work in coastal hydraulics, but somehow work in a research environment did not have the satisfying immediacy that I had learned to enjoy through my work in a Corps District office engaged in finding solutions to a lot of design and construction problems. Irene's experiences in Sacramento were sounding more like what we had encountered in Little Rock, and in less than a year I also transferred there.

In both the Missouri River Division and the Little Rock District my work was primarily related to design and construction. In Sacramento, however, I was assigned to the Planning Branch of the Engineering Division. There I encountered engineering studies different from those I had been familiar with up to that time. Not only was the scope and level of detail of the engineering work different from that associated with design and construction, but also the projects themselves were different. Urban flood protection, deep-draft navigation and irrigation were common purposes for projects in the Sacramento District and none of those had been major considerations in projects in Little Rock. It was time to begin learning again.

Also, while I was in Sacramento the Corps of Engineers planning process itself changed dramatically. Beginning with the passage of the National Environmental Policy Act (NEPA) in 1969, the Corps planning changed from one directed primarily toward producing the most cost-effective solution to a specified technical problem to one directed toward exploration of the economic, social and environmental consequences of a wide range of alternative solutions to water resources problems. Initially the requirements of NEPA were vague, and there was no guidance for preparing the environmental reports required by the law. The Corps addressed that problem by having its Districts prepare several environmental statements encompassing everything the planners thought to be pertinent and responsive to the law. I wrote three of them--one for a local flood protection project in an urban area, one for a large multiple-purpose reservoir project, and one for a deep-draft navigation project.

NEPA not only required the Federal agencies to consider environmental and social effects, it permitted them to provide measures for preservation and enhancement that previously could not have been addressed under existing laws. Efforts which the Corps had initiated in the 1960s to address concerns for fish and wildlife habitat, outdoor recreation, archaeology, public involvement in planning and other topics once considered beyond the Corps' purview received increasing emphasis in the revised planning process of the 1970s. Participating in the evolution of Corps planning from the relatively narrow focus of the 1950s and early 1960s to the broad, multiple objective

262

process of the 1970s was one of the most challenging assignments I encountered in more than thirty years as a federal employee.

Finally, in 1977--35 years after first setting foot in a Corps of Engineers office--I retired. Irene had had a serious heart attack in the early 1970s and had retired in 1975. I had a strong interest in art, weaving, and collecting textiles, and I had looked forward to retiring, but on making the final decision, I experienced an unexpected sense of having totally lost my identity. While I had been responsible for the environmentally controversial Morrison Creek project in Sacramento,* I sometimes said there were a number of people in Washington who didn't know me by my given name; they knew me only as "the Morrison Creek project manager." Now, retired, I suddenly found it hard to identify myself--even to myself--if I was no longer a hydraulic engineer.

For more than twenty years I had also been actively involved in the Hydraulics Division of the American Society of Civil Engineers (ASCE), serving on various committees and as chairman of the Division's Executive Committee in 1975 and 1976. After retiring I became active again. That too was a blessing, because at the time of Irene's unexpected death in 1979 I was chairman of the Society's Management Group responsible for the technical activities of all of its water-related divisions. My duties and responsibilities in that position helped fill the void created by the loss of the person who had been closest to me as both a friend and a professional colleague for more than thirty years.

By the fall of 1980 my term on the ASCE Management Group was approaching its end, and I was facing the question of what to do with the rest of my life. Again something fortuitous, and totally unexpected, happened. Emmett Laursen, who had been a fellow student at Iowa, called one day inviting me to the University of Arizona as a visiting professor for a semester or a year, or perhaps longer. "Maybe you'll like it," he said. I was terrified of the idea. I'd never wanted to teach. When I was young it was excruciating simply to stand up and give my name. Over the years I had overcome some of that youthful fear through participation in professional conferences and as a result of the many presentations I had been required to make to public audiences as part of my duties as project manager for the Morrison Creek project. However, those presentations all involved limited subject matter--material that I knew forward and backward. How could I stand in front of a class for an hour, several times a week, for weeks on end, and get all my facts in order and present them so that students could understand what had taken me a lifetime to learn? I didn't think it was possible! But Emmett was patient and persistent, and I finally agreed, reluctantly, to try teaching. It was probably the happiest choice I've made in my life. My students were mostly graduate students and many were foreign nationals. They needed not only the technical expertise I could

----

* The Morrison Creek project was an urban flood control and drainage project with a multitude of environmental and social consequences for the Sacramento metropolitan area. Because it was in the planning stage at the time of major changes in the Corps' planning process and because it exhibited important social and environmental impacts typical of water resources projects in urban areas, the Morrison Creek project was, in many ways, a test case for an emerging planning technology.

share with them; they also needed an extended family. I have found teaching to be the most rewarding thing I have done in my life.

In 1981 there were virtually no books suitable for use as texts for graduate classes in hydropower engineering, water resources planning, flow through dams and river engineering. I soon learned, too, that students (especially those for whom English is a second language) found it difficult to sift through a stack of handouts and identify the salient facts. This, in turn, led me to develop my class notes into one of the first books on water resources planning in 1984 and one on river engineering in 1985.

Margaret Mead once said that our society is changing so rapidly that everyone over the age of 25 is an immigrant in a strange land. I have learned more from my students than they have learned from me, and it is clear that the more things change, the more they are the same. In this life we can only rarely repay those who help us the most, but we can, in turn, help others, and such a chain becomes endless. Clare Boothe Luce's words to the effect that "the most important thing we do in life is to help each other" are confirmed by my experience.

My teaching and travels throughout the world sharpened my interest in developing water resources in lesser developed areas to improve living standards and public health. In recent years I participated in a symposium on drought, water management, and food production in Morocco and lectured there. I took part in a workshop in South Africa on water resource development in developing areas, and looked at sediment problems at hydroelectric plants in Transkei, one of the homelands. I participated in a symposium in Beijing on high dams, in a workshop in Wuhan on river navigation and the proposed Three Gorges Dam on the Yantze, and lectured there because some of our work on the Arkansas River was related to similar problems the Chinese have there.

I have always been interested in many things--reading, languages, sewing, cooking, travel, photography, art, weaving, bird-watching, and so on. There have never been enough hours in a day to pursue everything that interested me, and I recognize that I probably suffer from the Anglo-Saxon Protestant work ethic--if I'm not busy, I feel guilty! One's energy level diminishes with the years, and one eventually learns that it is impossible to "do it all and still be human." My social life was almost always limited to professional organizations and activities. When I was young, I thought everyone grew up and married; it was simply a matter of finding the right person, and at twenty that didn't appear to be impossible. Reality turned out to be different. Could I have accomplished what I have if I had been a married woman? I doubt if any American woman of my generation could have. Few American men of my generation seemed to be interested in intelligent women in a personal way; they may have been shy or simply intimidated to see us succeed. Once I entered engineering school my friends and mentors were all men, but those who influenced me the most up to that point were strong, kind, independent women, and all except my mother and my Greataunt Minnie were spinsters. I think young men today are much more appreciative and supportive of intelligent women.

If I had it to do all over again, would I make the same choices? Yes, I think so. It's been a richly rewarding life.

Percy Harold McGauhey
Vinton W. Bacon

National Academy of Engineering - **Memorial Tributes**

Election to the National Academy of Engineering is one of the highest tributes that can be paid to the accomplishments of an engineer in the United States. In 1979 the National Academy published a collection of Memorial Tributes to deceased members of the Academy. The tributes were prepared by colleagues of the deceased members.

Professor P. H. McGauhey, the subject of this tribute, is an outstanding example of a blend of professional expertise that is becoming increasingly rare in the engineering profession--an outstanding researcher, outstanding teacher, outstanding practitioner, and a warm and witty human being. There is no shortage of warm and witty human beings in the civil engineering profession, but the number of individuals with the talent, interest and ability to be simultaneously successful in research, practice and teaching is growing distressingly small.

At least part of the reason for the decline in the "all-around" expert civil engineer is a factor that Professor McGauhey himself spoke about on more than one occasion--the explosion of knowledge that makes it a full-time job for a researcher to stay abreast of the information needed to vigorously pursue solutions to problems at the "cutting edge" of a given area of expertise. The nature of the researcher's knowledge needs--to be familiar with as much knowledge as possible about a relatively small area of expertise--is diametrically opposed to the knowledge needs of the civil engineering practitioner--to know something about the many different areas of technical and non-technical subjects that must be considered and evaluated in finding solutions to today's real-world problems. In short, one can focus on improving one's breadth of knowledge or on one's depth of knowledge, but usually not on both.

Even the art of teaching has become more complex in recent years as more has been learned about differences in students' learning styles and more emphasis has been placed on one-on-one interactions with increasingly large numbers of students. Like the dedicated researcher and the dedicated practitioner, the dedicated teacher often finds little time to expend on efforts to become proficient in the other two areas.

Realizing that there are not likely to be many civil engineers like Professor McGauhey in the future makes it all the more important to recognize and celebrate his accomplishments.

P. H. (Mack) McGauhey, who died on October 8, 1975, was intimately known and deeply respected by professional engineers, educators, and governmental officials in the State of California and throughout the nation and world. His name was synonymous with sanitary engineering and water resources. There were few people facing perplexing, practical engineering problems who did not seek his help. Professor McGauhey was born on a homestead ranch on January 20, 1904, in Ritter, Oregon. The harshness of the eastern Oregon lands is reflected in his philosophy of life and in his verses, many of which appear in *Rimrock Ranch and Other Verses* and in *Oral History of the Sanitary Engineering Research Laboratory*, published by the Bancroft Library of the University of California, Berkeley.

Before getting to the real man and human being, let us look at his outstanding professional record, which was recognized by election to the National Academy of Engineering in 1973. In 1927 he received a Bachelor of Science degree in civil engineering from Oregon State University; a Civil Engineering degree in 1929 from Virginia Polytechnic Institute; and a Master of Science degree in hydraulic and sanitary engineering from the University of Wisconsin, Madison, in 1941. Utah State University honored him with a Doctor of Science honorary degree in 1971.

He served in faculty posts at Virginia Polytechnic Institute, at the University of Southern California and at the University of California, Berkeley, the last beginning in 1952. In 1957 he was appointed Director of the Sanitary Engineering Research Laboratory, Professor of Civil Engineering, and Professor of Public Health. In addition, he was appointed to the chairmanship of the Department of Civil Engineering.

It was in the latter capacity that he molded and led what has become one of the most respected sanitary engineering laboratories in the world. Professor McGauhey conducted pioneering investigations on a wide variety of subjects that included the composting and management of solid wastes, the economic evaluation of water, the treatment of waste by septic tanks and percolation fields, the eutrophication of natural waters, the fate of detergents in sewage treatment, and the use of the soil mantle as a waste management and water reclamation system. In each of these areas, Professor McGauhey became a world expert. What was so amazing, besides the diversity and excellence of his research, was his ability to bring his spirit of eternal optimism and his manner of meaningful compromise into the organization of his research, into the organization of the Sanitary Engineering Research Laboratory, and into the academic programs in civil engineering and public health. These qualities are reflected in the type of research that he undertook--he had the ability to hold together interdisciplinary research groups with the knack of allowing each investigator to contribute both toward the mutual objective of the group and toward fulfilling his own individual satisfactions.

He retired as Director of the Laboratory in 1969, but he was soon recalled by the Chancellor of the Berkeley campus to conduct a study of the role of the University in environmental studies.

In addition to the honors mentioned above, he received the Fuller Award of the American Water Works Association (1950), the Harrison Prescott Eddy Medal of the Water Pollution Control Federation (1960), the Distinguished Service Award of the National Clay Pipe Institute (1964), the Service Award of the California Water Pollution Control Board (1964), and the Gordon M. Fair Medal of the Water Pollution Control Federation (1969).

He served as a consultant for the State of California on the Lake Tahoe water management problems, the Department of Health, Education and Welfare, the Agency for International Development, the nation of Israel, the Ford Foundation, and many others. One of his last significant contributions was as

266

chairman of a three-person board of consultants that developed a Wisconsin Statewide Solid Waste Recycling Program, which resulted in the legislature creating an authority empowered to design and operate recycling systems--the first of its type in the nation.

But such creative achievements are expected of someone elected to the National Academy of Engineering. Virginians, Californians, and admirers around the world will always remember the warm, humble, helpful, philosophical and poetic man.

First and last, he was an educator. He had unique and old-fashioned views as to what an educator should be. Undoubtedly, these were formed by his early bleak schooldays in a one-room schoolhouse in eastern Oregon and by the fact he had to work his way through grade school, high school, and college. Such experiences would harden in philosophy even the softest of dispositions. His verse "Schoolhouse" expresses his ideas on the real purpose and meaning of education:

Schoolhouse

Its blackboard showed the sentence parsed--
Though feebly understood
And random truths there shone a while,
Then disappeared for good.

Yet stubborn minds perforce must yield
Beyond its battered door.
We went in poor and ignorant--
And came out only poor.

Adversity being the creator of character, Professor McGauhey had more than his share. Just when he was about to complete his doctorate, he contracted tuberculosis. He spent two years in a sanitorium and another year recuperating from surgery. From his verse "Sanitorium," a glimpse of this time emerges:

Sanitorium

Infirmary
Like patient oxen in their stalls
We lie benumbed of flesh and brain:
Each crack, each smear upon the walls,
Becomes the pattern of our pain.

Horizons
Slowly the restful arc where earth
Meets patterned skies we knew so well;
The far horizons of the mind
Are squared and shrunk to fit this cell.

267

Evening in Summer
>There is a hungriness that grips the heart
>When the last oblique rays of the dying sun
>Shatter like hopes against these ageless hills
>That wall us off from life.
>
>I see you there alone--yet cannot come
>To share your solitude
>When lengthening shadows of the evening grow--
>Suddenly--to a blackness that is night;
>Bearing on its restless wings
>The hot damp cloak of loneliness.

Education to him was the task of instilling useful and well-structured knowledge into recipients who were expected to work hard and doing this without unnecessary interference or ballyhoo from administrators. He had little time for professors who taught at 8:10 a.m. what came into their heads at 7:55 a.m.; nor did he have much sympathy for the professional student who spent too many years getting too few degrees. In his unpublished novel, aptly entitled "Phooey on Your Alma Mater," you can find these attitudes precisely stated:

>Sound advice and high purpose have not always been the
>considerations by which our institutions of higher learning
>are populated. A good long loaf at the old man's expense
>has always stood high among the reasons for congregating
>within ivy-covered walls. Nor has improvements in headwork
>always been the end result of the learning process in such
>an environment. A couple of generations ago, some colleges
>were so successful in converting their loafers into sots
>that many parents were thankful that poverty protected their
>sons from the moral strain of a college education.

On the "free-thinker" professor he states:

>Much of a University catalog is given over to a list of
>subjects along with descriptions indicating that the whole
>field of human knowledge is to be covered by Professor Van
>Beer in three hours per week for one semester. By this
>subterfuge the grouchy old professor can teach anything he
>pleases without fear that the accrediting committee will
>compare his course unfavorably with the same course at
>Harvard.

He gave equally short shrift to the elitist academician who felt that the dispensation of knowledge and the conduct of esoteric scholarly research had to be divorced from any semblance of everyday hard work. In his *Oral History of the Sanitary Engineering Research Laboratory*, he notes that the Richmond Field Station is some seven miles northwest of the Berkeley campus and remarks:

>Resistance of some faculty members to undertake such a long
>journey (from campus) was one of the problems of utilizing
>the (Field Station). I once explained this phenomenon on

268

the rationale that the (Field Station) was located on the
wrong side of the Campus. Thus it was not on the way to
Europe and hence (was) geographically inconvenient.

His feelings on the unreasonable world of academe were summed up, during
the so-called Free Speech Movement at Berkeley, on a scrap of paper found in
his desk drawer:

There is nothing so crude, lewd, or treasonable that the sub-
species of apes, that it [academe] tolerates as students, will not
attempt to force upon the community. There is nothing so
preposterous, unbecoming an academic community or even
common civilization, that its faculty will not, in its childlike
naivete, espouse in the name of Academic Freedom. There is
nothing so craven or absurd that its administration will not
embrace when the enemies of society and America have at the
gates of Sproul Hall.* There is no demand so preposterous, that
it will not be tolerated while faculty and students ponder how
to remove the cause of this blatant attack on society.

Mack said he was one person who "gave up church for Lent and never went
back"; and when he became Emeritus he would let you know that he had just
been "retarded from the University." His poetry conveys his wit and "pure
fun" humor:

Advice Is Worth Its Salt

I sought a friend (advice to borrow).
He said, "Go home and drown your sorrow!"
I felt so bad I shed a tear--
And found the salt improved my beer.

La Cucuracha

I picked up my glass and went for a drink,
There was something in it, and he jumped in the sink.
He got clean away--cause he saw me first.
Strange how a cockroach can quench one's thirst!

Professor McGauhey's poetry showed all his moods. There are things in these
lines that you rarely heard him say. The barely endurable pain that he must
have suffered almost constantly throughout his life only seeps through to the
outside world in his verses.

----

* The main Administration building on the Berkeley campus.

Dichotomy

Though 'gulfed in weariness by day
That makes him long for bed,
A man may come to dread the night--
When night holds things to dread.

Mack was a true example of the type of individual here now being paid homage: a descendent of pioneer Americans who took the promise of the American dream literally and who achieved it through the application of strenuous physical labor to a lifelong quest for education and excellence. In *Rimrock* he wrote his epitaph:

He did not lose his zest for life
Nor judge the race not worth the run.
But he would have judged his duty shirked
If he failed to do--what must be done.

Throughout his life, Percy Harold McGauhey was sustained by a loving and lovable wife called Margo.

Civil Engineer in Space

Utah State University Alumni Association **Outlook** - July-August 1981

Even civil engineers who view the entire globe as a potential jobsite don't commonly perceive outer space as a workplace. In fact, the space exploration program has created many new job opportunities for civil engineers. It is true that most of them are on earth--designing and constructing launch facilities, assisting in the structural design of spacecraft, and designing facilities for storing, recycling and disposing of waste products generated during space missions, to name a few. However, it is becoming clear that civil engineers will play increasingly important roles in future space missions and that some of those roles will require civil engineers to travel in space.

Two civil engineers have already been employed as astronauts by the National Aeronautics and Space Administration: Mary Cleave and James van Hoften. Van Hoften was the first civil engineer in space. He was aboard Challenger flight 41-C in April, 1984, and was responsible for replacing defective parts in the Solar Max satellite which had been inoperative for much of the three years it had been in orbit.

Mary Cleave was aboard Atlantis flight STS 61-B in November, 1985. She served as flight engineer during the ascent and re-entry and as operator of the mechanical arm located in the cargo bay of the shuttle. During that mission the crew deployed several satellites and conducted two six-hour spacewalks to demonstrate the feasibility of large-scale construction in space.

Dr. Cleave flew again on NASA flight STS-30 in May, 1989. She and the other members of the flight crew successfully launched the interplanetary probe **Magellan** *which is expected to produce radar imaging of the planet Venus in mid-1990.*

Environmental engineers like Mary Cleave are expected to play an important role in the space program. The discipline we now call environmental engineering was once called sanitary engineering. Sanitary engineers dealt primarily with providing safe drinking water supplies and developing facilities for collecting and treating sewage and garbage. Although their work was critically important to an urbanizing society, it was not attractive to many young people.

Today's environmental engineer is involved in one of the most challenging disciplines in the civil engineering career field. On earth, civil engineers working in the environmental discipline seek solutions to problems in water pollution, air pollution, and solid waste, nuclear waste and toxic and hazardous waste disposal. Their work probably requires the most diversified knowledge of any of the civil engineering disciplines. Many of the problems they are trying to solve are relatively new, which means that in many cases there are no tested solutions. The environmental engineer in space will be faced with the same types of problems, but with even more unknowns.

Mary Cleave and James van Hoften are among those leading us into the new frontier of the twentieth century, just as the civil engineers who built canals and railroads were among the leaders in the exploration of the western frontiers in the nineteenth century. This article describing some of Mary Cleave's early experiences as an astronaut appeared before she had journeyed into space.

In science fiction, there are women all over space. In actuality only one woman--from the Soviet Union--has gone into space. As of yet. Real life science is destined to catch its fictional counterpart as America's follow-up space shuttles escape earth's gravity.

Columbia, the first space shuttle, has made its first successful voyage, its second flight is scheduled in September, and others are planned through coming years. Women are very much part of the plans as astronauts and everything else.

Mary Cleave is one of them. The holder of a Ph.D. in civil and environmental engineering from Utah State University (USU) is one of a class of 21 astronauts now in training at the Houston space center. For the second shuttle flight her boss will be the person communicating with the orbiting craft and she says, "At that time I'll be his go-fer."

As to her own duties on a flight, Cleave says, "It could be anywhere from two to five years before I get on. You develop an attitude of patience."

"Hopefully," she adds with a grin, "it will be before I retire."

*   *   *

How does a native of Great Neck, New York, get to USU and from there into the country"s space program?

The first part is simply geography. "I went to school at Colorado State and then came to Utah because I liked the area," Cleave says, adding, with no intended pun on her current career, "I chose Utah State because of the down-to-earth people I met here."

She worked for the Utah Water Research Laboratory before deciding on the engineering degree. She heard about the astronaut program from a fellow graduate student and applied.

"I think the National Aeronautic and Space Administration (NASA) was looking for people who like to fly and who have a varied background," says the former high school biology teacher, who has a master's degree in microbial biology, experience with water research and the engineering degree. She has a pilot's license but hadn't flown for some time.

"My background is diverse because I could never decide what I wanted to be," she jokes.

She's decided now. NASA gives prospective astronauts one year of training before it requires them to decide. Though she is just now finishing her first year she made the decision long ago.

"Eventually, I'd like to manage a space station," she says.

In the meantime, rigorous training occupies Cleave and the 20 others in her class. As if to prove that astronauts are made, not born, NASA requires physical training similar to that experienced by the military's elite troops--the Navy Seals, Army Special Forces and Marine Recon groups. In fact, most of her counterparts are career military people.

The prospective astronauts also spend 15 hours each month flying a T-38, which is the same plane that escorted the *Columbia* in for its landing at Edwards Air Force Base.

According to Cleave the training and planning almost reach the point of overkill. "They go over and over it and tell you and tell you until you can respond automatically," she says.

Such automatic response is meant to save lives. Cleave says everyone in the space program is very careful and sensitive about the lives of astronauts. . . .

"They even assign us to spend some time with the people in tracking stations around the world and with other ground station crews," reports Cleave. "The reason for this is so the ground people will always realize they are dealing with humans rather than just machines in space."

Additionally, the astronauts attend many lectures and have a room full of books they must read on their own during the first year.

"To me, sitting at the desk is more grueling than the other training," she says.

The hours are somewhat erratic, she notes, but always long.

"I usually go in at 7:30 a.m. just so I can get a parking place," she laughs. "If we fly at night, we don't get home until 11:30, and we work most weekends to keep up. It amazes me to go in on a weekend and [find] everyone is there working just like a regular day. I'm not used to being around 20 other work-compulsive people."

And she doesn't have to be big and strong to keep up. Cleave is small. NASA requires astronauts to be in the 5' to 6'3" range. Mary Cleave just barely made it. She and the biggest person often serve as sort of guinea pigs for many of the tests; for instance, to see how fast a person can exit the orbiter or to see if the short person can reach the controls.

In the camaraderie that has developed among class members, nicknames and good-natured ribbing are standard procedure. Cleave's diminutive size and environmental engineering background have earned her the sobriquet "Sanitary Fairy" or sometimes "The Effluent Elf."

And what does the future hold for this space elf with the big background? Right now she just wants to go into space and believes that in spite of cutbacks this will happen.

Astronaut Mary Cleave

"Budget cuts are affecting some of what the shuttle is supposed to do in orbit but Defense Department spending will keep it in the air," she says. "The purely scientific experiments will come back, many because of private investment."

"I believe space is very important to our future. Being one of the first women has drawn a lot of attention to me but that isn't what matters. It's being on this frontier with so many exciting and dedicated people that makes me happy and proud."

Selected Additional Reading

Biographies and autobiographies of civil engineers are relatively rare, but the interested reader can often find substantial amounts of biographical information about individual civil engineers in books about civil engineering projects, particularly projects that are historically important. The books on this list are included for one of two reasons: they provide more information on a person identified in one of the readings in this book; or they are a general biographical reference. Through use of the biographical reference works one can identify specific civil engineers and the projects with which they have been associated and use that information as a starting point in a search for more extensive biographical information.

American Society of Civil Engineers Committee on History and Heritage. **A Biographical Dictionary of American Civil Engineers** (New York: American Society of Civil Engineers, 1972) 164 pgs.

Dorsey, Florence C. **Road to the Sea: The Story of James B. Eads and the Mississippi River** 340 pgs. (New York: Rinehart and Co., 1947)

Franzwa, Gregory M. and Ely, William J. **Leif Sverdrup** (Gerald, MO: Patrice Press, 1980) 387 pgs.

Hindle, Brooke. **David Rittenhouse** (Princeton, NJ: Princeton University Press, 1964)

Johnson, Jack T. **Peter Anthony Dey: Integrity in Public Service** (Iowa City, IA: The State Historical Society of Iowa, 1939) 246 pgs.

Jones, Helen H. **Rails from the West: A Biography of Theodore D. Judah** (San Marino, CA: Golden West Books, 1969) 207 pgs.

Lewis, Gene D. **Charles Ellet, Jr.: The Engineer as Individualist** (Urbana, IL: University of Illinois Press, 1968) 220 pgs.

Morris, M. D. **Civil Engineers in the World Around Us** (New York: American Society of Civil Engineers, 1974) 293 pgs.

Perkins, Jacob R. **Trails, Rails and War: The Life of General G. M. Dodge** (Indianapolis, IN: Bobbs-Merrill, 1929) 371 pp

Pursell Jr., Carroll W. **Technology in American: A History of Individuals and Ideas** (Cambridge, MA: The MIT Press, 1981)

Ratigan, William. **Highways Over Broad Waters: Life and Times of David B. Steinman** (Grand Rapids, MI: Eerdmans, 1959) 359 pgs.

Roysdon, Christine and Khatri, Linda A. **American Engineers of the Nineteenth Century: A Biographical Index** (New York: Garland Publishing Inc., 1978)

Sayenga, Donald. **Ellet and Roebling** (York, PA: American Canal and Transportation Center, 1983) 60 pgs.

Schuyler, Hamilton. **The Roeblings** (Princeton, NJ: Princeton University Press, 1931) 425 pgs.

Steinman, David B. **Bridges and Their Builders** (New York: G. P. Putnam's Sons, 1957) 379 pgs.

Steinman, David B. **The Builders of the Bridge** (New York: Harcourt, Brace & Jovanovitch, 1950) 420 pgs.

Stevens, John F. **An Engineer's Recollections** (New York: McGraw-Hill Publishing Co., 1935)

Stone, Irving. **Men to Match My Mountains** (New York: Doubleday & Co., 1956) 459 pgs.

Weigold, Marilyn. **Silent Builder** (Port Washington, NY: Associated Faculty Press, 1984) 188 pgs. (A biography of Emily Warren Roebling)

Wood, Richard G. **Stephen Harriman Long 1784-1854: Army Engineer, Explorer, Inventor** (Glendale. CA: A. H. Clark Co., 1966) 292 pgs.

Yager, R. **James Buchanan Eads: Master of the Great River** (Princeton, NJ: Van Nostrand, 1968)

CIVIL ENGINEERING PROJECTS

The selections included in this section have been chosen to illustrate the diversity of work in which civil engineers are engaged, the complexity and challenges inherent in the work, and the variety of ways which different individual civil engineers cope with the demands of their undertakings. Not all of the projects are well known, but each project was intended to serve a vital purpose. As is often the case in life, the most important projects are not necessarily the most well known.

The projects planned, designed, constructed, operated and maintained by civil engineers have become essential elements in the sustenance of modern American society. In today's world the ability of a society to conceive, support, finance and develop civil engineering projects is one of the most obvious indicators of the state of a nation's development. In a very real way, the extent of a nation's investment in its civil engineering projects is a measure of its stability and its willingness and ability to invest in its future. The importance of these projects to society as a whole is indicated by the fact that in progressive societies the impetus for investment in major civil engineering projects is institutionalized in the society's decision-making structure and not left to the whims of individual promoters.

The term "infrastructure" has been used increasingly in recent years as a descriptor for the transportation facilities, utilities, public buildings and other similar civil engineering projects that collectively make up the "built" part of our nation's environment. Despite the fact that individual projects might have been controversial at the time they were conceived or might have become controversial since they have been constructed, few people would question the necessity for and importance of these facilities in the aggregate. Despite their importance, the existence of these projects has been taken for granted by most Americans. Little thought is given to how they came into being. Even less thought is given to the need for maintenance unless that work is deferred so long that deterioration becomes a obvious threat to public safety (such as a decrepit bridge) or a persistent public nuisance (such as neglected potholes in urban roadways). And in recent years it seems that virtually no thought has been given to the fact that plans need to be made for the eventual replacement of many of these projects.

These readings, together with some in other sections of this book, clearly show that we are the beneficiaries of a past in which both government decision-making and the initiative of individual civil engineers contributed to the development of projects which have made it possible for us to enjoy one of the highest standards of living in the world. Recapturing the spirit of commitment to investment in the future that led to the development of these sometimes ordinary and sometimes spectacular, but always necessary, projects is a major challenge for the civil engineering profession in the years ahead.

Some of the projects described have been successful beyond even the wildest dreams of their creators. Others failed before they were even completed. Some have continued to provide useful service far beyond any reasonable expectation, and others were becoming obsolete while they were still under construction. Such is the nature of civil engineering.

Dealing as they do with projects with long physical lives, civil engineers must be capable of dreaming about the future, of designing to withstand loads imposed by forces which today might seem to exist only in the abstract; of thinking the unthinkable and searching for ways to prevent it from becoming reality; of facing and living with uncertainty. They must be prepared to balance reductions in risk against increases in cost; aesthetics versus economics; the value of human life and the cost of human suffering versus monetary expenditures; environmental quality versus economic development; and technical feasibility versus political reality. These judgments need to be made not only in the context of current needs and desires but also with some consideration of the needs and desires of future generations who will be required to live with the results of the decisions. They need to reflect the competition for scarce economic resources among (and impact upon) a wide variety of societal objectives, many of which deal with human services that are clearly beyond the professional concerns of the civil engineer. Finally, these judgments need to reflect what is implementable--not just feasible in the technical sense, but what is affordable, supported by the public at large, and politically achievable.

The projects described in this section reflect the needs and concerns of American society in every era of the nation's history and demonstrate both the response of the civil engineering profession to changing societal values and the ways in which those responses have shaped the nation as it exists today.

Engineering the Erie Canal
John Tarkov

American Heritage of Invention & Technology - Summer, 1986, Issue

In this brief description of the construction of the Erie Canal John Tarkov communicates the challenges encountered by men attempting to solve problems never solved before. He rightly concludes that learning and understanding the process of finding solutions to new problems is an inherent and important part of educating a person to become an engineer. Mastery of that process, he suggests, is demonstrated by repeated instances of creative problem solving in a variety of situations. Societal recognition of that mastery leads to recognition of one's qualification as an engineer.

Not every civil engineer gets the opportunity to apply his or her problem-solving ability to a project as visible and critical as the Erie Canal. And yet every civil engineer who practices engineering at the professional level engages in problem solving that in its own way is as original and creative as the engineering work on the Erie Canal.

Thomas Jefferson had a good eye for real estate on a grand scale. But when the notion of a canal linking the Great Lakes with the Hudson River near Albany, New York, was put before him in 1809 by two New York State legislators, he dismissed it out of hand. "Why, sir," he said, ". . . you talk of making a canal *three hundred and fifty miles long through a wilderness!* It is a little short of madness to think of it at this day!"

The idea of the Erie Canal tended to arouse that kind of skepticism wherever it was broached, and it has been difficult for later historians writing about the canal not to disparage its many early critics. The Erie Canal turned out, after all, to be a resounding success, perhaps the single most important public work ever built in the United States. But at the time, its opponents were merely being prudent. There were many more reasons to believe the Erie would fail than that it would succeed.

Canal building was hardly an advanced craft, much less a science, in early nineteenth-century America. The two most ambitious artificial waterways attempted prior to the Erie were the twenty-two-mile-long Santee Canal near Charleston, South Carolina, and the twenty-seven-mile-long Middlesex Canal linking the Merrimack River with the Charles River and Boston Harbor. Neither had much to commend it. Completed in 1800, the Santee took eight years to build and was a fiscal nightmare. It wasn't much of a canal either. Political cronyism had pushed its path away from natural water sources, so that eventually two-thirds of its channel lay bone dry. The Middlesex, finished in 1803 after nine years of work, didn't pay a dividend to its stockholders until 1819.

The Santee and the Middlesex were built at the approximate rate of two miles per year. If they were any yardstick, the Erie, with its 363-mile length and its significantly larger channel (forty feet wide and four feet deep) might well be open by the millennium.

The most serious impediment to progress on both the Santee and Middlesex works was the lack of trained engineers. In the early nineteenth century there was not a single native-born engineer in America. The Santee Canal had been engineered by a peevish and none-too-honest Swede named Christian Senf. The Middlesex began in American hands, but the local magistrate chosen to lay out the canal line quickly proved incompetent, and an Englishman named William Weston was called in as consulting engineer.

Of the handful of European engineers working in this country around 1800, Weston was the most active, but he was not a man of prodigious talent; many of the structures he designed for the Middlesex Canal simply fell apart. The historian Elting E. Morison writes of Weston: "Knowing not much, he knew a great deal more than anyone else and was in frequent demand." He was in demand once again in 1816, when--after years of discussion, debate, and political maneuvering--New York State seemed poised to proceed with its canal undertaking. The state's board of canal commissioners offered him seven thousand dollars to come from England and oversee construction of the Erie, but Weston declined on the grounds of his advancing age and his desire to stay with his family.

Benjamin Wright

At that point the canal commissioners discontinued their search for foreign expertise and instead appointed four residents of upstate New York--Benjamin Wright, James Geddes, Charles Broadhead, and Nathan S. Roberts--to be principal engineers on the canal. None of these men had ever even seen a canal before. Wright, Geddes, and Broadhead were judges. They knew surveying because such knowledge was useful to magistrates when they heard property cases. Roberts was a schoolteacher who had taught himself surveying at Wright's urging.

But simple surveying, the kind that goes into making boundaries, is of limited use when it comes to building a canal. Canal engineers must be able to measure elevations with a precision that allows for vertical errors measured in inches over horizontal stretches measured in dozens of miles. None of the four principal engineers appointed to building the Erie had ever taken a level before. But within a year they had taught themselves well enough so that when Geddes and Wright ran levels by different routes from Rome to Syracuse in the spring of 1818, enclosing a loop of one hundred miles, their final readings differed by less than two inches.

In all the many other details of canal building, they learned as they went, becoming engineers long after the title had been conferred on them. This was as true for the younger engineers on the canal as it was for men like Geddes and Wright. Virtually every American engineer of consequence during the first half of the nineteenth century learned his profession either on the Erie Canal or from an engineer who had been there. The Erie was truly, as a number of historians have said, America's first school of engineering. Men learned things there because they had to, and they learned them in whatever way they could: by mistakes, by watching and asking questions, and by accepting expert authority without regard to rank.

In 1818, for example, a minister and amateur mathematician named David Bates was serving as the resident engineer along a stretch of the canal east of Syracuse, and a young farmer named John Jervis was working as a target man in Bates's surveying party. Jervis had gained a rudimentary knowledge of leveling the previous year while felling trees for the canal and had buttressed this knowledge by studying two books on the subject. He ended up teaching Bates how to measure elevations.

Bates learned well from his subordinate. Later in life he was the principal engineer of the Ohio canal system. Jervis went on to become one of the greatest American engineers of the nineteenth century. Years later he wrote that on the Erie "the mechanical department of engineering was practically in its infancy. . . .The plan for a timber trunk for the aqueducts was prepared and submitted by a carpenter, Mr. Cady of Chittenango. This plan was adopted in nearly all the wood trunk aqueducts on the canal. At this day it stands as a well designed plan."

By 1819 local contractors and mechanics like Cady working on the first section of canal under construction--the ninety-four miles between the Seneca River and Utica--had invented three immensely valuable laborsaving devices. The land they were clearing was thick forest, and without their new machinery, the entire enterprise might have ground to a halt early on.

The first of the three inventions made it possible for one man to fell a tree of any size without using a saw or an ax. The worker would secure one end of a cable to the trunk of a tree some sixty feet above the ground and the other end to a roller turned by a gear with a crank. After anchoring the apparatus to the ground one hundred feet from the base of the tree, the worker would turn the crank. The tremendous leverage obtained by fastening the cable so high up made it only a matter of time and exertion before the tree crashed to the ground.

The stumps left behind could be extracted by another local invention. It rested on two huge wheels sixteen feet in diameter joined by an axle almost two feet thick and thirty feet long--in other words, a fair-sized tree. Midway along this axle-tree was a smaller wheel, fourteen feet in diameter, with its spokes firmly united to the axle barrel. A rope was fastened to the rim of this middle wheel, wound around it several times, and its loose end attached to a yoke of draft animals.

The middle wheel would be positioned almost directly over a stump, the two larger wheels braced, and a chain made fast to both the thick axle and the stump. When the team of horses or oxen pulled on the rope, the rotation of the wheel made the entire axle turn, winding the chain around it and gradually uprooting even fresh, green stumps. With this huge machine, seven men and a pair of horses could pull thirty or forty stumps in a day.

The third invention was a plow with a heavy piece of sharpened iron attached to it; when draft animals pulled the plow, the plate traveled below the ground, cutting through roots as thick as two inches so that they could be easily scraped away.

From almost the beginning, plowing and scraping were the preferred method of excavation on the Erie, since the continual traffic of men and animals packed and strengthened the banks in a way that shoveling and carting could not. Spades and wheelbarrows did have to be used when the ground was wet, but on the Erie new spades were designed to cut through roots more easily, and new wheelbarrows provided greater ease in carting dirt away.

All these means for clearing and shaping the land were in use by 1819. So was a well-organized system--based on the accountability of contractors--for letting many small excavation and construction contracts to private citizens along the canal's route (the state put up major structures, such as aqueducts and dams). There was an ample supply of local labor, supplemented with Irish immigrants shipped north by New York City's Tammany Hall. There was a corps of engineers whose diligence more than compensated for their inexperience.

And yet despite all that, the Erie Canal might have been a failure--even at that early stage--for lack of one vital commodity: material for building durable locks. If the locks were built of wood, they would rot in a few years. Good stone locks needed hydraulic cement for waterproof mortaring, but the only known sources of hydraulic cement were in Europe, so the cost of that would be prohibitive. The only solution was to build the locks by uniting stone to stone with ordinary mortar and applying a thin coating of imported hydraulic cement at the joints between them. It was a concession to the apparently inevitable: The Erie's locks were destined to fall apart fast. The only question was how fast.

That question became academic almost as quickly as it became critical. In 1818, quite by accident, contractors along the canal line discovered natural cement rock. It was also discovered by a Herkimer County physician named Andrew Bartow, who demonstrated its potential for the benefit of Benjamin Wright and Wright's chief assistant engineer, Canvass White. In a tavern in the village of Chittenango, Bartow mixed the pulverized rock powder with sand and placed a ball of it into a bucket of water. By morning the mixture had hardened to the point where it could be rolled across the floor like a stone.

Canvass White, easily the most gifted engineer on the canal, had spent the previous winter in England--at his own expense--studying existing canals and learning about hydraulic cement. By the start of the 1819 construction season, he had perfected the process for refining this local rock into true cement powder. By the time the canal was completed, more than four hundred thousand bushels of it had been used. It firmly held together every bit of masonry on the canal, from mundane little culverts to gigantic aqueducts--and, of course, good stone locks. (The gates were of wood.)

282

White's discovery exploded on the scene so quickly that his patent on the process was conveniently ignored by all the manufacturers. Eventually the state legislature considered awarding him ten thousand dollars in compensation. He was entitled to at least six times that in royalties, but even the attempt at partial reparation fell through. It was a typical outcome for White. Perhaps an engineering genius, he was luckless in his financial affairs. He died young, in 1834, leaving his widow little more than the furniture she was compelled to sell.

The discovery of native hydraulic cement came as the building of the Erie Canal was about to enter a more technically difficult phase. The middle section had been chosen as the starting point for construction in 1817 because it offered advantages to both the engineers and the pro-canal politicians. Politically the advantage of starting in the middle was twofold. Results, measured in navigable canal miles, could be effected quickly there and then used as leverage to obtain more state funds for further work. At the same time, as the middle section was completed, popular support for the rest of the canal would grow in the areas to the east and west.

Engineering a canal forty feet wide and four feet deep through the ninety-four miles of wilderness between Utica and the Seneca River was no mean technical feat, but the problems it presented were minor compared with the ones that lay in the later segments. If the canal was a school of engineering, the middle section offered the appropriate introductory course.

That course was completed in the fall of 1819, and the middle section was opened for navigation the following spring. Along the 270-odd miles of unfinished channel remaining, the most striking engineering problems (and solutions) were to be found in the 158-mile western portion, between the Seneca River and Buffalo, but the more difficult, if less spectacular, engineering had to be done in the east. There the canal dropped 419 feet in the 109 miles between Utica and the Hudson River; between Lake Erie and Utica the drop was only 146 feet over 252 miles.

The magnitude of the descent in the eastern section would have offered enough of a challenge by itself, but the difficulty was compounded by the inhospitable topography of the Mohawk River valley, through which eighty-six miles of the canal's line had to be laid out. The banks of the Mohawk were cramped by steep hillsides, which in places ended at the water's edge. This required construction of the canal channel in the river itself, supported on a masonry base and protected by high embankments. The canal's eastern section was a potential nightmare, and it fell to White, by now a principal engineer of the canal in everything but title, to make the best of it.

The key to his solution lay in his placement of locks, which automatically determined the location of the pound levels--the stretches of channel between the locks. In order to take advantage of the better line available on the north bank, he ran the canal across the Mohawk four miles below Schenectady on a 748-foot-long aqueduct. Twelve miles farther east, at the Cohoes Falls, he recrossed to the south bank via a 1,188-foot-long aqueduct. White saw the work through in three years. By the end of 1823, the Erie was open from Brockport, some twenty miles west of Rochester, all the way to the Hudson River at Albany.

At that point, about eighty miles of channel, between Brockport and Buffalo, awaited completion. Already standing were two of the western Erie's engineering triumphs: the Irondequoit Embankment and the Genesee River Aqueduct.

Not far east of Rochester, an unexceptional stream called Irondequoit Creek had carved out a valley and an engineering challenge. Taking the canal across it without adding about one hundred and fifty feet of up-and-down lockage was imperative. The only thing that made the task even remotely possible was the presence of several natural ridges that could carry the canal at least partway over the valley it would have to span. James Geddes had long advocated linking these ridges together with great earthwork embankments and running the canal across the top; the canal commissioners were hesitant to approve so bold a plan, but finally realizing that they had few real options, they authorized work to proceed as Geddes had proposed.

The Irondequoit Embankment, built entirely during the season of 1822, consisted of three natural ridges joined together by two man-made ridges, one 1,320 feet long and the other 231 feet. The canal ran along the narrow summit for 4,950 feet, passing 76 feet above Irondequoit Creek, which flowed through a 245-foot-long culvert. Since the valley's soil was unsuitable for such enormous earthworks, small mountains of earth had to be hauled in from elsewhere. Even so, there was no great confidence that the embankment would hold up; from its completion in October until the close of the 1822 season, the work was drained nightly.

A few miles farther west, in Rochester, a stone aqueduct carrying the canal over the Genesee River was completed in 1823; its combined span of 802 feet made it the second longest aqueduct on the canal. But impressive as the aqueduct and the embankment were, the engineering work that captured the most attention lay about sixty-five miles to the west, in Lockport, where the canal had to be lifted sixty feet up onto the Niagara Escarpment and where, for two miles, its channel would be blasted out of solid rock.

The work there began in 1822 and took three years. What stood at Lockport upon completion of the job were five double locks, one set of five for going up, a parallel set of five for going down. West of the locks, which quickly won popular renown as the Lockport Fives, the channel ran for seven miles through the Niagara ridge. To get through the most difficult portion--the two miles known as the Deep Cut--workers had to blast free and haul away nearly 1.5 million cubic yards of rock.

While progress was being measured in feet and inches at Lockport, the final section of the canal, between Lockport and Buffalo, was being built more quickly--but nevertheless to an extraordinary standard of care. To propel water down from Lake Erie, the fifty miles of channel from Buffalo to Lockport were sloped at exactly one inch per mile. By the summer of 1825, the job was done. A few final details remained; then, on October 26, 1825, the

The Erie Canal

Erie Canal formally opened amid statewide ceremony and celebration that lasted for weeks.

The ceremony and celebration ended with the year, but the effect of the canal on America had just begun. The Erie had cost the state about $7.9 million to build, but it attracted such a huge volume of commercial traffic that it paid for itself through toll revenues in less than ten years. Its awesome vitality as an avenue of commerce catapulted New York City into the position of pre-eminence that Philadelphia had always assumed would be its own. It turned western New York State from a wilderness into a prosperous country of farms, towns, and busy manufacturing cities. And as emigrants passed over the canal heading west, it had the same civilizing and nurturing influence on Ohio, Indiana, and the other states of the Old Northwest. The Erie appreciably advanced the timetable of American development.

But what if the canal had not been built? What if the task had proved too great and the work had been abandoned?

In his definitive **History of the Canal System of the State of New York**, written in 1905, Noble E. Whitford indulged in some fascinating speculative history. Without the Erie, he wrote, Canada would have been "enriched ... commercially and strategically almost in proportion as it would have tended to impoverish us." With the completion of the Welland Canal across the isthmus between Lake Erie and Lake Ontario in 1831, the already established tendency of Northwestern trade to gravitate up the St. Lawrence would have been greatly accelerated. Moreover, Canada would have gained strategic control of

285

the outlet to the Great Lakes, making it possible for the British government to translate that advantage into naval control of the lakes. All this at a time when their vital importance in the War of 1812 was still fresh in the memory.

Within the United States, the natural outlets for the northwestern trade were the Ohio and Mississippi rivers, and prior to the building of the Erie Canal, trade was drifting south along those routes just as it was drifting north to Montreal. Without the Erie Canal and the impetus it gave for the building of other canals and, later, east-west railroads, geographic expediency would have routed that trade north and south. "Chicago could hardly have become so great an emporium," wrote Whitford, ". . . and not a little of the commercial prestige of Boston, New York and Baltimore. . . would then, perchance, have descended upon New Orleans and Mobile and Galveston. More portentous still than this commercial alliance between the Northwest and the South is the consequent probability that out of it there would have grown racial sympathy and political kinship, with what effect upon the great issues which culminated in the Civil War or upon the present constituency of the American land and people, we can only conjecture."

But the canal had been built, and there was no conjecture as to its benefits. The men who built it found themselves in high demand as "canal fever" swept the country in the wake of the Erie's success. Benjamin Wright would still be referred to as the father of American engineering had he retired after the Erie, but he served as either the chief engineer or the consulting engineer for practically every major canal built in the United States for the next sixteen years, and for the Harlem and Erie railroads as well.

James Geddes was sixty-two when the Erie was done, but he continued canal work in Ohio and Maine before retiring. Canvass White served as the chief engineer of several canals, including the Delaware and Raritan. Nathan Roberts, who designed the famous Lockport Fives, was chief engineer on the Pennsylvania State Canal and at Muscle Shoals on the Tennessee River. After serving as chief engineer of the Ohio Canal System, David Bates moved on to success in railroad engineering. Before he too moved on to railroads, John Jervis's triumphs included the Delaware and Hudson Canal and New York City's Croton Aqueduct.

They did not build a perfect canal--no one does. The old Erie required a great deal of maintenance and repair, and it was alternately bedeviled by floods along streams that fed it or crossed its path and by low water due to leakage through its bed and banks. In 1836 the state began a twenty-six-year program of enlarging and improving the canal, guided by an authoritative, canal-long survey led by such men as John Jervis and Nathan Roberts. The survey found problems along every section of the canal, and some required substantial changes.

None of this detracts from the original accomplishment. That there was a profitable and maintainable canal in operation at all in 1837 speaks volumes about the kind of men who put it there. The few of them who had any technical knowledge at all had been nothing more than plain old-fashioned country surveyors before the Erie began. The rest had been farmers, craftsmen, merchants: the ordinary settlers of a wilderness. To do what they were asked to do, they had to reinvent themselves, and in reinventing themselves, they accomplished something unimagined and extraordinary. They invented the American engineer.

Granite Like Steel

An Excerpt from Chapter 6 of **A Great and Shining Road**
John Hoyt Williams

*The construction of the first transcontinental railroad was the most spectacular
engineering feat of the middle half of the nineteenth century. The Union
Pacific Company, working from Omaha to the west, struggled to cope with the
seemingly endless plains, the rugged terrain of the Rockies, and the fierce
attacks of Indians determined to resist the intrusion of the "iron horse". The
Central Pacific, working from Sacramento to the east, had only one major
physical obstacle: the Sierra Nevada mountain range. But what an obstacle.
This selection from John Hoyt Williams' book provides some insight into the
magnitude of the difficulties posed by the Sierras.*

*Of the Big Four (Leland Stanford, Collis Huntington, Mark Hopkins and
Charles Crocker) who founded the Central Pacific Railroad Company only
Crocker played an active role in the actual construction. He formed a
contracting company and was promptly awarded the contract to construct the
first segments of the railroad (conflicts of interest were not subjected to the
same degree of scrutiny then as they are now). The Central Pacific had hired
Samuel Montague, who had worked as an assistant to Theodore Judah during
the initial survey of the route over the Sierras, as chief engineer when Judah
died.*

*Crocker hired James Strobridge, a New Englander who had worked on
railroads in the East before coming to California as a prospector during the
Gold Rush, as foreman to oversee the actual construction. Labor was extremely
scarce in California and various proposals were put forward for using
Confederate prisoners of war, former slaves, Mexican immigrants and other
laborers who could be "induced" to work at tasks which men were not
voluntarily choosing to pursue. In the end, however, a substantial portion of the
labor force was made up of Chinese immigrants (called "Celestials" because of
their reference to their homeland as the "Celestial Kingdom").*

*The first fifty miles of construction from Sacramento east to the foothills of the
Sierra was relatively easy. Then, however, the task became considerably more
difficult.*

The summer and fall of 1865 must have been a period of numbing doubt and
strain for the leaders of both Pacific Railroads, for the problems they faced
must have appeared overwhelming. The Union Pacific, laying its first rails in
its three years of existence, was running out of time. It had just short of a
year to reach mile 100*, and it was in chaotic financial circumstances,

----

* Locations along the railroad were indicated by distance from the point of
origin. Mile 100 was 100 miles west of Omaha. According to the terms of the
agreement between the federal government and the railroad companies, the
companies were required to achieve certain milestones within specified time
frames in order to receive government financial incentives and, in some cases,
the right to continue working on the project.

bedeviled by lack of engineering leadership and cowed by the enervating shrill of war whoops in the Platte Valley.

The Central Pacific faced far worse. On September 1 its tracks reached Colfax [California], at mile 54, 2,242 feet high in what were still mere foothills. The next thirteen miles to Dutch Flat were murderous: an ascent of 1200 feet, broken terrain, two major bridges, numerous cuts, the granite monolith called Cape Horn to be circumvented, and the extremely difficult clearing of the right-of-way to be done. It would take a full year to reach mile 67. Far worse lay beyond Dutch Flat. From [there] to Cisco was only some twenty-five miles along the Central Pacific's projected line, but the elevation increased a remarkable 2,485 feet; there was far more blasting, cutting, and filling to be done, another precipitous gorge to be bridged, massive redwood stands to be cleared, numerous tight curves to be plotted, and the first three of the railroad's fifteen tunnels to be cut through the planet's most stubborn rock: Grizzly Hill Tunnel (mile 77), Emigrant Gap (mile 84), and Cisco (mile 92.25). At Cisco, the Big Four would still have thirteen miles and another 1,132 feet of elevation to conquer before reaching Summit, where, at mile 105.5, the worst of its tunnels--1,659 feet long--would have to be gouged out, followed by six more tunnels in less than two miles. Montague's talents had by now reduced the number of tunnels from Judah's eighteen to fifteen, but the task facing the Central Pacific was not merely improbable, it was unique in engineering annals. So rugged was the land after Colfax that the cost of grading alone on some portions would exceed $100,000 per mile. Tens of thousands of tons of granite and other durable stone would have to be chipped and blasted from the mountains every month. How ironic, for the Union Pacific, in a prairie swept clean by glaciers millennia before, desperately needed stone.

It would have been somewhat easier had it been possible for the Central Pacific to face its problems consecutively. Unfortunately, Crocker, Montague, Strobridge, Lewis Clement, and their host of Oriental workers would have to take on the worst problems simultaneously.

It is worth detailing the stark tableau faced by Jim Strobridge that summer and fall. With track laying assuming a low priority in the face of preparing the ground for iron, he and the omnipresent Charlie Crocker divided his growing pool of labor five ways. The largest of his work crews--some five thousand men and six hundred teams of draft animals--were sent ahead four miles east of Colfax to work on Cape Horn, an immense spur of granite rising some 3,800 feet above the American River. Another thousand or so were detailed to the backbreaking and dangerous work of clearing the right-of-way, and smaller teams of three hundred to four hundred men each were put to work farther east boring entrances for the first three tunnels.

----

* Lewis Clement was a Canadian engineer whose experience had been primarily in canal building. Because there was a shortage of trained railroad engineers on the West Coast Judah had hired him despite his lack of railroad experience. During the time Clement worked under Judah's supervision he had become an expert in railroad location and he was to become an expert in tunneling, bridge design and construction, and other aspects of railroad engineering before the Central Pacific work was completed.

The huge gang sent to Cape Horn, which soon resembled a giant anthill swarming with Celestials, faced what appeared to be the most impossible of a number of impossible tasks. Somehow they had to create a roadbed along the almost sheer sides of the granite monster, some 2,200 feet above the roiling river below--a sharply curving roadbed at that, whose curves would hug the monolith and, worse yet, ascend. For decades, trains going in either direction would stop at Cape Horn for a few minutes so travelers could marvel at the work accomplished and admire the splendid, vertiginous view.

In early September, Strobridge turned his Celestials loose on Cape Horn with their picks, drills, shovels, tiny wheelbarrows, and blasting powder. The "crumping" of explosives reverberated through the valley below as the Chinese--who either were not susceptible to acrophobia or possessed a singular wealth of fatalism--began to sculpt the mountain, great chunks of which were blasted or pried loose to tumble earthshakingly into the American River so far below. Hundreds of barrels of black powder were ignited daily to shear away the obdurate granite and form a ledge on which a roadbed could be laid; but no matter the volume of explosives, progress was too slow to suit Strobridge and his boss. While as many as half that work crew was engaged in building two massive retaining walls just above the emerging ledge (one a hundred feet long, the other two hundred feet), Montague suggested to Strobridge a new tactic, to which the Chinese headmen agreed. Beginning amidst the chill winds of late October, as snow swirled over the higher peaks in the distance, scores of Chinese were lowered by ropes from Cape Horn's summit to the almost vertical cliff face. There, nestled in flimsy-looking but strong woven baskets, the workers, sometimes swaying and swinging in the wind like ornaments on some bizarre outdoor Christmas tree, bored holes in the cold rock with their small hand drills. Dangling, they tamped in explosives that had been lowered to them, set and lit the fuses, signaled the men above by jerking a rope, and, wrote Thomas W. Chinn, "then scrambled up the lines while gunpowder exploded underneath." This was a hazardous business at best, and some of the Celestial acrophiles were not agile enough to escape the blasts or were hit by flying rock and followed the chunks of granite into the valley below. Notwithstanding the casualties there was no lack of volunteers, and to the surprise and relief of all, the basic work on Cape Horn was completed before winter's rather tardy fury forced a four-month halt to outside work. Track would be laid around Cape Horn the following May, well ahead of schedule. Most Cape Horn Chinese were shipped back to Sacramento for the winter, with a few score experienced rock men sent up the line to the tunnel facings.*

The basketmen faced danger every time they swung out on their lines from the summit of Cape Horn, but the crews clearing the right-of-way were even more exposed to injury or death. Their task was to clear an avenue a hundred feet wide on either side of the roadbed. At least twenty-five feet on each side had to be completely cleared and leveled, stripped of all rocks, obstructions, and vegetation. Past Colfax, this growth included some of the world's largest trees, timeless redwoods hundreds of feet high. How different from clearing

----

* The tunnel facing is the surface at which the work on tunnel construction is currently underway (or, in this case, about to begin).

the right-of-way in the Platte Valley, where the largest impediments were prairie-dog villages.

One three-hundred-man gang spent a full ten workdays clearing a single mile of right-of-way. The trees were felled (many of them shipped to sawmills to reappear later as ties and trestling), and then the stump and root complex had to be blasted from the soil. Some of the stumps were so massive and stubborn that ten barrels of powder were needed to break their grip on the earth. With every explosion come zinging chunks of rock and lethal redwood, tamarack, and pine splinters: mortal missiles fired back by nature at the army invading it. The road-clearing crews also had to contend with boulders and outcroppings, and often had to precipitate landslides to create a level and safe right-of-way. In any one week they used as much explosives as did Lee and McClellan at Antietam. Clearing a mile often cost Charlie Crocker $5,000 or more. When the snow accumulated to three feet that autumn, most of the road crews were sent down the line on furlough, until spring.

It was tunnel work, however, that was given the most attention, and cost the most money in 1865 and succeeding years. Those dark holes severely tested the ingenuity, patience, courage, and physical resources of every Central Pacific worker from Charlie Crocker down to the lowliest Chinese tea boy.

Immediately beyond Colfax an unusually hard form of volcanic granite and, estimated one observer, "even harder porphyritic rock" were encountered. Henry Root, who worked with Lewis Clement, now a tunnel engineer, recalled that "the rock was so hard that it seemed impossible to drill into it to a sufficient depth for blasting purposes." Another observer wrote that against such geological eccentricity "gunpowder seemed almost to have lost its force."

In the early fall, teams were scratching and blasting away at the portals of Grizzly Hill, Emigrant Gap, Cisco, and the more distant Summit tunnels, finding to their chagrin just how secure nature was in "her mightiest fortress." A distant, but knowledgeable observer, writing of the Central Pacific's assault on the Sierras, noted optimistically that the workers were "accustomed to pulverize quartz and reduce mountains in a manner which would have astonished Hannibal," and felt that the company would conquer the mountains in two and a half years, "if there [was] any virtue in gold, gunpowder or nitro-glycerine." There was very little gold in the Sacramentans' treasury and as yet no nitroglycerin, but by late 1865 as much blasting powder as the workers could handle was available.

While the Union Pacific faced no rock excavation at all in its first five hundred or so miles, the Big Four's chief tunnel engineer, John R. Gillis, found that his hardworking Chinese gangs at the Grizzly Hill facing (and elsewhere) were progressing an average of a pathetic seven *inches* in an exhausting day of dangerous labor. The best day's record at Summit Tunnel in two years of work, using powder, was twenty-seven inches on one facing. The Central Pacific had to order specially tempered steel for its drills in hundred-ton lots because the granite eroded the grooves so swiftly. Up to three hundred kegs of blasting powder a day was going up in smoke on Summit Tunnel alone by 1866, and the powder bill, according to engineer Clement, often ran from $53,000 a month to a high of $67,000 just for the tunnel work. Henry Root later explained that "more powder was used by the rock foremen than was economical," but they used it so lavishly because they were told that time, not money, was of the essence. The one benefit from all this blasting, chipping, and drilling was that the company found that it could use the small shards of granite for superior ballasting of the tracks, while the larger chunks shaped or not, could be sold to construction companies for good money. According to the *Railroad Record*, high-quality, construction-grade granite, formerly imported from China, was thus made available by the Central Pacific to American builders at an acceptable price.

Still, with only inches a day tallied by the Celestials and their picks and hand drills, the prospects were not cheering, for railroad tunnels, even if single-tracked (most of the company's were), are by nature large undertakings. On average, a Central Pacific railroad tunnel was sixteen feet wide at the bottom, sloping gently upward and inward to a height of at least nineteen feet. All tunnels were at least partially lined with stout timber--some, like Emigrant Gap and Tunnel Spur, were entirely timbered--and all were roofed with boarding almost three inches thick. The tunnels' side timbers (braces) were commonly twelve-by-twelve inch, or even twelve-by-sixteen inch, redwood--heavy, hard-to-maneuver supports that were held to the walls by long three-quarter-inch-thick iron bolts. The Central Pacific had twenty sawmills working full time on tunnel supports and crossties. To complicate matters, most of the Sierra tunnels were set on both curves and grades. It was nightmarish for the Chinese to labor in candlelight or by lantern in the increasingly foul, dusty air, and the engineers themselves had a hellish time plotting accurate lines under such conditions. Given the circumstances, it is remarkable that the tunnels run so mathematically true.

In view of the snaillike progress in the tunnels, the Big Four and their technical staffs made a number of decisions in late 1865. Work would continue on the tunnels throughout the winter, despite the weather conditions, from both their east and west facings. The very long Summit Tunnel would have to be pierced from above as well as from both ends so that four facings could be attacked at once. And newfangled explosives would have to be experimented with.

In late September, Crocker headed into the Sierras, which were already dappled with the season's first snows. The years of late, mild winters were no more. Up the line to Summit Tunnel churned Charlie Crocker, inspecting every foot of the way. He gave orders to establish permanent work camps on both sides of the Summit and ordered round-the-clock drilling, blasting, scraping, shoveling, and hauling by the Chinese. After all, he figured, there is no day or night within a gloomy tunnel. There would henceforth be three eight-hour shifts on each tunnel face, Crocker's only concession to those Chinese selected to work throughout the winter. As soon as the weather permitted, a steam power plant would be brought to the Summit to help haul the heavy detritus from the new shaft being sunk vertically, midpoint on the tunnel's line.

Some 90 percent of the Chinese work force was paid off for the winter, however, for there was little they could do in the frigid Sierras. Working outdoors was impossible because of the heavy snows, and since only a handful of men could work at one time on the constricted tunnel faces, a mere five hundred or so men were kept at work on the tunnels in gangs of approximately twenty men. Some of these would freeze to death that first Sierra winter, but that was only a harbinger of the next winter's toll.

291

Despite what was believed at the time and has been thought since, the Big Four did *not* recruit great numbers of Chinese from China itself. In fact, in 1866 and 1867--key years for the construction of the Central Pacific--more Chinese left California than entered--a reversal of traditional immigration patterns. Stanford was probably unaware of this odd exodus of Celestials just when they were most needed by his railroad, for early in 1866 he wrote to Crocker in the field: "We have the assurances from leading Chinese merchants, that under the just and liberal policy pursued by the Company, it will be able to produce, during the next year, not less than fifteen thousand laborers." Whether fifteen thousand were indeed made available, the Central Pacific hired only eight or nine thousand that year.

Tunneling, under wretched conditions, continued through the winter of 1865-66, even when fourteen feet of snow lay on the ground and drifts towered to thirty feet and more. While not an especially severe winter compared with those to follow, there were five feet of snow on the ground by the first of December. Other work stopped, although in the East Huntington, who tried to work eight months in advance of Crocker's needs (from pick handles to locomotives), was putting ten thousand tons of equipment in transit every thirty days or so.

Tunneling, clearing the right-of-way, grading, and track laying were not the only occupations of Central Pacific personnel. The company also employed about five hundred specialized artisans, mostly masons and carpenters, to build bridges (under master bridge man and architect Arthur Brown), company depots, and other buildings, and stone and brick culverts (ultimately, 375 miles of them) seated in hydraulic cement. The bridges were major engineering works in their own right, each one perched on solid granite piers. The first bridge, across the American River, had trestling nearly a mile long, for it not only spanned the seven-hundred-foot-wide river, but adjacent, spongy floodmarsh as well. Most of the trestling and actual bridgework was made from massive Puget Sound pine and redwood, but, as these deteriorated within a decade or so, they were to be replaced with fill (embankments) or iron-and-steel structures. The men who did such work were paid handsomely, for they were both in short supply and critical to the railroad's progress. This they realized, and they demanded and received--on time and in gold--wages of from three dollars to five dollars a day, with most employed year-round.

As the snows began in the Sierra that fall, another facet of mountain life became apparent. Even before the first heavy snowfall, patterns of drifting and avalanches had been recorded by Lewis Clement and other engineers. As the *Railroad Record* noted in November, "For several miles the track must be roofed, to slide off the snow." It was not clear at the time, however, that this would entail the labor of thousands, some $2 million in additional expenditure, and thirty-eight miles of heavily roofed and protected track.

In December, as the lonely, freezing tunnelers made the mountains shudder with their granite-shattering explosions, and while artisans plied their specialties in the company's expanding Sacramento shops, the leaders of the Central Pacific took stock. The railroad had fifty-four miles of working track to Colfax, less than twenty of which had been spiked in 1865. These few miles, however, according to the company's annual report, had cost an astounding $6 million, which, when added to the previous two years' outlay, came to a depressing $9 million.

Down in the Caisson

Chapter 9, **The Great Bridge**
David McCullough

*It is human nature to fail to appreciate the full measure of the hardships
endured by others, to minimize or even fail to recognize the problems solved by
others, and to dismiss as routine the accomplishments of another. As
susceptible as man is to these failings with regard to his contemporaries, he is
even more prone to such shortcomings when evaluating the achievements of the
past. To some extent this is due to a lack of familiarity with the past, but it is
also due to a kind of arrogance that seems to be based on the implicit assumption that old problems were simply not as difficult as those we face today.*

*This, of course, is far from true. David McCullough's description of the
difficulties encountered in sinking the caissons for the Brooklyn Bridge
provides ample evidence that problems encountered by civil engineers in the
nineteenth century were as challenging in every respect as the most difficult
problems faced by modern engineers. Those who have not experienced being
faced with the responsibility for making a major decision without any
experience or guidance to rely upon cannot fully appreciate Washington
Roebling's misgivings and hesitancy regarding the use of explosives in the
caisson. However, many engineers could identify with Roebling's approach to
solving the problem.*

*McCullough's account of the building of the Brooklyn Bridge reminds us that
the bridge is not only a landmark, not only a symbol, not only an inspiration,
but also a tremendous technical achievement. This excerpt calls to mind the
fact that some of the most astounding aspects of that achievement, as in the case
of many civil engineering projects, are buried--out-of-sight forever.
Descriptions such as this help assure that these achievements, and the ordeals
undergone by those who created them, remain alive in memory as well as in
reality.*

In all the thousands of years men had been building things, no one had ever
attempted to sink into the earth so large a structure as the Brooklyn caisson
and there were not very many places where the job would have been more
difficult than the Brooklyn side of the East River.

Roebling and his assistants thought they had learned quite a lot about the
ground they had to penetrate while dredging the site, but as he commented
with his usual dispassion, "The material now became sufficiently exposed to
enable us to arrive at the conclusion that it was of a very formidable nature,
and could only be removed by slow, tedious, and persistent efforts." Compared
to this everything before had been child's play. Now that which had looked so
reasonable on paper was turning out to be quite a different matter in practice.
Indeed, so bad was the first month of excavation inside the caisson, so
painfully slow and discouraging, that it began to look as though the whole idea
for the foundations had been a terrible mistake, that they would have to give
up and try again some other way or some other place.

There was never any public awareness of such feelings, which was just as well. There was, for that matter, very little real awareness on the part of the public of what actually went on inside the caisson, the work being entirely concealed.

The best over-all view of the site was still from the deck of the ferry. So every day thousands of people on their way to and from New York got a splendid, close-up look at the three towering boom derricks swinging blocks of limestone into place and at the squads of men swarming about the masonry work or through the adjacent yards, every last man appearing to know just what was expected of him. There were half a dozen different steam engines sending up columns of black smoke and everywhere a bewildering clutter of tackle, hand tools, nail kegs, and tar barrels, stacks of lumber and great heaps of coal, sand, and stone. How anything orderly or rational might emerge from such seeming chaos was something for ordinary men to ponder in dismay.

Still, seen from above, the work did not appear all that different from other big construction projects. The activity around the gigantic new Post Office being built in New York, for example, was every bit as confusing and impressive to watch. All this was lit by the same light of day and the men appeared no different from other mortals, breathing the same good air. But down in the caisson, everyone had heard, things were different. That was the part of the work that had the most fascination and of course the fact that it was hidden away where no one could see it, except for a relative few, made the fascination that much greater.

The newspapers sent reporters down soon enough. By July better than two hundred workers were going down every day and naturally they had their own stories to tell. So as a result a picture began to emerge, of a strange and terrifying nether world at Brooklyn's doorstep, entered only by men of superhuman courage, or by fools, and as sometimes happens with ideas that grow in the imagination, it was not so very far from the truth.

Probably the most vivid description was one given by E. F. Farrington, Roebling's master mechanic, a plain, blunt, practical man ordinarily. There would be rumors later about who actually was doing Farrington's writing for him, or at least dressing up his literary style, but there is no doubting the authenticity of the image.

> Inside the caisson everything wore an unreal, weird appearance. There was a confused sensation in the head, like "the rush of many waters." The pulse was at first accelerated, then sometimes fell below the normal rate. The voice sounded faint unnatural, and it became a great effort to speak. What with the flaming lights, the deep shadows, the confusing noise of hammers, drills, and chains, the half-naked forms flitting about, with here and there a Sisyphus rolling his stone, one might, if of a poetic temperament, get a realizing sense of Dante's inferno. One thing to me was noticeable--time passed quickly in the caisson.

Even the air lock was an unnerving experience for most men the first time they went down. For some it was also an extremely painful experience. The little iron room was abundantly lighted by daylight through glass set in the ironwork overhead. But once the attendant had secured the hatch with a few turns of a windlass, the common sensation was that of being enclosed in an iron coffin. Then a brass valve was opened. "An unearthly and deafening screech, as from a steam whistle, is the immediate result," wrote one man, "and we instinctively stop our ears with our fingers to defend them from the terrible sound. As the sound diminished we are sensible of an oppressive fullness about the head, not unaccompanied with pain, somewhat such as might be expected were our heads about to explode." (For many the sensation did not pass and they were said to be "caught in the lock.") Then the sound stopped altogether, the floor hatch fell open by itself, and the attendant pointed to an iron ladder leading into the caisson. The immediate wish of most men at this point, whether they showed it or not, was to get back out into the open air just as fast as humanly possible. But once the ladder had been negotiated and three or four minutes has passed, most men also found they felt reasonably steady.

The initial view of the caisson interior was generally something of a shock, once the eyes had adjusted to the light. The six big chambers looked something like vast cellars from which a flood had only recently receded. Every post and partition, every outside wall, and the entire ceiling were covered with a slimy skim of mud. Every man in the place wore rubber boots and got about on planks laid from one section to another and between the planks the muck and water were sometimes a foot deep or more. Most days the work force would be concentrated in a few locations, leaving some of the huge chambers as dark and silent as subterranean caves.

Where there was light it came from calcium lamps, limelights as they were also called, which threw steaming, blue-white, luminous jets into the corners where the men worked, or from squat sperm candles that blazed like torches at the end of iron rods planted alongside the plank walkways. "The subject of illuminating a caisson in a satisfactory manner is rather a difficult problem to solve," Roebling remarked in his report to the directors of the Bridge Company. At first the candles had burned with such vigor in the compressed air and sent up such clouds of smoke that the air had become intolerable. This had been overcome somewhat by reducing the size of the wick and of the candle and by mixing alum with the tallow and drenching the wicks in vinegar. Even so Roebling worried about the quantities of floating carbon the men were breathing into their lungs.

Kerosene lamps had to be ruled out from the start. They smoked even worse than candles, and with fire a constant hazard in such a charged atmosphere, Roebling did not want the risk of spilled oil. So he had hit upon the idea of limelights, of the kind ordinarily used for stage lighting or nighttime political rallies. He had the gas--a combination of compressed oxygen and coal gas-- piped into the caisson, put burners in every chamber, and found two lamps per chamber would do the job. One small explosion had singed the beard off an attendant, but other than that the system had worked most satisfactorily. Ordinary street gas would have been about five times less expensive, but when that had been tried, the heat inside the caisson had built up to the point where no one was able to take it.

The air as it was, besides being heavy and dank, was uncomfortably warm. On the way from the compressors it passed through a cooling spray of water. Even so, winter or summer, regardless of the time of day or the weather outside, the temperature inside stayed 80 degrees or more and the air was so saturated with water that under the best conditions the chambers seemed continuously shrouded in mist. Visitors who did not have to exert themselves in any way soon found they were wringing-wet with perspiration.

Most of the people who visited the caisson--newspapermen, local politicians, an artist from *Harper's Weekly*, editors from some of the professional journals --came out with their clothes thoroughly mud splattered and quite relieved to have the experience behind them. Many of them also expressed open amazement that men could actually work in such a place day after day.

The first load of rock and mud was hauled out of the caisson by clamshell dredge buckets on July 5. Most of the effort inside was spent removing the sharp-edged boulders that threatened to damage the frames and shoe as the caisson began to come down on them with crushing force. Boulders under the water shafts were the most serious initial problem, for if the caisson were to settle suddenly, the shafts might be blocked shut or badly damaged. And there was no way to get the boulders out of there except to chip away laboriously hour by hour, by hand, with long steel bars and sledge hammers.

In the middle chambers the ground was nearly all traprock, packed like gravel and joined by what Roebling described as a natural cement made of decomposed fragments of green serpentine rock. Every boulder was coated with this unyielding substance, upon which a steel-pointed pick had virtually no effect. Only by driving in steel-pointed crowbars with heavy sledges were the men able to make the slightest headway.

In chambers No. 1 and 2, those nearest the ferry slip, there was clay and gravel between the rocks, which made the going easier, while in Nos. 5 and 6, those at the upstream end of the caisson, there was a gummy blue clay that extended down forty feet, just as indicated by earlier soundings. This made the digging there relatively easy, of course, but it also meant that the caisson would have to go down at least forty feet--or beyond the clay. As Roebling said, no better foundation could have been wished for than what they were finding in chambers No. 3 and 4, but only if it had extended all over. And with the nature of the material so vastly irregular from one chamber to the other, lowering the caisson uniformly seemed practically an impossibility.

*   *   *

The idea of driving the cutting edge of the caisson through such material by building weight overhead had to be abandoned at the start. The pressure needed to do that would crush the cast-iron shoe and smash the bearing frames. . . . Instead, every boulder, every rock of any size, had to be removed before the shoe or frames began bearing down on them. And all such work had to be done by probing underwater . . .

Just finding the boulders under the shoe, let alone removing them, was an unbelievably tough and disagreeable task. The full perimeter of the cutting

----

* The shoe was the outer perimeter of the caisson. It had been "armored" to serve as a cutting edge as the caisson descended.

edge was 540 feet. This added to the five frames, each 102 feet long, brought the caisson's full bearing surface to 1,050 lineal feet, or a distance greater than the length of three football fields, every inch of which had to be probed beneath with a steel sounding bar twice daily with each shift. Whenever a new shift came down, the work accomplished in the preceding eight hours had to be carefully explained; and since most of the trouble spots discovered would be underwater, there was no way simply to point them out--the information had be to written down or memorized. "Moreover," as Roebling wrote,"a settling of the caisson of six inches or a foot would bring to light an entirely fresh crop of boulders in new positions, and very often half without and half within the caisson."

*　　*　　*

. . . Through July and on into early August, the rate of descent had been less than six inches a week, and the boulders, instead of diminishing in number, as had been expected, became more plentiful. It was a hopeless rate of progress Roebling reported to his directors. At this rate it would take nearly two years to sink just the one caisson.

*　　*　　*

To step up the pace, Roebling organized a special force of forty men who worked at boulders exclusively, from eleven at night until six in the morning, when the regular shift came on. . . .

But when the caisson had reached a depth of some twenty feet, or approximately half the distance Roebling intended to sink it, the boulders became so large and numerous that there was no choice left but to begin blasting.

The idea of using powder on the boulders had, of course, been considered from the start. It would have saved all kinds of time and effort obviously, and as things grew increasingly difficult and frustrating inside the caisson, the men were more than ready to give it a try, whatever the supposed risks involved. But Roebling had held off. In such a dense atmosphere, he reasoned, a violent concussion might rupture the eardrums of every man inside. Smoke from the explosions might make the air even more noxious and certainly more unpleasant than it already was. The doors and valves of the air locks might be damaged.

His greatest fear, however, was the possible effect on the water shafts. The two immense columns of water that stood above the work chambers and every man in the caisson were held there in a critical balance only by the pressure inside the chambers. The margin of safety was just two feet of water--the distance from the surface of each pool and the bottom edge of each shaft. An explosion inside the caisson, Roebling explained, might suddenly depress the level of the pool and allow the air to escape underneath. A water shaft might blow out, in other words. All the compressed air would escape in one sudden blast and almost certainly with the following immediate consequences: with the work chambers instantly deflated, so to speak the full weight of the caisson would come down all at once, smashing blocks and frames and outer edges. The impact might be so great as to crush every interior support and everyone inside; and in the early stages of the work, the river would have rushed in and drowned everyone. What the effect might be on top was anybody's guess, but it was realistic to assume that all that water bursting out of a shaft would be about the same as a major explosion.

Still, Roebling knew, such prospects, however sobering, were all hypothetical. There was no past experience to go by. So whether he was right or not remained to be seen. With luck, he might be wrong. He decided to find out.

He began by firing a revolver with successively heavier charges in various parts of the caisson. When it was clear this was perfectly safe and causing no adverse effects, he set off small charges of blasting powder, fired by a fuse, gradually working these up in magnitude until they were on the order of what was needed to get on with the work. The concussions bothered no one especially, nor did they have any noticeable effect on the air locks or water shafts. "The powder smoke was a decided nuisance," Roebling said. "It would fill the chambers for half an hour or more with a thick cloud, obscuring all the lights." But this he alleviated greatly by switching to fine rifle powder.

The results were spectacular. With a little practice the work moved ahead as never before. A long steel drill would be hammered into the rock to make a hole for the blasting charge and the charge would be tamped in and set off. "As many as twenty blasts were fired in one watch," Roebling reported, "the men merely stepping into an adjacent chamber to escape the flying fragments." The hard crystalline traprock split more easily than the tough gneiss or rotten quartz boulders. Invariably the traprock broke neatly into three equal-sized boulders. The caisson now began descending twelve to eighteen inches a week, instead of six.

Care was taken to guard against fires igniting in the yellow-pine roof and the men did their best not to injure the shoe with the charges they set off beneath it. But the shoe by this time was in such shape that a little more damage hardly mattered. The armor plating was bent and torn, the shoe itself cracked or badly crushed in dozens of places.

One convenient method for disposing of a boulder lodged beneath the shoe was to drill straight through to the other side, plant a charge at the far end, then shoot the boulder bodily into the caisson. Some boulders encountered now were up to fourteen feet in length and five feet in diameter.

For the people of New York and Brooklyn all such activity was considered somehow removed from reality. The whole concept of an enormous wooden chamber descending below the river was a little difficult for many to understand and the men who went in and out of it seemed a breed apart. There was simply something quite unnatural about all this. "For night is turned into day and day into night in one of these bridge caissons," wrote the *Herald*; "and when the steam tugs, with their red and blue lights burning from their wooden turrets go creeping along the bosom of the river like monstrous fireflies, then do these submarine giants delve and dig and ditch and drill and blast . . . The work of the buried bridgebuilder is like the onward flow of eternity; it does not cease for the sun at noonday or the silent stars at night. Gangs are relieved and replaced, and swart, perspiring companies of men follow each other up and down the iron locks, with a dim quiet purpose . . ."

The sheer physical exertion inside the caisson was as great as ever, the work every bit as unnerving as it had been. And the deeper they went, the more the men felt the discomforts of the compressed air.

The work went on around the clock, except for Sundays, with three shifts of eight hours each. The first shift went down at six in the mornings, the second shift at three in the afternoon, the third, the special night gang, went down about eleven. Most men stayed in the caisson the full eight hours, taking their dinner pails down with them. Work in such an atmosphere brought on an uncommonly fierce appetite, they said, and the standard meal consisted of great slabs of bread and cheese or beef, washed down with beer.

The two day shifts were composed of 112 men each, while the night shift Roebling kept to roughly forty picked men. So the full force working inside the caisson came to about 264. Up on the surface there were two shifts to operate the dredging gear and two shifts to dispose of the material brought up from below. In addition, there were people to run the compressors and hoisting engines, blacksmiths, mechanics, men to look after the gas for the lighting below, a carpenter's force of some twenty-five men, and thirty men working on the masonry, bringing the total force aboveground to something like a hundred.

But the number of those inside the caisson who had been with the work from the start was quite small comparatively. According to the time books, a total of 2,500 different individuals worked in the Brooklyn caisson from start to finish. This means then that men were quitting in droves--at a rate of about a hundred a week on the average, or, to put it another way, every week about one man in three decided he had had enough of building the Great Bridge and walked off the job, never to return again.

There were notable exceptions, of course. One man named Mike Lynch went down with the very first shift to go into the caisson and would be the last man to come out. He not only never lost an hour's time during the ten months he worked in the caisson, but he made a day's extra pay in overtime. "He is strictly temperate and regular in all his habits," William Kingsley notes, "and is none the worse for his long service in compressed air."

That the turnover was so great is not surprising.

Amenities provided by the management were very few--about what was customary. Two unpainted frame sheds had been put up in the yard, with rows of pegs and hooks inside for the men to hang their clothes. (The temperature inside the caisson was such that most men went down wearing nothing but pants and a pair of company boots.) In front of the sheds were sets of washtubs, with hot and cold water. And that was about the sum of the comforts provided aboveground.

Inside the caisson itself there were generally a few dry spots where a man could eat his midday meal. And against one wall stood what was considered by all the world's most extraordinary toilet. It was described in one of Roebling's official reports as a pneumatic water closet and consisted of a wooden box with a lid and a large iron pipe that passed up through the timber roof. The box was kept about half full of water, and whenever its contents were to be discharged, a valve was opened and the pressure from within the caisson would blast everything instantly overhead in the form of a fine mist. This particular device was not installed until the work had progressed some little time, however, and until then the pools beneath the water shafts, or any convenient corner, had sufficed for the same purpose. When he came to describe the general working conditions, Roebling would note that the sense of smell was almost entirely lost in the "made air," as he called it. "This" he said, "is a wise provision of nature, because foul odors certainly have their home in a caisson."

For an ordinary laborer the pay was two dollars a day. But after the caisson reached a depth of twenty-eight feet, it was decided to revise that. The bad air, the increased unpleasantness over all, and the widespread feeling that the deeper down they went the more hazardous the work, all called for a . . . hike in wages, the management decided. So from that point on the pay was $2.25 a day.

Men kept quitting just the same, but for every one who did, there were at least a dozen anxious to take his place, most of them Irish, German, or Italian immigrants who were desperate for work of any kind, and many of them, like those who had gone into the Eads caisson* were thinly clothed and undernourished.

But for all the talk and worry there had been over caisson sickness, and for all the growing fear of it as pressure inside the caisson increased steadily, only a few so far had experienced any ill effects.

One pound of air pressure equals two feet of tidewater, so for every two feet the caisson was lowered, one pound had to be added to the pressure. Gauges in the engine room indicated the height of the tide and the pressure of the air. The greatest the pressure would ever be in the Brooklyn caisson was twenty-three pounds per square inch above normal atmospheric pressure, or nearly ten pounds less than it had been inside the Eads caisson the day James Riley fell dead. In St. Louis several more had died miserably, but there had been only mild symptoms in Brooklyn. A little paralysis in the legs was all. Only three or four men had been bothered in the slightest and none of the engineering staff so far.

Like Eads, Roebling noted that the ones who did have trouble were all new to the job. His way of alleviating their discomfort was to send them right out of the caisson. Now that he had seen something of the problem first hand and had spent as much time under compression as anyone on the job, Roebling was convinced that Eads's system of shortening the hours was the best possible prevention and said he would follow that same system in the New York caisson. The thing to do, he said, was to "reduce the period that the human system is in contact with the exciting cause." The increased quantity of oxygen inhaled under pressure was what did the damage, he thought. "That the system struggles against this abnormal state of affairs," he said, "is shown by the fact that the number of inhalations per minute is involuntarily reduced from thirty to fifty per cent. It follows, therefore, that the shorter the period of exposure to compressed air the less the risk."

But any change in the schedule would wait until the New York side, since the Brooklyn caisson was not going deep enough to produce anything like what

----

* James Eads, who was building a bridge across the Mississippi River at St. Louis, was also using caissons on his project.

was happening in St. Louis, where Eads had had a special hospital ship fitted up and had hired a fulltime physician who prescribed special diets and set down strict rules about rest. Eads's men by this time were permitted to work in the caisson only an hour at a time.

But the men in the Brooklyn caisson were having their troubles all the same. The work was a hazard to the health, it was agreed, and far more exhausting than anything any of them had ever done before. Collingwood said a full day inside would leave him feeling worn-out and in ill temper for days. And when the weather turned cold in the late fall, dozens of men began coming down with severe colds and bronchitis, caused by the abrupt drop in temperature inside the air lock. Every time they "locked out" at the end of the day, hot, tired, and dripping wet, the men would experience a sudden temperature drop, from at least 80 to 40 degrees. Roebling had steam coils installed in the air locks to keep the temperature the same as in the chambers below, but the men still had to face the chill open air once they emerged from the locks.

A hacking cough also became common among those who had been on the job any time. Candle smoke and the blasting were said to be the cause. Those who had been going down the longest could spit black and would still be able to do so several months after the work was finished.

But what plagued everyone most was the thought of all that weight bearing down overhead and the river outside and the unspoken fear that sometime, sooner or later, something was going to go wrong and they would all be drowned like rats or suffocate or be crushed to death. And then just to confirm how very tenuous was the balance upon which they were all trusting their lives, there occurred what would afterward be called "The Great Blowout."

It happened at about six in the morning and on a Sunday, when only a few men were about, a fact the pious took to be more than a matter of coincidence. Eads had his men working seven days a week, it was noted, while Roebling kept the Sabbath. This was a sign, it was said, and thanks were given through Brooklyn and nowhere more fervently than in the Irish neighborhoods near the Navy Yard. Heaven only knew how many would have been left widows, people were saying, had it happened any other day of the week.

All at once in the very still early morning there had been a terrific roar. The few who actually saw the thing go off said it looked more like a volcano than anything else. It was as though the river had exploded, sending a column of water, mud, and stones five hundred feet into the air and showering yellow water and mud over ships and wharves and houses for blocks around. The column was seen from a mile off and the noise was so frightful that people began pouring out of their doors and rushing pell-mell up Fulton Street. The whole neighborhood was on the run. Roebling described it as a stampede. "Even the toll-collectors at the ferry abandoned their tills," he said.

Nobody was inside the caisson at the time and only three men were on top of it. One of them, a yard watchman, said later that the current of air rushing *toward* the blowing water shaft was so powerful it knocked him off his feet, ruining his Sunday suit. He had been struck by a stone after that and could remember no more. One of the other men leaped into the river, while the third tried to bury himself in a coal pile.

Then in an instant it was all over and everything was as silent as before. Both doors of the air locks fell open and for the first and only time the submerged caisson was flooded with daylight. The quiet lasted but briefly. Within minutes there was another rush of people heading down Fulton to see what had happened.

Roebling, Collingwood, and one or two of the others from the work force were on the scene almost immediately. They turned hoses into the open water shaft, closed the air locks, and in about an hour had a head of water thirty-one feet high back in the shaft and fifteen pounds of pressure back in the work chambers. When it was time to go down to take a look at the damage, Roebling led the way. "The first entry into the caisson was made with considerable misgiving," he wrote. But incredibly none of the disastrous consequences he had feared had occurred, as he reported later to the Board of Directors:

> The total settling that took place amounted to ten inches
> in all. Every block under the frames and posts was abso-
> lutely crushed, the ground being too compact to yield; none
> of the frames, however, were injured or out of line. The
> brunt of the blow was, of course, taken by the shoe and
> sides of the caisson. One sharp boulder in No. 2 chamber
> had cut the armor plate, crushed through the shoe casting,
> and buried itself a foot deep into the heavy oak sill, at
> at the same time forcing in the sides some six inches. In
> a number of places the sides were forced in to that amount,
> but in no instance were they forced outward. The marvel
> is that the airtightness was not impaired in the least.

His caisson had withstood the staggering blow of 17,675 tons dropped ten inches. By the way certain boltheads were sheared off, he could tell that the sides of the caisson--nine solid courses of timber--had been compressed two inches, such had been the impact. In the roof, however, there was not a sign of damage except for the slightest sag near the water shafts, where the support from the frames was the least.

With a little figuring Roebling concluded that once all the settling had stopped and before the compressed air was built up again inside the chambers, the caisson was carrying a total weight of twenty-three tons per square foot. This was an astonishing revelation. As nerve-racking as the whole episode had been, it had demonstrated just how large a margin of safety Roebling had built into the structure, since its ultimate load, once the bridge was built, would be only five tons per square foot. So he had built the caisson at least four times as strong as it needed to be.

*   *   *

Trusting matters to take care of themselves was something this extremely competent young man had seldom done in his life. He had had the contrary attitude drummed into him since childhood and from here on, more than ever, he would insist on the strictest attention to every detail and to safety precautions especially, and he would come down hard on anyone caught taking such matters lightly. The thing to fight against, he told the men, was the kind of carelessness that comes from familiarity with the job.

The Reversal of the Chicago River

From **Landmarks in American Civil Engineering**
Daniel L. Schodek

The role of the civil engineer in protecting public health is not as obvious today as it once was, although the health and well-being of almost every urban dweller in the United States as well as most other developed nations is dependent on water supply treatment and wastewater treatment facilities designed, constructed and operated by civil engineers (to say nothing of solid waste disposal facilities and air pollution control facilities).

The reversal of the Chicago River, described here, is undoubtedly one of the most monumental undertakings ever attempted in the name of public health. Strangely enough, the Metropolitan Sanitary District of Greater Chicago is again engaged in a mammoth project (the Chicago Tunnel and Reservoir Plan) to provide additional capability to protect the water quality of Lake Michigan. This plan involves excavation of underground tunnels in solid rock to convey to surface reservoirs combined storm and sanitary sewer flows that cannot be treated by the existing treatment plants during periods of high runoff. The excess flow will be diverted into the reservoirs for storage until the treatment facilities have the capacity to treat it, at which time is to be pumped into the sewers for conveyance to the treatment plants. It is estimated that the project will be under construction for ten or twelve years and cost almost four <u>billion</u> dollars.

The reversal of the Chicago River, carried out in the last decade of the nineteenth century, represents an innovative engineering solution to problems of water supply and pollution in an era of rapid American urban development. The direction of flow of the water that carried Chicago's wastes was changed so that instead of entering Lake Michigan it flowed into the Des Plaines River and thence into the Illinois [River], the Mississippi [River] and finally the Gulf of Mexico, thus preserving Lake Michigan as a source of clean water for the city of Chicago. The project, which involved the construction of a 28-mile channel through a glacial moraine and bedrock ridge, was unprecedented in scale and introduced earth-moving techniques later employed elsewhere. More earth was moved in this undertaking than in the digging of the Panama Canal.

The rapid expansion of Chicago as a gateway to the West in the nineteenth century resulted in a growing pollution problem. From the time of its incorporation in 1837, Chicago relied on Lake Michigan as its sole source of fresh water. A supply system of intake cribs* constructed one to two miles from the shoreline in the lake and connected to pumping stations through underground tunnels furnished the city's burgeoning population with water. From 1856 waste disposal was accomplished through a sewer system that fed into the Chicago River--which, though virtually stagnant most of the year,

----

 * A crib is a framework constructed around an intake to eliminate or reduce the introduction of unwanted suspended material in a water supply intake.

carried untreated sewage into Lake Michigan and the intake cribs during the spring floods. This unfortunate interface between the two systems resulted in periodic outbursts of typhoid fever, amoebic dysentery, and cholera.

In an attempt to relieve the increasing domestic sewage and industrial waste problem, work was done in 1871 to deepen the existing Illinois and Michigan Canal and create a gravity flow away from the lake toward the Des Plaines River, thus carrying away from the lake a portion of the untreated sewage. The canal had been constructed much earlier in the century in response to demand for a southern outlet to the Great Lakes--a possibility because of the unique geology of the area. At the time of the formation of the Great Plains, a vast waterway connected Lake Michigan with the Gulf of Mexico through the Chicago, Des Plaines, Illinois, and Mississippi rivers. A slight tilting of the lake plateau, however, occurring approximately 10,000 years ago, virtually closed this southern outlet, leaving a shallow water trail and large swamp in its place. The explorers Joliet and Marquette, first passing through the region in 1673, noted the importance of a southern waterway to the Gulf of Mexico in their reports. After American independence the northern outlet from the Great Lakes through the St. Lawrence River was controlled by the British, and demand for a southern outlet grew. Construction of the Illinois and Michigan Canal began in 1836 and, although its planners took advantage of the residual pathway that already existed, proceeded with great difficulty, owing to underlying bedrock formations and a slumping national economy. Work was finally completed on a modified "shallow cut" design in 1848, thus realizing the first man-made southern connection to the Gulf of Mexico.

The deepening of this canal in 1871 provided only partial relief of the waste problem, as changing water levels rendered the gravity flow unreliable. Moreover, the growth of towns along the Des Plaines River produced further problems as sewage from Chicago polluted their drinking water supply. In 1876 a pumping station and conduit from Lake Michigan were built to flush fresh water down the canal, but this proved of little help; additional pumping installations in 1881 brought only a marginal improvement. Outbreaks of water-borne disease continued to afflict the metropolis.

The crisis escalated in August 1885, when 6.19 inches of rain fell in a two-day period in and around Chicago. Flood waters overwhelmed the pumping stations and sewer pipes, sending the scourings of catch basins* and inlets into the river and lake. An immense mass of sewage and bacteria spread across the lower portion of Lake Michigan, fouling the intake crib for the city's water supply system. Within months almost 12 percent of Chicago's 250,000 inhabitants died of cholera and other diseases. A drainage and water supply commission was established in January 1886 to develop a comprehensive, long-term solution to Chicago's water problems. The group's report, submitted the following year, recommended that a sanitary and ship canal be constructed connecting Lake Michigan at Chicago with the Des Plaines River at Lockport; that this canal be of sufficient size to permit a flow of 24,000 cubic feet of water per minute for every 100,000 inhabitants of Chicago, based on an

----

* Catch basins are storage areas excavated or constructed to trap debris for the purpose of preventing obstruction of storm sewers during periods of storm runoff.

expected future population of 2.5 million; and that the canal be connected to the Chicago River and deep enough to permanently reverse the river's flow. These recommendations were formally adopted in November 1889 by the Illinois legislature, and the Sanitary District of Chicago was created. This autonomous agency, known today as the Metropolitan Sanitary District of Greater Chicago, embraced an area of 185 square miles and had its own elected officials and independent taxing authority.

The 28-mile channel between Lake Michigan and Lockport that engineers accordingly devised for the Sanitary District brought about the second reversal of the Chicago River in 10,000 years. The design relied on dilution and natural biological processes to render the sewage harmless to downstream communities. (Treatment plants had to be added in later years.) In addition to diverting contaminants from Lake Michigan, the Sanitary and Ship Canal provided drainage margins to allow for storm runoff under flood conditions, thus preventing a recurrence of the 1885 disaster. Navigation facilities for barges were also provided, so that this waterway could take the place of the obsolete Illinois and Michigan Canal. Finally, land acquisition along the channel's route, authorized under the Sanitary District's charter, created sites for future industrial development.

Ingenious construction techniques, particularly with regard to earth moving, were employed during the eight-year course of the project, which involved canal building, river dredging, bridge and road building, and other activities. More cubic yardage of rock and earth was displaced than in any other single earth-moving project to date. Fifteen miles of the route were dug through solid rock; almost 8 miles through earth, mostly clay; and an additional 5 miles through combined earth and rock. At the project's completion in January 1900 a force of 8,500 men had been working without halt for eight years, blasting and removing 29,559,000 cubic yards of earth and 12,261,000 cubic yards of rock. The Lockport control structure contained sluice gates* and the Bear Trap Dam, by which the amount of water flowing through the channel was regulated. Horse-drawn graders removed and leveled the first 13 miles of earth excavation, beginning at the Chicago end. This completed section measured 160 feet wide at the canal's bottom and 225 feet at the top. The final 15-mile section to Lockport, cut through rock, was accomplished with dynamite and steam-operated cranes and steam shovels mounted on rails in the trench. These digging techniques, particularly the manner in which steam shovels were semiautomatically operated, were later adopted in the Panama Canal dig. The channel section in the rock cut had a profile of uniform 160-foot width at top and bottom. The depth throughout was 24 feet.

Extensive dredging of the Chicago River was required in order to bring about its reversal. In addition, a 13-mile section of the Des Plaines River was rerouted, the former riverbed being employed as a portion of the new channel. Finally, new bridges were constructed to maintain existing roadways across the new canal.

----

* Gates for controlling the flow of water through the structure.

On January 17, 1900, the gates at Lockport were opened for the first time. A great crowd gathered along the river's edge in Chicago, full of doubters as to the predicted reversal. But the Chicago River did indeed change direction, and sewage stopped flowing into Lake Michigan. In the following decades Daniel Burnham's Chicago Plan of 1909, as constructed, created one of the world's most beautiful and inviting urban shorelines along the formerly polluted lake's edge. By 1922 the death rate from typhoid fever was down to about 1 per 100,000 of the city's inhabitants.

Minor modifications were incorporated over the years. In 1910 the 8-mile North Shore Channel was built, connecting Lake Michigan at Evanston, just north of Chicago, to the Chicago River. The new channel supplied additional lake-water inflow and provided an outlet for drainage for the city of Evanston and neighboring North Shore towns. The 18-mile Calumet Sag Channel was constructed between 1910 and 1920, connecting the main canal to the Calumet River and reversing the latter's flow as well. During the 1920s the dilution method of sewage treatment proved inadequate for the expanding metropolitan area, and treatment plants were subsequently constructed. Supported by these plants, the Sanitary and Ship Canal and its subsidiaries continue to operate today as originally conceived.

A Disaster in the Making
John Tarkov

American Heritage of Invention & Technology - Spring, 1986 Issue

Civil engineers understand that their job is to produce a project that satisfies the client's requirements at the lowest possible cost without compromising the public safety and wellbeing. If cost and safety were not considerations there would be no need for engineers; anyone can design and construct a bridge if the bridge doesn't have to be safe or if there is no limit on how much money can be spent to make it safe. Balancing the requirements for safety and economy with the need for functionality and aesthetics is the essence of the responsibility of a professional civil engineer.

Like many other things in life, the work of civil engineers, when done well, goes virtually unnoticed. However, unlike most other endeavors, when the work of a civil engineer is done improperly the failures are spectacular and cannot be concealed. This article describes one such failure. Like many other disasters, the occurrence of this one was the result of many errors, not just a single error. The nature of the civil engineering design process--a process that has evolved over the years--is for all design calculations to be checked and rechecked by independent reviewers. When the process is short circuited, as it was in this case, a simple error--whether an error in judgment or arithmetic--slips into the system. That error alone might not be enough to cause a failure, but when aggravated by other errors totally beyond the control of the designer--errors in, say, fabrication or in construction, or defects in material--the simple design error becomes an integral part of a major disaster.

One of the morals of the story of the Quebec Bridge failure is that failure has no respect for fame or eminence. Those who deliberately or inadvertently circumvent the checks and balances built into the engineering design process for detecting and correcting human error, whether the rawest rookie or the most grizzled veteran, run the risk of becoming the victim of a professional calamity.

At five-thirty on the afternoon of August 29, 1907, a steelworker named Ingwall Hall was perched high on the partially constructed south cantilever arm of the Quebec Bridge, a few miles from Quebec City. The bridge was to have a span of eighteen hundred feet when completed--the longest in the world. The first whistle signaling the end of the workday had just blown, and Hall was waiting out the few minutes before the final whistle that would send the men on the structure home for the night.

Instead of the final whistle, the workers heard a loud report, like a cannon shot. Two compression chords* in the south anchor arm of the bridge had failed, either by the rupture of their latticing or by the shearing of their lattice rivets, and as the distress of mortally tortured steel spread through the entire superstructure, the nineteen thousand tons of the south anchor and cantilever

----

* Chords are structural members; compression chords are structural elements that are designed to withstand a compression [or crushing] load as opposed to a tension [or stretching] load.

arms and the partially completed center span thundered down onto the banks of the St. Lawrence River and into the water the bridge had been designed to cross. One eyewitness likened the collapsing columns to "ice pillars whose ends were rapidly melting away."

Swallowing water and fighting the river's sudden turbulence, Hall had to struggle in order to breathe. After a few long minutes, a rescue boat reached Hall, and he was dragged aboard. He had lost two fingers, but of the eighty-six men on the bridge when it went down, he was one of only eleven who survived.

No bridge collapses quickly. Just as the safe completion of a bridge is measured in years, the failure of a bridge can be reckoned in the same way. Though the chaotic physical dismemberment of the south arm of the Quebec Bridge took no more than fifteen seconds, the more orderly prelude to the catastrophe began long before.

It began in the summer of 1897, when the consulting engineer Theodore Cooper attended the annual convention of the American Society of Civil Engineers in Quebec City. A former director of the society, Cooper was one of the most respected bridge builders of the time. He made an excursion to the proposed site of the Quebec Bridge and within a week expressed an interest in giving the Quebec Bridge Company the benefit of his expertise.

Cooper's tender of interest was hardly unbidden. The Quebec Bridge Company had been sounding out American bridge engineers as consultants because its own chief engineer, Edward A. Hoare, had never worked on a bridge with a span longer than three hundred feet. Cooper was a proud, confident man, fiercely devoted to his calling. He had been graduated as a civil engineer from the Rensselaer Institute (now Rensselaer Polytechnic) in 1858 at the age of nineteen. Enlisting in the Navy in 1861, he served as an assistant engineer of the gunboat *Chocura* for the last three years of the Civil War, then moved on to a teaching post at the United States Naval Academy. After a tour of duty in the South Pacific, he resigned from the Navy in July 1872. In May of that year, Capt. James Eads appointed Cooper the inspector of steel manufacturing for Eads's most important engineering work, the St. Louis Bridge.

If the Navy laid the groundwork for Cooper's career, the St. Louis Bridge launched it along a high trajectory. Captain Eads moved Cooper up quickly, placing him in charge of erection at the bridge, which was the most ambitious use of the cantilevered method of erection yet attempted. Cooper performed his duties admirably--once going without sleep for sixty-five hours during a crisis, another time wiring Eads at midnight to warn him that the arch ribs were rupturing, a potentially disastrous condition that was remedied by following the instructions Eads immediately wired back. Upon completion of the work in 1874, Cooper found himself much in professional demand. By 1879, after resigning as the superintendent of Andrew Carnegie's giant Keystone Bridge Company in Pittsburgh, Cooper was able to set up as an independent consulting engineer in New York.

The projects he undertook there were notable and prestigious. His works included the Seekonk Bridge in Providence, the Sixth Street Bridge in Pittsburgh, and the Second Avenue Bridge in New York. He moved through the most rarefied atmosphere of his profession, but unlike his mentor Eads, he never oversaw a truly heroic masterwork. The Quebec Bridge, viewed in that light, was irresistible to Cooper. He said the bridge would be his last work. It would stand as the crowning achievement to an elegant career.

Almost two years would go by before Cooper's affiliation with the Quebec Bridge Company became formal. The financially troubled company had a history of moving slowly--or not at all. Incorporated by an Act of Parliament in 1887, it had accomplished virtually nothing in its first eleven years. In March 1899 officials of the company met with Cooper in New York and arranged for him to review the bids for the long-awaited bridge contracts. All prospective contractors' plans and tenders were sent to him, as well as clear instructions on how he should proceed. He was especially urged to keep in mind the weak financial position of the Quebec Bridge Company.

The Quebec Company had been in close touch with the Phoenix Bridge Company of Phoenixville, Pennsylvania, since 1897, and the Phoenix Company had already submitted preliminary plans for the bridge. Now that the bidding was open, the Quebec Company's desire to give the Phoenix Company the contract for the superstructure was barely concealed.

On April 14, 1899, John Sterling Deans, the chief engineer of the Phoenix Bridge Company, wrote to Edward Hoare, his counterpart in Quebec: "Dear Mr. Hoare--Mr. Szlapka [Phoenix's chief design engineer] and I were with Cooper the greater part of yesterday, and you will be glad to learn that there was not a single vital or important criticism or mistake found in our plans. . . . Mr. Cooper, however, somewhat upset me, by making the following remark, which of course I understood was entirely personal and without any full knowledge of the situation. He said: 'Well, Deans, I believe that all of the bids will probably overrun the amount which the Quebec Bridge Company can raise, and that the result will be. . . . that all of the bids will be thrown out and a new tender asked on revised specifications and plans.' Mr. Cooper undoubtedly desires to be perfectly fair, but. . . does not fully understand the situation. I trust, therefore, that you will give his report the most careful scrutiny, and get it in the right shape before it is submitted."

There were more collegial letters between Phoenixville and Quebec, and both Deans and Hoare stayed in close touch with Cooper. Later Cooper would maintain that no pressure had been brought to bear on him. In any event, on June 23, 1899, he sent his findings to the Quebec Bridge Company. "I therefore hereby conclude and report," he wrote, "that the cantilever superstructure plan of the Phoenix Bridge Company is the 'best and cheapest' plan and proposal."

Those three words--"best and cheapest"--became a touchstone for Cooper in his approach to the bridge. His subsequent letters to Quebec and Phoenixville are seasoned with references to the fiscal consequences of major design decisions. None of the parties involved ever placed costs before safety outright, but their aim was clearly to build a bridge that could bear the twin loads of its own mechanical burden and the Quebec Bridge Company's financial burden.

The Quebec Company had no cause to be dissatisfied with Cooper's scrupulous concern for its ledger books--and had every reason to be confident of his ability

309

to oversee the building of a good bridge. On May 6, 1900, Cooper was
appointed the company's consulting engineer for the duration of the work. He
had become, finally, the master builder on a project of historic magnitude.

Five days before his formal appointment on May 1, Cooper exercised his
authority by recommending that the span of the bridge be lengthened from
sixteen hundred feet to eighteen hundred feet. His explanation for this major
design change revealed an attentiveness to both engineering and expense.
Piers constructed in deeper water would be subject to the heavy ice floes of
the main channel. Closer to shore, they would be less vulnerable--and quicker
to build, speeding up the completion of the entire work by at least one year.
The change would also make the bridge the world's longest. To keep down
the increased cost of steel in the superstructure for an eighteen-hundred-foot
span, Cooper recommended another major design change: modified specifications that would allow for higher unit stresses.

His recommendations were approved at Quebec almost as a matter of course.
And then, for the next three years--as work proceeded on the substructure,
the anchorages, and the approach spans--practically nothing was done to
prepare for the engineering difficulties posed by the eighteen-hundred-foot
span and the higher allowable stresses.

Once again, money was the root of inaction. Short of funds as usual, the
Quebec Bridge Company was making no promises to anybody about its
capacity to pay for the bridge's superstructure once the preliminary work was
done. For all the goodwill between Phoenixville and Quebec, the Phoenix
Bridge Company was politely declining to enter into a contract until payment
might be assured.

And so, while the huge size of the bridge cried out for preliminary tests and
research studies, none were conducted during the long slack period between
1900 and 1903. It was not in the interests of the Phoenix Bridge Company to
go out-of-pocket on research costs it might never recoup, and it was plainly
impossible for the Quebec Bridge Company to provide the funds. An unspoken
assumption became necessary instead: Theodore Cooper's experience and
authority were sufficient to confer success upon the untested work.

Then in 1903 the Canadian government guaranteed a bond issue of $6.7
million to pay for the work. With that, the torpor enveloping the project
turned into humming activity. Phoenix and Quebec entered into serious
contract discussions while design engineers and draftsmen struggled to meet
the urgent demand for detailed drawings.

Three years of opportunity for deliberate preparation had been lost. In the
rush to provide drawings so that the steel for the bridge could be fabricated

----

    * Unit stresses are a measure of the load carried by a structural member.
Increasing the load carried by a member without increasing the size of the
member causes an increase in unit stress. Structural members are designed
so that the unit stress does not exceed the allowable stress for that particular
member. The allowable stress is determined by the type of material, e.g.
steel, concrete, etc.)

with little loss of time, there was no recomputation of assumed weights for the bridge under the revised specifications. It was an oversight of critical importance, and Theodore Cooper did not intervene. He decided to accept the theoretical estimates of weight that the Phoenix Bridge Company had provided.

During the three languid years that preceded the project's lurch into progress, Cooper visited the site of the bridge three times. His third visit, in May 1903, when he was sixty-four, would be his last. After that, he would decline requests that he come to Quebec. His health was poor, he said, and his physician had advised him not to travel. From that point on, he would oversee the construction of the world's longest spanning bridge from his office in New York.

Cooper's health may indeed have been fragile, but he was hardly an invalid, commuting almost daily to his office at the foot of Manhattan Island from his home on West Fifty-seventh Street. The only specific references to illness in his letters to Quebec and Phoenixville cite "the grippe" and "fatigue" as his reasons for not being able to be there.

In fact, he had never much appreciated being there in person as a consulting engineer. He regarded on-site visits as unproductive and largely devoted to atmospherics. From his earliest days in private practice, Cooper had insisted on a clause in his contract that limited his on-site responsibilities to a maximum of five days a month. When the Quebec Bridge Company's secretary, Ulric Barthe, at one point brought Cooper's attention to that understanding, Cooper replied that the five days were not an obligation but a limit, implying that it was a limit not to be abused. With his health now weakened, the five-day limit became academic.

The question of his health also caused Cooper to offer what amounted to a pro forma resignation in 1904. On a visit to New York, S. N. Parent, the president of the Quebec Bridge Company, asked Cooper when he might see him in Quebec again. Cooper's answer was never. He then asked to be relieved of his responsibilities, but Parent would not hear of it. A short time later Cooper made the same offer to John Deans of the Phoenix Bridge Company, who also refused to treat it seriously. The matter was laid to rest, and Cooper refrained from pressing it. Feeling, as he later said, "a pride and a desire to see this great work carried through successfully, I took no further action."

In the summer of 1903, while Cooper was still well enough to travel, his pride in the great work took him to Ottawa. He was incensed. Collingwood Schreiber, the chief engineer of the Department of Railways and Canals, had suggested that the department hire its own consulting bridge engineer to review and correct the detailed drawings of the Quebec Bridge--after Cooper had seen them--and then submit them to Schreiber for final approval. Robert Douglas, an engineer in Schreiber's department, had reviewed Cooper's new specifications for the eighteen-hundred-foot-span bridge and had criticized the high unit stresses. "Considering that the American government in several cases appointed four or five engineers to consider and determine unit stresses of unexampled magnitude," Douglas would say later, "I thought that this matter was too important to be left to the judgment of Mr. Cooper." But confidence in Cooper was the byword just then, and foresight was at a premium.

311

Upon learning of Schreiber's proposal, Cooper wrote angrily to Quebec: "This puts me in the position of a subordinate, which I cannot accept." His brisk discussions with Schreiber in Ottawa yielded a decidedly one-sided compromise, much to the relief of Cooper's worried colleagues in Quebec and Phoenixville. It was agreed that plans and specifications would pass from Cooper to Schreiber for final approval; as it would turn out in practice, Schreiber's initials could just as well have come from a rubber stamp.

"I think," Cooper wrote to Hoare upon his return from Ottawa, "this will allow us to go on and get the best bridge we can, without putting metal where it will do more harm than good." By now, whether he wanted it or not initially, Cooper had attained virtually absolute authority over the engineering of the Quebec Bridge. He would say later that the burden had been imposed upon him by the circumstances, and that it was an onerous one. But in 1903 he had journeyed in haste to Ottawa to block Schreiber's attempt to have drawings independently reviewed; in 1904 he had quickly acceded to the protestations of Deans and Parent that he not resign; and in 1905 he insisted that a young, recently graduated engineer be installed at Quebec to serve, in effect, as his eyes and ears on the bridge.

The young engineer's name was Norman McLure, and though nominally he answered to both Cooper and Hoare, he was in fact Cooper's personal representative, communicating with Cooper frequently. McLure's intelligence, energy, and loyalty suited Cooper well. He was well trained and well recommended, and he had enough technical competence to keep Cooper accurately informed and to execute Cooper's instructions, but not nearly enough experience to act without Cooper's authority.

The practical effect of all this, after the contract between Quebec and Phoenix was signed and erection of the superstructure got under way late in the summer of 1904, was to leave the day-to-day, hands-on building of the most technically ambitious bridge project in the world to a group of men utterly unprepared to grasp the scope of the work. No one at the site knew enough about what he was doing to act with authority. Everything of import was referred to Cooper.

Work on the superstructure proceeded uneventfully at first. The few difficulties that occurred were minor. The first sign of potentially serious trouble surfaced in 1906.

The best opportunity for the critical computation of weights, during the waiting period from 1900 to 1903, had long since been missed, but early in 1905 the shop drawings of the south anchor arm were practically complete, and it would have been possible to recompute the weight of the arm to within a few percentage points of its actual weight. Neither the Phoenix Bridge Company nor Theodore Cooper bothered to do it--now for the second time.

On February 1, 1906, they began to pay the price. Cooper received a report from E. L. Edwards, the Phoenix Bridge Company's inspector of materials, revealing that the actual weight of steel put into the bridge had far exceeded the original estimated weight. (By June the projected weight for the complete structure would have to be raised from sixty-two to seventy-three million pounds.) Cooper concluded that the increase in the already high stresses, due to the error reported by Edwards, was between 7 and 10 percent.

312

By this time the south anchor arm, tower, and two panels of the south cantilever arm had been fabricated, and six panels of the anchor arm were already in place. Cooper decided that the increase in stresses was safe, and he permitted work to continue. The only alternative would have been to start building the bridge all over again.

By the summer of 1907 the consequences of allowing the bridge's actual dead load* to go uncalculated for so long began to show up on the structure itself, in the lower chord compression members--the lower outside horizontal pieces running the length of the bridge.

On June 15 McLure wrote to Cooper: "In riveting the bottom chord splices of [the] south anchor arm, we have had some trouble on account of the faced ends of the two middle ribs not matching . . . This has occurred in four instances so far, and by using two 75-ton jacks we have been partly able to straighten out these splices, but not altogether."

Cooper replied: "Make as good work of it as you can. It is not serious. It would be well. . . in future work to get the best results in matching all the members before the full strains [forces] are brought upon them."

When work on the central, suspended span began in July--as the span crept out over the river--the rapidly increasing stresses on the compression members farther back became intolerable. The instability of built-up, latticed compression members in a major work under construction was poorly understood then, so key portions at the ends of the Quebec Bridge's weight-bearing lower chords were still unriveted, even as the stresses upon them grew insupportable with the steady outward advance of the span.

By early August the end details of the compression chords began to show signs of buckling. On August 6 McLure reported to Cooper that lower chords 7-L and 8-L of the south cantilever arm were bent. Cooper was troubled. He wired back with instructions, and with the almost plaintive question: "How did bend occur in both chords?"

On August 12 McLure informed him that the splice between lower chords 8-L and 9-L was now bent as well. Cooper's concern grew, but it was not shared in Phoenixville. Chief Engineer Deans insisted that chords 7-L and 8-L had already been bent when they left the shop. McLure insisted that they only began to show deflection after being installed on the bridge. The debate over chords 7-L and 8-L occupied the greater part of August. Meanwhile work continued, and the stresses on the lower chords grew.

On August 20 chords 8-R, 9-R, and 10-R showed distortion. On August 23 the joint between chords 5-R and 6-R showed a half-inch offset. The bend in chord 8-R was increasing. The bridge was collapsing with glacial slowness, but no one--not even Cooper, for all his concern in the face of the Phoenix Bridge Company's almost cavalier attitude--appreciated fully what was happening.

----

* Dead load is the term engineers use to describe the weight of the structure itself.

313

On August 27 the crisis should have been obvious to all. A week before, chord 9-L of the south anchor arm had been only three-quarters of an inch out of line. On the morning of August 27 McLure measured it again. The deflection was now two and one-quarter inches. McLure wrote to Cooper immediately. Had he been more experienced, he might have sent a telegram, the way a younger Theodore Cooper had once wired Captain Eads at midnight years before.

As word of what had happened to chord 9-L of the anchor arm swept the bridge, gusts of anxiety swept along with it. By the end of the day B. A. Yenser, the Phoenix Company's general foreman on the bridge, decided to suspend work, saying that he feared for his own life and the lives of the men under his charge. The next morning he changed his mind and ordered work to continue. Chief Engineer Hoare of the Quebec Company endorsed this decision--there is some evidence that he may have requested it. He saw no immediate danger, and he was afraid that stopping work then might mean that it would not resume until spring.

Officials of the Phoenix Bridge Company continued to insist that all the bends detected in the lower chord members had been present before installation. They made no effort at all to explain how the deflection of chord 9-L had grown by an inch and a half in the past week.

Fear was everywhere on the bridge on August 28, while the men in charge at the site were paralyzed by a vacuum of authority. Hoare, the Quebec Company's responsible engineer on the project, was technically unqualified--and thus unable--to take command. After much discussion, he dispatched McLure to New York to brief Cooper in person.

Shortly before 11:30 a.m. on August 29, Theodore Cooper arrived at his Manhattan office and found Norman McLure waiting for him. McLure's letter of August 27 had arrived as well. Cooper read it, spoke briefly with McLure, and at 12:16 p.m. he sent a terse telegram to Phoenixville that read: "Add no more load to bridge till after due consideration of facts. McLure will be over at five o'clock."

Cooper was unaware that work was still going on at Quebec. He was under the impression, based on McLure's letter, that construction had stopped two days before. In his haste to catch a train to Phoenixville, McLure neglected to wire Cooper's decision to Quebec as he had promised to do, and so work continued through the afternoon.

Cooper's telegram reached Phoenixville at about 3:00 p.m. John Deans read it --and disregarded it. The workers stayed on the bridge. When McLure arrived at five o'clock, Deans and Peter Szlapka met with him. They agreed to meet again in the morning, when a letter from Phoenix's field engineer at Quebec was due to arrive. The letter would support the Phoenix Company's position that the chords had left Phoenixville slightly bent but serviceable. Almost precisely as the meeting adjourned, chords 9-L and 9-R of the anchor arm buckled, and the Quebec bridge collapsed.

The members of the Royal Commission of Inquiry investigating the collapse wrote in their 1908 report, "We are satisfied that no one connected with the work was expecting immediate disaster, and we believe that in the case of Mr.

Cooper his opinion was justified. He understood that erection was not proceeding; and without additional load the bridge might have held out for days."

John Deans was excoriated for his abysmally poor judgment during the final crisis, and the Quebec Bridge Company was criticized for appointing the unqualified Edward Hoare as the responsible engineer at the site. But the brunt of the blame was placed on the shoulders of Theodore Cooper and Peter Szlapka. Cooper had examined and approved Szlapka's design for the bridge. "The failure," said the commissioners, "cannot be attributed directly to any cause other than error in judgment on the part of these two engineers . . . . A grave error was made in assuming the dead load for the calculations at too low a value. . . . This error was of sufficient magnitude to have required the condemnation of the bridge, even if the details of the lower chords had been of sufficient strength."

The second Quebec Bridge was completed in 1917. Weighing two and a half times more than its ill-fated predecessor, it has stood without any additional reinforcement since the day it opened. It did undergo a calamity of its own, however. In 1916 its prefabricated central span dropped into the river while being raised into place, killing eleven.

Theodore Cooper

Theodore Cooper's career ended with the collapse of the first Quebec Bridge. He testified twice before the Royal Commission, speaking candidly and with some bitterness toward both the Phoenix and Quebec bridge companies. His testimony brought forth a fusillade of countercharges from officials at Phoenixville and Quebec. With that last tremor of the tragedy behind him, Cooper retired from public life. He died at home on August 24, 1919, at the age of eighty.

Several months after the disaster, a party of engineering students from McGill and Laval universities made an excursion to the site of the ruins that had been the first Quebec Bridge.

There was little they could learn from the tortured steel; the Royal Commission had already pronounced it of strictly limited value to its own investigation.

What lessons the debris contained had to be gleaned on levels other than the purely technical. And if the twisted metal spelled out anything to the young engineering students as they made their way around it, the message was this: Any great bridge builder is by nature a figure of hubris. Here is what happens when hubris goes insufficiently checked by deliberation and exquisite care in the face of the little known. You may not think this could happen to you, but it can. It can happen to anyone who dares to build. It happened here to the best of them.

The Charles River Basin

From **Landmarks in American Civil Engineering**
Daniel L. Schodek

The recognition that civil engineers have received for their contributions to the American landscape has generally been limited to their contributions to the design and construction of beautiful structures. It is seldom recognized that civil engineers, in their search for solutions to transportation, or housing, or sanitation, or water supply, or flood control problems have made major contributions to the quality of the nation's natural environment. For each of the widely publicized situations where civil engineers have been involved in the creation of "eyesores" there are literally dozens, if not hundreds, of beautiful spectacles created by civil engineers. Many of them are structures such as the Golden Gate or Brooklyn Bridge, or dams such as Hoover and Grand Coulee, or buildings such as the Louisiana Superdome or the Empire State Building, but there are also hundreds of projects that have resulted in great scenic beauty and recreational opportunity. The Charles River project is one such project, but there are hundreds of miles of urban floodways, thousands of miles of scenic highway, and tens of thousands of public use areas for everything ranging from wilderness camping to outdoor sports at reservoir projects in every part of the nation. All of these environmental assets are attributable to the efforts of America's civil engineers.

Boston is among the most beautiful of American cities--a fact attributable in large part to the presence of the Charles River Basin, which separates Boston and Cambridge. The broad fresh-water basin provides not only a focus for the city but a recreational resource as well. Sailboats and crew boats use its waters; Boston's nineteenth-century fabric of brick buildings, softened by Storrow Park, flanks one side, and the campus of MIT overlooks the other. The stately Longfellow Bridge, with its arches and towers, crosses the basin and defines it. This incalculable urban asset is the result of the damming of the Charles River, completed in 1910, to flood its tidal flats.

Prior to 1903 salt-water tidal inflows from Boston Harbor filled the estuary of the Charles River, alternating with outflows that left broad, unsightly, and often malodorous mudflats along each shoreline. By the late 1800s most of the marshy territory adjacent to the river had been filled in and built upon, bringing ever-increasing portions of the populace face to face with twice-daily displays of estuarial muck and leading to growing public pressure for a dam across the Charles that would form a basin of sufficient depth to continuously cover the flats. In fact, a dam had been proposed as early as the beginning of the nineteenth century. Various more recent proposals favored half-tide dams, full-height dams, a fresh-water basin, a salt-water basin, and several possible dam locations.

In 1891, at his inaugural, Boston's Mayor Matthews spoke of "the opportunity for making the finest water park in any city in this country . . . an imitation of the plan adopted by the city of Hamburg . . . We should dam up the stream . . . and lay out a series of boulevards along the basin thus created." A group named the Committee on the Charles River Dam became the primary public force favoring a dam. Its opponent was the Beacon Street Committee, which represented the shoreline landowners, primarily along Beacon Street, and local wharf interests.

The issue came to a head in 1893-1894. The Massachusetts legislature had funded an investigation of the condition of and the means for improving "the beds, shores, and waters of the Charles River between the Charles River Bridge and the Waltham Line." The conclusion of this study was in favor of a dam and a fresh-water basin. Howls of protest arose from the Beacon Street Committee, which, although motivated chiefly by commercial concerns, cited the prediction of some experts that the dam would create a malodorous lake, with huge, floating masses of excrement, unaffected any longer by daily tidal cleansing. The appalling vision took sufficient hold of the public imagination to prevent any action on the mudflat problem until 1901, when once again a background study was funded, this time under the guidance of an extremely dedicated and effective civil engineer, John Ripley Freeman.

John Ripley Freeman

Freeman had already had twenty-five years of civil engineering experience when asked to advise the Committee on the Charles River Dam. He spent ten of them working on hydraulics projects in New England. He also knew many of the people involved in the basin controversy. The chairman of the dam committee was Henry Pritchett, president of MIT [Massachusetts Institute of Technology], Freeman's alma mater. Freeman himself was a member of the MIT Corporation. Freeman engaged many other experts in his work, including Charles T. Main, an MIT classmate, and the eminent civil engineer Hiram Mills.

Assisted by specialists in a number of fields, Freeman conducted his researches throughout 1902, assembling enough data to discount all the pseudoscientific proclamations that had fueled the opposition to the dam. The work began with new surveys of the basin area, in order to reflect recent dredgings, but eventually involved the following: the continual monitoring of air and water temperatures at selected points; the charting of the paths and velocities of currents; the mapping of channels; the measure of flow rates of sewer discharges, under normal conditions and during storms; chemical and bacterial analyses of fresh- and salt-water samples from various points and under varying conditions of water motion, oxygenation, and temperature; the inspection of harbor bottom borings with reference to potential structural

318

integrity, as well as to thickness of silt deposits subject to tidal scouring; and the collection of other relevant information.

Freeman's work was remarkable for its thoroughness and for its attention to what would today be termed the probable "environmental impacts" of the proposed dam. He found, for example, that fresh water in the area was less likely to precipitate sludge and give off odors than salt water, and that the rate of pollution abatement in oxygenated water was not reduced by lack of water movement. He also found that the existing tidal flows mainly moved the same water back and forth and thus slowed the dilution of contaminants. He even found that the elimination of tidal inflows would not seriously affect air temperatures in the area, thus responding to opponents' fears that there would be a loss of cooling influence on hot days associated with influxes of sea water.

In sum, Freeman demonstrated that the dam and basin would in actuality function positively against pollution, malaria, and storm flooding and would have no substantial effect on summer temperatures or the harbor channels. Freeman was thus able to raise the quality and broaden the scope of the investigation to such an extent that his eventual support for the dam was based on rock-solid scientific refutation of the objections raised by various dam opponents. His report was made available to the committee and thence to the legislature in early 1903.

Freeman's final recommendations included many specific suggestions. He recommended that the dam be built so as to create a basin in the lower Charles River and that it be built at the site of the aging Craigie Bridge, so that the dam and the much-needed new bridge, by being virtually one and the same structure, could render substantial benefits at reduced costs. His original plans also included a park area. Further, he recommended that the dam be full rather than half height to take advantage of fresh water's greater antipollution effect, as well as to maximize recreational boating safety; and that a constant basin level be established to protect adjoining properties from high-tide storm sewage overflows and subsequent cellar floodings and to expedite navigation in the basin area irrespective of tide. Other recommendations were that marginal conduits approximately 16 feet in diameter be placed near the shorelines, terminating via tide gates below the dam (in order to further protect the basin from any excessive discharges from contiguous waterways--the Fens basin, the Broad Canal, the Lechmere Canal, and various sewer lines); and finally that various other ancillary measures be undertaken to help the basin exert positive effects on the entire surrounding system of waterways.

The projected costs of the dam and basin were only slightly higher than the costs of providing similar benefits by any other means, such as heavy dredging of the estuary. By filling it with water, such dredging could be limited to the amount of material required to build shore embankments. Because fresh water had been found to be preferable for pollution control, expensive tidal sluices were unnecessary. Most important, because the dam would incorporate a replacement for Craigie Bridge and could be built for about the same cost as simply replacing the bridge--which had to be done in any case-- the basin represented a substantial bargain.

A large lock to facilitate commercial shipping was included in the plan. Given that the railroads were taking over most of the loads, this appears to have been done at least partly to mollify the Beacon Street Committee.

Enabling legislation having been passed in 1903, construction began under the able direction of Frederic P. Sterns. The project was completed in 1910. Its chief component, the Charles River Dam, included a roadway, a large lock for the passage of boats, sluice gates, and overflow conduits. A large recreational area was part of the dam complex design.

Studies were made for dams of various widths, from one a little greater than 100 feet wide to one having a park of several acres on the upstream side. It was decided to continue the 488-foot width required by the lock nearly across the river. This made an area to be filled in the river of about 6.9 acres, of which 5.7 acres would be used for park purposes and 1.2 acres for the roadway.

The first contract for construction was signed on January 14, 1905, and included, among other things, the lock, sluices, harbor and basin walls, and the earth filling between. The award was made to the Holbrook, Cabot, and Rollins Corporation of Boston. Later contracts were let for the dam, pile driving in the Broad and Lechmere canals, sections of the embankment, and other work. In order to construct the lock, a cofferdam was built enclosing about 3.5 acres. The water was then pumped out and piles driven to receive the concrete. By January 1, 1907, a total of 9,969 foundation piles had been driven.

A critical part of the construction was the creation of a shutoff dam capable of acting quickly to change the water level in the basin. This device was necessary because the swift flow of the tide and of the river would otherwise prevent earth fill from being deposited in such a way that it would remain in place. The shutoff dam was composed of six rows of round piles, cross-braced, running across the river. Six-inch yellow pine sheeting was driven between the middle rows and cut off slightly below mean low water in Boston Harbor. Earth was eventually filled in on both sides of the sheeting and surrounded by riprap to prevent scour. The driving of the piling was difficult, because of the swift flow of the river, but it was accomplished successfully.

The method used for shutting off the river involved a series of gates. There were 82 of them, running in grooves built on heavy uprights. The gates were 6-inch by 8-inch timber, braced lengthwise and diagonally, forming solid pieces, 10 feet wide by 15 feet high. Across the top of the uprights were placed timbers, and the gates were held to these by ropes. On October 20, 1908, the ropes were cut, and in 7 seconds all of the gates were dropped into place. Rubber hose had been nailed to the bottom edge of each gate so that the joint between the gate and the sheet piling might be as close as possible. As soon as the river was closed off, the uprights projecting above the round piling were cut off and removed, and large shovels and dredges began at once to pile earth against the shutoff. This filling was continued until the structure became an earth dam with a wooden core wall. The finished dam consisted of two concrete retaining walls (faced with granite) with earth fill between.

The discharge of the river was provided for by eight sluices, each 7.5 feet wide by 10 feet high and controlled by a sluice gate operated by electricity. The roofs of the sluices and the lock for small boats and launches were made of concrete reinforced by steel beams.

The lock consisted of a reinforced-concrete structure with expansion joints, placed on piles and including the foundations for a Scherer rolling lift bridge at the roadway. Also included were recesses into which the great steel lock gates are drawn when the lock is opened, bollards, and electrically operated capstans to aid vessels passing through the lock. The lock gates were made of structural steel beams, plates, and girders and were supported on two 4-wheel trucks running on heavy steel rails at right angles to the center line of the lock. As the gates were to be operated in winter, they were heated by steam pipes, where necessary, to keep ice from forming against their sides and between the bearing surfaces.

The Charles River Basin

The Charles River Basin was an immediate success. Some repairs and alterations have been required over the years. In 1955 a hurricane caused flood flows exceeding those designed for, leading to construction of a new dam that was dedicated on May 24, 1978. The new dam is located a half-mile below the original one. Three smaller locks, instead of one large one, brought increased passage efficiency through the new dam, but craft still have to clear the bottleneck created by the old dam's lock. A major improvement consisted

321

in the new dam's contribution to water quality in the basin, which had suffered increasing salinity due to sea water intrusion through the lock. Because this salt water sinks to the bottom, it tends to stay in the basin. The dam included a pumping station drawing from 20 feet below the surface, thus reversing most of the saline intrusion at the locks. In addition, bubblers were installed upstream to promote the flushing of deep saline pockets and to destratify the basin water, with consequent water quality improvement. Increased storm sewage overflows were addressed by a new chlorination and detention plant between the two dams.

It is a measure of Freeman's achievement that water quality in the basin has only recently been considered below desirable standards, following a half-century of increasing pollutive strain. The quality and depth of his pioneering investigations swept objections firmly aside and helped provide Boston with an urban jewel comparable to Frederick Law Olmsted's "Emerald Necklace," which rings the city with some of the loveliest waterways and parklands in the country. Though less informed by the older Olmsted's nostalgic sense of the landscape, which was brought to bear on the design of Boston's Fenway, than by scientific and commercial concerns, the basin nonetheless contributes to Boston's environment at every level, from pollution control to the provision of a beautiful river park, a recreational resource of irreplaceable value, and a source of civic pride.

"The Greatest Liberty Ever Taken With Nature"

Excerpts from Chapter 19, The Path Between the Seas
David McCullough

One of the aspects of civil engineering that goes almost totally unnoticed by the layman is what civil engineers call "geotechnical engineering." Geotechnical engineering is that part of civil engineering which deals with the compaction, excavation and removal of soil and rock during a construction project, and with the planning and analysis that are necessary for these activities to be carried out efficiently and economically. The geotechnical work often goes unnoticed because it occurs out of sight of the casual observer and because the excavated material often becomes an integral part of the finished project. Excavated areas are sometimes filled with the structure that is being constructed, or, as in the case of the Panama Canal, with water. In either event the true magnitude of the geotechnical effort is frequently not apparent once the project is completed.

In this selection from David McCullough's **Path Between the Seas** *the magnitude and difficulty of geotechnical work becomes obvious. The Culebra Cut, the excavation that carried the Panama Canal across the continental divide, was one of the largest earth and rock excavations ever attempted. Without successful completion of the Cut there would be no Panama Canal. Mr. McCullough describes not only the technical dimensions of the problems encountered but also the human suffering and frustration which were endured in the execution of this mammoth project.*

Within less than a year after [Chief Engineer Colonel George] Goethals took charge, several major changes were made in the basic plan of the canal, and with a sweeping reorganization, beginning in early 1908, he installed his own entirely new regime. The widespread impression was that the plan was firm, that this at last was the canal that was to be built, and that these were the men who would build it. The widespread impression was correct.

The changes, each very important, were as follows:

--The bottom width of the channel through Culebra Cut was to be made half again wider, from two hundred to three hundred feet. Thus it was to be more than four times as broad as the French canal[*] would have been at that point.

--The width of the lock chambers was enlarged, primarily to satisfy the Navy. The locks would be 110 feet wide (rather than 95 feet) to accommodate the largest battleship then on the drawing boards, the Pennsylvania, which had a beam of 98 feet. (The largest commercial vessel then being built was the Titanic, with a beam of 94 feet.) So each lock chamber was to be 110 feet by 1,000 feet.

----

    * References to the "French canal" are references to the unsuccessful attempt by the French to construct a canal across the isthmus.

--On the Pacific side, where heavy silt-bearing currents threatened to clog the entrance to the canal, the engineers now planned a tremendous breakwater that would reach three miles across the tidal mud flats to Naos Island.

--When trestles began sinking in the mud at the site of the Sosa Dam, a major change had to be made in the placement of the Pacific locks. Previously, there was to have been one lock at the south end of Culebra Cut, at Pedro Miguel, then an intermediate-level lake and another set of two locks close to the Pacific shore, at Sosa Hill. In the new arrangement, the Pedro Miguel complex remained unchanged, but the dam and second set of locks were pulled back from Sosa Hill--back from the Pacific--to a new site at Miraflores. Consequently the terminal lake (called Sosa Lake on the old plan) was greatly reduced in area and the first flight of locks at the Pacific end was now to be as far inland as were the Gatun Locks. From the military viewpoint this was regarded as a far better solution, since the Pacific locks would now be far less vulnerable to bombardment from the sea, a point Goethals had made to [Secretary of War William Howard] Taft as early as 1905, following their tour of the area. The possibility of bombardment from the air had not been considered then, nor was it now late in 1907, since the world had as yet to catch up to the achievements at Kitty Hawk.

With his reorganization Goethals did away with all the old departments first established by Wallace and carried on by Stevens*. Under that system the work had been portioned off according to specific types of activity--excavation and dredging, labor and quarters, and so forth. Now everything was simply divided into three geographic units--an Atlantic Division, a Central Division, and a Pacific Division--each run by one overall chief who was responsible for virtually everything within the district other than sanitary and police activities. It was a scheme very like that used by the French, with the fundamental difference that none of the work was to be done by contract, except for the lock gates. Stevens' contract plan had been dropped at the time Goethals took over.

The Atlantic side was to be run solely by Army men, with Major Sibert as division head assisted by several other engineering officers. Forty-seven years old, large, headstrong, full of ambition and good humor, William Sibert was cut from much the same pattern as John Stevens, with whom he was one day to collaborate on a book about the canal. Sibert's civilian clothes fit him badly, he chewed on unlit cigars, and he spoke his mind. His relations with Goethals, strained from the start, were to become more and more unpleasant.

Born on a farm in Alabama, Sibert had finished at West Point in 1884, worked on the famous Poe Lock at the Soo Canal and ran a railroad in the Philippines. But for the past six years, assigned to river and harbor work at Pittsburgh, he had built more than a dozen locks and dams on the Allegheny, Monongahela, and Ohio rivers. His experience in such work was second to none, a point neither he nor Goethals would lose sight of.

----

   * John F. Wallace and John Stevens were the two Chief Engineers who preceded Goethals in Panama. Both resigned--Wallace to take a higher paying and safer position in the United States and Stevens for reasons that he never fully explained.

The Central Division was assigned to Major Gaillard, but his highly competent executive officer was a civilian, a lean, red-haired Bostonian named Louis K. Rourke, who had been running things very well in Culebra Cut for nearly two years.

David Du Bose Gaillard (pronounced Ge-*yard*) was a South Carolinian. He was a year older than William Sibert and a close friend. As cadets at the Military Academy they had been roommates and were known as David and Goliath. Still slim and youthful-looking, Gaillard had had a solid if unspectacular career in the Corps of Engineers and like Goethals had been singled out for the initial General Staff. "Sibert's experience on locks and dams makes his assignment to that work very necessary," Goethals explained to his West Point son, ". . . so Gaillard had to take the Cut."

Like Goethals, these and the other engineering officers who were to serve in Panama considered themselves part of an honored tradition; and this, it should be emphasized, gave to their whole mode of operation a very different tone from that of the previous regime. It was not that they were necessarily superior technicians to the railroad people who preceded them, but that their entire training and experience had been directed toward large construction works in the national interest. They were engineers of the state, no less than those who had come out from France to build the de Lesseps canal. Even their training had been patterned after that of the Ecole Polytechnique* from the time Sylvanus Thayer instituted the sweeping academic reforms at West Point that were to make him "Father of the Military Academy." It was Thayer in the 1820's who, after observing the program of the famous French school, made engineering the heart of the curriculum at West Point and instilled the mission to construct into the academic program. "We must get up early, for we have a large territory," a cadet once explained to a visitor in the 1850's; "we have to cut down the forests, dig canals, and make railroads all over the country." And that had remained the prevailing spirit. Only the top men from each class qualified for the Engineers.

But the Goethals regime did not consist solely of Army people, the common view again notwithstanding. Indeed, the only division head that he personally appointed was a civilian, Sydney B. Williamson, who had been a young assistant at Muscle Shoals when Goethals constructed the high-lift lock. He and Williamson had worked well together then and on several subsequent projects, and their trust in each other was total. Williamson was put at the head of the Pacific Division and all his subordinate engineers were to be civilians. So naturally the lines were drawn: if the Army was to build the Atlantic locks and the civilians the Pacific locks, then it would be a test to see which group was the most resourceful and competent. A sharp rivalry ensued, just as Goethals anticipated.

Meantime, Rear Admiral Harry Harwood Rousseau, who at thirty-eight was the youngest member of the canal commission, was given responsibility for the design and construction of all terminals, wharves, coaling stations, dry docks, machine shops, and warehouses. Lieutenant Frederick Mears, aged twenty-nine, was put in charge of relocating the Panama Railroad, a large and very

----

* The most prestigious French school of science and engineering.

difficult task. To build the forty-odd miles of the new line would take five years and cost nearly $9,000,000.

Two further resignations were announced, those of Joseph Ripley, who had been Stevens' choice for lock design, and Jackson Smith, whose competence Goethals recognized but whose manner had become more than Goethals was willing to tolerate. As a result Smith's Department of Labor and Quarters was broken up and Major Carroll A. Devol was named Chief Quartermaster of the Zone, with responsibility for labor, quarters, and supplies. The personal choice of Secretary Taft, Devol had been in charge of the Army transport service in San Francisco in 1906 at the time of the earthquake and had managed the distribution of all supplies to the stricken city, an enormous and ably handled operation for which the Army was wholly responsible and for which the Army was to get too little credit.

Up until now all the design work on the locks had been handled in Washington, but with Ripley's departure, Goethals transferred the design staff to the Isthmus and installed still another Army officer, Lieutenant Colonel Harry Foote Hodges, at its head.

Everything considered, Hodges was probably Goethals' most valuable man, as well as the sort journalists and historians could readily overlook. Born in Boston, class of '81 at the Academy, he was small, fussy, humorless, quite unspectacular in manner and appearance. With his sharp little face and large, dark, intense eyes, he looked not unlike a bright mouse. Like Sibert, he had spent several valuable years working with Colonel Poe on the Soo, and like Sydney Williamson, he was Goethals' personal choice. Hodges, henceforth, had overall responsibility for the design and erection of the lock gates, all the tremendous conduits and valves beneath the walls and floors of the locks, every intricate mechanism required. He had, that is, the most difficult technical responsibility in the entire project, upon which depended the canal's success. When Goethals was away from the Isthmus, Hodges would serve as acting chief engineer. According to Goethals, the canal could not have been built without him.

For anyone to picture the volume of earth that had to be removed to build the Panama Canal was an all but hopeless proposition. Statistics were broadcast-- 15,700,000 cubic yards in 1907, an incredible 37,000,000 cubic yards in 1908-- but such figures were really beyond comprehension. What was 1,000,000 cubic yards of dirt? In weight? In volume? In effort?

The illustrative analogies offered by editors and writers were of little help, since they were seldom any less fantastic. The spoil from the canal prism, it was said, would be enough to build a Great Wall of China from San Francisco to New York. If the United States were perfectly flat, the amount of digging required for a canal ten feet deep by fifty-five feet wide from coast to coast would be no greater than what was required at Panama within fifty miles. A train of dirt cars carrying the total excavation at Panama would circle the world four times at the equator. The spoil would be enough to build sixty-three pyramids the size of the Great Pyramid of Cheops. . . .

The material taken from Culebra Cut alone, exclaimed one writer toward the completion of the work, would make a pyramid topping the Woolworth Building by 100 feet (the Woolworth, at 792 feet, was then the world's tallest building), while the total spoil excavated in the Canal Zone would form a pyramid 4,200 feet high, or more than seven times the height of the Washington Monument.

If all the material from the canal were placed in one solid shaft with a base the dimension of a city block, it would tower nearly 100,000 feet--nineteen miles-- in the air.

But who could imagine such things? Or how many could also take into account the smothering heat of Panama, the rains, the sucking mire of Culebra, none of which was less troublesome or demoralizing than in times past. For however radically systems or equipment were improved upon, however smoothly organized the labor army became, the overriding problem remained Panama itself--the climate, the land, the distance from all sources of supply. At the bottom of Culebra Cut at midday the temperature was seldom less than 100 degrees, more often it was 120 to 130 degrees. . . .

In any one day there were fifty to sixty steam shovels at work in the Cut, and with the dirt trains running in and out virtually without pause, the efficiency of each shovel was more than double what it had been. Along the entire line about five hundred trainloads a day were being hauled to the dumps. . . .

Perhaps as extraordinary as anything that can be said is that the work could not have been done any faster or more efficiently in our own day, despite all technological and mechanical advances in the time since, the reason being that no present system could possibly carry the spoil away any faster or more efficiently than the system employed. No motor trucks were used in the digging of the canal; everything ran on rails. And because of the mud and rain, no other method would have worked half so well.

The "special wonder of the canal" was Culebra Cut. It was the great focus of attention, regardless of whatever else was happening at Panama. The building of Gatun Dam or the construction of the locks, projects of colossal scale and expense, were always of secondary interest so long as the battle raged in that nine-mile stretch between Bas Obispo and Pedro Miguel. The struggle lasted seven years, from 1907 through 1913, when the rest of the world was still at peace, and in the dry seasons, the tourists came by the hundreds, by the thousands as time went on, to stand and watch from grassy vantage points hundreds of feet above it all. Special trains had to be arranged to bring them out from Colon and Panama City, tour guides provided, and they looked no different from the Sunday crowds on the Boardwalk at Atlantic City. Gentlemen wore white shoes and pale straw hats; ladies stepped along over the grass in ankle-length skirts and carried small, white umbrellas as protection from the sun. A few were celebrities: Alice Roosevelt Longworth, Lord Bryce, President Taft, and William Jennings Bryan (who "evinced more general excitement than anyone since T.R.") "He who did not see the Culebra Cut during the mighty work of excavation," declared an author of the day, "missed one of the great spectacles of the ages--a sight that no other time, or place was, or will be, given to man to see." Lord Bryce called it the greatest liberty ever taken with nature.

A spellbound public read of cracks opening in the ground, of heartbreaking landslides, of the bottom of the canal mysteriously rising. Whole sides of mountains were being brought down with thunderous blasts of dynamite. A visiting reporter engaged in conversation at a tea party felt his chair jump half an inch and spilled a bit of scalding tea on himself.

To Joseph Bucklin Bishop, writing of "The Wonderful Culebra Cut," the most miraculous element was the prevailing sense of organization one felt. "It was organization reduced to a science--the endless-chain system of activity in perfect operation."

> On either side were the grim, forbidding, perpendicular walls of rock, and in the steadily widening and deepening chasm between--the first man-made canyon in the world--a swarming mass of men and rushing railway trains, monster-like machines, all working with ceaseless activity, all animated seemingly by human intelligence, without confusion or conflict anywhere . . . The rock walls gave place here and there to ragged sloping banks of rock and earth left by the great slides, covering many acres and reaching far back into the hills, but the ceaseless human activity prevailed everywhere. Everybody knew what he was to do and was doing it, apparently without verbal orders and without getting in the way of anybody else . . .
> Generally, the more the observer knew of engineering and construction work, the higher and warmer was his appreciation.

Panoramic photographs made at the height of the work gave an idea of how tremendous that canyon had become. The columns of coal smoke that towered above the shovels and locomotives--"a veritable Pittsburgh of smoke"-- were blue-black turning to warm gray; exposed clays were pale ocher, yellow, bright orange, slate blue, or a crimson like that of the soil of Virginia; and the vibrant green of the near hills was broken by cloud shadow into great patchworks of sea blue and lavender.

The noise level was beyond belief. On a typical day there would be more than three hundred rock drills in use and their racket alone--apart from the steam shovels, the trains, the blasting--could be heard for miles. In the crevice between Gold Hill and Contractors Hill, where the walls were chiefly rock, the uproar, reverberating from wall to wall, was horrible, head-splitting.

For seven years Culebra Cut was never silent, not even for an hour. Labor trains carrying some six thousand men began rolling in shortly after dawn every morning except Sunday. Then promptly at seven the regular work resumed until five. But it was during the midday break and again after five o'clock that the dynamite crews took over and began blasting. At night came the repair crews, men by the hundreds, to tend the shovels, which were now being worked to the limit and taking a heavy beating. Night track crews set off surface charges of dynamite to make way for new spurs for the shovels, while coal trains servicing the shovels rumbled in, their headlights playing steadily and eerily up and down the Cut until dawn. And though it was official Isthmian Canal Commission policy that the Sabbath be observed as a day of rest, there was always some vital piece of business in the Cut that could not wait until Monday.

Among the most fascinating of the surviving records of the work is a series of Army Signal Corps films made down in the Cut. Watching these rare old motion pictures (now in a collection at the National Archives), seeing the trains cut back and forth across the screen, seeing the dynamite go off and tiny human figures rush about through clouds of dust and smoke, one senses too how extremely dangerous it all was. At one point, when a shovel suddenly swings, Goethals can be seen to jump nimbly out of the way.

Bishop and those others who described the spectacle from the cliffs above had very little to say about such hazards. But year after year hundreds of men were being killed or hideously injured. They were caught beneath the wheels of trains or struck by flying rock, crushed to death, blown to bits by dynamite. "Man die, get blow up, get kill or get drown," recalled one black worker; "during the time someone asked where is Brown? He died last night and bury. Where is Jerry? He dead a little before dinner and buried. So on and so on all the time."

Construction of the canal would consume more than 61,000,000 pounds of dynamite, a greater amount of explosive energy than had been expended in all the nation's wars until that time. A single dynamite ship arriving at Colon carried as much as 1,000,000 pounds--20,000 fifty-pound boxes of dynamite in one shipload--all of which had to be unloaded by hand, put aboard special trains, and moved to large concrete magazines built at various points back from the congested areas.

At least half the labor force was employed in some phase of dynamite work. Those relatively few visitors permitted to walk about down in the Cut saw long lines of black men march by with boxes of dynamite on their heads, gangs of men on the rock drills, more men doing nothing but loading sticks of dynamite into the holes that had been drilled. The aggregate depth of the dynamite holes drilled in an average month in Culebra Cut (another of those statistics that defy the imagination) was 345,223 feet, or more than sixty-five miles. In the same average month more than 400,000 pounds of dynamite were exploded, which meant that all together more than 800,000 dynamite sticks with their brown paper wrappings, each eight inches long and weighing half a pound, had been placed in those sixty-five miles of drill holes, and again all by hand.

Difficulty was had at first in determining how much dynamite to use in a single shot, depending on the depth of the holes, the spacing of the holes, and the character of the rock, which could be anything from basalt to the softest

----

* The drills themselves were of two types, a well drill that could bore a hole five inches in diameter to a depth of one hundred feet and a smaller tripod drill that could bore a three-inch hole to a depth of thirty feet. These drills were all powered by compressed air fed into the Cut through some thirty miles of pipe from big compressors at Rio Grande, Empire, and Las Cascadas. The elaborate compressed-air system was another of those advances that distinguished the American effort from that of the French.

329

shale. The foremen responsible for the loading and tamping learned by trial and error. Different grades of powder were tried, different kinds of fuses and methods of firing.

Premature explosions occurred all too often as the pace of work increased. "We are having too many accidents with blasts," Goethals noted in June 1907. "One killed 9 men on Thursday at Pedro Miguel. The foreman blown all to pieces." Several fatal accidents were caused when shovels struck the cap of an unexploded charge. Another time a twelve-ton charge went off prematurely when hit by a bolt of lightning, killing seven men. Looking back years later, one West Indian remembered, "The flesh of men flew in the air like birds many days."

The worst single disaster occurred on December 12, 1908, at Bas Obispo. More than fifty holes had been loaded with some twenty-two tons of dynamite. The charges had been tamped, the fuses set, but none of the holes had been wired since the blast was not scheduled until the end of the day. As the foreman and one helper were tamping the final charge, the whole blast went off, by what cause no one was ever able to determine. Twenty-three men were killed, forty injured.

As time went on the men became extremely proficient and accidents became comparatively rare considering the volume of explosives being used and the numbers of laborers involved. Still, more men would be killed, and very often, as at Bas Obispo, there would be too little left of them to determine who they were.

The shovels in the Cut set records "never anticipated," as Goethals noted, and in the eyes of most beholders they became something more than mere machines. They had personality and gender--usually feminine, yet they were also likened to Theodore Roosevelt--and accounts of their prodigious feats of strength, as well as their agility, acquired a kind of mythical quality. The Canal Record's full-page reports on their performance were read as avidly as baseball scores.

The peak was in March 1909, when sixty-eight shovels, the largest number ever used at one time in the Cut, removed more than 2,000,000 cubic yards, ten times the volume achieved by the French in their best month. The record for a single shovel was set in March 1910, when a ninety-five-ton Bucyrus (No. 123), working twenty-six days, excavated 70,000 cubic yards. More astonishing is the realization that the vast rift in the earth at Culebra was dug entirely by what, comparatively speaking, was a mere handful of machines. The volume removed from the Cut was 96,000,000 cubic yards. So even allowing for replacements, the average shovel dug well over 1,000,000 cubic yards, despite the worst kind of punishment year in, year out. No machines had ever been subjected to such a test and their record was a tribute to the men who designed and built them.

The shovels were deployed along the entire nine miles of the Cut, but in one section just to the north of Gold Hill they were stacked one above another at seven different levels, while seven parallel tracks carrying the dirt trains were kept constantly busy. "There were any amount of . . . trains, which were going in every direction," noted a young English tourist in her diary; "they must be very well arranged." In fact about 160 trains a day were running in and out of the Cut, and the degree of planning needed to handle such traffic can be further appreciated when it is taken into account that most of the track had to be shifted--removed, replaced, relocated--time and again. There were 76 miles of construction track within the nine-mile canyon, while in the Central Division as a whole there were 209 miles, not counting the Panama Railroad. In any one year well over a thousand miles of track had to be shifted about within that area just to keep the work moving in the Cut. And to complicate the problem further still, the bottom of the Cut, the main work level, kept steadily contracting in width the deeper the Cut became.

No one part of the operation--not the drilling, the blasting, the shoveling, the dirt hauling--could ever be permitted to interfere or disrupt another. So consequently every move was the result of very careful study. All shovels, every mile of track, every one of the hundreds of rock drills in use, were located daily on a map at division headquarters at Empire. Careful estimates were made as to the progress of each individual steam shovel, when it would have to be repositioned, when tracks would have to be shifted, what effect such moves would have on the disposition of drilling and blasting crews. So neatly was everything coordinated, so smooth were communications, that at the close of each day locomotive crews, as an example, had only to check the assignment boards at the roundhouses to see exactly what they were to do the day following.

Traffic in and out of the Cut was directed from towers at either entrance by yardmasters who kept in telephone contact with the various dumps and with a half-dozen small towers strung out along the line of excavation. The yardmasters, who took their orders from the chief dispatcher at Empire, directed the passage of each loaded train to a particular dumping ground and ordered the right of way for the train when it hit the main line of the Panama Railroad. When the empty trains returned, it was the yardmaster again who distributed them to the shovels.

The dumping grounds--the other end of the system--were located anywhere from one to twenty-three miles from the Cut. Sixty-odd locations were used in the course of excavation, and though much of the spoil[*] was simply gotten rid of--that is, put to no useful purpose--a very considerable part of it served to build earth dams, to build embankments on the new line of the railroad, and to create the huge new Naos Island breakwater at the Pacific end. To keep the flooding Chagres [*River*] from backing up into the Cut as the great trench deepened, an earth dike was thrown across the north end, at Gamboa, seventy-eight feet above sea level.

All the dumps were carefully engineered, with tracks on several terraces. At each dump was another yardmaster who reported the arrivals and departures of trains and his "readiness for spoil," who ordered the distribution of loaded trains to the several dumping tracks, and who, in addition, directed the movements of the Lidgerwood unloaders[**] as well as two additional pieces of

----

[*] Spoil is the term used to describe excavated material that can not be used elsewhere in the project.

[**] A mechanical unloading device used to unload an entire train.

331

equipment that had since come into use: the dirt spreader and the track shifter.

Both devices were of vital importance to the efficiency of the entire system, since the least delay at the dumping end at once decreased progress in the Cut. The dirt spreader was a railroad car with big steel blades mounted on either side, these operated by compressed air. Once a train had been unloaded, its spoil dumped beside the tracks, the spreader came through, pushed by a locomotive, and did the job of several hundred men working with shovels. The track shifter, an even cruder-looking piece of equipment, was the creation of William Bierd, former head of the Panama Railroad, who had built the first one in the shops at Gorgona shortly before Goethals' arrival. It was a huge crane-like contraption that could hoist a whole section of track--rails, ties, and all--and swing it in either direction. And since the tracks at the dumps had to be shifted constantly, to keep pace with the loads being delivered, it was an extremely valuable adjunct. Bierd's own creation could shift track about three feet, but subsequent models, built after he resigned, could reach as much as nine feet. With one such rig, fewer than a dozen men could move a mile of track in a day, a task that would have taken not less than six hundred men working by hand.

The largest of the dumps were at Tabernilla (fourteen miles beyond the north end of the Cut), Gatun Dam (the most distant location), Miraflores, and La Boca, the largest, which had been renamed Balboa. Some of the dumps covered as much as a thousand acres, and in the rainy season they became great seas of mud, with tracks slipping and sinking five or six feet. At Tabernilla, more than 16,000,000 cubic yards of spoil were simply dropped in the jungle. At Balboa, 22,000,000 cubic yards were deposited, with the result that 676 acres were reclaimed from the Pacific as a site for a new town.

By far the most troublesome of the dumps was the Naos breakwater, where, as at Gatun Dam, spoil from the Cut was dumped from a huge trestle, this one being extended slowly across the mud flats of the bay. At first everything went as hoped. But then the soft bottom sediments began to give way beneath the heavier material being poured on top. Overnight whole sections of trestle and track would vanish into mud and everything would have to stop until they were replaced. In some areas the vertical settlement exceeded a hundred feet, while the slippage sideways was three times worse. In time not a single foot of the long trestle remained where it had been to start with. By 1910 well over 1,000,000 cubic yards of spoil had been dumped into the breakwater and still it was a mile short of Naos Island. To reach the island, ultimately, would require 250,000 cubic yards of earth and rock from Culebra, which was ten times what had been originally estimated.

"Culebra Cut was Hell's Gorge," one steam-shovel man would write, recalling the heat and dust and noise. Nor were the rains any less of a problem than in times past. In 1908 and again in 1909, the years of the heaviest work, well over ten feet of rain fell. To check the torrential runoff, to reduce the chance of landslides, Goethals did what the French had done: he had diversion channels dug parallel to the Cut. But he greatly expanded on their plan. The channel on the east side of the Cut, known as the Obispo diversion, ran for a distance of five and a half miles and had a minimum width of fifty feet. To build this ditch, and another similar to it on the opposite side, the so-called Camacho diversion, required another 1,000,000 cubic yards of excavation. And very possibly they were a mistake, as Goethals himself later conceded, since they were dug too close to the Cut and water seeping from them below ground may have been the cause of several of the more disastrous slides.

All technical problems at Panama were small problems compared to the slides in the Cut. The building of the great dam at Gatun, for so long the most worrisome part of the plan, turned out to be one of the least difficult tasks of all. A tremendous man-made embankment simply grew year by year at Gatun, extending a mile and a half across the river valley, a ridge of earth that was to be fifteen times as wide at its base as it was high. At the eastern end were the beginnings of the Gatun Locks; in the center were the beginnings of what was to be the dam's giant concrete spillway. Two big outer walls of "dry" spoil were built first as a base for the embankment. These toes, as they were called, were nearly half a mile apart--the river, meantime, having been turned into an old diversion channel built by the French--and into the space between them was pumped hydraulic, or "wet," fill, a solution of blue clay, which when dry would create a core almost as impervious as concrete. There was no lack of controversy over the project as time went on (much of it stirred up by Philippe Bunau-Varilla, who was convinced that Goethals did not know what he was doing), and once, on November 20, 1908, a section about two hundred feet long slipped sidewise and sank nearly twenty feet at the point where the dam crossed the old French canal. In the face of a storm of criticism and alarm in the newspapers, Goethals insisted that the situation was not serious and as it turned out he was perfectly correct. The damage was repaired; the work went on.

The slides, however, were a wholly different matter. The first occurred early in the fall of 1907, or just as Goethals was beginning to feel he had things under control.

The Cucaracha slide, located on the east bank of the Cut just south of Gold Hill, was the slide that had given the French such grief. On the night of October 4, 1907, after days of unusually heavy rain, Cucaracha "started afresh." Without warning, an avalanche of mud and rock plunged into the bottom of the Cut, destroying two steam shovels, obliterating all track in its path. And for days afterward that same part of the slope, about fifty acres in area, kept moving down and down, slipping anywhere from ten to fifteen feet a day. "It was, in fact, a tropical glacier--of mud instead of ice," Major Gaillard noted in an article for Scientific American, "and stakes aligned on its moving surface and checked every 24 hours by triangulation, showed a movement in every respect similar to stakes on moving glaciers in Alaska upon which the writer has made observations in 1896." After ten days, when the slipping stopped, 500,000 cubic yards of mud had been dumped into the canal.

In 1910 Cucaracha let go twice again, burying shovels, track, locomotives, flatcars, and compressed-air lines. The entire south end of the Cut was bottled up for months. Within a year Gaillard reported that the worst of the slides were over, but in fact they were still to come. From 1911 on, as the Cut grew very much deeper, the slides occurred season after season and grew increasingly worse. "No one could say when the sun went down at night what the condition of the Cut would be when the sun arose the next morning," Bishop wrote. "The work of months and years might be blotted out by an avalanche of earth or the toppling over of a small mountain of rock." There were slides at Las Cascadas, La Pita, Empire, Lirio, East Culebra--twenty-two slides all together. Cucaracha was almost never still. It took three months to dig out the rock and mud dumped into the Cut by slides in 1911. In 1912 more than a third of the year, four and a half months, was spent removing slides. On one day more than a hundred trains would roll out of the Cut; the next day there would be none, because a monstrous slide had occurred.

Steam shovels were buried so deep in mud that only the tips of their cranes were left protruding. Hundreds of miles of track disappeared or were twisted into crazy roller-coaster patterns. In one bizarre instance a shovel and track were picked up by a landslide and were deposited unharmed halfway across the floor of the Cut.

On some of the terraced slopes the ground crept ever so slowly, barely inches a day, which was never enough to do any serious damage, but for two years gangs of men had to be kept constantly at hand, day after day, moving the track back to where it belonged.

At another place a slow but relentless slide kept perfect pace with the steam shovel working at its base. The shovel never had to move; as much as it dug, the slide replenished.

For the engineers the problem was not merely the size of the slides. They were also confronted with a type nobody had anticipated. Those slides that had beset the French, like the comparative few experienced by Wallace and Stevens, were normal, or gravity, slides--Cucaracha being the largest and most destructive example. As explained earlier, they nearly always occurred in the rainy season, when a top layer of soft, porous material slid from the sloping plain of underlying rock, "like snow off a roof," as one American said. But the new variety, and much the worst, were what geologists classified as structural break and deformation slides. They were due not to sliding mud, but to unstable rock formations, the height of the slopes, and, in part, to the effects of heavy blasting. As the Cut deepened, the underlying rock formations of the slopes lost their lateral support and were unable to withstand the enormous weight from above. It was as if the flying buttresses had been removed from the wall of a Gothic cathedral: the exposed wall of the Cut simply buckled outward under its own load and fell. Rains and saturation actually had little to do with such slides. In fact, some of the most horrendous happened during the dry season.

The first signs of trouble were huge cracks in the ground running along the rim of the Cut, anywhere from a few feet to a hundred yards back from the edge. The next stage might come weeks or months later, or it might take years. A settling or outward tilt of big blocks, whole sections of the slope, would commence. Then the whole slope would give way, sometimes in an hour or two, sometimes over several days.

The worst of such slides occurred in front of the town of Culebra, on the west bank of the Cut, where huge cracks in the ground began appearing in 1911. By the summer of 1912, "the large and annoying Cucaracha" had put an additional 3,000,000 cubic yards in the path of the canal, but the slide on the west bank at Culebra had deposited more than twice that amount. Thirty buildings in the town of Culebra had to be moved back from the brow of the Cut.

Before long some seventy-five acres of the town broke away and fully half of all the buildings had to be dismantled and removed to save them from being carried over the edge. Ultimately these breaks, all occurring in the dry season, dumped 10,000,000 cubic yards into the Cut, while on the opposite side another 7,000,000 cubic yards fell away, with the result that the top width of the Cut at that point was increased by a quarter of a mile.

The slides "seem to be maneuvered by the hand of some great marshal and sent forth to the fray in every way calculated to put the canal engineers to discomfiture," declared the <u>National Geographic Magazine.</u> "Now they are quiescent, attempting to lull the engineers into a false security . . . now they come in the dead of night, spreading chaos and disrupting everything in whatever direction they move . . ." To many of the workers it seemed the task would go on forever. "I personally would say to my fellow men," recalled one Barbadian, "that . . . my children would come and have children, and their children would come and do the same, before you would see water in the cut, and most all of us agree on the same."

Often wisps of smoke would trail from the moving embankments. Once cracks in the surface below Culebra issued boiling water. When Gaillard arrived to investigate the matter, he took a Manila envelope from his pocket and held it over one of the vents in the earth. In seconds the paper was reduced to ashes. The explanation, according to the geologist who was summoned, was "oxidation of pyrite," but the terrified workers were convinced that they were cutting into the side of a volcano.

The most uncanny of all effects, however, was the rising of the floor of the Cut. Not merely would the walls of the canal come crashing down, but the bottom would rise ten, fifteen, even thirty feet in the air, often quite dramatically. Gaillard on one occasion grew concerned as a steam shovel appeared to be sinking before his eyes, but looking again he realized it was not that the shovel was descending, but that the ground where he stood was steadily rising--about six feet in five minutes, "and so smoothly and with so little jar as to make the movement scarcely appreciable."

This phenomenon, diabolical as it seemed, had a simple explanation. It was caused by the weight of the slipping walls of the Cut acting upon the comparatively soft strata of the exposed canal floor. The effect was exactly that of a hand pressed into a pan of soft dough--the hand being the downward pressure of the slides, the rising dough at the side of the hand being the bottom of the canal.

The slides attracted worldwide attention and inspired all kinds of suggestions as to how the problem might be solved, very few of which were practical. The most popular remedy was to plaster the sides of the Cut with concrete, and this was actually tried in one particularly troublesome area, but without success. The concrete crumpled and fell along with everything else as soon as the slide resumed its downward progress.

To check the deformation slides considerable excavation was also done along the uppermost portions of the slopes in an effort to decrease the pressure on the underlying strata. But by and large there was still only one way to cope

The aftermath of one of the numerous slides in the Culebra Cut

with the problem and that was the same as it had been since the time of the French--to work for an angle of repose, to keep cutting back at the slopes, to keep removing whatever came down, until the slides stopped. And no one honestly knew how long that might take. By late 1912 at Cucaracha and at Culebra, the chief trouble spots, the angle of inclination was about one on five (one foot vertical to five horizontal). Still the ground kept moving.

Fifteen thousand tourists came to watch the show in 1911 and in 1912 there were nearly twenty thousand. "You are now overlooking the world-famous Culebra Cut," exclaimed the tour guides at the start of their standard spiel. There was more tonnage per mile moving on the tracks below, the visitors were informed, than on any railroad in the world. But meanwhile a big clubhouse at the town of Culebra was being dismantled and removed ("in order to lighten the weight upon the west bank of the canal at this point"), and on January 19 Cucaracha broke loose once again. It was one of the worst slides on record. It spilled the whole way across the Cut and up the other side. All traffic was blocked at that end; for the sixth or seventh time, the slide had wiped out months of work.

Gaillard was practically in shock, according to one account, and Goethals was hurriedly called to the scene. "What are we to do now?" Gaillard asked. Goethals lit a cigarette. "Hell," he said, "dig it out again."

336

"It Is a Work of Civilization"

Excerpts from Chapter 21, The Path Between the Seas
David McCullough

In the final chapter of his book, **Path Between the Seas**, *author David McCullough conveys the feeling of accomplishment civil engineers realize from their involvement in projects such as the Panama Canal. Although some might believe that such a feeling of accomplishment is limited to those who are involved in major projects such as the Canal, most civil engineers would agree that one of the rewarding aspects of the profession is the frequency with which one experiences the realization that his or her professional efforts are contributing to the welfare of society. The opportunity to observe one's fellow man use and benefit from the fruits of one's labors is an intangible reward that contributes materially to the positive feelings most civil engineers have about their work.*

"It is hard for me to transmit to you the feeling we all possessed toward the work," Robert Wood* said. "Rarely can man see his own work, but we saw it physically . . . year by year . . ."

They saw it in the deepening of Culebra Cut; in the rise of docks and warehouses; in the new railroad; in the fortifications being built at Toro Point and Margarita Island in Limon Bay and on the islands of Perico, Flamenco, and Naos in the Bay of Panama. (The giant 16-inch guns being installed were the largest, heaviest weapons in the possession of the United States and had a range of twenty miles.) They saw it in the hydroelectric plant built adjacent to the spillway of Gatun Dam, and in the dam itself, which in its final stage looked as if it had been there always, more like a huge glacial moraine than anything else. It was all measurable progress.

And there was the lake. It had begun its rise with the closing of the West Diversion channel in 1910, when the dam was still incomplete. In the time since, as the water inched steadily up the long, sloped inner face of the dam, as the Chagres [River] gathered and spread inland mile by mile, the realization of what a very different kind of canal it was to be, the conception of a long arm of fresh water suspended in the jungle, began to take hold and with good effect.

Popular interest at home mounted proportionately, the nearer the dream seemed to fulfillment. In the last years of construction, hundreds of articles appeared in magazines and Sunday supplements under such titles as "The Spirit of the Big Job" or "Realizing the Dream of Panama," "Great Work Nobly Done," "The Greatest Engineering Work of All Time," "Our Canal." In 1913, in anticipation of the projected grand opening, close to a dozen different books were published about Panama and the canal.

But it was also in the closing years of the task that the great locks took form for all to see and they were the most interesting and important construction

----

* Wood was one of the first American Army officers assigned to work on the Canal.

feats of the entire effort. They were the structural triumphs at Panama. In their overall dimensions, mass, weight, in the mechanisms and ingenious control apparatus incorporated in their design, they surpassed any similar structures in the world. They were, as was often said, the mighty portals of the Panama Gateway. Yet they were something much more than monumental; they did not, like a bridge or a cathedral, simply stand there; they worked. They were made of concrete and they were made of literally thousands of moving parts. Large essential elements were not built, but were manufactured, made in Pittsburgh, Wheeling, Schenectady, and other cities. In a very real sense they were colossal machines, the largest yet conceived, and in their final, finished form they would function quite as smoothly as a Swiss watch. They were truly one of the engineering triumphs of all time, but for reasons most people failed to comprehend.

To build all the locks took four years, from the time the first concrete was laid in the floor at Gatun, August 24, 1909. Most impressive of all was their size, and especially if seen during the last stages of construction before the water was turned in. Visitors who stood on the dry floor of a single lock chamber when all of it was still open to the light felt as though they had suddenly lost their sense of scale. Each individual chamber was a tremendous concrete basin closed at both ends with steel gates. The walls, one thousand feet long, rose to eighty-one feet, or higher than a six-story building. The impression was of looking down a broad, level street nearly five blocks long with a solid wall of six-story buildings on either side; only here there were no windows or doorways, nothing to give human scale. The gates at the ends, standing partly open to the sky, were like something in a dream.

The artist Joseph Pennell, having climbed down to the floor of an empty lock chamber at Pedro Miguel, found the shapes of gates and walls towering above him so "stupendous" that he was almost unable to draw. Walter Bernard, editor of Scientific American, returned from a visit to the Isthmus to write an article on "The Mammoth Locks" in which he conceded that it is impossible even to consider the subject "without drifting into the superlative mood." Another visitor would recall "the feeling that follows a service in a great cathedral."

To build the Great Pyramid or the Wall of China or the cathedrals of France, blocks of stone were set one on top of the other in the age-old fashion. But the walls of the Panama locks were poured from overhead, bucket by bucket, into gigantic forms. And within those forms there had to be still other forms to create the different culverts and tunnels, the special chambers and passageways, required inside the walls. Everything had to be created first in the negative, in order to achieve the positive structure wanted.

Moreover, the creation of the building material itself was a "science" requiring specific, controlled measurements and a streamlined system of delivery from mixing plant to construction site. Timing was vital.

Concrete--a combination of sand, gravel, and portland cement (itself a mixture of limestone and clay)--had been known since the time of the Romans, but was used very little as a building material until the late nineteenth century and then mainly for subbasements and floors. Dry docks and breakwaters were

338

built of reinforced (or ferro) concrete--concrete in which metal rods are added--
and in the early 1900s several major buildings were built of the same material
in Europe and the United States, as well as silos, some small bridges, and a
Montgomery Ward warehouse in Chicago. George Morison drew up plans for
his concrete bridge over Rock Creek Park in Washington, and by 1912 a
tremendous concrete railroad bridge, the Tunkhannock Viaduct, was under
way near Scranton, Pennsylvania. Nothing even approaching the size of the
Panama locks had yet been attempted, however, and not until the building of
Boulder Dam in the 1930s would any concrete structure equal their total
volume. The largest amount of concrete ever poured in a day anywhere else
was about 1,700 cubic yards. At Gatun alone the daily average was nearly
double that.

"No structure in the world contains as large an amount of material," William
Sibert wrote proudly of the great flight of locks at Gatun. With their approach
walls, they measured nearly a mile from end to end. The volume of concrete
poured was more than 2,000,000 cubic yards--enough, somebody figured, to
build a solid wall 8 feet thick, 12 feet high, and 133 miles long. Taken
together, the locks at the other end of the canal, at Pedro Miguel and
Miraflores, were larger still, with a volume of some 2,400,000 cubic yards.

The lock chambers all had the same dimensions (110 by 1,000 feet) and they
were built in pairs, two chambers running side by side in order to
accommodate two lanes of traffic. The single flight at Gatun consisted of
three such pairs. There was one pair at Pedro Miguel and two at Miraflores,
making six pairs (twelve chambers) in all.

The chambers in each pair shared a center wall that was sixty feet wide from
bottom to top. The width of the side walls was forty-five to fifty feet at the
floor level, but on the outside they were constructed as a series of steps, each
step six feet high, starting from a point twenty-four feet from the base level.
So at the top, the side walls were only eight feet wide.

The floors of the chambers were solid concrete, anywhere from thirteen to
twenty feet thick.

Once completed, the stepped backs of the side walls would be filled in, covered
entirely with dirt and rock. And the locks, once they were in use, would never
be less than half full of water. So their size would appear nowhere near so
overwhelming.

Seen during construction they were a fantasy of huge, raw-looking concrete
monoliths, of forms of sheet steel that looked like colossal, blank theatrical
flats, of monstrous cranes and soaring cableways--roads shunting here and
there. The swarms of workers at the lock sites appeared lost beside the rising
shapes and the incredible array of mechanical contrivances. The noise was
shattering.

At Gatun big square buckets of concrete, nearly six tons to a bucket, were
swung through the air high above the locks, dropped to position, and dumped,
all by means of a spectacular cableway. Eighty-five-foot steel towers stood on
either side of the locks (four on each side) and the cables stretched across a
span of some eight hundred feet. The towers were on tracks, so they could be
moved forward as the work progressed.

Sand and gravel were brought up the old French canal in barges and were stockpiled near a mixing plant. Then sand, gravel, and portland cement were fed into the plant (a battery of eight concrete mixers) by a little automatic railroad, the cars running in and out on a circular track. Another small railroad carried the buckets of wet concrete from plant to cableway, two buckets on two flatcars pushed by one of the French locomotives. At the cableway two empty buckets would descend from overhead, the two full buckets would be snatched up, delivered through the air at a speed of about twenty miles per hour, then returned to repeat the cycle.

The advantage of such an overhead delivery system was that the work area could be kept free of everything except the essential forms within which the concrete was poured. As fast as a bucketload was deposited, men knee-deep in wet concrete would spread it out.

All the locks were constructed in thirty-six foot sections, each a single monolith that took about a week to build to its full height. The big steel forms, also on tracks, would then be moved ahead to the next position.

At Pedro Miguel and Miraflores, where the terrain was not so open or spacious as at Gatun, division head Williamson* and his civilian engineers decided to use cantilever cranes rather than cableways, cranes so enormous in size that they could be seen rising above the jungle from miles distant. Some of these were in the shape of a gigantic T. Others looked like two gigantic T's joined together and were known as "chamber cranes," because they stood within the lock chambers, their long cantilever arms reaching out over the center and side walls. All the cranes moved on tracks and were self-propelling.

The T-shaped variety were the "mixing cranes." One arm of the T hoisted sand and gravel and cement from stockpiles to mixing plants located in the base of the T. The other arm transferred buckets of fresh concrete to the chamber cranes that in turn swung the buckets to the desired position. The complete operation was about as mechanized as it could possibly have been and to the average onlooker a very weird, unearthly sight to behold. The operator of a chamber crane, the man who guided the concrete to its destination, sat alone in a tiny box hanging from the delivery arm of the crane, nearly a hundred feet off the ground.

Five million sacks and barrels of cement were shipped to Panama to build the locks, dams, and spillways, all of it from New York on the *Ancon* and the *Cristobal*, and an idea of what such quantities amounted to is imparted by a single budgetary statistic: an estimated $50,000 was saved in recovered cement after Goethals issued a directive requiring the men to shake each sack after it was emptied.

By latter-day standards the engineers were novices in the use of concrete. Numerous discoveries had still to be made about the critical water-cement

----

* Sydney Williamson, the civilian engineer Colonel Goethals had selected to be supervisor of the Pacific Division of the Canal.

ratio in the "mix design" and the susceptibility of the material to environmental attack. To build anything so large as the concrete locks at Panama was an unprecedented challenge, but what was built had also to hold up in a climate wherein almost everything, concrete included, could go to pieces rapidly. Yet, however comparatively crude the level of theoretical technology may have been regarding the material, the results were extraordinary. After sixty years of service the concrete of the locks and spillways would be in near-perfect condition, which to present-day engineers is among the most exceptional aspects of the entire canal.

The design and engineering of the locks, the results of years of advance planning, can be attributed largely to three men: Lieutenant Colonel Hodges[*] and two exceptionally able civilians, Edward Schildhauer and Henry Goldmark. Schildhauer, slight of build, clean-shaven, very businesslike, was an electrical engineer and still in his thirties. Goldmark, who with his starched collars and thin, well-brushed hair looked like a corporation lawyer, had responsibility for designing the lock gates.

The fundamental element to be reckoned with and utilized in the locks--the vital factor in the whole plan and all its structural, mechanical, and electrical components--was water. Water would lift and lower the ships. The buoyancy of water would make the tremendous lock gates, gates two to three times heavier than any ever built before, virtually weightless. The power of falling water at the Gatun spillway would generate the electrical current to run all the motors to operate the system, as well as the towing locomotives or "electric mules." The canal, in other words, would supply its own energy needs.

No force would be required to raise or lower the level of the water in the locks (and thus to raise or lower the ship in transit) other than the force of gravity. The water would simply flow into the locks from above--from Gatun Lake or Miraflores Lake--or flow out into the sea-level channels. The water would be admitted or released through giant tunnels, or culverts, running lengthwise within the center and side walls of the locks, culverts eighteen feet in diameter, as large nearly as the Pennsylvania Railroad tubes under the Hudson River. At right angles to these main culverts, built into the floor of each lock chamber, were smaller cross culverts, fourteen to a chamber, these about large enough to admit a two-horse wagon. Every cross culvert had five well-like openings into the floor, which meant there were all together seventy such holes in each chamber, and it was from these that the water would surge or drain, depending on which valves were opened or shut.

The valves in the large culverts were immense sliding steel gates that moved on roller bearings up and down in frames in the manner of a window. There were two gates to each valve and they weighed ten tons apiece. To fill a lock, the valves at the lower end of the chamber would be closed, those at the upper end opened. The water would pour from the lake [Gatun Lake] through the large culverts into the cross culverts and up through the holes in the chamber

----

* Hodges was assigned the overall responsible for the design and construction of the lock gates.

floor. To release the water from the lock, the valves at the upper end would be shut, those at the lower end opened.

The reason for having as many as seventy wellholes in the chamber floor was to distribute the turbulence of the incoming water evenly over the full area and thereby subject chamber and ships to a minimum of disturbance. It was the engineers' intention to be able to raise or lower a ship in a chamber in about fifteen minutes.

Of all the moving parts in the system, the largest and most conspicuous were, of course, the lock gates, or "miter gates," as they were known, which swung open like double doors and closed in the form of a flattened V. The leaves of the gates weighed many hundreds of tons apiece and were the largest ever erected. Their construction was begun at Gatun in May 1911. As structures they were relatively simple and posed no special challenge, except, again, for their magnitude. A skin of plate steel was riveted to a grid of steel girders in exactly the manner of a steel ship's hull--or of a modern airplane wing, which they much resembled in vastly enlarged form. and being both hollow and watertight, they would actually float, once there was water in the locks, and thus the working load on their hinges would be comparatively little.

The leaves were all a standard sixty-five feet wide and seven feet thick. They varied in height, however, from forty-seven to eighty-two feet, depending on their position. The highest and heaviest (745 tons) were those of the lower locks at Miraflores, because of the extreme variation in the Pacific tides.

During construction, inspectors went down inside the gates through a system of manholes to check every rivet, an extremely uncomfortable task with the sun beating on the outer steel shell. All imperfect rivets were cut out and replaced and the watertightness of the shell was tested by filling the gates leaves with water.

As a safety precaution there were also to be duplicate gates throughout. One set of double doors was backed by another, in the event that the first set failed to function properly or was rammed by a ship. And since each lock chamber (except the lower locks at Miraflores) had its own set of intermediate gates, the complete system consisted of 46 gates (92 leaves), the total tonnage of which (sixty thousand tons) was almost half again greater than that of a ship such as the *Titanic*.

The purpose of the intermediate gates was to conserve water. While the locks were built to accommodate ships as large as the *Titanic* or the *Imperator*, or larger, each lock chamber could be reduced in size, by closing the intermediate gates, if the ship in transit was not one of the giants and could be accommodated by a chamber of six hundred feet or less. And of all the oceangoing ships in the world at that time, approximately 95 percent were less than six hundred feet long.

To lift a great merchant liner, or any ship of more than six hundred feet, to the level of Gatun Lake would require an expenditure from the lake of 26,000,000 gallons of water, the equivalent of a day's water supply for a major city. For a complete lockage through the canal, for one ocean-to-ocean transit, the expenditure would be double that amount, all of it fresh water and all washed out to sea.

342

The most obvious and frequently emphasized differences between the French and American efforts at Panama, between failure and success at Panama, were in the application of modern medical science, the methods of financing, and the size of the excavation equipment. But it should also be understood that the canal that was built was very different from what could have been built by anyone thirty years earlier. It was not only a much larger canal than it would have been (the locks were nearly twice as large as those designed by Eiffel, which measured 59 by 590 feet); it was constructed differently and of different materials. And its means of operation and control were altogether different. "Strongly as the Panama Canal appeals to the imagination as the carrying out of an ideal," wrote one astute editor, "it is above all things a practical, mechanical, and industrial achievement."

As a ship approached the entrance to the locks, its path would be blocked by a tremendous iron "fender" chain stretched between the walls. The chain would be lowered (into a special groove in the channel floor) only if all was proceeding properly--that is, if the ship was in proper position and in control of the towing locomotives. If the ship was out of control and struck the chain, then the chain would be payed out slowly by an automatic release until the ship was brought to a stop, short of the lock gates. (A 10,000-ton ship moving at five knots could be checked within seventy feet.) The length of the chain was more than four hundred feet and its ends were attached to big hydraulic pistons housed in the lock walls. There were pumps to supply water for the pistons and more electric motors to run the pumps.

If by some very remote chance a ship were to smash through the fender chain, the safety gates would still stand in the path, the apex of their leaves pointed toward the ship. To break through the safety gates would take a colossal force, and it was almost inconceivable that the forward motion of any ship could be that great, having just encountered the fender chain. But in the event that this too occurred, there was still one further safeguard.

The most serious threat to the locks would be from a ship out of control as it approached the upper gates, a ship, that is, about to go down through the locks and out of the canal. For if the upper gates were destroyed, then the lake would come plunging through the locks.

So on the side walls at the entrance of each upper lock, between the fender chain and the guard gates, stood a big steel apparatus that looked like a cantilever railroad bridge. This was the emergency dam. It was mounted on a pivot and in a crisis it could be swung--turned electrically--across the lock entrance in about two minutes' time. From its underside a series of wicket girders would descend, their ends dropping into iron pockets in the concrete channel floor. The girders would descend, their ends dropping into iron pockets in the concrete channel floor. The girders would form runways down which huge steel plates would be dropped, one after another, until the channel was sealed off. It was an ungainly contraption, but it worked most effectively.

343

Under normal procedure a ship would be controlled by the towing locomotives all the way through the locks, with four locomotives to the average-sized ship, two forward pulling, two aft holding the ship steady. At no time in the locks would a ship move under its own power.

Like nearly every detail of the locks, the towing locomotives were the first of a kind. Presently they would become one of the most familiar features of the canal. They were designed by Schildhauer to work back and forth on tracks built into the top of the lock walls and to move a ship from point to point at about two miles an hour or less. But they also had to negotiate the 45-degree incline between the locks.

For the still young, still comparatively small General Electric Company the successful performance of all such apparatus, indeed the perfect efficiency of the entire electrical system, was of the utmost importance. This was not merely a very large government contract, the company's first large government contract, but one that would attract worldwide attention. It was a chance like none other to display the virtues of electric power, to bring to bear the creative resources of the electrical engineer. The canal, declared one technical journal, would be a "monument to the electrical art." It had been less than a year since the first factory in the United States had been electrified.

. . . the chief virtue of electricity was in the degree of control it afforded. Things could be made to happen--stop, start, open, close--with the mere press of a button or the turning of a few simple switches on a central control board. And so it was to be at Panama, and with one other extremely important feature. In this operation, things could be made to happen only as they were supposed to, in exactly the prescribed sequence.

Though the fundamental principles were much like those developed for railroad switchboards, no comparable control system had been produced heretofore. Again credit for the basic conception belongs to Edward Schildhauer, but otherwise it was a wholly joint effort. "No specifications could have been more exacting or explicit as to the results to be accomplished," wrote one of the engineers at Schenectady, "or have given a wider range as to the method of their accomplishment . . . . It was the single aim of all concerned to produce something better, safer and more reliable than anything before undertaken." A special department was set up at the General Electric works, wherein picked employees concentrated solely on the Panama project. Company engineers were sent to the Isthmus to become thoroughly familiar with all aspects of the problem; Schildhauer and members of his staff came to Schenectady. The result was an unqualified success.

The operation of each flight of locks was to be run from the second floor of a large control house built on the center wall of the uppermost lock. From there, with an unobstructed view of the entire flight, one man at one control board could run every operation in the passage of a ship except the movement of the towing locomotives.

344

Each control board was a long, flat, waist-high bench, or a counter, upon which the locks were represented in miniature--a complete working reproduction. The board at Gatun was sixty-four feet in length and about five feet wide. There were little aluminum fender chains that would actually rise in place or sink back out of sight on the board as a switch was turned. The lock chambers were represented by slabs of blue marble. There were aluminum pointers placed in the same relative positions as the lock gates and these opened or closed as the actual lock gates opened and closed. There were upright indicators showing the positions of the rising stem valves, and there were still taller upright indexes showing the level of the water in the chambers to within half an inch.

Everything that happened in the locks--the rise and fall of the fender chains, the opening and closing of the gates--happened on the board in the appropriate place and at precisely the same time. So the situation in the locks could be read in an instant on the board at any stage of the lockage.

More than half a century later the same control panels would still be in use, functioning exactly as intended, everything as the engineers originally devised. "They were very smart people," a latter-day engineer at Miraflores would remark. "After twenty-one years here I am still amazed at what they did."

Once, just before the canal was completed, the Commission of Fine Arts sent the sculptor Daniel Chester French and the landscape architect Frederick Olmsted, Jr., son of the famous creator of New York's Central Park, to suggest ways in which the appearance of the locks and other components might be dressed up or improved upon. The two men reported:

> The canal itself and all the structures connected with it impress one with a sense of their having been built with a view strictly to their utility. There is an entire absence of ornament and no evidence that the aesthetic has been considered except in a few instances. . . . Because of this very fact there is little to find fault with from the artist's point of view. The canal, like the Pyramids or some imposing object in natural scenery, is impressive from its scale and simplicity and directness. One feels that anything done merely for the purpose of beautifying it would not only fail to accomplish that purpose, but would be an impertinence.

Consequently nothing was changed or added. The canal would look as its builders intended, nothing less or more.

For all practical purposes the canal was finished when the locks were. And so efficiently had construction of the locks been organized that they were finished nearly a year earlier than anticipated. Had it not been for the slides in the Cut, adding more than 25,000,000 cubic yards to the total amount of excavation, the canal might have opened in 1913.

The locks on the Pacific side were finished first, the single flight at Pedro Miguel in 1911, Miraflores in May 1913. Morale was at an all-time high.

Asked by a journalist what the secret of success had been, Goethals answered, "The pride everyone feels in the work."

"Men reported to work early and stayed late, without overtime," Robert Wood remembered. ". . . I really believe that every American employed would have worked that year without pay, if only to see the first ship pass through the completed Canal. That spirit went down to all the laborers."

The last concrete was laid at Gatun on May 31, 1913, eleven days after two steam shovels had met "on the bottom of the canal" in Culebra Cut. Shovel No. 222, driven by Joseph S. Kirk, and shovel No. 230, driven by D.J. MacDonald, had been slowly narrowing the gap all day when they at last stood nose to nose. The Cut was as deep as it would go, forty feet above sea level.

In the second week in June, it would be reported that the newly installed upper guard gates at Gatun had been "swung to a position halfway open; then shut, opened wide, closed and . . . noiselessly, without any jar or vibration, and at all times under perfect control."

On June 27 the last of the spillway gates was closed at Gatun Dam. The lake at Gatun had reached a depth of forty-eight feet; now it would rise to its full height.

Three months later all dry excavation ended. The Cucaracha slide still blocked the path, but Goethals had decided to clear it out with dredges once the Cut was flooded. So on the morning of September 10, photographers carried their gear into the Cut to record the last large rock being lifted by the last steam shovel. Locomotive No. 260 hauled out the last dirt train and the work crews moved in to tear up the last of the track. . .

Then on September 26 at Gatun the first trial lockage was made.

A seagoing tug, *Gatun,* used until now for hauling mud barges in the Atlantic entrance, was cleaned up, "decorated with all the flags it owned," and came plowing up from Colon in the early-morning sunshine. By ten o'clock several thousand people were clustered along the rims of the lock walls to witness the historic ascent. There were men on the tops of the closed lock gates, leaning on the handrails. The sky was cloudless, and in midair above the lower gates, a photographer hung suspended from the cableway. He was standing in a cement bucket, his camera on a tripod, waiting for things to begin.

But it was to be a long, hot day. The water was let into the upper chamber shortly after eleven, but because the lake had still to reach its full height, there was a head of only about eight feet and so no thunderous rush ensued when the valves were opened. Indeed, the most fascinating aspect of this phase of the operation, so far as the spectators were concerned, was the quantity of frogs that came swirling in with the muddy water.

With the upper lock filled, however, the head between it and the middle lock was fifty-six feet, and so when the next set of culverts was opened, the water came boiling up from the bottom of the empty chamber in spectacular fashion.

The central control board was still not ready. All valves were being worked by local control and with extreme caution to be sure everything was just so. Nor were any of the towering locomotives in service as yet. Just filling the locks took the whole afternoon. It was nearly five by the time the water in the lowest chamber was even with the surface of the sea-level approach outside and the huge gates split apart and wheeled slowly back into their niches in the walls.

The tug steamed into the lower lock, looking, as one man recalled, "like a chip on a pond." Sibert, Schildhauer, young George Goethals, and their wives were standing on the prow. "The Colonel" and Hodges were on top of the lock wall, walking from point to point, both men in their shirt sleeves, Goethals carrying a furled umbrella, Hodges wearing glossy puttees and an enormous white hat. The gates had opened in one minute forty-eight seconds, as expected.

The tug proceeded on up through the locks, step by step. The gates to the rear of the first chamber were closed; the water in the chamber was raised until it reached the same height as the water on the other side of the gates ahead. The entire tremendous basin swirled and churned as if being stirred by some powerful, unseen hand and the rise of the water--and of the little boat--was very apparent. Those on board could feel themselves being lifted, as if in a very slow elevator. With the water in the lower chamber equal to that in the middle chamber, the intervening gates were opened and the tug went forward. Again the gates to the stern swung shut; again, with the opening of the huge subterranean culverts, the caramel-colored water came suddenly to life and began its rise to the next level.

It was 6:45 when the last gates were opened in the third and last lock and the tug steamed out onto the surface of Gatun Lake. The day had come and gone, it was very nearly dark, and as the boat turned and pointed to shore, her whistle blowing, the crowd burst into a long cheer. The official time given for this first lockage was one hour fifty-one minutes, or not quite twice as long as would be required once everything was in working order.

That an earthquake should strike just four days later seemed somehow a fitting additional touch, as if that too were essential in any thorough testing-and-providing drill. It lasted more than an hour, one violent shudder following another, and the level of magnitude appears to have been greater than that of the San Francisco quake of 1906. The needles of a seismograph at Ancon were jolted off the scale paper. Walls cracked in buildings in Panama City; there were landslides in the interior; a church fell. But the locks and Gatun Dam were untouched. "There has been no damage whatever to any part of the canal," Goethals notified Washington.

Water was let into Culebra Cut that same week, through six big drain pipes in the earth dike at Gamboa. Then on the afternoon of October 10, President Wilson pressed a button in Washington and the center of the dike was blown sky-high. The idea had been dreamed up by a newspaperman. The signal, relayed by telegraph wire from Washington to New York to Galveston to Panama, was almost instantaneous. Wilson walked from the White House to an office in the Executive Building (as the State, War, and Navy Building had been renamed) and pressed the button at one minute past two. At two minutes past two several hundred charges of dynamite opened a hole more than a hundred feet wide and the Cut, already close to full, at once became an extension of Gatun Lake.

In all the years that the work had been moving ahead in the Cut and on the locks, some twenty dredges of different kinds, assisted by numbers of tugs, barges, and crane boats, had been laboring in the sea-level approaches of the canal and in the two terminal bays, where forty-foot channels had to be dug several miles out to deep water. Much of this was equipment left behind by the French; six dredges in the Atlantic fleet, four in the Pacific fleet, a dozen self-propelled dump barges, two tugs, one drill boat, one crane boat, were all holdovers from that earlier era. Now, to clear the Cut of slides, about half this equipment was brought up through the locks, the first procession from the Pacific side passing through Miraflores and Pedro Miguel on October 25.

The great, awkward dredges took their positions in the Cut; barges shunted in and out, dumping their spoil in designated out-of-the-way corners of Gatun Lake, all in the very fashion that Philippe Bunau-Varilla had for so long championed as the only way to do the job. Floodlights were installed in the Cut and the work went on day and night. On December 10, 1913, an old French ladder dredge, the *Marmot*, made the "pioneer cut" through the Cucaracha slide, thus opening the channel for free passage.

The first complete passage of the canal took place almost incidentally, as part of the new workaday routine, on January 7, when an old crane boat, the *Alexandre La Valley*, which had been brought up from the Atlantic side sometime previously, came down through the Pacific locks without ceremony, without much attention of any kind. That the first boat through the canal was French seemed to everyone altogether appropriate.

The end was approaching faster than anyone had quite anticipated. Thousands of men were being let go; hundreds of buildings were being disassembled or demolished. Job applications were being written to engineering offices in New York and to factories in Detroit, where, according to the latest reports, there was great opportunity in the automobile industry. Families were packing for home. There were farewell parties somewhere along the line almost every night of the week.

The first oceangoing ship to go through the canal was a lowly cement boat, the *Cristobal*, and on August 15 the "grand opening" was performed almost perfunctorily by the *Ancon*. There were no world luminaries on her prow. Goethals again watched from shore, traveling from point to point on the railroad. The only impressive aspect of the event was "the ease and system with which everything worked," as wrote one man on board. "So quietly did she pursue her way that . . . a strange observer coming suddenly upon the scene would have thought that the canal had always been in operation, and that the *Ancon* was only doing what thousands of other vessels must have done before her."

There were editorials hailing the victory of the canal builders, but the great crescendo of popular interest had passed; a new heroic effort commanded world attention. The triumph at Panama suddenly belonged to another and very different era.

Of the American employees in Panama at the time the canal was opened only about sixty had been there since the beginning in 1904. How many black workers remained from the start of the American effort, or from an earlier time, is not recorded. But one engineer on the staff, a Frenchman named Arthur Raggi, had been first hired by the Compagnie Nouvelle in 1894.

Goethals, Sibert, Hodges, Schildhauer, Goldmark, and the others had been on the job for seven years and the work they performed was of a quality seldom ever known.

Its cost had been enormous. No single construction effort in American history had exacted such a price in dollars or in human life. Dollar expenditures since 1904 totaled $352,000,000 (including the $10,000,000 paid to Panama and the $40,000,000 paid to the French company). By present standards this does not seem a great deal, but it was more than four times what Suez had cost, without even considering the sums spent by the two preceding French companies, and so much more than the cost of anything ever before built by the United States government as to be beyond compare. Taken together, the French and American expenditures came to about $639,000,000.

The other cost since 1904, according to the hospital records, was 5,609 lives from disease and accidents. No fewer than 4,500 of these had been black employees. The number of white Americans who died was about 350.

If the deaths incurred during the French era are included, the total price in human life may have been as high as twenty-five thousand, or five hundred lives for every mile of the canal.

Yet amazingly, unlike any such project on record, unlike almost any major construction of any kind, the canal designed and built by the American engineers had cost less in dollars than it was supposed to. The final price was actually $23,000,000 below what had been estimated in 1907, and this despite the slides, the change in the width of the canal, and an additional $11,000,000 for fortifications, all factors not reckoned in the earlier estimate. The volume of additional excavation resulting from slides (something over 25,000,000 cubic yards) was almost equal to all the useful excavation accomplished by the French. The digging of Culebra Cut ultimately cost $90,000,000 (or $10,000,000 a mile). Had such a figure been anticipated at the start, it is questionable whether Congress would have ever approved the plan.

The total volume of excavation accomplished since 1904 was 232,440,945 cubic yards and this added to the approximately 30,000,000 cubic yards of useful excavation by the French gave a grand total, in round numbers of 262,000,000 cubic yards, or more than four times the volume originally estimated by Ferdinand de Lesseps for a canal at sea level and nearly three times the excavation at Suez.

The canal had also been opened six months ahead of schedule, and this too in the face of all those difficulties and changes unforeseen seven years before.

Without question, the credit for such a record belongs chiefly to George Goethals, whose ability, whose courage and tenacity, were of the highest order.

That so vast and costly an undertaking could also be done without graft, kickbacks, payroll padding, any of the hundred and one forms of corruption endemic to such works, seemed almost inconceivable at the start, nor does it seem any less remarkable in retrospect. Yet the canal was, among so many other things, a clean project. No excessive profits were made by any of the several thousand different firms dealt with by the I.C.C. [Isthmian Canal Commission]. There had not been the least hint of scandal from the time Goethals was given command, nor has evidence of corruption of any kind come to light in all the years since.

Technically the canal itself was a masterpiece in design and construction. From the time they were first put in use the locks performed perfectly.

Ten years after it opened, the canal was handling more than five thousand ships a year; traffic was approximately equal to that of Suez. . . By 1939 annual traffic exceeded 7,000 ships.

But in the decades following the Second World War, that figure more than doubled. Channel lighting was installed in 1966 and nighttime transits were inaugurated. Ships were going through the canal at a rate of more than one an hour, twenty-four hours a day, every day of the year. Many of them moreover--giant container ships, bulk carriers--were of a size never dreamed of when the canal was built: the 845-foot *Melodic*, the 848-foot *Arctic*, the 950-foot *Tokyo Bay*, the largest container ship in the world at the time she made her first transit in 1972. Traffic in the canal by the 1970's was beyond fifteen thousand ships a year, annual tonnage was well beyond the 100,000,000 mark. Tonnage in 1915 had been 5,000,000.

But fundamentally, and for all general appearances, the canal remains the same as the day it opened and its basic plan has been challenged in only one respect. It has been argued that the separation of the two sets of locks at the Pacific end was a blunder, that it would have been a more efficient canal had the Pacific locks been built as a unit at Miraflores, just as at Gatun. But those who have had the most experience with running the canal in recent years do not regard the Pacific arrangement as a limiting factor, and indeed various tests run by the Panama Canal Company through the years indicate that Gatun is actually more of a bottleneck. With certain improvements, the engineers believe the capacity of the present canal could be increased to about twenty-seven thousand ships a year.

The creation of a water passage across Panama was one of the supreme human achievements of all time, the culmination of a heroic dream of four hundred years and of more than twenty years of phenomenal effort and sacrifice. The fifty miles between the oceans were among the hardest ever won by human effort and ingenuity, and no statistics on tonnage or tolls can begin to convey the grandeur of what was accomplished. Primarily the canal is an expression of that old and noble desire to bridge the divide, to bring people together. It is a work of civilization.

350

The Red Queen

Chapter 2, **Cadillac Desert**
Marc Reisner

The story of how water was brought from California's Owens Valley to Los Angeles, as told by Marc Reisner, is full of the material that makes good stories--heroes and villains, tragedies and triumphs, insidious plots and righteous causes, and genius and stupidity. Some say that Reisner unjustly impugns the reputations of well-intentioned men concerned with finding and developing a water supply to satisfy the needs of a growing metropolis, while others insist that he has only documented the nefarious deeds of a group of selfish and unscrupulous scoundrels masquerading as civic leaders. Regardless of which view is correct, the story he tells is fascinating, and engineers play pivotal roles.

This selection, longer than most in this collection, is included because it illustrates the range of economic, social, cultural, political, institutional and environmental considerations that attend major civil engineering projects. It is not realistic to think that civil engineers can foresee and understand all such problems, or, even if that were possible, that civil engineers could magically produce solutions that would satisfy all of the interested parties. It is, however, important for civil engineers to understand that such concerns exist, even if they cannot be fully identified during the planning and development of a project, and that consideration of these concerns as they are identified is part of the job of the civil engineer. In fact, it is the understanding of a highway as more than just a slab of concrete; the water supply project as more than just a system of pipes and reservoirs; the airport as more than just runways, a terminal and some associated parking; the skyscraper as more than just a system of columns, beams and interior and exterior walls; the wastewater treatment system as more than a network of sewer pipes connected to a chemical and biological processing plant; in general, the understanding of the role things play in society rather than a mere understanding of the things themselves that distinguish the truly professional civil engineer from the person who is merely earning a living by working within the profession.

While Los Angeles moldered, San Francisco grew and grew. The city owned a superb natural harbor--the best on the Pacific Coast, one of the best in the world. When gold was struck in the Sierra Nevada foothills, 150 miles across the Central Valley, San Francisco became the principal destination of the fortune seekers of the world. The names of the camps suggested the potency of the lure: New York-of-the-Pacific, Bunker Hill, Chinese Camp, German Bar, Georgia Slide, Nigger Hill, Dutch Corral, Irish Creek, Malay Camp, French Bar, Italian Bar. Those who found their fortunes were inclined to part with them in the nearest haven of pleasure, which was San Francisco. Those who did not discovered that they could do just as well providing the opportunities. With oranges going for $2 apiece at the mines, and a plate of fresh oysters for $20 or more, it was a bonanza for all concerned.

In 1848, the population of San Francisco was eight hundred; three years later, thirty-five thousand people lived there. In 1853 the population went past fifty thousand and San Francisco became one of the twenty largest cities in the United States. By 1869, San Francisco possessed one of the busiest ports in the world, a huge fishing fleet, and the western terminus of the transcontinental railroad. It teemed with mansions, restaurants, hotels, theaters, and whorehouses. In finance it was the rival of New York, in culture the rival of Boston; in spirit it had no competitor.

Los Angeles, meanwhile, remained a torpid, suppurating, stunted little slum. It was too far from the gold fields to receive many fortune seekers on their way in or to detach them from their fortunes on the way out. It sat forlornly in the middle of an arid coastal basin, lacking both a port and a railroad. During most of the year, its water source, the Los Angeles River, was a smallish creek in a large bed; during the few winter weeks when it was not-- when supersaturated tropical weather fronts crashed into the mountains ringing the basin--the bed could not begin to contain it, and the river floated neighborhoods out to sea. (For many years, Santa Anita Canyon, near Pasadena, held the United States record for the greatest rainfall in a twenty-four-hour period, but it may be more significant to state that the twenty-six inches that fell in a day were nearly twice the amount of precipitation that Los Angeles normally receives in a year.) Had humans never settled in Los Angeles, evolution, left to its own devices, might have created in a million more years the ideal creature for the habitat: a camel with gills.

The Spanish had actually settled Los Angeles long before they ever saw the Golden Gate. It was more convenient to Mexico and, from an irrigation farmer's point of view, it was a more promising place to live. By 1848, the town had a population of sixteen hundred, half Spanish and half Indian, with a small sprinkling of Yankees, and was twice the size of San Francisco. A decade later, however, San Francisco had grown ten times as large as Los Angeles. By the end of the Civil War, when San Francisco was the Babylon of the American frontier, Los Angeles was a filthy pueblo of thirteen thousand, a beach for human flotsam washed across the continent on the blood tide of the war. . . .

*   *   *

They came by ship, they came by wagon, they came by horse. They came on foot, dragging everything they could in a handcart, but the real hordes came by train. In 1885, the Atchison, Topeka, and Santa Fe Railroad linked Los Angeles directly with Kansas City, precipitating a fare war with the Southern Pacific. Within a year, the cost of passage from Chicago had dropped from $100 to $25. During brief periods of mad competition, you could cross two-thirds of the continent for a dollar. If you were asthmatic, tubercular, arthritic, restless, ambitious, or lazy--categories that pretty well accounted for Los Angeles' first flood of arrivals--the fares were too cheap to pass up. Out came Dakota farmers with hopes and wills blasted by blizzards and droughts. Out came farmers who despaired at the meager profits they made growing wheat. *You could grow oranges.* Out came Civil War veterans looking for an easy life, failures looking for another chance, and the usual boom-town complement of the slick, the sharp, and the ruthless.

The first boom began in the early 1880s and culminated in 1889, when the town transacted $100 million worth of real estate--in today's economy, a $2 billion year in Idaho Falls. Fraud was epic. Hundreds of unseen, paid-for-lots were situated in the bed of the Los Angeles River, or up the nine-thousand-foot summits of the San Gabriel Range. The boom was, predictably, short-lived. In 1889, a bank president, a newspaper publisher, and the town's most popular

352

minister all fled to Mexico to spare themselves jail terms, and a dozen or more victims took their own lives. By 1892, the population had dropped by almost one-half, but the bust was followed quickly by an oil boom, and enough fortunes were being made (the original Beverly Hillbillies were *from* Beverly Hills, then a patch of jackrabbit scrub overlying an oil basin) to pack the arriving trains again. Los Angeles soon drew close to San Francisco in population and was crowing with glee. "The 'busting of the boom' became but a little eddy in the great stream," enthused the Los Angeles *Times*, "the intermission of one heartbeat in the life of...the most charming land on the footstool of the Most High...the most beautiful city inhabited by the human family." Only one thing stood in the way of what looked as if it might become the most startling rise to prominence of any city in history--the scarcity of water.

The motives that brought Harrison Gray Otis, Harry Chandler, and William Mulholland to Los Angeles were the same that would eventually bring millions there. Otis came because he had been an incontrovertible, if not quite an ignominious, failure. He was born in Marietta, Ohio, and as a young man held a series of unspectacular jobs--a clerk for the Ohio legislature, a foreman at a printing plant, an editor of a veterans' magazine. His one early taste of glory came during the Civil War, in which he fought on the Union side, acquired several wounds and decorations, and ultimately rose to the rank of captain. *Captain* Harrison Gray Otis. He liked the title well enough to think himself deserving of a sinecure, and after the war he drifted out to California in search of one.

*   *   *

The city was still small when Otis arrived, but it was already served by several newspapers, one of which, the *Times and Mirror*, was owned by a small-time eastern financier named H.H. Boyce. Boyce was looking for a new editor, and, though the pay was a miserable $15 a week, Otis took the job. Perhaps because he was fuming about the pay, or perhaps because he knew that time was running out, Captain Otis then made one of the bolder decisions of his life. He took all of his savings and, to help offset the low pay, convinced Boyce to let him purchase a share in the newspaper. Privately he was thinking that someday, perhaps, he could force H.H. Boyce out.

Harry Chandler came to Los Angeles for his health. He grew up in New Hampshire, a cherubic child with cheeks like Freestone peaches. His falsely benign appearance, which stayed with him all his life, made him a popular boy model among advertisers and photographers. But cherubic Harry was a rugged individualist and a ferocious competitor, and if there was money involved he would rarely pass up an opportunity or a dare. While at Dartmouth College, he accepted someone's challenge and dove into a vat of starch--a display that nearly ruined his lungs. Advised by doctors to recuperate in a warm and dry climate, he bought a ticket to Los Angeles. Arriving there, he moved from flophouse to flophouse because none of his fellow tenants could endure his hacking cough. When he was thoroughly friendless and nearly destitute, Harry met a sympathetic doctor who suffered from tuberculosis and owned an irrigated orchard near Cahuenga Pass, at the head of the San Fernando Valley. Would Harry like a job picking fruit?

The work was hard but invigorating. Before long, Harry felt almost cured. The work was also surprisingly lucrative. The doctor was as uninterested in money as Harry was interested, and let him sell a large share of what he picked. In his first year, Harry made $3,000. It was a small fortune, and inspired in Harry an awed faith in the potential of irrigated agriculture and, most particularly, agriculture in the San Fernando Valley. With the proceeds, Harry began to acquire newspaper circulation routes, which, at the time, were owned independently of the newspapers and bought and sold like chattel. Before long, he was a child monopolist, owning virtually all the routes in the city.

By 1886, Harrison Gray Otis had finally managed to hound H.H. Boyce out of the Los Angeles *Times and Mirror*. . . . [and] acquired a new circulation manager and guiding light, whose name was Harry Chandler, and in 1884 Harry Chandler acquired a new father-in-law, whose name was Harrison Gray Otis.

William Mulholland came to Los Angeles more or less for the hell of it. He was born in 1855 in Dublin, Ireland, where his father was a postal clerk. At fifteen, he signed on as an apprentice seaman aboard a merchant ship that carried him back and forth along the Atlantic trade routes. By 1874 he had had enough, and spent a couple of years hacking about the lumber camps in Michigan and the dry-goods business in Pittsburgh, where his uncle owned a store. It was in Pittsburgh that Mulholland first heard about California. He had just enough money to get to Panama by ship, and after landing in Colon, he traversed the isthmus on foot and worked his way north aboard another ship, arriving in San Francisco in the summer of 1877. Being back on a ship had renewed Mulholland's taste for the sea, and, after a brief failure at prospecting in Arizona--where he also fought Apaches for pay--he decided to ship out at San Pedro, the port nearest Los Angeles. He had ten dollars to his name. Anxious to make a little extra money, he joined a well-drilling crew. "We were down about six hundred feet when we struck a tree. A little further we got fossil remains. These things fired my curiosity. I wanted to know how they got there, so I got hold of Joseph Le Conte's book on the geology of the country. Right there I decided to become an engineer."

In his official photograph for the Los Angeles Department of Water and Power, which was taken when he was nearly fifty, Mulholland still looks young. He is wearing a short-brimmed dark fedora and a dark pinstripe suit; a luxuriant silk cravat circumnavigates a shirt collar that appears to be made of titanium; from a thick, bushy mustache sprouts a lit cigar. The face is supremely Irish: belligerence in repose, a seductive churlish charm. Once, in court, Mulholland was asked what his qualifications were to run the most far-flung urban water system in the world, and he replied, "Well, I went to school in Ireland when I was a boy, learned the Three R's and the Ten Commandments--most of them--made a pilgrimage to the Blarney Stone, received my father's blessing, and here I am." He began his engineering career in 1878 as a ditch-tender for the city's private water company, clearing weeds, stones, and brush out of a canal that ran by his house. One day Mulholland was approached by a man in a carriage who demanded to know his name and what he was doing. Mulholland stepped out of his ditch and told the man that he was doing his goddamned job and that his name was immaterial to the quality of his goddamned work. The man, it turned out, was the president of the water company. Learning this, Mulholland went to the company office to collect his pay before being fired. Instead, he was promoted.

The Sierra Nevada blocks most of the weather fronts moving across California from the Pacific, so that a place on the western slope of the range may receive eighty inches of precipitation in a year, while a place on the east slope, fifty miles away, may receive ten inches or less. The rivers draining into the Pacific from the West Slope are many and substantial, while those emptying into the Great Basin from the East Slope are few and generally small. The Owens River is an exception. It rises southeast of Yosemite, near a gunsight pass that allows some of the weather to come barreling through, heads westward for a while, then turns abruptly south and flows through a long valley, ten to twenty miles wide, flanked on either side by the Sierra Nevada and the White Mountains, which rise ten thousand feet from the valley floor. The valley is called the Owens Valley, and the lake into which the river empties--used to empty--was called Owens Lake. Huge, turquoise, and improbable in a desert landscape, it was the shrunken remnant of a much larger lake that formed during the Ice Ages. Due to a high evaporation rate and, for its size, a modest rate of inflow, the lake was more saline than the sea, but it supported two species of life in the quadrillions: a salt-loving fly and a tiny brine shrimp. The soup of shrimp and the smog of flies attracted millions of migratory waterfowl, a food source whose startling numbers were partially responsible for inducing some of the valley's first visitors to remain. "The lake was alive with wild fowl," wrote Beveridge R. Spear, an Owens Valley pioneer. "Ducks were by the square mile, millions of them. When they rose in flight, the roar of their wings...could be heard...ten miles away....Occasionally, when shot down, a duck would burst open from fatness which was butter yellow."

The greater attraction, however, was the river. When whites arrived in the 1860s, Paiute Indians who had learned irrigation from the Spanish were already diverting some of the water to raise crops. In traditional pioneer fashion, the whites trumped up some cattlerustling charges against the Indians, which appear to have led to the murder of a white woman and a child. The pious Owens Valley citizens then murdered at least 150 Paiutes in retaliation, driving the last hundred into Owens Lake to drown. They then took over the Indians' land, borrowed their irrigation methods, and began raising alfalfa and pasture and fruit. By 1899, they had established several ditch companies and had put some forty thousand acres under cultivation.

The huge new silver camp at Tonopah, Nevada, consumed most of what the valley grew. With prosperity, several thriving towns sprang up: Bishop, Big Pine, Lone Pine, Independence. The irrigated valley was postcard-pretty, a narrow swath of green in the middle of the high desert, with 14,495-foot Mount Whitney, the highest peak between Canada and Mexico, looming over Lone Pine and the river running through. Mark Twain came to visit, and Mary Austin, who was to become a well-known writer, came to live. But the entrance that most excited the valley people was that of the United States Reclamation Service (later renamed the Bureau of Reclamation). The Service was an unparalleled experiment in federal intervention in the nation's economy, and was being watched so closely by skeptics in Congress that it could not afford to have any of its first projects fail. To Frederick Newell, the first Reclamation Commissioner, the Owens Valley looked like a place where he could almost be guaranteed success. The people were proven irrigation farmers--a rarity in the non-Mormon West; the soil could grow anything the climate would permit; the river was underused; and there was a good site for a reservoir. Sixty thousand additional acres were irrigable, and all of them could be gravity-fed. In early 1903, just a few months after the Service was created, a team of Reclamation engineers was already trooping around the valley, gauging streamflows and making soil surveys. Sixty thousand new acres would even make it worthwhile to run a railroad spur to Los Angeles. Los Angeles, everyone thought, was going to make the Owens Valley rich.

Fred Eaton thought differently. Eaton had been born in Los Angeles in 1856; his family had founded Pasadena. Most of the Eaton men were engineers, and when they looked around them it seemed that half of what they saw they had built themselves; it gave them an overpowering sense of pride-in-place. Fred had gone into hydrologic engineering, which is to say that he pretty much taught it to himself, and by the time he was twenty-seven, he was superintendent of the Los Angeles City Water Company. As San Francisco had bloomed into pseudo-Parisian splendor, Fred Eaton had chafed. When Los Angeles finally began to take on the appearance of a place with a future, he had been intensely proud. But he was one of the few people who understood that this whole promising future was an illusion. With artesian pressure still lifting fountainheads of water eight feet into the air, no one believed that someday the basin would run out of water. Few understood that the occasional big floods in the Los Angeles River were testimony to the *absence* of rain: that the basin was normally so dry there wasn't enough ground cover to hold the rain when it fell. The annual flow of the Los Angeles River (that which ran aboveground) represented only a fifth of 1 percent of the runoff of the state, and because of the pumping the flow was dropping fast, from a hundred cubic feet per second (cfs) in the 1880s to forty-five cfs in 1902. If growth continued, the population and the water would fall hopelessly out of balance. Everyone was living off tens of thousands of years of accumulated groundwater, like a spendthrift heir squandering his wealth. No one knew how much groundwater lay beneath the basin or how long it could be expected to last, but it would be insane to build the region's future on it.

There was no other source of water nearby. Deserts lay on three sides of the basin, an ocean on the fourth. The nearest large rivers were the Colorado and the Kern, but to divert them out of their canyons to Los Angeles would require pumping lifts of thousands of feet--an impossibility at the time. It would also require a Herculean amount of energy.

But there was, 250 miles away, the Owens River. It might not be quite sufficient for the huge metropolis forming in Eaton's imagination, but it was large enough; there was water for at least a million people. Indeed, Eaton was one of the few Los Angeleans who knew the river even existed. Its distance from Los Angeles was staggering, but its remoteness was overshadowed by one majestically significant fact: Owens Lake, the terminus of the river, sat at an elevation of about four thousand feet. Los Angeles was a few feet above sea level. The water, carried in pressure aqueducts and siphons, could arrive under its own power. Not one watt of pumping energy would be required. The only drawback was that the city might have to take the water by theft.

During their years together at the Los Angeles City Water Company, Fred Eaton and Bill Mulholland became good friends, thriving on each other's differences. Eaton was a western patrician, smooth and diffident; Mulholland an Irish immigrant with a musician's repertoire of ribald stories and a temperament like a bear's. Eaton thought so much of Mulholland that he groomed him to be his successor, and when Eaton left the company in 1886 to pursue a career in politics and seek his fortune, Mulholland was named superintendent. In the years that followed, Fred Eaton would become messianic about the water shortage he saw approaching. The only answer, he told Mulholland, was to get the Owens River. At first, Mulholland found the idea preposterous: going 250 miles for water was out of the question, and Mulholland didn't much believe in surface-water development anyway. Damming rivers meant forming reservoirs, and in the heat and dryness of California, reservoirs would evaporate huge quantities of water. It made more sense to slow down the rainfall as it returned to the ocean and force more of it into the aquifer. Mulholland preached soil and forest conservation thirty years before its time. He wanted to seed the whole basin, and when he said that the deforestation of the mountainsides would reduce the basin's water supply, everyone thought he was slightly nuts. He had his men filling gullies and installing infiltration galleries and checkdams* all over the place. Everything he did, however, was nullified by the basin's growth.

By 1900, Los Angeles' population had gone over 100,000; it doubled again within four years. During the same period, the city experienced its first severe drought. Even with lawn watering prohibited and park ponds left unfilled, the artesian pressure, as Eaton had predicted, began to drop. Gushes became gurgles, then dried up. Pumps were frantically installed. By 1904, the pressure was low enough to prompt Mulholland to begin shutting irrigation wells in the San Fernando Valley, which lay across the Hollywood Hills and fed both the aquifer and the river. The farmers were furious, and Mulholland began spending a lot of time in court. The Los Angeles City Water Company was eventually taken over by the city, and Mulholland was retained in command. (The city didn't have much choice in the matter. Mulholland was such a seat-of-the-pants engineer that the plan of the entire water system resided mainly in his head; the most elemental schematics and blueprints did not exist.) In late 1904, the newly created Los Angeles Department of Water and Power issued its first public report. "The time has come," it said, "when we shall have to supplement the supply from some other source." With that simple statement William Mulholland was about to become a modern Moses. But instead of leading his people through the waters to the promised land, he would cleave the desert and lead the promised waters to them.

There is a widely held view that Los Angeles simply went out to the Owens Valley and stole its water. In a technical sense, that isn't quite true. Everything the city did was legal (though its chief collaborator, the U.S. Forest Service, did indeed violate the law). Whether one can justify what the city did, however, is another story. Los Angeles employed chicanery, subterfuge, spies, bribery, a campaign of divide-and-conquer, and a strategy of lies to get the water it needed. In the end, it milked the valley bone-dry, impoverishing it, while the water made a number of prominent Los Angeleans very, very rich. There are those who would argue that if all of this was legal, then something is the matter with the law.

It could never have happened, perhaps, had the ingenuous citizens of the Owens Valley paid more attention to a small news item that appeared in the Inyo *Register*, the valley's largest newspaper, on September 29, 1904. The

----

* Infiltration galleries and checkdams are measures designed to increase the amount of precipitation that infiltrates into the groundwater reservoir.

item began: "Fred Eaton, ex-mayor of Los Angeles, and Fred [sic] Mulholland, who is connected with the water system of that city, arrived a few days ago and went up to the site of the proposed government dam on the [Owens] River." The person who took them around, the story continued, was Joseph Lippincott, the regional engineer for the Reclamation Service. It wasn't so much this small piece of news that should have aroused the valley's suspicions. It was the fact that Lippincott had already taken Eaton around the valley twice before.

The valley had no particular reason to distrust J.B. Lippincott, although a search into his background would have dredged up a revelation or two. As a young man out of engineering school, he had joined John Wesley Powell's Irrigation Survey, the first abortive attempt to launch a federal reclamation program in the West, but had lost his job soon thereafter when Congress denied Powell funding. Embittered by the experience, Lippincott migrated to Los Angeles, where, by the mid-1880s, he had built up a lucrative practice as a consulting engineer. In 1902, when the Reclamation Service was finally created, its first commissioner, Frederick Newell, immediately thought of Lippincott as the person to launch its California program. He had a good reputation, and he understood irrigation--a science few engineers were familiar with. The post, however, meant a substantial cut in salary, and Lippincott insisted on being allowed to maintain a part-time engineering practice on the side. Newell and his deputy, Arthur Powell Davis (who was John Wesley Powell's nephew), were a little wary; in a fast-growing region with little water, a district engineer with divided loyalties could lead the Service into a thicket of conflict-of-interest entanglements. The centerpiece of the Service's program in California was to be the Owens Valley Project, and there were already rumors that Los Angeles coveted the valley's water. One of the Service's engineers, in fact, had raised this issue with Davis; with Lippincott, a son of Los Angeles, in charge, a collision between the city and the Service over the Owens River might leave the city with the water and the Service absent its reputation. But the Service's early leadership, unlike those who succeeded them, suffered from a certain lack of imagination. "On the face of it," Davis scoffed, "such a project is as likely as the city of Washington tapping the Ohio River."

The only person who seemed suspicious when Lippincott began showing Eaton and Mulholland around the Owens Valley again and again was one of his own employees, a young Berkeley-educated engineer named Jacob Clausen. His apprehensions had been aroused during Eaton's second visit, when Lippincott and Eaton had ridden up to the valley from Los Angeles by way of Tioga Pass and Clausen, at Lippincott's request, had met them at Mono Lake. On the way down the valley, Lippincott insisted that they stop at the ranch of Thomas Rickey, one of the biggest landowners in the valley. Rickey's ranch was in Long Valley, an occluded shallow gorge of the Owens River, hard up against the giant Sierra massif, which contained the reservoir site the Reclamation Service would have to acquire in order for its project to be feasible. Eaton had told Clausen that he wanted to become a cattle rancher and was interested in buying Rickey's property if he was willing to sell. As they visited the ranch, however, he seemed much more interested in water than in cattle. Clausen understood the dynamics of the Owens Valley Project-- the streamflows, the water rights, the interaction of ground and surface water--better than anyone, and Lippincott asked him to explain to Eaton how the project would work. Eaton hung on his every word, and that, Clausen was

358

to testify later, "was exactly what Lippincott wanted." The two Los Angeleans were good friends, and Eaton had been the first to dream of Los Angeles going to the Owens Valley for water. Was it so farfetched, Clausen would remember thinking to himself, to believe that Lippincott was out to help Los Angeles steal the valley's water?

If Clausen's suspicions were aroused, those of his high superiors remained utterly dormant, even though they would soon have equal reason to suspect Lippincott of being a double agent for Los Angeles. In early March of 1905, Lippincott had sent his entire engineering staff to Yuma, Arizona, on the Colorado River, to move the Yuma Irrigation Project forward at a faster pace. Work on the Owens Valley Project had been held up by winter and by the delayed arrival of a piece of drilling equipment which was on order. During the hiatus, the Reclamation Service received a couple of applications for rights-of-way across federal lands from two newly formed power companies in the Owens Valley. Each was interested in building a hydroelectric project, and Lippincott had to decide which, if any, of the plans could coexist with the Reclamation project. Unable or unwilling to look into the matter himself, Lippincott might have waited for one of his engineers to return later in the spring, but he wanted to dispose of the issue, so he decided to appoint a consulting engineer to look into the matter for him. And though there were dozens of engineers in Los Angeles and San Francisco among whom he could have chosen, he decided to turn to his old friend and professional associate Fred Eaton.

The news that Lippincott had hired Fred Eaton to decide on a matter that could affect the whole Owens Valley Project left his superiors stunned, but their response, typically, was one of bafflement rather than anger. "I fail to understand in what capacity he is acting" was the only response Arthur Davis managed to give.

Eaton himself had no questions about the capacity in which he was acting, though the public face he presented was very different. With his letter of introduction from Lippincott and an armload of freshly minted Reclamation maps, he strode into the government land office in Independence, claiming to represent the Service on a matter of vital importance to the Owens Valley Project. For the first three days, however, his investigations had nothing to do with the hydroelectric plans. Poring over land deeds in the office's files--deeds to which he might have had no access as a private citizen--Eaton jotted down a wealth of information on ownership, water rights, stream flows--things Los Angeles had to know if and when it decided to move on the Owens Valley's water. Handsome and charming, Eaton even managed to get the land office employees to help him, unaware that the information they were digging out had nothing to do with the matter that had allegedly brought Eaton there. . . .

* * *

On the 6th of March, exactly three days after Lippincott had hired Eaton as his personal representative in the matter of the power company applications, the city of Los Angeles had quietly hired its own consultant to prepare a report on the options it had in its search for water. The report had taken only a couple of weeks to prepare--most of the information was in Mulholland's office, and the conclusion was foregone anyway--and the consultant had received an absurdly grandiose commission of $2,500, more than half his annual salary. It was not so much a commission as a bribe. The money, however, was well spent: the name of the consultant was Joseph B. Lippincott.

*   *   *

. . .the $2,500 contract accepted by Joseph Lippincott from Los Angeles was, if not exactly illegal, an apparent violation of the most basic ethical standards for government officials. Newell had let Lippincott off with another fatherly lecture, but everyone in the Reclamation Service had heard about it, and since the Service had been created as an answer to the epic graft and fraud associated with the General Land Office, some of Lippincott's associates were furious with him. By July of 1905, Newell realized that the whole thing might blow up in his face; he had to do something to contain the damage. As a result, he decided to appoint a panel of engineers to review the conflict between the Reclamation project and the water needs of Los Angeles and decide whether the Owens Valley Project should move forward, be put on hold, or be abandoned. Newell felt that Lippincott, as the senior engineer most familiar with the project, should sit on the panel. To his and Lippincott's astonishment, several Reclamation engineers said they would refuse to sit next to him. Lippincott now realized that he, too, would have to mount a damage-control operation in a hurry. On July 26, the night before the panel was scheduled to convene, he dashed off a telegram to Eaton that read, "Reported to me and publicly accepted that you had represented yourself as connected with Reclamation and acting as my agent in Owens Valley. As this is entirely erroneous and very embarrassing to me, please deny publicly or the Service will be forced to do so." The truth of Lippincott's denial can best be judged by Fred Eaton's reaction, which was incendiary. He received the telegram in the federal land office in Independence, where he was still trying to masquerade as Lippincott's agent. After reading it he felt compelled to vent his spleen on the nearest person available, agent Richard Fysh. "Eaton said he had a telegram from Mr. Lippincott and it was a damned hot one," said Fysh later in a deposition, "and he, Eaton, did not like it a little bit, as it put him in a wrong light."

Newell's panel of engineers was convened in San Francisco on July 27. After two days of hearing divided opinions (Clausen testified in favor of continuing, Lippincott in favor of abandonment), the panel reached a unanimous verdict. The Owens Valley Project should not be sedulously pursued, they recommended; the needs of Los Angeles had become too great an issue. But neither should it be formally abandoned until a more persuasive case could be made for doing so. Los Angeles would have to demonstrate that it had absolutely no choice but to go to the valley for water, and it would have to prove that it had the resources to carry out such a gigantic undertaking on its own. Such a recommendation, the panel added, was of course based on the assumption that the Reclamation project was still feasible.

Which, unbeknownst to anyone but Eaton and a select handful of Los Angeles officials, it was not. Four months earlier, after completing his consultant duties for Lippincott, Eaton had gone back to see the stubborn Thomas Rickey, who held the key piece of land in the valley--the land the city had to have in order to block the federal project--but who had refused to sell. In Eaton's hand was his recommendation that Rickey's hydroelectric company be allowed to usurp its competitor's claim on the main power sites on the river. That, Eaton thought, was the sweetener that would surely make Rickey sell.

After hours of pleading and cajoling, however, the rancher still held out. In disgust, Eaton finally stood up, roughly shook Rickey's hand, and stomped out the door. As he was standing at the railroad depot, waiting for the train that would take him back to Los Angeles, Rickey raced up in his carriage. He had had a sudden change of heart; for $450,000, he told Eaton, he would sell him an option clear on the ranch, including the Long Valley reservoir site.

Eaton's jubilation was so great he couldn't restrain himself. He ran to the telegraph office and shot off a cryptic message to Mulholland. "The deal is made," he wired. All it had required was "a week of Italian work."

Los Angeles now had most of what it needed, but Mulholland still wanted some additional water rights in order to kill the Reclamation project once and for all. Within hours of receiving Eaton's telegram, he was frantically organizing an expedition of prominent Los Angeleans to the Owens Valley, using the pretext that they were investors interested in developing a resort. The group included Mayor Owen McAleer and two prominent members of the water commission. For them to see the river firsthand was crucial, Mulholland reasoned, because he and Eaton would need more money to buy the last water rights they wanted, and the city could not legally appropriate money toward a project that hadn't even been described, let alone authorized. A group such as this could easily free up some money in the Los Angeles business community if they fathomed how much water there was.

It went exactly as planned. The group arrived in the valley on the cusp of spring, when even small tributaries of the Owens River were overflowing; days after they returned, Eaton and Mulholland had all the money they needed. They requisitioned an automobile and raced off to the valley by the shortest route, across the Mojave Desert--probably the first time anyone crossed it by car. After a week of frantic, furtive buying, the two men returned. "The last spike has been driven," Mulholland announced to the assembled water commissioners. "The options are all secured."

Like all the other newspaper publishers in the city, Harrison Gray Otis had been operating under a self-imposed gag rule. Although the publishers knew what was going on, not a word of Mulholland and Eaton's stealthy grab of water options had appeared in the papers. However, on July 29, the same day the Reclamation panel reached its verdict, Otis could no longer contain himself. Under a headline that read, "Titanic Project to Give the City a River," the whole unauthorized story spilled out in the Los Angeles *Times*.

Otis seemed to take particular satisfaction in the way Fred Eaton had hoodwinked the greedy but guileless rubes in the Owens Valley. "A number of the unsuspecting ranchers have regarded the appearance of Mr. Eaton in the valley as a visitation of providence," the *Times* chortled. "In the eyes of the ranchers he was land mad. When they advanced the price of their holdings a few hundred dollars and he stood the raise, their cup of joy fairly overflowed...The farmer folk in the Owens River Valley think that he had gone daffy on stock raising. To them he is a millionaire with a fad." The paper even admitted that the town of Independence, whose neighboring ranchers had been made offers they couldn't refuse, was faced with financial ruin, but it refused to let such a fact spoil its enjoyment of a good joke. The paper also recalled in excruciating detail Joseph Lippincott's career as a double agent, apparently thinking it was doing him a favor. "In the consummation of the great project that is to supply Los Angeles with sufficient water for all time, great credit is given to J.B. Lippincott," it said, "Without Mr. Lippincott's interest and cooperation, it is declared that the plan never would have gone through...*Guided by the spirit of the Reclamation Act*...he recognized the fact that the Owens River water would fulfill a greater mission in Los Angeles than if it were to be spread over acres of desert land...Any other government engineer, a nonresident of Los Angeles and not familiar with the needs of this section, undoubtedly would have gone ahead with nothing more than the mere reclamation of the arid lands in view" (emphasis added). It was praise that was to damn Lippincott for the rest of his life.

There was nothing quite as revealing in the *Times*'s story, however, as its very lead sentence: "The cable that has held the San Fernando Valley vassal for ten centuries to the arid demon," it gushed in a spasm of metaphorical excess, "is about to be severed by the magic scimitar of modern engineering skill."

There was something very strange about that sentence. All along, the Owens River had been portrayed as a matter of life or death to the city of Los Angeles. No one had ever said a word about the San Fernando Valley.

*   *   *

The first sign something was afoot came in the weeks following the *Times*'s disclosure of Mulholland and Eaton's daring scheme, when Otis's newspaper took time out from its usual broadsides to laud the future of the San Fernando Valley, an encircled plain of dry, mostly worthless land on the other side of the Hollywood Hills. "Go to the whole length and breadth of the San Fernando Valley these dry August days," the paper editorialized on August 1. "Shut your eyes and picture this same scene after a big river of water has been spread over every acre, after the whole expanse has been cut up into five-acre, and in some cases one-acre, plots--plots with a pretty cottage on each and with luxuriant fruit trees, shrubs and flowers in all the glory of their perfect growth...." Again on October 10, a so-called news story began, "Premonitory pains and twitches: The San Fernando Valley has caught the boom. It appears just about ready to break...."

What was odd about this was that there was as yet no guarantee--at least one publicly offered by Mulholland--that the San Fernando Valley was going to receive any of the Owens Valley water. In the first place, the route of the aqueduct had not yet been disclosed; it might go through the valley, but then again it might not. Secondly, the voters had not even approved the aqueduct, let alone voted for a bond issue to finance it. Mulholland had been saying that the city had surplus water sufficient for only ten thousand new arrivals. If that was so, and if the city was expected to grow by hundreds of thousands during the next decade, where was this great surplus for the San Fernando Valley to come from? In those days, the valley was isolated from Los Angeles proper; it sat by itself far outside the city limits. In theory, the valley couldn't even *have* the city's surplus water, assuming there was any--it would be against the law.

The truth, which only a handful of people knew, was that William Mulholland's private figures were grossly at odds with his public pronouncements; it was the same with his intentions. Despite his talk of water for only ten thousand more people, there was still a big surplus at hand.

(During the eight years it would take to complete the aqueduct, in fact, the population of Los Angeles rose from 200,000 to 500,000 people, yet no water crisis occurred.) The crisis was, in large part, a manufactured one, created to instill the public with a sense of panic and help Eaton acquire a maximum number of water rights in the Owens Valley. Mulholland and Eaton had managed to secure water rights along forty miles of the Owens River, which would be enough to give the city a huge surplus for years to come. But Mulholland was not saying that he would *use* any of the surplus; in fact, he seemed to be going out of his way to assure the Owens Valley that he would not. For example, the proposed intake for the aqueduct had been carefully located downstream from most of the Owens Valley ranches and farms, so that they could continue to irrigate; Mulholland would later tell the valley people that his objective was simply to divert their unused and return flows.

In truth, Mulholland planned to divert every drop to which the city held rights as soon as he could. Like all water-conscious westerners, he lived in fear of the use-it-or-lose-it principle in the doctrine of appropriative rights. If the city held water rights that went unused for years, the Owens Valley people might successfully claim them back. But where would he allow the surplus to be used?

Privately, Mulholland planned to lead the aqueduct through the San Fernando Valley on its way to the city. In his hydrologic scheme of things, the valley was the best possible receiving basin; any water dumped on the earth there would automatically drain into the Los Angeles River and its broad aquifer, creating a large, convenient, nonevaporative pool for the city to tap. It provided, in a word, free storage. That it was free was critically important, because Mulholland, intentionally or not, had underestimated the cost of building the aqueduct, and to build a large storage reservoir in addition to the aqueduct would be out of the question financially. Even had it been feasible, Mulholland was deeply offended by the evaporative waste of reservoirs; he was much more inclined to store water underground.

Mulholland had an even more important reason for wanting to include the San Fernando Valley in his scheme. Under the city charter, Los Angeles was prohibited from incurring a debt greater that 15 percent of its assessed valuation. In 1905, that put its debt limit at exactly $23 million, which was what he expected the aqueduct to cost. But the city already had $7 million in outstanding debt, which left him with a debt ceiling too low to complete the project. After coming this far--securing the water rights, organizing civic support--he wouldn't have the money to build it!

Mulholland, however, was clever enough to have thought of a way out of this dilemma. If the assessed valuation of Los Angeles could be rapidly increased, its debt ceiling would be that much higher. And what better way was there to accomplish this than to *add to the city*? Instead of bringing more people to Los Angeles--which was happening anyway--*the city would go to them*. It would just loosen its borders as Mulholland loosened his silk cravat and wrap itself around the San Fernando Valley. Then it would have a new tax base, a natural underground storage reservoir, and a legitimate use of its surplus water in one fell swoop.

Anyone who knew this, and bought land in the San Fernando Valley while it was still dirt-cheap, stood to become very, very rich.

*   *   *

On November 28, 1904--just six days after Joseph Lippincott was paid $2,500 to help steer his loyalties in the direction of Los Angeles--a syndicate of private investors had purchased a $50,000 option on the Porter Land and Water Company, which owned the greater part of the San Fernando Valley--sixteen thousand acres all told. Innocent enough. But the investors had then waited to consummate their $500,000 purchase until March 23, 1905--*the same day* that Fred Eaton had telegraphed the water commission that the option on the Rickey ranch in Long Valley was secured. On that day, as anyone who had access to Mulholland's thinking knew, Los Angeles was all but guaranteed 250,000 acre-feet of new water--an amount that would leave the city with a water surplus for at least another twenty years. And the only sensible place to use the surplus water was in the San Fernando Valley.

*   *   *

On September 7, 1905, the bond issue passed, fourteen to one.

To the Los Angeles *Times*, it was a "Titanic Project to Give the City a River." to the Inyo *Register*, it was a ruthless scheme in which "Los Angeles Plots Destruction, Would Take Owens River, Lay Lands Waste, Ruin People, Homes, and Communities." That sensational headline actually belied the feeling in the valley somewhat. Few people thought, at first, that things would be so bad. A number of the ranchers had made out well selling their water rights, and they would be able to keep their water for years, until the aqueduct was built. The city had bought up nearly forty bank miles of the river and would probably dry up the lower valley, but the upper valley, except for Fred Eaton's purchase of the Rickey estate, had been left mostly intact. When Eaton moved up from Los Angeles as promised and began his new life as a cattle rancher, the valley people were reassured. After a while, they even began to fraternize with him.

Mulholland, meanwhile, had begun his own campaign to mollify the people of the valley, a campaign in which he was joined, somewhat more bellicosely, by the Los Angeles *Times*, which featured headlines such as "Ill-feeling Ridiculous" and "Owens Valley People Going Off at Half-Cock." Inyo County's Congressman, Sylvester Smith, was an influential member of the House Public Lands Committee, and since the city would have to cross a lot of public land it would have to deal with him. Meanwhile, Theodore Roosevelt, the bugaboo of monopolists, had just been elected to a second term. He would never let the Owens Valley die for the sake of . . . Harrison Gray Otis, and [his] cronies in the San Fernando Valley syndicate. On top of all this, the Owens was a generous desert river, with a flow sufficient for two million people. It was laughable to think of Los Angeles growing that big, so even under the worst of circumstances there would be water enough for all. The reasoning was very sensible, the logic very sound, and it was fatefully wrong.

There was one person who knew that it was. She was Mary Austin, the valley's literary light, who had published a remarkable collection of impressionistic essays entitled **Land of Little Rain** that won her recognition around the world. In the course of her writing she had spent long hours with the last of the Paiutes, the Indians who had lived in the valley for centuries until they were instantly displaced by the whites. The Paiutes showed her

364

what no one else saw--that order and stability are the most transient of states, that there is rarely such a thing as a partial defeat. In a subsequent book, a novella about the Owens Valley water struggle called **The Ford**, she wrote about what happens when "that incurable desire of men to be played upon, to be handled," runs up against "that Cult of Locality, by which so much is forgiven as long as it is done in the name of the Good of the Town." Mary Austin was convinced that the valley had died when it sold its first water right to Los Angeles--that the city would never stop until it owned the whole river and all of the land. One day, in Los Angeles for an interview with Mulholland, she told him so. After she had left, a subordinate came into this office and found him staring at the wall. "By God," Mulholland reportedly said, "that woman is the only one who has brains enough to see where this is going."

No sooner had the city gotten the aqueduct past the voters than it faced the more difficult task of getting it past Congress. Most of the lands it would traverse belonged to the government, so the city would have to appeal for rights-of-way. The Reclamation project, though moribund, was still not officially deauthorized, which was, at the very least, a nuisance to the city. But deauthorization could prove to be even worse, because tens of thousands of acres that the Service had withdrawn would return to the public domain and be available for homesteading. Homesteading in California was another name for graft; half of the great private empires were amassed by hiring "homesteaders" to con the government out of its land. If the withdrawn lands went back to the public domain, every available water right would be coveted by speculators for future resale to the city. Mulholland seemed to believe that the city would never require more water, but others, notably Joseph Lippincott, thought him wrong. The withdrawn lands had to be kept off-limits at all costs.

The instrument for achieving this wishful goal was a bill introduced at the behest of Mulholland's chief lawyer, William B. Matthews, by Senator Frank Flint of California, a strong partisan of Los Angeles and urban water development in general. The bill would give the city whatever rights-of-way it needed across federal lands and hold the withdrawn lands in quarantine for another three years, which would presumably give the city enough time to purchase whatever additional water or land it might need. Flint's bill reached the Senate floor in June of 1906, and flew through easily. Its next stop, however, was the House Public Lands Committee, where it crashed into Congressman Sylvester Smith. Smith was an energetic and charming politician, a former newspaper publisher from Bakersfield with a sense of public duty and enough money to maintain an ironclad set of principles. The idea of Harrison Gray Otis . . . becoming vastly richer . . . on water abducted from his district inflamed his well-developed sense of outrage. Smith knew what he was up against, however, and realized that his best defense was to appear utterly reasonable. As a result, he said that he was willing to acknowledge the city's need for more water, that he was willing to let it have a substantial share of the Owens River, and that he was willing to grant the aqueduct its necessary rights-of-way. He was not willing, however, to do any of this in the way the city wanted. He suggested a compromise. Let the Reclamation Service build its project, including the big dam in Long Valley--a dam that could store most of the river's flow. The water could then be used first for irrigation, and because of the valley's long and narrow slope, the return flows would go back to the lower river, where they could be freely diverted by Los Angeles. The city would sacrifice some of the water it wanted, the valley would sacrifice some irrigable land. It was, Smith argued, an enlightened plan: sensible, efficient, conceived in harmony. It was the only plan under which no one would suffer. He would add only two stipulations: the Owens Valley would have a nonnegotiable first right to the water, and any surplus water could not be used for irrigation in the San Fernando Valley.

Smith's proposal was obviously anathema to the San Fernando land syndicate, and to the city as well. The chief of the Geologic Survey doubted that it would work, and even if it did, for the West's largest city to settle for leftover water from a backwater oasis of fruit and cattle ranchers was, to say the least, humiliating. The city might have to beg for extra water in times of drought or go to court to try to condemn it. If the Owens Valley held on to its first rights and expanded its irrigated acreage, Los Angeles might soon have to look for water again, and the only river in sight was the Colorado, a feckless brown torrent in a bottomless canyon which the city could never afford to dam and divert on its own. Smith's proposal led directly to one unthinkable conclusion: at some point in the relatively near future, Los Angeles would have to cease to grow.

What was William Mulholland's response? He took a train to Washington, held a summit meeting with Smith and Senator Flint, and decided to do what any sensible person would have done: he accepted the compromise.

If it was a smokescreen, as it appears to have been, it was a brilliant move. (Mulholland seems to have been a far better political schemer than he was a hydrologist and civil engineer). For one thing, it put Sylvester Smith off guard, making him believe that the reconciliation he wanted to effect was a success. For another, it gave Los Angeles some critical extra time to plead its case before the two people who might help the city get everything it wanted: the President of the United States, Theodore Roosevelt, and the man on whom he leaned most heavily for advice--Gifford Pinchot.

Pinchot was the first director of Roosevelt's pet creation, the Forest Service, but that was only one of his roles. He was also the Cardinal Richelieu of TR's White House. Temperamentally and ideologically, the two men fit hand in glove. Both were wealthy patricians (Pinchot came from Pittsburgh, where his family had made a fortune in the dry-goods business); both were hunters and outdoorsmen. Though their speeches and writings rang of Thomas Jefferson, at heart Pinchot and Roosevelt seemed more comfortable with Hamiltonian ideals. Roosevelt liked the Reclamation program because he saw it as an agrarian path to industrial strength, not because he believed--as Jefferson did--that a nation of small farmers is a nation with a purer soul. Pinchot espoused forest conservation not because he worshiped nature like John Muir (whom he privately despised) but because the timber industry was plowing through the nation's forests with such abandon it threatened to destroy them for all time. Roosevelt was a trust-buster, but only because he feared that unfettered capitalism could breed socialism. (For evidence he only had to look as far as Los Angeles, where Harrison Gray Otis was whipping labor radicals into such a blind, vengeful froth that two of them blew up his printing plant in 1910 and killed twenty of their own.) The conservation of Roosevelt and Pinchot was utilitarian; their progressivism--they spoke of "greatest good for the greatest number"--had a nice ring to it, but it also happens to be the progressivism of cancer cells.

366

On the evening of June 23, Senator Frank Flint left his offices on Capitol Hill for a late meeting with the President. It was a hot and muggy night, and Roosevelt seemed in an irritable mood. Behind him, however, stood a man who seemed a model of coolness and decorum, Gifford Pinchot. Flint, who had just received an intensive coaching from Matthews and Mulholland, began a passionate appeal.

Smith's so called compromise, he said, was nothing less than capitulation. Los Angeles had agreed only in despair; it was going to run out of water any day and it couldn't afford to be filibustered to death in Congress. Smith's prohibition on using surplus water in the San Fernando Valley left the city no choice but to leave any surplus in the Owens Valley or dump it in the ocean. In the first case, water rights the city had purchased at great expense might revert to the valley under the doctrine of appropriative rights; in the second case, the city would violate the California constitution, which forbade "inefficient use" of water. The real estate bust of 1889 had depopulated the city by one-half. Imagine what a water famine would do! All of the city's actions in the Owens Valley had been legitimate. It had paid for its water, fair and square, and it wanted to let the valley survive. But there was only so much water, and it was a hundredfold--thousandfold, said Smith--more valuable to the state and the nation if it built up a great, strong, progressive city on America's weakly defended western flank instead of maintaining a little agrarian utopia in the high desert.

It was a rousing speech--the kind of speech that Roosevelt liked to hear. It was, in fact, just the kind of speech *he* would have made.

Roosevelt turned to his other visitor. "What do you think about this, Giff?"

"As far as I am concerned," Pinchot answered coolly, "there is no objection to permitting Los Angeles to use the water for irrigation purposes."

It was as simple as that. Roosevelt did not even bother to call in the Interior Department's lawyers or the Geologic Survey's hydrologists to ask whether Flint's argument was sense or nonsense. He never invited Sylvester Smith to give his side of the argument. He didn't even tell Smith or his own Interior Secretary, Ethan Hitchcock, about his decision; they found out about it secondhand a day and a half later. Hitchcock, a wealthy, principled man in the style of Sylvester Smith, had been profoundly embarrassed by the two-faced behavior of his employee J.B. Lippincott, and had been looking for a way to make amends to the Owens Valley. Flabbergasted and infuriated by the President's decision, Hitchcock raced over to the White House, where Roosevelt refused to hear him. Instead, he forced him to suffer the humiliation of helping him draft a letter explaining "*our* attitude in the Los Angeles water supply question." As Hitchcock stood by, impotent and enraged, Roosevelt wrote, "It is a hundred or a thousandfold more important to state that this water is more valuable to the people of Los Angeles than to the Owens Valley." The words could have come right out of William Mulholland's mouth.

*   *   *

Roosevelt's support for Flint's bill was only the beginning of the aid and comfort he was to give to the most powerful city on the Pacific Coast. When the Reclamation Service officially annulled the Owens Valley Project in July of 1907, the hundreds of thousands of acres it had withdrawn were not returned to the public domain for homesteading, on Roosevelt's orders--just as Mulholland wished. It was a decision without precedent, and its result was that the handful of rich members of the San Fernando syndicate could continue using the surplus water in the Owens River that thousands of homesteaders might have claimed instead. Ethan Hitchcock had promised that such a decision, which he already foresaw when Roosevelt closed ranks behind Los Angeles, would be made over his dead body, but Roosevelt spared his life by firing him first. And when the city, immensely satisfied with the result, asked Pinchot whether he couldn't go a step further, the chief of the Forest Service decided to include virtually all of the Owens Valley in the Inyo National Forest.

The Inyo National Forest! With six inches of annual rainfall, the Owens Valley is too dry for trees; the only ones there were fruit trees planted and irrigated by man, some of which were already dying for lack of water. This didn't seem to bother Pinchot, nor did the fact that this action appears to have been patently illegal. The Organic Act that created the Forest Service says, "No public forest reservation shall be established except to improve and protect the forest...or for the purpose of creating favorable conditions of water flow, and to provide a continuous supply of timber for the use and necessities of the United States; but it is not the purpose of these provisions...to authorize the inclusion...of lands more valuable for the mineral therein, *or for agricultural purposes*, than for forest purposes" (emphasis added). The valley's irrigated orchards were infinitely more valuable than the barren flats and scattered sagebrush that characterized the new national forest, so Pinchot's action was incontrovertibly a violation of the legislation that put him in business. He lamely countered that he was simply acting to protect the quality of Los Angeles' water; but since much of the treeless acreage he included in the Inyo National Forest lay *below* the intake of the aqueduct, it was a flimsy excuse. As a formality, Pinchot was obliged to send an investigator to the Owens Valley to recommend that he do what he had already made up his mind to do. He sent three before he found one who was willing to go along. "This is not a government by legislation," lamented Sylvester Smith on the Senate floor, "it is a government by strangulation."

In July of 1907, with the reclamation project in its grave and the Owens Valley imprisoned inside a national forest without trees, Joseph Lippincott resigned from the Reclamation Service and immediately went to work, at nearly double his government salary, as William Mulholland's deputy. He remained utterly unchastised. "I would do everything over again, just exactly as I did," he said as he departed.

The one thing that no one seems to have thought about in all this was that the people of Owens Valley were only human, and there was just so much they could take.

The aqueduct took six years to build. The Great Wall of China and the Panama Canal were bigger jobs, and New York's Catskill aqueduct, which was soon to be completed, would carry more water, but no one had ever built anything so large across such merciless terrain, and no one had ever done it on such a minuscule budget. It was as if the city of Pendleton, Oregon, had gone out, by itself, and built Grand Coulee Dam.

The aqueduct would traverse some of the most scissile, fractionated, fault-splintered topography in North America. It would cover 223 miles, 53 of them in tunnels; where tunneling was too risky, there would be siphons whose acclivities and declivities exceeded fifty-grade [a slope of fifty percent]. The city would have to build 120 miles of railroad track, 500 miles of roads and trails, 240 miles of telephone line, and 170 miles of power transmission line. The entire cement-making capacity of Los Angeles was not adequate for this one project, so a huge cement plant would have to be built near the limestone deposits in the grimly arid Tehachapi Mountains. Since there was virtually no water along the entire route, steampower was out of the question and the whole job would be done with electricity; therefore, two hydroelectric plants would be needed on the Owens River to run electric machinery that a few months earlier had not even been invented. The city would have to maintain, house, and feed a work force fluctuating between two thousand and six thousand men for six full years. And it would have to do all this for a sum equivalent, more or less, to the cost of one modern jet fighter.

The workers would have to supply their own hard-shelled derby hats, since hard hats did not yet exist, and even if they had the city couldn't afford them. They would live in tents in the desert without liquor or women--although both were available nearby and ended up consuming most of the aqueduct payroll. They would eat meat that spoiled during the daytime and froze at night, since the daily temperature range in the Mojave Desert can span eighty degrees. Nonetheless, the men would labor on the aqueduct as the pious raised the cathedral at Chartres, and they would finish under budget and ahead of schedule. If you asked any of them why they did it, they would probably say that they did it for the chief.

The loyalty and heroics that Mulholland inspired in his workers were a perpetual source of wonder. For six years he all but lived in the desert, patrolling the aqueduct route like a nervous father-to-be pacing a hospital waiting room--giving advice, offering encouragement, sketching improvised solutions in the sand. In sandstorms, windstorms, snowstorms, and terrifying heat, his spirits remained contagiously high. Pilfering, which can add millions to the cost of a modern project, was almost unknown. Although the pay was terrible--Mulholland simply couldn't afford anything more--he initiated a bonus system that shattered records for hard-rock tunneling. (The men were in a race with the world's most illustrious tunnelers, the Swiss, who were digging the Loetchberg Tunnel at the same time.)

Throughout the entire time, Mulholland showed the better side of a complex and sometimes heartless character. If he wandered through a tent city and discovered that a worker's wife had just had a baby, he would stop long enough to show her the proper way to change a diaper. He would sit down and eat with the men and complain louder than anyone about the food. In lieu of newspapers, his wit was breakfast conversation. Once, when a landslide sealed off a tunnel with a man still inside, Mulholland arrived to check on the rescue effort.

"He's been in there three days, so I don't suppose he's doing so well," said the supervisor, a mirthless Scandinavian named Hansen.

"Then he must be starving to death," said Mulholland.

369

"Oh, no, sir," said the supervisor. "He's getting something to eat. We've been rolling him hard-boiled eggs through a pipe."

"Have you?" said Mulholland archly. "Well, then, I hope you've been charging him board."

"No, sir," said the flustered Hansen. "But I suppose I should, eh?"

And Los Angeles loved Mulholland even more than the men, because its reward would be infinitely greater than theirs--to the thirsty city, he was Moses. And he was that greater rarity, a Moses without political ambition. When a move was afoot a few years later to run him for mayor, Mulholland dismissed it with a typical bon mot: "I would rather give birth to a porcupine backwards than become the mayor of Los Angeles." But nothing that William Mulholland ever said or did quite matched the speech he gave when, on November 5, 1913, the first water cascaded down the aqueduct's final sluiceway into the San Fernando Valley. It had been a day of long speeches and waiting, and the crowd of forty thousand people was restless. Mulholland himself was exhausted; his wife was very ill, and he had slept only a few hours in several nights. When the white crest of water finally appeared at the top of the sluiceway and cascaded toward the valley, an apparition in a Syrian landscape, Mulholland simply unfurled an American flag, turned toward the mayor, H.H. Rose, and said, "There it is. Take it."

It was the high point of Mulholland's life and career.

Very little of the water that was, according to Theodore Roosevelt, a hundred or a thousandfold more important to Los Angeles than to the Owens Valley would go to the city for another twenty years. All through the teens and early twenties, the San Fernando Valley used three times as much aqueduct water as the city itself, the vast part of it for irrigation. During one particularly wet year, every drop of the copious flow of the aqueduct went to irrigate San Fernando Valley crops; the city took nothing at all. Understandably, this news enraged the people of the Owens Valley. For Los Angeles to take their water to fill their washtubs and water glasses was one thing. For it to turn their valley back to desert so that another desert valley, owned by rich monopolists, could bloom in its place was quite another.

The teens and early twenties, however, were extraordinarily wet years-the same wet years that caused the Reclamation Service to overestimate dramatically the flow of the Colorado River--and there was water enough for everyone. The irrigated acreage in the San Fernando Valley rose from three thousand acres in 1913--the year both the completion of the aqueduct and the annexation of the valley occurred--to seventy-five thousand acres in 1918. Even so, the Owens Valley lost few of its orchards and irrigated pasturelands, and the new railroad to Los Angeles and the silver mine at Tonopah fed in enough wealth to allow the town of Bishop to build a grand American Legion Hall and Masonic Temple, those cathedrals of the rural nineteenth century.

The same uncharacteristically engorged desert river that was keeping the Owens Valley green was responsible, in Los Angeles, for the most transfixing change. Santa Monica Boulevard, once a dry dusty strip, became an elegant corridor of palms; in Hollywood, where the motion picture industry had risen up overnight, outdoor sets resembled New Guinea; and since most Los

Angeleans were immigrants from the Middle West, every bungalow had a green lawn. The glorious anomaly of a fake tropical city with a mild desert climate brought people from everywhere. . . .

*   *   *

In the West, drought tends to come in cycles of about twenty years, and the next drought arrived on schedule. The years of 1919 and 1920 were a premonition; rainfall was slightly below average, it rose back to average--a measly fourteen inches--in 1921 and went slightly over that in 1922. Then it crashed. Ten inches in 1923; six inches in 1924; seven inches in 1925. In Florida, a seven-inch rainstorm may occur two or three times a year, but Los Angeles was trying to look like Florida, and grow even faster, on a fifth of its precipitation, and when the drought struck it keep going on a tenth. Mulholland had expected 350,000 people by 1925, but had 1.2 million on his hands instead. The city was growing fifteen times faster than Denver, eleven times faster than New York. And though the city at its core had become a metropolis, Los Angeles County led the nation in the value of its agricultural output. All of this agriculture depended on irrigation, which, together with the phenomenal urban growth, depended on a river draining Mount Whitney two hundred miles away.

As the drought intensified, the Owens River moved perilously close to overappropriation. The problem was not only that the river was small, but also that no carryover storage existed--nothing but some small receiving reservoirs around the basin and the snowfields in the Sierra. The Los Angeles Aqueduct was essentially a run-of-the-river project. If the river didn't run, the city collapsed.

If the city and the Owens Valley were to continue sharing the river, carryover storage would have to be built; otherwise, one place or the other would lose its water during a drought. Mulholland, of course, knew this, but still refused to build the dam at Long Valley. He blamed it on the city's fragile finances, but that was a poor excuse; the real reason was that he and his old friend Eaton had had a nasty falling-out.

Fred Eaton had not even bothered to attend the dedication of the aqueduct in 1913, though its existence was owed mainly to him. He had bought the initial water rights the city needed with his own money, taking a considerable risk; had the voters failed to approve the bond referendum, he would have been drowning in both unusable water and debt. The city had paid him quite adequately for the right, but it had not made him a multimillionaire. Originally, Eaton had hoped to operate the Owens Valley end of the aqueduct as a private concession, which could have made him incredibly rich, but Fredrick Newell and Roosevelt had dashed that dream, insisting that the project be municipally owned from end to end. Eaton had also had some bad luck in the cattle business, and had to switch ignominiously to chickens. He was sixty five years old; it was time things finally went right. The one item of real value Eaton owned was the reservoir site on the ranch he had purchased from Thomas Rickey. Ideally, a dam built at the site ought to be 140 feet high, the approximate depth of the gorge; that would create a reservoir large enough to provide for both the city and the valley during all but the worst droughts. A damsite of such importance to the city--a site which, if developed, would drown a good portion of his ranch--was worth a lot of money, as far as

371

Eaton was concerned. When Mulholland asked him what his price was, Eaton said $1 million. Mulholland, who seemed personally indifferent to money (though he was reputedly the highest-paid civil servant in California), laughed him off. Time and time again he asked Eaton to accept a reasonable offer-- $500,000, perhaps, or a little more--and each time his offer was more angrily refused. By 1917, the two old friends were no longer on speaking terms.

As the drought intensified, Mulholland begged the city fathers to end their abject deification of growth. The only way to solve the city's water problem, he grumbled aloud, was to kill the members of the Chamber of Commerce. When he was ignored, he began to regulate irrigation practices in the San Fernando Valley. First he forbade the irrigation of alfalfa, a low-value, water-demanding crop; then he prohibited winter planting. When these measures proved inadequate, he swallowed his disdain for surface storage and began building reservoirs in the basin--first the Hollywood Reservoir, then a much larger dam in San Francisquito Canyon, a deep fissure in the shaky, shaly topography of the Santa Paula hills.

With the tens of thousands of people pouring in each year, everything was a stopgap measure. By the early 1920s, Mulholland was already lobbying for an aqueduct from the Colorado River. This, however, put him on a collision course with Harry Chandler, who owned 860,000 acres in Mexico that relied on the Colorado and who was so greedy that, despite his enormous wealth, he put the interests of his Mexican holdings above the welfare of the city he had created out of whole cloth. Chandler's opposition, together with fierce feuding among the Colorado River Basin states, kept the Boulder Canyon Project Act, which would create the storage reservoir that any Colorado River aqueduct would need, bottled up for years. Frustrated at every turn, Mulholland reached the end of his tether sometime in 1923. The only answer, he decided, was to do what Mary Austin had predicted the city would ultimately do--dry the Owens Valley up.

*   *   *

[The] master strategist . . . was sixty-nine years old and a changed man. Thirty years earlier, Mulholland had spent his idle hours in a cabin at one of the city's outlying reservoirs, reading the classics and planting poplars. When the city had first talked about tapping the Owens River, his concern about the valley's welfare led him to suggest that the city plant millions of trees which the residents could sell for firewood to the barren mining camps in Nevada-- until someone informed him that so many trees would suck up enough groundwater to bleed the river dry. In his later years, however, the William Mulholland who had read Shakespeare and quoted Alexander Pope was hardly recognizable. No person ever put his imprint on an agency as strongly as Mulholland left his on the Los Angeles Department of Water and Power, and that agency was now using secret agents, breaking into private records, and turning neighbors into mortal foes. And, worst of all, Mulholland was ignoring a solution that would have satisfied everyone--a dam at Long Valley--out of petty niggardliness and almost fanatical pride.

In 1980, there were few people still alive who remembered Mulholland, but one who did was Horace Albright, the director of the National Park Service under Herbert Hoover. Albright could no longer remember the year--he was eighty-two--but it was probably 1925 or 1926, and he was a young park superintendent invited to attend a testimonial dinner for Senator Frank Flint, the man who had engineered the dubious federal decisions that allowed the Owens Valley aqueduct to be built. Albright was seated at Mulholland's table, a couple of chairs away, and midway through dinner he felt a rough tap on his shoulder.

"You're from the Park Service, aren't you?" Mulholland demanded more than asked.

"Yes, I am," said Albright. "Why do you ask?"

"Why?" Mulholland said archly. "Why? I'll tell you why. You have a beautiful park up north. A majestic park. Yosemite Park, it's called. You've been there, have you?"

Albright said he had. He was the park's superintendent.

"Well, I'm going to tell you what I'd do with your park. Do you want to know what I would do?"

Albright said he did.

"Well, I'll tell you. You know this new photographic process they've invented? It's called Pathe. It makes everything seem lifelike. The hues and coloration are magnificent. Well, then, what I would do, if I were custodian of your park, is I'd hire a dozen of the best photographers in the world. I'd build them cabins in Yosemite Valley and pay them something and give them all the film they wanted. I'd say, 'This park is yours. It's yours for one year. I want you to take photographs in every season. I want you to capture all the colors, all the waterfalls, all the snow, and all the majesty. I especially want you to photograph the rivers. In the early summer, when the Merced River roars, I want to see that.' And then I'd leave them be. And in a year I'd come back, and take their film, and send it out and have it developed and treated by Path . And then I would print the pictures in thousands of books and send them to every library. I would urge every magazine in the country to print them and tell every gallery and museum to hang them. I would make certain that every American saw them. And then," Mulholland said slowly, with what Albright remembered as a vulpine grin, "and then do you know what I would do? I'd go in there and build a dam from one side of that valley to the other and *stop the goddamned waste!*"

"It was the tone of his voice that surprised me," Albright said. "The laughingly arrogant tone. I don't think he was joking, you see. He was absolutely convinced that building a dam in Yosemite Valley was the proper thing to do. We had few big dams in California then. There were hundreds of other sites, and there were bigger rivers than the Merced. But he seemed to want to shake things up, to outrage me. He almost *wanted* to destroy."

It was the same tone, the same bitter and unreasoning quarrelsomeness, that Mulholland displayed when a reporter from the *Times* asked him why there was so much dissatisfaction in the Owens Valley. "Dissatisfaction in the valley?" said Mulholland mockingly. "Yes, a lot of it. Dissatisfaction is a sort of condition that prevails there, like foot and mouth disease." It was the same

373

unreasoning rage that made him say, when his war of attrition against the Owens Valley had finally caused events to take a drastic turn for the worse, that he half regretted the demise of so many of the valley's orchard trees, because now there were no longer enough live trees to hang all the troublemakers who lived there.

*   *   *

. . . On May 21, 1924, a group of men . . . "stole" three cases of dynamite, and blew a large section of the aqueduct to smithereens. From that moment on, William Mulholland refused to refer to anyone in the Owens Valley by any other name than "dynamiter." . . . But the dynamitings continued. When the Department of Water and Power released a report that recommended "destroying all irrigation"--those were the exact words--in the valley, and it turned out that the main author was Joseph P. Lippincott, the response was a fresh series of blasts. . . . The Ku Klux Klan, sensing a perfect battle stage between "Hollywood"--which was to say, cities, big business, liberalism, and Jews--and the small-town, revanchist values it cherished, was sending recruiters into the valley and getting good results. Even Fred Eaton, after holding himself aloof, finally entered the fray against the city of which he had been mayor. "Wherever the hand of Los Angeles has touched Owens Valley," he wrote in a letter to the editor, "it has turned back into desert."

Joseph Lippincott, whose one admirable quality may have been prescience, had said twenty years earlier that the Owens Valley was doomed as soon as Los Angeles obtained its first water right. Mulholland, however, kept insisting blindly that the valley could live on--he didn't say how--even as he turned life there into a kind of hell.

*   *   *

On November 16, 1924, as the drought continued to hold Los Angeles in a deadly grip, a caravan of automobiles rumbled slowly southward through the town of Independence. . . . The cars turned toward the Alabama Hills, a small range of barren rises at the foot of the Sierra escarpment. Weaving through the hills was the Owens River aqueduct, and somewhere along its course were the Alabama Gates. In wetter times, the gates had turned floodwaters in the aqueduct onto the desert to keep them from straining the capacity of the siphons below. They hadn't been used in years, but they still worked. When the caravan arrived at the gatehouse, a hundred men got out of the cars, walked up to the spillway, and turned the five huge wheels that moved the weirs. For the first time in many years, the Owens River flowed back across the desert into Owens Lake.

The effect of the seizure was electrifying. Mulholland was in a murderous rage. He dispatched two carloads of armed city detectives to take back the gates, but news of their imminent arrival prompted the local sheriff to go down to meet them. "If you go up there and start trouble," he told the detectives, "I don't believe you will live to tell the tale." They never went. Mulholland, in the meantime, secured a court injunction against the seizure, but when the papers were served to the men at the gates they threw them into the water.

*   *   *

Meanwhile, Mulholland's public relations department was flooding the state with a booklet "explaining" the Owens Valley crisis. "Never in its history has the Owens Valley prospered and increased in wealth as it has in the past twenty years," it said. And it was true, as long as you looked at only the first nineteen of those years; in the twelve subsequent months, the city had almost brought the valley to its knees. Shops and stores were closing for lack of business--thousands of people had already moved out--but Mulholland dismissed pleas for reparations out of hand. If business was down, he said, the shopkeepers could move, too.

The first order to shoot to kill came on May 28, 1927, a day after the No Name Siphon, a huge pipe across a Mojave hill, lay in shards, demolished by a tremendous blast of dynamite. As city crews hauled in 450 feet of new twelve-foot pipe, another blast destroyed sixty feet of the aqueduct near Big Pine Creek. On June 4, another 150 feet went sky high. In response, a special train loaded with city detectives armed with high-velocity Winchester carbines and machine guns rolled out of Union Station for the Owens Valley. Roadblocks were erected on the highways; all cars with male occupants were searched; floodlights beamed across the valley as if it were a giant penitentiary. Miraculously, though the Owens Valley water war had gone on for more than twenty years, though it had turned violent during the past three, there were still no corpses. . . . Then, a few months later, came the collapse of the Saint Francis Dam.

*   *   *

By refusing to pay Fred Eaton the $1 million he wanted for his reservoir site, Mulholland had left himself short of water storage capacity. It was a serious situation to begin with, and it was compounded by the drought, the dynamitings, and the phenomenal continuing influx of people. His power dams were also running day and night, spilling water into the ocean before it could be reused. The water he had obtained at such expense and grief was being wasted. As a result, he turned to the dam he had under construction in San Francisquito Canyon, and, ignoring the advice of his own engineers, decided to make it larger.

The reservoir behind the enlarged Saint Francis Dam reached its capacity of 11.4 billion gallons in early March of 1928, and immediately began to leak. Few dams fail to leak when they are new, but if they are sound they leak clear water. The water seeping around the abutment of the Saint Francis Dam was brown. It was a telltale sign that water was seeping through the canyon walls, softening the mica shale and conglomerate abutment.

It was also a sign that William Mulholland chose, if not exactly to ignore, then to disbelieve. After all, it was *his* dam. Would the greatest engineering department in the entire world build an unsafe dam? To reassure the public, Mulholland and his chief engineer rode out to the site on March 12 for an inspection. The last of the season's rains was falling, and muddy water was running from a nearby construction site. After a perfunctory look, Mulholland decided that the site was the source of the mud, and pronounced the dam safe. On the same night, at a few minutes before midnight, its abutment turned to Jell-O, and the reservoir awoke from its deceptive slumber and tore the dam apart.

375

There are few earthly phenomena more awesome than a flood, and there is no flood more awesome than several years' accumulation of rainfall released over the course of an hour or two. The initial surge of water was two hundred feet high, and could have toppled nearly anything in its path--thousand-ton blocks of concrete rode the crest like rafts. Seventy-five families were living in San Francisquito Canyon immediately below the dam. Only one of their members, who managed to claw his way up the canyon wall just before the first wave hit, survived. Ten miles below, the village of Castaic Junction stood where the narrow canyon opened into the broader and flatter Santa Clara Valley. When the surge engulfed the town, it was still seventy-eight feet high. Days later, bodies and bits of Castaic Junction showed up on the beaches near San Diego.

The flood exploded into the Santa Clara River, turned right, and swept through the valley toward the ocean. It tore across a construction camp where 170 men were sleeping, and carried off all but six. A few miles below, Southern California Edison was building a project and had erected a tent city for 140 men. At first, the night watchman thought it was an avalanche. As it dawned on him that the nearest snow was fifty miles away, the flood crest hit, forty feet high. The men who survived were those who didn't have time to unzip their canvas tents, which were tight enough to float downstream like rafts. Eighty-four others died.

When the flood went through Piru, Fillmore, and Santa Paula it was semisolid, a battering ram congealed by homes, wagons, telephone poles, cars, and mud. Wooden bridges and buildings were instantaneously smashed to bits. A woman and her three children clung to a floating mattress until it snagged in the upper branches of a tree. They survived. A rancher who heard the deluge coming loaded his family in his truck and began to dash to safety. As he stopped by his neighbors' house and ran to the door to warn them, the flood arrived and swept his family out to sea. A four-room house was dislodged and floated a mile downstream without a piece of furniture rearranged; when the dazed owners came to inspect it, they found their lamps still upright on their living-room tables. A brave driver trying to outrace the flood could not bring himself to pass the people waving desperately along the way; his car held fourteen corpses when it was hauled out of the mud. The flood went on, barely missing Saticoy and Montalvo, and, at five o'clock in the morning, went by Ventura and spent itself at sea.

Hundreds of people were dead, twelve hundred homes were demolished, and the topsoil from eight thousand acres of farmland was gone. William Mulholland, whose career lay amid the ruins, was still alive, but as he addressed the coroner's inquest he bent his head and murmured, "I envy the dead." "After a feeble effort to put the blame on "dynamiters," he took full responsibility for the disaster.

But the great city his aqueduct had created was, for the moment at least, willing to forgive him, "Chief Engineer Mulholland was a pitiable figure as he appeared before the Water and Power Commission yesterday," the Los Angeles *Times* reported on March 16. "His figure was bowed, his face lined with worry and suffering. . . . Every commissioner had the deepest sympathy for the man who has spent his life for the service of the people of Los Angeles . . . his Irish heart is kind, tender, and sympathetic."

Nine separate investigations eventually probed the collapse of the Saint Francis Dam. No one is even sure how many lives were lost, but a likely total is around 450; it would become one of the ten worst disasters in American history. The precise cause of the collapse was never officially determined, but when an investigator dropped a piece of the rock abutment into a glass of water, it dissolved in a few minutes. It was also learned that Mulholland had ordered the reservoir filled fast--a violation of a cardinal engineering rule-- because he didn't want Owens River water to go to waste.

The city took full responsibility for all losses and paid most of the claims without contest, which cost it close to $15 million. For much less than that, Mulholland not only could have bought the Long Valley site, but built the dam, too.

In the ensuing months, in hearing after hearing, Mulholland was dragged through an agonizing reappraisal of his career. It was learned that two other dams in whose design and construction he participated as a consultant eventually collapsed, and a third had to be abandoned when partially built. He was a bold engineer, an innovative engineer; he was also a reckless, arrogant, and inexcusably careless engineer. His fall from grace was slow, awful, and complete. By the time he wearily resigned, in November of 1928, at the age of seventy-three, his reputation was sullied beyond redemption. His wit and his combativeness vanished in retirement, and even in the company of his perfervidly loyal children he often lacked the energy to speak. He told them, "The zest for living is gone."

The city finally settled with Fred Eaton . . . for $650,000. A few weeks later, the two old and broken men moved to heal their twenty-year rift. Lost in despondency at home, Mulholland received a message that Eaton, who had since returned to Los Angeles, would like to see him. Without a word, he got his hat and strode out the door. Eaton had suffered a stroke; he needed a cane to walk, and he looked ancient. "Hello, Fred," said Mulholland as he approached Eaton's bedside. Then both of them broke down and wept.

The dam in Long Valley was ultimately built, and the reservoir that formed behind it, which was named Lake Crowley in honor of a priest who devoted the latter part of his life to healing the rift between city and valley, was, in its day, one of the largest in the country. By then, however, all hope of fruitful coexistence had died. On a map, the Owens Valley was still there, but it had ceased to exist as a place with its own aspirations, its own destiny. By the mid-1930's, Los Angeles was landlord of 95 percent of the farmland and 85 percent of the property in the towns. In the town of Independence, the Eastern California Museum, which tells the story of the battle largely from the valley's side, sits on land leased from the city.

Los Angeles leased some of the land back to farmers for a while, but the unpredictability of the water supply discouraged most of those who tried to carry on. There might be enough for twenty or thirty thousand acres in wetter years; then there might be enough for only three or four thousand. As the city grew, the river became utterly appropriated; when that happened, the Department of Water and Power sank wells and began depauperating the aquifer, as would happen--as is happening--in so many places in the West. The last of the ranchers quit in the 1950s and the economy shifted to tourism; most of those who remain now pump gas, rent rooms, or serve lunch to the skiers and tourists driving through on Highway 395. By the 1970s even that tenuous existence was threatened; the aquifer was so drawn-down that desert plants which can normally survive on the meagerest capillary action of groundwater began to die; and the valley went beyond desert and took on the appearance of the Bonneville Salt Flats. When the winds of convection blow, huge clouds of alkaline dust boil off the valley floor; people now live in the Owens Valley at some risk to their health. The city has refused every request that it limit its groundwater pumping, just as it has refused to stop diverting the creeks that feed Mono Lake to the north--another casualty of its unquenchable thirst. Some sporadic dynamitings began to occur again in the 1970s, and reporters arrived eager to cover the "second Owens Valley War," but the war was long since over--there was nothing left to win.

*   *   *

Between the arrival of William Mulholland and his death, Los Angeles grew from a town of fifteen thousand into the then most populous desert city on earth. Today it is the second-largest, barely surpassed by Cairo. Its obsessive search for more water, however, was never to end. While Lake Crowley was filling, the city was already completing its aqueduct to the Colorado River, whose construction almost precipitated a shooting war with Arizona, a rival as formidable as the Owens Valley was weak. And though the first Colorado River aqueduct was supposed to end its water famines forever--as was the Owens River aqueduct--the city was soon planning a second Colorado River Aqueduct and plotting to seize half of the Feather River, six hundred miles away, at the same time. No sooner had it managed to do all of that than the city fathers were secretly meeting with the Bureau of Reclamation, mapping diversions from rivers a thousand miles distant in Oregon and Washington. Like the Red Queen, Los Angeles runs faster and faster to stay in place.

*   *   *

The Owens River created Los Angeles, letting a great city grow where common sense dictated that one should never be, but one could just as well say that it ruined Los Angeles, too. The annexation of the San Fernando Valley, a direct result of the aqueduct, instantly made it the largest city in the world in terms of geographic size. From that moment, it was doomed to become a huge, sprawling, one-story conurbation, hopelessly dependent on the automobile. The Owens River made Los Angeles large enough and wealthy enough to go out and capture any river within six hundred miles, and that made it larger, wealthier, and a good deal more awful. It is the only megalopolis in North America which is mentioned in the same breath as Mexico City or Djakarta--a place whose insoluble excesses raise the specter of some majestic, stately kind of collapse. In **The Water Seekers**, Remi Nadeau, a city historian, says, "They brought in so much water for so many people that few cared any more whether Los Angeles grew at all . . . . Indeed, one might say that . . . . they have brought in too much water. For if California now has enough water to more than double in population, then much of California is doomed to be insufferable."

That, in any event, is the way it appears some days from atop Mulholland Drive.

The Holland Tunnel

From **Landmarks in American Civil Engineering**
Daniel L. Schodek

Like many other types of civil engineering projects the tunnel used for vehicular traffic is rarely appreciated, even by other civil engineers, for the remarkable engineering feat it is. Once in existence it is taken for granted, and seldom is any thought given to the problems encountered in the design and construction of the project. This brief description of some of the design and construction problems encountered in the development of one of America's most extensively used vehicular traffic tunnels calls to mind the extent of the true creative genius and technical expertise necessary for such a project to become a reality.

When completed in 1927, the Holland Tunnel, with its unprecedented length of over 8,500 feet, was a bold step forward in vehicular tunnel construction. Its 29-foot-diameter twin tubes had been shield-driven through extremely difficult river bottom conditions; its unique ventilating system for drawing off exhaust-laden air was a major innovation. Linking Manhattan and Jersey City, it was originally called the Hudson River Vehicular Tunnel but was later renamed the Holland Tunnel as a memorial to its builder, Clifford M. Holland.

The tunnel was initiated by two state-appointed bodies, the New Jersey Interstate Bridge and Tunnel Commission and the New York Bridge and Tunnel Commission, who began their deliberations in 1906 and were considering a vehicular tunnel by 1913. The original idea of building a bridge was replaced after investigations indicated that its cost would be prohibitive. The all-weather nature of a tunnel was attractive, in addition to its lower cost. Authority was granted in 1919 by the two states for the commissions to proceed with the construction of a vehicular tunnel between a point in the vicinity of Canal Street on the island of Manhattan and a point in Jersey City.

The able Clifford Holland, a young tunnel engineer for New York's Public Service Commission who was in charge of the construction of all subway tunnels under the East River, was chosen by the commissions to design and build the tunnel. Chief Engineer Holland took office in July 1919 and gathered a staff consisting largely of individuals who had worked on the East River subway tunnels. A number of traffic-engineering studies determined the dimensions and the best location for the tunnel. The planned separation of entrance and exit plazas was unique at the time. Eventually the planners settled on two tubes, each of which would accommodate two traffic lanes in a single direction. Each tube would have a diameter of 29.5 feet, the largest in the United States at the time. The north tube would be 8,558 feet long, portal to portal, and the south tube 8,371 feet.

Holland faced many design and construction difficulties with a subaqueous tunnel of this length. A principal design problem was the ventilation system. With an expected constant stream of motor vehicles passing through, the tunnel would soon become lethal if some way of removing the fumes were not devised. Holland studied this problem very carefully, beginning by attempting to ascertain what quantities of exhaust were to be dealt with, what were the components of the exhaust, and what would be their effects on tunnel occupants. He initiated a series of studies conducted by the U.S. Bureau of

Mines, Yale University, and the University of Illinois. The goal was to provide a tunnel as safe as any roadway.

*   *   *

The final design consisted of four vent buildings with a total of 84 fans (42 blowers and 42 exhaust fans). Fresh air was to be drawn in and blown by the blower fans into a duct running beneath the roadway. The air would enter the tunnel proper through a series of slots spaced 10 to 15 feet apart, located above the curb on each side. The steady stream of fresh air emanating from these louvers would mix with the exhaust fumes from tunnel traffic, and the mixture would be drawn up through grills in the tunnel ceiling by powerful exhaust fans in the ventilation buildings, where it would then be discharged into the open air. A complete air change would be accomplished in 90 minutes.

As for the method of constructing the tunnel itself, Holland considered trench, caisson, and shield techniques. The need to reduce interference with river traffic, as well as the silty composition of the river bottom, influenced his choice of the shield method rather than an alternative such as dredging a trench, floating tube segments over the trench, and sinking them in place (a process that would necessarily obstruct river traffic). By now the shield-driving technique was an accepted method of tunnel construction and known to be particularly suited to the subsurface conditions encountered in this case.
    . . . The shield used in Holland's time was a steel cylinder whose forward end acted as a cutting edge and whose rear end overlapped the tunnel lining of inserted cast-iron rings. Inside the shield, hydraulic jacks pushed against the tunnel lining to drive the shield forward. As the shield thrust forward, the encountered material was either pushed aside or admitted into the shield through special openings and then removed. In subaqueous tunnels, compressed air prevented the entry of water into the shield.

Construction began on October 12, 1920. The first shield was built on the New York side, and tunneling was begun on October 26, 1922. Work progressed from both shores. Each shield was some 30 feet in diameter and a little over 16 feet long. The upper half had a 2.5-foot projecting hood. The shields were divided into 13 compartments. The thirty 10-inch jacks used in each shield had a total thrust available to drive the shield forward of about 6,000 tons. The cast-iron rings that lined the tunnel consisted of 2.5-foot-wide, 6-foot-long segments, bolted together. These were put in place by a hydraulic erector. A special grout lining was introduced under high pressure into the void created by the difference between the diameter of the shield and that of the rings. Hemp rope and lead wire were used to make junctures watertight.

The work was difficult. Holland was ever-present. Sadly, he did not live to see the completion of the great tunnel. He died on October 19, 1924, at the age of forty-one, and was succeeded by his assistant, Milton H. Freeman. Less than two months later, on December 7, 1924, the south tube headings were joined. Freeman himself died on March 24, 1925. Ole Singstad, Holland's engineer of design, succeeded Freeman and carried the work to its conclusion. On November 13, 1927, the tunnel was opened to traffic. In 1928, the first full year of operation, it handled 8,744,600 vehicles and became an indispensable part of the transportation network in the area.

"It is a Marvel of Engineering Skill"

An Excerpt from **Friends in High Places**
Laton McCartney

Leon McCartney's critical study of the Bechtel Corporation, one of the world's largest engineering and construction companies, includes this interesting account of the events leading to the formation of the consortium of companies which constructed Hoover Dam and describes some of the highlights of the construction of the dam itself.

Hoover Dam was an engineering marvel. It would have been a marvel just on the basis of its size alone. However, a number of other aspects of the project make it interesting to civil engineers: the use of a construction consortium, the accelerated construction schedule and the attendant use of shifts to permit work around the clock; the technical innovations associated with material handling and large-scale concrete placement; the development of facilities for housing workers; and the adoption of the project by the nation as a symbol of American "know how" and "can do" in a time when much of American society was struggling to find anything which could serve as a basis for optimism about the future.

The weather was hot; the air moist and sticky; the jobsite, deep in the jungles of Cuba, remote as any could be. But if the conditions bothered Henry J. Kaiser, he did not show it. Seated on a campstool, arms waving, voice booming, round body fairly bursting with enthusiasm and effervescence, he was describing to his friend Dad Bechtel[*] what he called "the mightiest project of them all."

It was two years since the two men had seen each other. Kaiser had accepted in 1928 a $20 million contract to build 200 miles of roadway through the interior of Cuba. And yet during the months he had been working on the project, which involved the employment of 6,000 people and the erection of no fewer than 500 bridges, Kaiser had been obsessed by one extraordinary notion concerning the American West: the building of a dam across the raging Colorado River. "I lay awake nights thinking about it," he would say later. "I lay in my sweltering tent and dreamed it over and over."

Kaiser was not the first to have had such thoughts. Engineers had been talking for more than twenty years of damming the Colorado, harnessing its power to produce electricity and irrigate the West. But though studies had been made and plans drawn up, nothing had come of them. Nothing, that is, until 1930, when a former engineer from California named Herbert Hoover decided, as president, that it was worth a go. Under his aegis, funds for the

----

[*] Warren A. Bechtel, called "Dad" by his friends, was founder of what was then called the W. A. Bechtel Company. It eventually became the Bechtel Corporation. His friend, Henry J. Kaiser was, like Bechtel, a self-educated man. Kaiser founded companies that were, over the years, involved in construction, shipbuilding, aircraft manufacturing, automobile manufacturing, nuclear engineering, and aluminum production, among other things.

project had been allocated, and the U. S. Department of Reclamation had announced that it would soon be accepting construction bids. It was then that Kaiser had summoned Bechtel to Cuba.

As he heard Kaiser reveal his plans, though, Bechtel was less than optimistic. Unlike his sometime partner, whose dam-building experience was limited to a pair of small projects, Bechtel knew at first hand how demanding an undertaking like Boulder could be. For while Kaiser was building roads in Cuba, the W. A. Bechtel Company had, among its other endeavors, put up Bowman Dam, the second-largest rock-fill* dam in the world. The work, managed by Dad's oldest son, Warren junior, had proved enormously taxing, not the least because of the dam's location, high in the Sierra Nevada. Cut off from the rest of the world for almost half the year by deep mountain snows, the site had required the Bechtels to import a large herd of beef cattle simply to feed the crew. A complete hospital had also been necessary, along with a slaughterhouse and self-contained work camp. Hellish as building Bowman had been, Bechtel knew it was nothing compared with what would be required to erect Boulder. When at last Kaiser paused for breath, Bechtel allowed cautiously, "It sounds a little ambitious."

Kaiser merely smiled. "Dad," he said, "problems are only opportunities in work clothes."

As the days went on, and Kaiser kept talking, Bechtel found himself gradually being swayed. Building Boulder would be the capstone to his career. More important, it would expose his sons to an endeavor beside which everything else would pale. In his ebullient fashion, Kaiser had told him that people someday would view Boulder as they did the pyramids of Egypt or the Great Wall of China, and that the dedication plaque at its base would list the W. A. Bechtel Company as one of its builders. Kaiser didn't have to say anything more. Dad was in.

With the decision made, the two men hurried back to the United States-- Kaiser heading east to begin raising additional capital, Bechtel returning to California to sniff out what other bids were being made. He soon discovered that more than a few other builders were interested in Boulder as well, chief among them an old competitor from Boise, Idaho, named Harry J. Morrison.

A tall, sparely built man with a fondness for singing cowboy songs beside construction-site campfires, Morrison was the boss of the Boise-based Morrison-Knudsen Construction Company, and a most formidable rival. He had already come to California in quest of financial backing from Leland Cutler, a prominent San Francisco banker and an old friend and schoolmate of Herbert Hoover's. Cutler declined to put up any cash, and he told Morrison that whoever obtained the Boulder contract would be required to put up a $5 million surety bond.** It was a staggering sum, especially in the midst of an

----

* A rock-fill dam is a dam constructed of loosely-placed rock or stone.

** A surety bond is a performance guarantee provided by the contractor to the owner. It is intended to compensate the owner for costs incurred if the contractor fails to fulfill the contract.

economic depression, and Cutler had suggested that Morrison consider bringing in partners. As a possibility, he mentioned another builder who had expressed an interest in Boulder, Felix Kahn of San Francisco's MacDonald and Kahn.

Since that conversation, Morrison had been busily trying to round up backers. In addition to $500,000 of his own, he had one important lure: the participation of Frank T. Crowe, a whipcord-tough former Department of Reclamation superintendent, who was then counted as the premier dam-builder in the country. Twenty years before, as a young engineer fresh out of the University of Maine, Crowe had drawn up the original estimates for Boulder, and the dam had loomed large in his imagination ever since. "I was wild to build this dam," he told a reporter, years after Boulder's completion. "I had spent my life in the river bottoms and Boulder meant a wonderful climax--the biggest dam built by anyone anywhere."

Walker Young, Frank Crowe, Charles Shea, "Dad" Bechtel and R. F. Walter at Hoover Dam construction site in 1933.

With Crowe's help, Morrison made his first approach to the Wattis brothers of Utah Construction, a firm with which both he and Bechtel had worked in the past. W. H. Wattis was 76, and his hand shook so badly he could barely hold a pen; but both he and his ill and sour-tempered brother, E. O., were enthusiastic about Morrison's proposal. The Wattises' problem was lack of cash. Though they could match Morrison's $500,000, they could not handle the entire surety bond on their own, much less the $40 million they estimated that Boulder would eventually cost. Moreover, they were not happy at the prospect of taking in partners. "If we can't do it on our own," W. H. griped, "the hell with it." Morrison managed to mollify them, but only after promising them that any outsiders brought in would be "our kind"--if not other Mormons, then clean-living Christian folk.

Returning to San Francisco, Morrison went next to Charles A. Shea, a fiercely independent 47-year-old Irish-American with the build of a bantamweight boxer and a disposition to match. Though not precisely the sort of clean-cut associate the Wattises had had in mind, Shea had a reputation as the best tunnel-builder in the Bay area. Also, he was rich, as evidenced by the permanent suite he kept in the Palace Hotel. As Morrison made his pitch, Shea paced the room, hands thrust deep in his pockets, a cigar stuck in the corner of his mouth. When Morrison finished, Shea not only agreed to come in for $500,000, but to bring along his friends at Pacific Bridge for an equal amount.

The next candidate on Morrison's list was the builder Cutler had suggested: Felix Kahn. A rabbi's son and a University of Michigan graduate, Kahn was one of five brothers, all of whom were highly successful engineers. His partner was Alan MacDonald, a hot-tempered, outspoken Kentuckian with mechanical and electrical engineering degrees and a penchant for antagonizing bosses. Prior to teaming up with Kahn, he had been fired from fifteen consecutive jobs.

This odd couple were one of San Francisco's most successful building concerns and had put up a number of the city's largest structures, including the Mark Hopkins Hotel. When Morrison approached them, they quickly agreed to ante up $1 million--a sum more than sufficient to dispel any concern the Wattises might have had about working with a fiery Scot and a Jew. "That $1 million," Kahn wryly noted, "made me one of the family right away."

On the east coast, meanwhile, Kaiser had gotten Warren Brothers, the Boston construction company from which he'd subcontracted the Cuban work, to put up $500,000. He and Bechtel also were prepared to write checks for half a million dollars each, but, even so, they fell far short of the funding needed for the surety bond. Consequently, Bechtel suggested joining up with Morrison, the Wattises and their associates. When Kaiser agreed, Bechtel went to see the elder Wattis, who had recently learned he had hip cancer and was becoming increasingly concerned that he would die before Boulder was completed. Worried at the prospect of leaving his ailing brother to shoulder the entire financial load, and respectful of Bechtel's abilities--not to mention the $1.5 million he, Kaiser and Warren Brothers were offering--W. H. proved happy to have them. All that remained was sorting out the details of the partnerships and coming up with an acceptable bid.

On a mid-February morning in 1931, just two weeks before the deadline for submission of bids to the Department of Reclamation's Denver office, the contractors and their coterie of lawyers, accountants and engineers convened at the Engineers Club in San Francisco. Until the last moment, the participation of a number of the builders had been in doubt. Hospitalized now, W. H. had lately been threatening to pull out, while Morrison had had to scramble to borrow funds to meet his $500,000 pledge. Pacific Bridge, which Shea had brought in, had also found itself short, and was able to participate only after its president, Gorrill Swinert, sold 40 percent of the company's stock to the firm's attorney. Warren Brothers, once one of the most successful builders in the country, was in even tougher shape. Owing to the Depression, the company was teetering on bankruptcy, and had asked Bechtel and Kaiser to come up with its promised $500,000 ante. This meant that Bechtel and Kaiser, having between them contributed $1.5 million, had nearly 30 percent equity in the consortium, more than any other member of the group.

As Kaiser mounted the podium and called the meeting to order, the air was charged with a special electricity. Collectively, those present had done much to change the face of the American West; and until this moment, most had never met. Kaiser quickly got them down to business. The first order was comparing various estimates of Boulder's costs. Morrison and Crowe stood by their projection of $40 million, a vast sum in the Depression era. Remarkably, given the magnitude of the project and the host of variables involved, two of the other companies present came in with estimates within $700,000 of that figure.

Next the consortium turned to the task of naming itself. Someone suggested "Continental Construction" or "Western Construction," neither of which elicited much enthusiasm. Then Kahn had an inspiration. Why not, he suggested, "Six Companies"? Though there were actually eight companies in the room, his idea won enthusiastic approval. Six Companies, as the Californians present were well aware, was the name of the council the Chinese tongs in San Francisco used to arbitrate their differences. If Six Companies was good enough for the tongs, Kahn joked, it was good enough for this group.

Once the name was chosen, Kaiser proposed that the organization be run along military lines, under the control of one commander-in-chief--presumably himself. His notion was voted down almost as soon as it was voiced. The builders were an independent lot and not at all willing to take orders from any individual. Instead, it was decided that Six Companies would function under a board of directors. By the time the group adjourned to the Palace Hotel for a working lunch, an organizational chart had been drawn up. W. H. Wattis, who was fading fast and couldn't be present, was named president, Dad Bechtel first vice-president, and E. O. Wattis second vice-president. Charlie Shea was put in charge of field construction, working closely with Harry Morrison, while Felix Kahn was designated to handle finances. To no one's surprise, Frank Crowe was the unanimous choice for general superintendent.

One crucial piece of business--exactly how much to bid on Boulder--was put off until later. The construction business was a tight-knit, competitive, gossipy fraternity, ever alert for intelligence on upcoming projects. One offhand remark by anyone about Six Companies' bid could, conceivably, doom the entire project. Accordingly, it was not until forty-eight hours before the bid deadline that the builders met to settle on a final figure.

Gathering at the hospital bedside of W. H. Wattis, Bechtel, Kaiser and the other builders listened as Frank Crowe, who had built a scale model of the dam for the occasion, reviewed each phase of the project and its attendant costs. When he had finished, the partners totaled the numbers, then added 25 percent. If all went well, this would be their profit. As soon as the decision was reached, Crowe bolted from the room and rushed to catch the train east to Denver.

He arrived on March 3, checked into the Cosmopolitan Hotel and spent the rest of that day and evening double-checking and triple-checking his numbers. The next morning, he formally submitted Six Companies' bid. It totaled $48,890,000--$5 million below the next-lowest bid, $10 million below the highest and only $24,000 above the government's own estimates. Six Companies had won the right to build the world's largest dam. Within hours of the announcement of the awarding of the contract, its competitors were predicting it would go broke in the process.

On the face of it, there was much to suggest that they might be right. Emanating from the deep winter snows of the Rockies in Colorado and Wyoming, then cascading 1,700 miles to the Gulf of California, the Colorado is one of the world's mightiest rivers. Over the course of millions of years, it carved out the Grand Canyon. By the time it reaches Black Canyon, another of its creations, 270 miles downstream, it is roiling with red mud and moving at a rate of between 100,000 and 200,000 cubic feet a second. Dr. Elwood Mead, the Department of Reclamation's chief engineer, after whom the

Boulder-created Lake Mead would later be named, likened its power at this juncture to "the force of a railway train." It was at the Black Canyon site, hard by the Arizona-Nevada border, 30 miles southeast of Las Vegas, that Six Companies proposed to put up its dam.

The Black Canyon site had been chosen for two reasons. One was the comparative shallowness of the river's bedrock. The other, even more important consideration was the relative narrowness--1,000 feet at the top, closing to 370 feet at the river's bottom--of the Canyon's gorge. Here, Six Companies planned to plug the Colorado with a mammoth concrete ledge, one that would extend 140 feet below the river bed and more than 700 feet above it--about the height of the Empire State Building, and nearly twice the height of any dam ever built.

It was a colossal, and in the opinion of many, insurmountable, undertaking, not merely because of the vast amount of materials that would be required, or the harshness of the desertlike conditions, but because of the power of the river itself. Even in the later summer and fall, when its water was low and sluggish, the Colorado could be capricious and deadly. During the preliminary work on Boulder, drill barges were frequently wrecked by a sudden rush of water down the canyon, and on more than one occasion, it seemed the entire project would be swept away. "The Colorado is a wild river," Frank Crowe said. "One day, it rose 40 feet in 40 minutes. It became a wall of yellow mud that kept rising and rising until I thought it was going to wash all of us right out of the canyon."

Making working conditions even more harrowing were the extremes of weather. Sudden storms washed out roads and blew down workers' tents. Always fierce, the heat during the summer of 1931 averaged 12 degrees above normal, with temperatures at river level often hitting 120 to 130 degrees. With the heat refracting off the canyon walls, workers felt as if they were roasting in a huge oven. Such were the temperatures that gasoline tanks exploded by spontaneous combustion. The man who unthinkingly picked up a crowbar with his bare hands usually came away with a second-degree burn. During the first few months of building, heat prostration alone caused several deaths per week.

Yet thanks to the Depression, there was no shortage of workers eager to sign on. By the time Six Companies set up operations at Black Canyon in June 1931, upwards of 10,000 men from all over the country had converged on the Las Vegas area, lured by the newspaper stories about Boulder and by Six Companies' announcements of job openings. As a result, Crowe was quickly able to assemble a work force, picking and choosing from an enormous labor pool that had gathered almost overnight. Eventually, a thousand of the men would be housed in Boulder City, a company town with paved streets and shade trees that Six Companies was building seven miles from the damsite. Boulder City, however, would not be completed until 1932, and until then, the men lived in tents.

Getting a fast start on the project was critical for Six Companies, since in negotiating the Boulder contract, Crowe and his bosses--led now by Bechtel, who had been elected president after W. H. Wattis succumbed to cancer in September 1931--had built in performance incentives. The more quickly the work went, the fatter would be Six Companies' financial rewards in the form of paybacks and bonuses.

By July, when Crowe announced that Six Companies had begun "highballing" the project--going all out--the first of the building materials had started to arrive. Altogether, in building the dam Six Companies used 45 million pounds of reinforced steel, 8 million tons of sand, 840 miles of pipe and more concrete than had been needed for all fifty of the previous Department of Reclamation dams *combined*. On some days, Six Companies would take delivery of 60 railroad cars of cement and other building materials. Forty-two railroad cars were required simply to bring in parts for each of the 2-million-pound bulkhead gates used to open and close the tunnels Crowe's workers were digging through both sides of the gorge.

As materials kept arriving, conveyed by Six Companies' own truck fleet and twenty-nine-engine railroad, some of Crowe's crews began bringing in power lines from California to provide the electricity needed to drive much of the heavy equipment. Other workers, meanwhile, started laying hundreds of miles of railroad track and roads, while still others began spanning sections of the gorge with steel bridges and a network of cableways.

Once all these preparations were in place, workers began blasting and drilling thousands of feet of earth and silt from the canyon walls to find the bedrock that would ultimately anchor the dam. The so-called "high-scalers" hung like mountain climbers from ropes extending down the rim of the gorge and chopped away at loose rocks, the staccato clatter of their drills interrupted from time to time by the rumble of dynamite blasts. The army of workers below had to be on constant watch for falling rocks or, on occasion, the plummeting body of a comrade who had slipped or misstepped.

After the canyon walls had been cleared of silt, Crowe and his men began what was the most difficult phase of construction: the laying dry of the entire riverbed, which, in turn, required diverting the Colorado's course. Crowe's plan was to drill two mile-long tunnels, one on the Nevada side of the river, the other on the Arizona side. Each would have the approximate diameter of the Lincoln Tunnel and would start several thousand feet above the damsite, emerging nearly half a mile below it. The tunnels were the key to the project, and by midsummer, work on them had progressed to the point where Crowe could report to Bechtel and the other Six Companies partners that Boulder was well ahead of schedule.

The partners were pleased, and so was Crowe, who was rapidly living up to his reputation as the best dam-builder in the business. But though the work was going smoothly, there was trouble on the horizon: not from the Canyon or the River, but from the partners themselves.

They were, in their personalities, and ways of doing things, a highly disparate bunch, and though they were united in wanting to build Boulder, each had distinct notions of how it ought to be done. The result, during the first few months of construction, was a welter of conflicting orders to Crowe, who became sufficiently exasperated to begin seriously considering whether to quit. The problem was finally resolved when the partners agreed to leave the running of Boulder to an executive committee of four: Bechtel, Kaiser, Shea and Kahn. As Six Companies' president, Bechtel headed the committee, and ensconced in the fine Spanish-style mansion the partners built for themselves high above the damsite, he received regular progress reports from Crowe during a weekly pinochle game. It was Shea, however, who provided the day-to-day liaison with Crowe, straightening out difficulties before they reached the field superintendent. Kahn, meanwhile, added legal affairs and oversight of the workers' feeding and housing to his financial responsibilities. The fourth executive-committee member, Henry J. Kaiser, was designated chief lobbyist and dispatched to Washington. The posting was fine with Kaiser, who had suffered heat prostration during one tour of the damsite and was never to visit it again thereafter.

Soon after its formation, the four-man executive committee was confronted with a crisis that threatened to shut Boulder down. Under intense pressure from his profit-conscious bosses, Crowe had been driving his men mercilessly. "Some of the carpenters were working so fast, they'd put handfuls of nails in the cuffs of their pants, [so] they wouldn't have to keep going back to the keg for more," one Boulder veteran recalled. "One foreman was so tough, we used to say he had three crews: the one working with him today, the one he had coming on tomorrow, and the one he had just fired." By August, the daytime temperatures at Boulder hadn't dropped below 98 degrees for a solid month, and the workers were ready to walk off the job. Led by a group of "muckers"-- tunnel shovelers, whose salaries had been cut when they were displaced by mechanical shovels--the workers compiled a number of grievances. Among their complaints were the primitive sanitary conditions and the fact that they were being charged half of their $4-a-day salary for meals and the privilege of living in a Six Companies tent. They demanded that their pay be raised to match the $5.50 to $6 per day workers were making elsewhere in the Southwest, and that a number of safety improvements be made, including the provision of ice water on the canyon floor, where 13 men had already died of heat prostration. Unless Six Companies complied, the workers said they would strike.

Crowe was not intimidated. Rejecting the workers' demands, he reported to his superiors, "We are six months ahead of schedule . . . and we can afford to refuse concessions which would cost us $2,000 daily or $3 million in the seven years we are allowed to finish the work."

In the face of Crowe's intransigence, the ranks of the dissidents, who had originally numbered no more than a few hundred, began to swell, and by August, totaled 1,400, two-thirds of the Boulder work force. Still unmoved, Crowe announced on August 10 that Six Companies was firing the entire group. They were to be given three days' pay and were to pack up and leave the damsite immediately.

To ensure that they did, and to quell possible rioting, the government sent in troops from Fort Douglas, Utah. Meanwhile, state and federal officials were brought in to search the workers' cars for firearms and liquor. In what would be the first of a series of bitter strikes at Boulder, the laborers, many of them members of the American Federation of Labor or the Industrial Workers of the World, refused their severance checks, took up pickets and set up their own camp in the desert.

The unions rallied quickly to their cause. "We feel it's a crime against humanity to ask men to work in that hell-hole of heat at Boulder Dam for a mere pittance," the AFL wired the U. S. Secretary of Labor, William Doak. The Hoover administration . . . was unsympathetic, and aside from halfheartedly pressing Six Companies to better the sanitary conditions, it did nothing. Crowe, meanwhile, stood fast. Cut off from support, with no prospect of victory, the strike finally collapsed. Six Companies agreed to rehire the dissidents, though at only their original salaries, and with no guarantees whatever of improving their plight.

With the crew back on the job, work at Boulder resumed in earnest, helped along by a drought that sent river levels to record lows, thus making it easier to control the Colorado and prepare the gorge for the laying of the dam. To accelerate the pace of construction even more, Six Companies brought in the most modern equipment on the market. Giant aluminum-bodied trucks roared back and forth across the construction site, transporting as much as 50 tons of material in a single load, while jumbo drilling rigs, fitted with as many as thirty different drills, attacked the canyon walls like crazed mechanical monsters, eating through 15 and 20 feet of solid rock in a matter of minutes.

By the beginning of 1932, Crowe was so far ahead of schedule that Six Companies had recouped its initial $5 million surety bond, and pocketed an additional $1 million in contract incentives. And more was yet to come, from savings on building materials--whose cost had fallen through the floor, thanks to the Depression--as well as from peculiarities in the contract, which allowed, for instance, an $8 charge for every cubic yard of earth excavated, when, thanks again to the Depression, it was costing Six Companies a third less.

In April, though, the project hit an unexpected snag when, unaccountably, Herbert Hoover, himself an engineer and a product of the West, failed to ask Congress for sufficient funds to carry on the work. In a panic, Kaiser mounted a whirlwind lobbying effort to get a deficit appropriation passed by Congress. As Kaiser buttonholed lawmakers, pressing on them the importance Boulder would have in relieving unemployment, Bechtel told reporters in San Francisco that unless funds were soon forthcoming, work on Boulder would be delayed as much as a year, forcing Six Companies to lay off nearly half of its 3,400-man payroll.

Dad's dire prediction, coupled with Kaiser's smooth salesmanship, eventually secured Boulder its congressional appropriation. . . .

*   *   *

At Boulder, meanwhile, the breakneck pace was continuing unabated. On April 10, 1935, the final phase of construction got under way, with the pouring of the first concrete. Crowe presided over the operation like a symphony conductor, orchestrating the movement and synchronization of each piece of equipment and the dozens of work crews. Blended at Six Companies' own mixing plants after it was brought to Boulder by train, the concrete was loaded into an unending succession of huge dump buckets, each of which was immediately placed on a waiting Six Companies train and transported to the damsite. There the individual buckets were unloaded and hung from hooks on one of four cable systems. The systems, in turn, functioned like marvelously efficient ski lifts, bearing bucket after bucket to the appropriate vertical column of the dam where the [concrete] was finally poured. As the columns filled, the [concrete] was cooled in summer and heated in winter by steel tubing inserted into it and carrying, depending on the season, cold or hot water--this to prevent the contraction that would have occurred had the concrete been allowed to harden on its own.

Every few months, Six Companies poured half a million more [cubic] yards of concrete, until, by the summer of 1935, all 3.25 million yards had been poured and the wedge that would contain the Colorado was in place. "A remarkable record," one of Crowe's engineers wrote, in describing the dam pouring for the Smithsonian Institution. "Twelve hundred men with modern equipment had in 21 months built a structure whose volume is greater than the largest pyramid of Egypt, which, according to Herodotus, required 100,000 men 20 years to complete."

Hoover Dam

There remained a big mopping-up job, the construction of a power plant and the eventual building of a $100 million aqueduct; but Six Companies' work on Boulder was essentially complete. On September 30, 1935, President Roosevelt, accompanied by [Secretary of Interior Harold] Ickes and a retinue of federal officials, reporters and governors and congressmen from the Colorado River Basin states, arrived at Boulder City. At precisely 11:00 A.M. Pacific Time, the president dedicated the dam in a ceremony broadcast live over both major national radio networks. Even Harold Ickes could not but be impressed. "It is a marvel of engineering skill," he recorded in his diary. "We were all struck with the wonder and marvel of the thing."

The Tallest Building: The Empire State

Chapter 5 from **Wonders of the Modern World**
Joseph Gies

Although it is no longer the tallest building in the world, the Empire State Building is probably as well known, if not more well known, than the buildings which have surpassed it. From an engineering perspective it is significant not only for its size, but also for the fact that it was constructed at a pace which seems unbelievable today. Like many other "firsts", once it had been completed and occupied it became an attraction by simply being there, and not because of the wonder of how it came into being. Joseph Gies, writing at a time when the Empire State Building was still the world's tallest, reminds his readers of the many marvels inherent in creating such a structure. Given today's constraints, his account of the construction period seems like a fairy tale.

Few Americans are unable to name the world's tallest building. Yet in 1930, almost no one thought that the Empire State Building's soaring title was more than a momentary thing. The skyscraper race had lent an invigorating note to the booster spirit. The Woolworth Building, Farmer's Trust, Bank of Manhattan, Chrysler, Empire State--what would be next? The record had risen from sixty floors to seventy to eighty to a hundred-plus. As late as October of 1930, Al Smith was still keeping the ultimate height of the Empire State Building a secret, though the basic eighty-six floors were almost completed. The sixteen-story capping tower was in fact an alteration made at the last minute by John J. Raskob, a du Pont partner and Al Smith's backer in this enterprise as in the 1928 presidential election. "It needs a hat," was Raskob's comment when shown the scale model of the original eighty-six-floor structure. The hat would lift it decisively above the new, not quite completed Chrysler Building. So numerous were the dizzy towers of the new age that many New Yorkers expressed concern lest their island sink under the weight of steel and concrete, an apprehension engineers relieved by pointing out that the material excavated for a skyscraper's foundation weighed more than the structure itself.

How tall could a building be built? The newspapers of 1930 printed widely varying calculations, but the American Institute of Steel Construction gave an authoritative-sounding 2,000 feet, nearly two hundred stories.

No one doubted that such heights would soon be achieved. The next step was already publicized--an anonymous cigar magnate was planning a 1,600-foot-tall building, over one hundred and fifty floors, and there were plenty of projects on slightly less grandiose scales.

That was in 1930. Then came 1931; the tower was topped out far above Fifth Avenue and Thirty-Fourth Street, and the opening ceremonies held, with Al Smith, Raskob, Mayor Walker, and many top hats. Tenants began moving in. Then they stopped moving in. Vast office spaces looking down majestically on midtown Manhattan lay empty and silent. Nobody needed them. The Empire State Building was a bust. No more plans for bigger, taller buildings were published. The anonymous cigar magnate was not heard from. The skyscraper race had suddenly ended before the finish lines. It was thirty years and one world war later before anybody started talking again about a building higher than the Empire State.

The skyscraper rush started with steel, for a skyscraper is a steel skeleton curtained by brick and glass. The curtain is of no structural significance. It can be anything--stone, wood, aluminum, or cardboard. The moment steel became available, skyscrapers were possible, and as soon as tall buildings could be built, they went up--and up and up.

Originally buildings had been held to heights of three, four, or five stories by the limited endurance of their stairclimbing tenants. Then the elevator, first run by steam on a screw shaft and later greatly improved by the development of the electric motor and a suspension design, pushed building heights to nine, ten, even twelve stories. But as their solid-masonry walls rose, they grew thicker at the base, until the first-floor walls were taking up an appreciable amount of the available space and costing a small fortune.

A remarkable signpost of the future had been built as early as 1851 in London, where Joseph Paxton is supposed to have gotten the idea for a *cage* construction from seeing his young daughter stand on a lily pad. Utilizing the umbrellalike rib design of the lily pad, Paxton gave the London Exposition its distinguishing feature, the celebrated Crystal Palace. It was thirty years before the Crystal Palace principle was adapted to commercial buildings, but in 1883-1885, a structure on the northeast corner of LaSalle and Adams Streets in Chicago, the present site of the Field Building, introduced a revolution. This was the Home Insurance Building, and its ten stories, carried by a metal skeleton partly imbedded in its exterior walls, were the creation of William Le Baron Jenney, a California Forty-Niner who had helped build the railroad across Panama and had risen to the rank of major on Sherman's staff during the Civil War. Coming home early one night from his downtown-Chicago office, legend has it that Jenney startled his wife, who jumped to her feet, dropping the book she was reading on a bird cage. What the cage was doing on the floor is not clear, but Jenney, it is said, observed the success with which it withstood the shock of the book and decided then and there that a cage construction would support the floor loadings of a building. Though the story may be apocryphal, Jenney's conception was a bold one, and the Home Insurance Building is generally considered the original skyscraper, although it was never, even after the addition of an eleventh and twelfth floor, the tallest building in the world. By the time its extra floors were put on, the Tacoma Building, also in Chicago, had outstripped Jenney's creation in height and design. In the Home Insurance Building, the outside walls, 4 feet thick on the ground floor, had supported themselves, with the metal columns and girders carrying only the floor loads. In the Tacoma, on the northeast corner of LaSalle and Madison, built by Holabird and Roche in 1887-1888, the skeleton carried nearly all the loads, including those of the brick walls fronting the street. Back and court walls were still self-supporting.

The skeletons of these two Chicago trail blazers were composed of both iron and steel. Jenney used wrought iron for the beams of the lower floors and steel on the upper, possibly because steel first became available during the building's very construction. The columns were of cast iron. This was the first use of steel in a building, though not the first in a structure, James Eads' St. Louis Bridge having been completed a decade earlier.

The French architect-engineer Eugene Emanuel Viollet-le-Duc was the first to perceive the almost limitless possibilities of skeleton-frame construction in metal. Leroy S. Buffington, a Minneapolis architect, after reading Viollet-le-Duc's theories designed a twenty-eight story building in 1888, which he nicknamed a "cloudscraper." Buffington's cloudscraper was never built, but the next year another engineer familiar with Viollet-le-Duc's theories built a skeleton of iron that astounded the whole world. This was Alexandre Gustave Eiffel, one of Europe's leading bridge engineers, who was invited by the French government to create a symbol for the Paris Exposition of 1889, the centenary of the French Revolution.

The conception of the Eiffel Tower was breathtaking; it was an even 300 meters high, 984 feet. Nothing remotely like it had ever been made by man. The Great Pyramid rose only 481 feet. The spire of the Rouen Cathedral, for six centuries the highest thing in France, measured only 495 feet. The dome of St. Peter's in Rome was only 457 feet. All these soaring wonders had taken decades, even centuries to build, had employed thousands of laborers, squandered treasure houses of money, and used up mountains of building materials. Eiffel threw up his airy creation in a few months, with a small crew of a new kind of laborer, and for a trifling cost--$1,200,000. Most remarkable of all, its 984-foot height took a total of less than 7,000 tons of material. Here indeed was a building revolution.

The Eiffel Tower was strictly a demonstration. Now the question arose, how tall could actual buildings framed by metal cages go? The next year, 1890, saw the twenty-one-story Masonic Temple in Chicago capture the title of "tallest building in the world." The Masonic's skeleton was entirely steel, both beams and columns. New York hastened to get into the skyscraper business (exactly when this word came into use is not known) with several interesting buildings. Two of the most famous, long landmarks of Manhattan, are the Flatiron Building, erected in 1902, and the Times Building (recently remodeled as the Allied Chemical Building), in 1904. Both have odd, trapezoidal floor plans, dictated by the pie-shaped real-estate slices Broadway strews along its diagonal path as it crosses Manhattan avenues, Fifth at Twenty-Third, site of the Flatiron, and Seventh at Forty-Second, site of the Times. The resulting slenderness of the two buildings, plus the absence of scientific data on wind stresses, caused the New York engineers to take special precautions. Triangular "gusset plates" were inserted as braces, four to each joint of horizontal beam and vertical column.

The twenty-story, 286-foot-high Flatiron (whose nickname may be a puzzle to modern housewives who have never seen an old-fashioned flatiron) and the twenty-two-story, 375-foot-high Times building were soon surpassed, first by the thirty-two-story, 468-foot-high City Investing Company Building on Wall Street, then the forty-seven-story, 612-foot-high Singer Building, and climactically, in 1913, by the memorable Woolworth Building. There was something peculiarly appropriate about the Woolworth Building as a symbol of America. American business, mass-production minded and consumer oriented, cornucopia of inexpensive merchandise for every man, woman, child, and dog, was especially well represented, in the eyes of European intellectuals of that day, by the ten-cent store. And in 1913, a year that might be taken as the end of the nineteenth century, the headquarters of the ten-cent chain rose above the hubbub of lower Manhattan Island, sixty stories high, 746 feet up from the ground. This was just 28 years after Jenney's Chicago Home Insurance

Building and 24 years after Gustave Eiffel's Paris tower had lighted the way. The skyscraper was here for sure, not an idea or a skeleton, but a building full of people working at desks and typewriters, stacked up in offices, floor by floor, layered one above the other so high that people, horses, wagons, and automobiles below, everyone said, looked like ants.

Many problems had to be solved before the Woolworth Building could be built: plumbing, electricity, heating, ventilation, fireproofing, building maintenance, window washing, mail delivery, long-distance elevators, and many more. The facility with which steel girders could be riveted together into a building skeleton, and the amazing capacity of the structure to withstand stresses, put considerable pressure on all kinds of fabricators, engineers, designers, and production superintendents. There was one problem that nobody could immediately solve. If you filled a square mile of New York with sixty- and seventy-story buildings, where would all the people fit when they went down into the street during their lunch hours? Or at nine o'clock in the morning coming in, or six o'clock going home? New York streets were already crowded in 1913, especially in the downtown areas where the skyscraper boom was on. Since that time, they have become steadily more crowded and no real solution is yet in sight. A negative, but effective, palliative was a 1916 zoning ordinance specifying setback design for high-rise buildings. At the same time, the elevator core was becoming an inhibiting factor. For a thirty-story building, at least two banks of elevators are needed, one to service the lower floors and one the upper, and two banks of, say, six elevators each take up much valuable space. The express elevators in a real skyscraper steal space from all the floors they pass by, and the higher the building the more space they steal. Worst of all, the space loss is largest on the ground floors, which command the best rents.

The problems involved in high-rise buildings are such that one is prompted to ask, why build skyscrapers in the first place? The obvious answer--to economize on ground space in a crowded metropolis--is too simple. There was plenty of room in lower Manhattan at the time the Woolworth Building was built, and throughout the skyscraper boom of the 1920's there was still room in lower and midtown Manhattan. Even today, if all buildings on the island were of uniform height, they would only have to be about eight stories high to accommodate all the office and residential space presently used. The real reasons behind what might be called the First Skyscraper Boom were apparently three.

First, despite the telephone and such later innovations as teletypewriters and closed-circuit television, a good deal of business must be transacted face to face. For this reason, businesses tend to congregate by industry, the financial industry in Wall Street, the communications industry in midtown Manhattan, and so on. This effect of convenient concentration can best be achieved by very large buildings.

Secondly, the shape of most city blocks, notably in New York, does not lend itself to large-based buildings. A large building fronting on Fifth Avenue must necessarily be tall. The obviousness of this is a little deceptive. The biggest commercial building in the world is still the Merchandise Mart in Chicago, with 4 million feet of floor space. The Merchandise Mart is only eighteen floors high, plus a tower, but it has an enormous frontage of 724 feet along Kinzie Street and 577 feet along the Chicago River. The only larger office building is the Pentagon, built on open land in Arlington County, Virginia, whose five-floor pentagons, nested one inside the other, contain 6 1/2 million square feet. All these human-inhabited structures are dwarfed by one at Cape Kennedy designed for habitation by the Saturn V rocket and enclosing 125 million cubic feet in its four assembly bays. It is beyond all comparison the world's biggest building, but by no means the world's tallest; it is only 526 feet high.

Finally, transportation terminals promote concentration. Grand Central Station is hemmed in by a circle of giants--Chanin, Lincoln, Chrysler, Graybar, Pan Am--somewhat to its aesthetic disadvantage, but from the point of view of New York Central and New Haven railroad commuters, natural enough.

Yet when all is said and done, there is something rather mystifying about the skyscraper. If New York were to be rebuilt from scratch on a rationally planned basis, there might be none. The skyscraper has been called a product of a competitive and anarchic economic society, and it is interesting to note that the boom of the 1920's had strong overtones of status rivalry. The Bank of Manhattan Building, under construction in 1929, kept a very jealous eye on its Lexington Avenue rival, known, before purchase of plans and site by Walter P. Chrysler, as the "Reynolds Building." Originally designed to rise only forty-seven floors, the Bank of Manhattan Building was gradually revised upward to sixty-three, and ultimately to seventy-one, with several modifications coming after construction had actually begun. The giant at 40 Wall Street was topped out with a 50-foot flagpole to carry it to 925 feet, which still left it short of the Eiffel Tower. Chrysler had his architects design a distinctive, if slightly bizarre, lid for his champion, a "finial tower" of spiraling steel plates, which added another 185 feet to the building's basic 845 feet and enabled it to surpass the Eiffel. The plate sections of the tower were lifted one by one up along the outside of the building, over the top of the sixty-sixth floor, and set down inside, in the middle of the sixty-fifth floor. There they were put together and the whole assembly hoisted in one 27-ton piece and its supports placed under it. With sixty-six true floors and six "penthouse" floors, the Chrysler Building is seventy-two stories high.

Al Smith's original announcement had described the Empire State Building as "close to 1,000 feet high," and though there was some talk of making the Chrysler Building taller, it was soon conceded that the Empire State's basic eighty-six floors would put its summit out of reach. *Engineering News-Record* greeted the new building with the somewhat restrained comment that greater convenience for tenants and more floor space could be achieved by other designs but that the new building would at least test the utility of such slender-tower types. Less technical circles were far more enthusiastic. Despite the market crash and the "slump" as it was optimistically labeled, "Al Smith's new building" was the talk of New York in the winter of 1929.

The first step was one that had already become a commonplace of American urban life, though it still awed Europeans--the demolition of a perfectly good old building to make room for the new one. In this case, the old building was not only sound but distinguished; it was the Waldorf-Astoria, America's most famous hotel. Five months--from October 1929 to the end of February 1930--were required to take its 15,000 tons of iron and steel apart, reduce the masonry to rubble, and haul it all away in sixteen thousand truckloads. Well before the last truck had rolled away, the second step was under way. This was excavation. Contrary to what many New Yorkers think, Manhattan is by no means a solid piece of bedrock. Foundation problems on New York buildings have been numerous. Very often engineers have had to resort to the bridgebuilders' tool, the pneumatic caisson, to get through soft ground. To make it a little harder, the soft ground is frequently water bearing. The Empire State site proved very good. On the other hand, the hole in the ground had to be large. Excavation of 9,000 cubic yards of earth and 17,398 cubic yards of rock, carried out by power shovel and hand labor between January 22 and March 17, created a space 55 feet deep, enough for a basement and subbasement. On April 7, with a goodly turnout of sidewalk superintendents on hand, the first steel columns were erected.

By this time, steel erection had been thoroughly mastered. Yet every building presented the same problems in slightly varying form, and the bigger the building the bigger the problems. They may be summarized under four headings: 1. Steel supply (fabrication and delivery of structural members); 2. Plant layout (number and position of derricks and hoisting engines); 3. Steel-handling methods; 4. Erection procedure (setting, fitting up, riveting).

As is commonly done on supersized jobs, two contractors were brought in for the steel erection, McClintic-Marshall and American Bridge Company, a subsidiary of United States Steel. The construction was divided between them in a rather unusual way--in alternate layers, each consisting of from one to four "lifts," each lift being the steel members of two floors. On the first ten floors, where the building ran its full width from Thirty-Third to Thirty-Fourth Streets, the contractors' layers were single lifts of two floors each. As the building rose through successive setbacks, the layers were increased.

McClintic-Marshall and American Bridge set up a huge joint supply depot at Bayonne, New Jersey, across the bay. By April 7, the big column sections--H beams and I beams --were stacked in great tidy piles. Movement to the site was carried out with military precision, for there was no room at the busy Fifth Avenue intersection for any overflow of materials and no patience with delays either. Lighters picked up the steel at Bayonne just two days ahead of erection and lugged it across the bay and around the corner of the island into the East River, where it was unloaded at piers at Twenty-Third Street and at Nineteenth Street. The heaviest members were unloaded at the former, the lighter ones at the latter. Trucks ferried the beams uptown to Thirty-Third Street, then west to the Fifth Avenue site. Thirty-Third Street rather than Thirty-Fourth was used primarily because the sidewalk of the latter is exceptionally broad, and in the early stages of construction, when the derricks were down in the bottom of the excavation hole, they could not reach a truck on Thirty-Fourth Street. Other materials were unloaded inside the building itself to keep the dust out of the street. Rubbish was dumped down a steel-plate chute reaching every floor, a truck standing always at the bottom of the chute.

A continuous, almost split-second delivery schedule having been achieved, the next step was plant layout. There were nine derricks in the basement, four at

----

* The terms "H beam" and "I beam" are used to describe two different types of steel beams. The terms derive from the shape of the beams' cross.sections, similar to the letter "H" in one case and the letter "I" in the other.

the corners and five in the middle. The four corner derricks were of lighter capacity--rated at 20 tons, though capable of lifting a good bit more. The five middle derricks, rated at 30 tons, could actually lift 50. The biggest loads that had to be moved actually weighed 44 tons, but these, the half-sections of the bottom columns in the core of the building, did not have to be lifted very high. Sidewalk superintendents might take note of the fact that derricks and hoists sometimes do have accidents. Just a few months before work began on the Empire State Building a hoist operator on the Chrysler Building was knocked out by a falling brick and dropped a load of steel.

The Empire State derricks were powered by electric hoists initially positioned in the basement. As work progressed, their hoist lines had to reach higher and higher to the floors being constructed above them. For this purpose, holes were left in the concrete floor slabs, which were being placed as soon as steel erection had gone up to the next floor. This was done not only in the interests of speedy concrete erection, but to facilitate the steelwork. Even the unfinished top floors were given temporary planking to provide firm footing for the riveting and erection crews.

Within the building, an industrial-railway setup was used, with 4,000 feet of straight portable track, 360 of curved, thirty-one turntables and six portable switches. Twenty-four dump cars and twenty-four platform cars ran on the constantly shifted tracks. Material, brick or stone, for example, was dumped from truck to hopper, fed from hopper to dump car, hauled to a material hoist in the center of the building, lifted to the right floor, pushed off on the track, and hauled to the spot where it was needed. This automation represented an enormous speed advantage over the old-fashioned wheelbarrow of earlier skyscrapers.

When steel reached the sixth floor, where the first, and major, setback was designed, the hoist engines were lifted to the second floor. There they remained until the derricks were on the twenty-fifth floor, when they were trussed up and lifted on the outside of the building to that floor. When the steel reached the fifty-second floor, the three smaller, 80-horsepower hoists were moved up again to the fifty-second. But the two biggest, 100-horsepower hoists stayed down on the twenty-fifth because the setback at the fifty-second floor put the derricks out of reach of the unloading trucks on Thirty-Third Street. Consequently, a relay station had to be set up on the twenty-fifth floor and the steel sent up in two moves.

In putting the steel pieces together, three steps were involved. First, the columns and beams were maneuvered into proper position by the steel connectors, the topmost workers in a high-iron gang. Then they were "fitted up" by bolting, and finally, they were riveted solidly together by the riveting gangs. To minimize the number of bolts needed, and consequently the time taken, the riveting gangs were organized to follow closely on the erection crews, working just one floor below them. In the first weeks of construction, an average of two lifts--four floors--a week was maintained. At the end of each lift, the five derricks had to be hoisted to the next platform, two stories higher. This was a 3-hour job.

Until the twenty-sixth floor, three hundred ironworkers were kept busy. After the twenty-fifth-floor setback, the smaller working area could only accommodate two hundred and fifty. Each riveting gang consisted then as today of a riveter, a heater, a bucker-up, catcher, and a helper, or "punk." The heater, bucker-up, catcher, and riveter passed red-hot rivets, in a popular metaphor, like a baseball infield. The heater toasts the rivet glowing red on his little stove, then tongs it 20, 30 feet or farther to the catcher, who fields it neatly with his bucket (nowadays a metal mitt) and extends it to the bucker-up, who "bucks" it into the hole while the riveter, from the other side, flattens its nose with a rattling volley from his gun.

While the ironworkers lifted the big pieces from the street and riveted them together into a cage, the concrete, stone-facing, and metal-trim crews went to work a few floors below. These specialized crews laid what ultimately came to over 62,000 cubic yards of concrete reinforced by 3 million square feet of steel-wire mesh. They affixed 300 tons of stainless-steel window trim and 450 tons of aluminum *spandrels,* the facing between the top of one window and sill of the one above. The rest of the curtain, covering most of the vast steel skeleton, was blocked in with 200,000 cubic feet of Indiana limestone. Finally, 10 million bricks were laid, mostly in places where they don't show.

In applying the limestone curtain, the builders made use of experience on previous skyscrapers where breaking, or *spalling*, of stone had occurred. Sometimes temperature drops caused steel members to contract enough to put additional pressure on the stone, or steel members were compressed by unexpectedly heavy floor loadings, or temperature changes affected the facing but not the interior structural members. Occasionally, wind stresses produced cracking. Settling and even vibrations from trains and subways occasionally led to damage.

For the Empire State Building, a specially designed "cowing pressure-relieving joint" was used on each floor. A corrugated sheet-lead filler enclosed in a sheet-lead envelope, the joint is placed at corners where vertical piers and horizontal girders meet and acts like a spring, flexing with whatever overpressure or underpressure may be given by changes in loading, temperature expansion and contraction, and wind force. All the Empire State's relieving joints together can absorb a vertical compression of 6 inches.

The ironworkers formed the vanguard, working always on the open top floor in sun, rain, and wind. Closely behind came the battalions of stonemasons, bricklayers, concrete layers, and trim crews. Still farther down worked the rear echelon of this vertically advancing army, a vast number of specialists. The ten service elevators, two of which were salvaged from the Waldorf, were replaced by the seventy-four permanent elevators, whose 1,200 feet-per-minute speed in their 7 miles of shafts prompted newspaper columnist Bugs Baer to comment that riding them would make one realize how a post card felt in a mail chute. Heating crews installed sixty-seven hundred radiators. Telephone men put in 17 million feet of wire. Plumbers installed 51 miles of pipes. Ventilation crews put in blowers and ducts capable of pulling over 1 million cubic feet of fresh air per minute into the building. (Air conditioning installations came later.) Every one of these installations established size records, most of which remain unbroken.

At one point, thirty-five hundred men were employed. The peak payroll was $250,000, which works out to an average of about $70 a week, not a figure to make present-day New York construction workers long for the good old days. But by the standards of 1930, the pay was good and considerable pains were taken with the men's well-being. Water-supply and sanitary equipment were of course installed on a temporary basis, and cafeterias were set up as needed. This was relatively easy since each floor was completed as rapidly as possible after the steelwork. The first cafeteria was on the third floor; the next on the ninth, then on the twenty-fourth, then the forty-seventh, and finally the ironworkers were munching sandwiches and drinking hot coffee on the sixty-fourth floor, setting a record for the world's highest restaurant. Elevators also increased in number and height as the job progressed. (The seventy-four elevators are capable of handling fifteen thousand passengers per hour, despite which a team of Polish mountain climbers preferred to use the stairs, climbing the 1,860 steps in 21 minutes.)

The command post for the army of workers was in a long wooden shed built out over the Fifth Avenue sidewalk. There, John W. Bowser, veteran superintendent for Starret Brothers and Eken, the main contractors, directed four divisions: construction, accounting, administration, and project management. This last was the general staff of the task force, receiving the plans from the engineers and architects, and translating them into action in steel and concrete.

Bowser and his assistants, while proud to be building a building taller than anyone else's, were prouder still to be building faster than anyone else ever had. They were right. Although the Empire's height record is due at long last to be broken, its speed record remains untouchable. The last rivet was heated, tossed, caught, and rattled home just twenty-three weeks after the first truckload of steel was delivered at Thirty-Third Street. The masonry job was complete in 8 months. In one 10-day period in the fall, the building jumped fourteen floors--steel, stone, concrete, and everything else. The architects actually had trouble providing detailed designs to keep up with the men; steel was up thirty stories before final plans were ready for some of the lower floors. Some of the steel was rolled into beams, shipped to the Bayonne depot, barged to the East River, trucked to Thirty-Third Street, and hoisted in the miniature railroad cars to the working floor in a total of 18 hours, incredible though that may sound.

The steel was topped out 12 days ahead of a closely figured schedule. The exterior limestone was completely set 13 days ahead of schedule. Yet no overtime was necessary, and the men worked only a regular 5-day week. They put in a total of 7 million man-hours over a period of 1 year and 45 days. It was a fantastic accomplishment, though still another record held by the job is even more fantastic. This is the cost. Since costs normally are always on the rise, the common story of engineering achievements is that they cost from a little to a lot more than their original estimates. This situation can be taken for granted, but the Empire State, launched on the very morrow of the market crash and driven to its breakneck completion while the country sank into the depression, showed an astonishing reversal of form. Figured to cost around $50,000,000, it ended up at just $40,948,900, of which $24,718,900 was for the building work, the rest for acquisition, demolition, and so on.

New York was proud of the shining new queen of skyscrapers, though her reputation was tarnished for some time through no fault of her builders. The depression left her half-empty for several years, and the words "white elephant" were murmured, outside of Al Smith's hearing. However, with the arrival of the affluent society, the Empire State Building more than vindicated

Al Smith. Today, she has fifteen thousand tenants, many of whom can see the world's most magnificent view on sunny days, and an impenetrable gray wall on foggy ones.

The observation tower, it will be recalled, was an eleventh-hour decision of John J. Raskob. Without it, the building would have topped the Chrysler by only a few feet. Nobody in 1930 could think of a good reason for adding 200 feet to the building's height, but Raskob insisted, and the tower was added, theoretically as a mooring mast for dirigibles. These sausage-shaped monsters are so forgotten today that it comes as a bit of a shock to recall that they were perfectly serious competitors of the airplane in 1930. During this era, the *Graf Zeppelin* made several trips across the Atlantic as a regular commercial service. However, even if the lighter-than-air ship had proved to be the transportation of the future, it is extremely doubtful that the Empire State Building could have contributed to its success. The whipping winds 1,250 feet above Manhattan made delicate maneuvering of a dirigible impossible, but Raskob's "hat" was added--600 tons of steel erected and riveted into sixteen stories in 17 days of January 1931, in the teeth of treacherous north winds accompanied by freezing rain and fog. A four-story 47-foot-high base was capped by a 105-foot shaft and a circular top piece 53 feet high and 32 feet in diameter. The top two stories were given additionally heavy bracing because to the end the fiction was maintained that Count Zeppelin's giant namesake would tie up there.

An attempt was even made--once--to use the tower for its stated purpose. Not a dirigible, but a blimp--a smaller, balloonlike lighter-than-air craft--was flown up to the mooring mast, but a gust of wind put an end to the experiment and very nearly to the experimenters, who included the usual quota of celebrities. Valiantly striving to execute its balancing act, the blimp released water ballast, which did not enable it to achieve the desired stability, but which bewildered shoppers on Fifth Avenue, who were drenched by rain from a cloudless sky. That was the end of the attempt to make the observation tower a hitching post.

The beacon atop the Empire State's tower has served as a valuable guidepost for aircraft--though fears that a plane might crash into the building were justified one foggy day in 1945, when an Air Force B-25 bomber struck the seventy-eighth and seventy-ninth floors. Eleven persons in the building were killed, twenty-seven injured. Structural damage was slight.

A lucrative and important, though quite unforeseen, use for the tower was found, and surprisingly early. In 1931, only 6 months after the building's inauguration, an experimental station was opened on its top for a new electronic toy known as television. Owned by NBC, the station presented its first commercial show 10 years later, in 1941. When commercial television really became a fact after World War II, NBC agreed to share the Empire State tower with other broadcasters, and in 1949 the interested parties agreed to build a tower on top of the tower, a 222-foot-high needle that would accommodate five TV stations (since grown to eight) and six FM radio channels. A lightning rod was thoughtfully added--it's been struck thousands of times-- as well as red aircraft-warning lights. Work on the antenna was begun on July 27, 1950, and completed on May 1, 1951, not a particularly gaudy record in comparison with those of the original builders. The cost was $3 million, or nearly an eighth the expense of the original 102 stories.

The tower has another function, and a very valuable one. It is close to being New York's premier tourist attraction. Visitors can get a magnificent view of city and harbor from the lower, eighty-sixth-floor, Observatory (visibility 40 miles), and a positively awe-inspiring one from the circular, glass-enclosed summit Observatory on the top floor (visibility 80 miles). Celebrities have their pictures taken by news photographers; ordinary tourists, by each other. Many find it a thrill to record their voices this high up, and others consider it a particularly romantic place to kiss a girl. Many a motion picture has included a scene filmed in one of the two observatories.

Among the questions most frequently asked of the guides by visitors is one that is peculiarly difficult to answer: Does the Empire State Building sway in the wind? Engineering Professor J. Charles Rathbun of the City College of New York made a study in the 1930's and reported in the 1940 *Transactions* of the American Society of Civil Engineers that the building has two separate motions in the wind--it "deflects" and "vibrates." The deflection of the top of the building amounts to as much as 10 inches in a strong wind of 55 to 90 mph. At the same time, it is also vibrating, that is, moving back and forth at a rate of six to eight times per minute, by up to 2 inches per vibration.

The antenna brought the total height of the building up to 1,472 feet, emphatically reaffirming the world's record, 440 feet beyond the reach of the Chrysler Building, and almost equally beyond the tip of the Eiffel Tower, which itself has acquired a TV antenna. However, the Empire State Building is officially listed at merely the basic 1,250 feet. Why this modesty? Because if TV antennas are to be considered parts of structures, they why should they not be considered as structures by themselves? In Columbus, Georgia, there is a TV mast 1,749 feet tall, and in Fargo, North Dakota, a brand-new one breaks 2,000 feet. Even the Russians are reported to be building one in Moscow far overtopping the pinnacle of Fifth Avenue.

The Empire State Building

The Queen of Skyscrapers is in imminent danger of losing her true title, that of being the tallest inhabited building in the world. The wonder is that she has managed to keep it so long. Initially, she ruled through absence of competition as the depression drove real-estate promoters to cover. Postwar affluence brought a construction boom in New York that put the original skyscraper boom in the shade. Few of the new breed of skyscrapers were really tall, most of them stopping at a mere thirty or forty stories. One of them was monstrous (aesthetically as well as otherwise, according to some critics)--the Pan Am Building, which captured the title of "world's largest commercial office building," with its 2,700,000 square feet ranking only behind the Pentagon and the Merchandise Mart. But its squat fifty-nine stories rise only 830 feet above Park Avenue.

The interesting challenger for the title of world's tallest building is the Port of New York Authority's World Trade Center near the lower tip of Manhattan. A magnificent civil engineering and structural conception, it will consist of immense twin towers one hundred and ten stories high, rising 1,350 feet and topping the Empire State at last.

Designed by the distinguished architect Minoru Yamasaki of Detroit, the towers will have many unusual features, including an elevator transfer system via "sky lobbies" at the forty-first and seventy-fourth floors, where passengers will switch from expresses to locals as in the subway. However, the most remarkable feature by far lies in the basic structural design. It is the work of the Seattle firm of Worthington, Skilling, Helle and Jackson, and it carries out a truly amazing revolution, or perhaps counter-revolution. The Seattle firm has already built a modest forerunner, the thirteen-story IBM Building in Pittsburgh, on the new principle, which calls for bearing walls--a return to the ancient method of the earliest masonry buildings before Joseph Paxton and Major Jenney changed everything, but these bearing walls are steel. The World Trade center towers will have walls of closely spaced parallel vertical columns of stainless steel. Columns around the elevator core will help to carry the floor loadings, which otherwise will be unsupported and consequently completely free of obstructing columns. Vast amounts of space will be saved, and office layouts will be completely uncluttered by interior structural elements. Fifty thousand people will work in the freely organized spaces.

And the Empire State Building will at long last lose its championship--in height, that is. Would anybody want to make a bet that the World Trade Center will be completed in 1 year and 45 days? As for cost, early estimates, sure to be revised, upward, indicated $350 million for the new champion, nine times the cost of the Empire State Building. No question about it, Al Smith had something.

A Bridge That Speaks For Itself
Margaret Coel

American Heritage of Invention & Technology - Summer, 1987 Issue

Margaret Coel's article on the construction of the Golden Gate Bridge provides an interesting summary of some of the technical problems encountered in the construction of the bridge. As is so frequently the case in engineering work, a common problem--bridging a channel--requires an uncommon solution, a solution that involves solving problems that are unique to that particular site or structure. One of the aspects of civil engineering which appeals to many civil engineers is that there are very few problems that can be resolved with "off the shelf" solutions. Each situation presents its peculiarities, requiring the engineer to utilize experience and judgment. Each solution leans heavily on solutions to similar problems in other settings, but each solution incorporates an element of creativity that leaves the civil engineer with the feeling of accomplishment that only comes with solving a problem that's never been solved before.

On May 28, 1937, President Franklin D. Roosevelt pushed a telegraph key at the White House and sent a message around the world: The Golden Gate Bridge was now open. It spanned the mile-wide entrance to San Francisco Bay, hanging from skyscraper-tall towers, and it defied engineering experts who had once called it impossible.

The bridge crossed one of the most treacherous channels on earth. Named the Golden Gate by the "Pathfinder," John C. Fremont, in 1846, the 5-mile-long channel connects the Pacific Ocean to the 245-mile-long bay, and through it ocean currents rush into the bay while bay tides ebb and flow out, setting up crosscurrents powerful enough, according to one historian, "to swing a ship halfway around." To make matters worse, fog blankets the area for days on end, and the wind often gusts at forty to seventy miles per hour. Nonetheless, people called for a bridge almost from the day gold seekers arrived on the peninsula and founded San Francisco. To reach Marin County, across the channel, they had to take one of the ferries that began operating in the 1860's. In 1872 Charles Crocker, the railroad speculator who drove the Central Pacific Railroad across the Sierra Nevada, announced that for his next engineering feat he would bridge the Golden Gate. Nothing came of it. Considering nineteenth-century metallurgy and technology, a bridge really was impossible.

Almost half a century later James Wilkins, an engineer turned journalist, began a new campaign for a bridge. His newspaper articles caught the imagination of the San Francisco city engineer Michael Maurice O'Shaughnessy, the first practicing engineer to believe the bridge could be built. In 1917 O'Shaughnessy sought the opinion of several bridge engineers. One was Joseph Baermann Strauss, five feet tall, 120 pounds, and described by one historian as a "grim, Teutonic Chicagoan." Then forty-seven years old, Strauss was one of the leading bridge builders of the day. In the twenty-five years since he had earned an engineering degree from the University of Cincinnati, he had built four hundred bridges on six continents, including the Arlington Memorial across the Potomac, the Longview on the Columbia, and a bridge across the Neva in Leningrad. Among his inventions was a mechanized counterweight that revolutionized the building of drawbridges.

Joseph Strauss

Years later Strauss remembered his first meeting with O'Shaughnessy. "He turned to me and said, 'Why don't you design a bridge for the Golden Gate? Everybody says it can't be done.' If I had realized this was the ruddy old Irishman's stock challenge to every bridge builder he met, I might have laughed it off with the rest of them."

Instead, Strauss said it could be done, for twenty-five to thirty million dollars. In his view a bridge across the Golden Gate was no more impossible than any other bridge he had built. Quoting a fellow engineer, he said, "I'll build a bridge to Hell, if they'll give me enough money to do it." Armed with a recent U.S. Coast and Geodetic Survey channel study, Strauss started drafting a plan. By June 1921 he had completed a design for a cantilever-suspension bridge with a mass of girders eighty stories high--a design that one critic has compared to "two grotesque steel beetles crawling out from either shore." There would also be masonry toll plazas modeled after the Arc de Triomphe, glass elevators in the towers, and gilded wrought-iron gates.

Appalled at the idea of such an ugly bridge in such a beautiful setting, O'Shaughnessy kept the design under wraps for eighteen months. Even Strauss later called it "an awkward looking structure." When San Franciscans finally viewed sketches in the newspapers, they began questioning the wisdom of building any bridge at all.

Strauss, seized with the idea of a great bridge on a great harbor, persisted. He sought support in the counties north of San Francisco, promoting his bridge to civic groups and anyone else who would listen, even though he was, in the words of one listener, the "world's worst speaker." Said his stenographer, "He would talk in complex engineering terms, without bothering to explain them." Nevertheless, northern Californians got the idea: The bridge would increase land values and usher in prosperity. At their urging, the California legislature, in May 1923, created the Golden Gate Bridge and Highway District, with power to levy property taxes and build the bridge. By now, San Franciscans, sensing economic benefits, were also rallying behind the bridge.

A decade of hurdles lay ahead before construction could begin, however. Legal challenges were mounted by taxpayers, majority owner of the ferry service, which carried fifty million passengers and four million cars every year. Not until 1931 did the California Supreme Court uphold the bridge district's legality. One full year was spent obtaining approval from the War Department, which had final say over bridges on navigable waters. It also owned the military reserves over which Strauss intended to build the approaches--Fort Point in San Francisco and Fort Baker in Marin County.

Still more time was consumed by a "War of the Professors." After making a preliminary channel survey, one deep-sea diver reported that the floor was "soft as plum pudding." Even though this same diver also claimed to have spotted "mermaid caverns," the geologist Bailey Willis, a Stanford professor emeritus, declared the floor unstable. No bridge could stand on pudding stone, he said, especially in earthquake-prone San Francisco.

"Bunkum," replied the project's geologist, Andrew Lawson, a University of California professor emeritus. Calling it "disconcerting to be caught between professors," Strauss ordered hundred-foot-deep holes drilled every twenty-five feet over ten acres of the channel floor. Informed that rock samples proved that the floor was stable enough to support the bridge even in earthquakes, Willis advised, "Wait and see."

At each hurdle Strauss explained and defended the bridge. When San Francisco's Joint Council of Engineers stated it would cost $112 million, he explained why it would cost only about $30 million. "They clung to their figures," he said later. "I stuck to mine." When, during a hearing before the War Department, an attorney argued that the bridge would interfere with shipping, Strauss countered that the largest ships in the world could pass easily under it. It would have a maximum clearance of 236 feet and a minimum of 220, he explained.

"You mean," said the attorney, "the clearance will be sixteen feet greater at low tide than at high tide?"

"No," replied Strauss. "What I mean is that cables will lengthen on hot days and lower the bridge sixteen feet."

"Oh," said the attorney, "so you are building a rubber bridge."

Strauss even defended the bridge to O'Shaughnessy, who, discouraged by long delays and stiff opposition, withdrew his support. "How long will your bridge last?" the skeptical city engineer asked.

"Forever," said Strauss.

The bridge district finally passed a thirty-five-million-dollar bond issue in 1930, during the worst banking period in American history. there was no market for the bonds. In 1932 a desperate Strauss paid a call on A. P. Giannini, the Italian immigrant who had founded and was chairman of the Bank of America. "I recounted the long fight against overwhelming odds," remembered Strauss.

Said Giannini, "We'll take the bonds."

While clearing hurdles, Strauss also worked on a new design. Plans for the cantilever-suspension bridge had reflected the consensus around 1920 that suspension bridges should be rigid and must not exceed two thousand feet. Since then advances in metallurgy had produced stronger structural steel, making longer suspension bridges possible, and new engineering theories

405

allowed for greater flexibility. By 1929, the thirty-five-hundred-foot George Washington Bridge, the first superspan, was under construction. With all his experience Strauss had never built a suspension bridge. After formally appointing him as chief engineer for the Golden Gate Bridge, in 1929, the bridge district hired three suspension-bridge consultants: Charles E. Derleth, Jr., dean of engineering at the University of California, and Othmar H. Ammann and Leon Moisseiff, principal engineers on the George Washington Bridge.

Stung by criticisms of his earlier design, Strauss was determined to give San Francisco a beautiful bridge. With that goal in mind, he hired two men with a strong sense of design: Clifford E. Paine, who became his able assistant chief engineer, and Irving Morrow, a young architect whose ideas were considered so radical he had never been commissioned to design a major building. Later Strauss's critics claimed he had had little to do with the bridge's final appearance, a claim his assistant rebutted. Strauss made the final decisions, Paine said. "He was the guiding spirit and genius behind the Golden Gate."

In 1930 Strauss submitted new plans for the Golden Gate Bridge. It would be by far the longest single-span suspension bridge in the world, with a center span of 4,200 feet. (Not until the 4,260 foot Verrazano-Narrows Bridge was built in 1961 was that span exceeded.) In total length it would extend 1.5 miles, including the 1,125-foot side spans and the approaches.

Its two cables would be the largest and longest ever constructed; its two towers, the highest. It would be flexible enough not only to rise and fall sixteen feet, depending upon temperature and traffic, but to swing sideways almost twenty-eight feet under maximum wind loads. It would weigh a hundred million pounds and support more than four times that amount. Its steel could fill a freight train twenty miles long, its concrete build a twenty-five-foot square block two miles high, its lumber frame five thousand houses, and its wire circle the earth three times. Yet it would seem light and graceful, welding architecture and engineering in a seamless whole. The high towers, curving cables, and long roadway were all, in Strauss's words, "basic essentials of the engineering."

When large crowds turned out for the groundbreaking on February 26, 1933, construction had been under way for six weeks. Impatient to begin, Strauss had sent bulldozers to dig twelve-story-deep craters for the anchorages on both shores. Each crater was filled with 128 million pounds of concrete. Buried in the concrete were the 134-foot-long steel rods that would hold the cables. To ensure that the cables couldn't yank the anchorages into the sea, crater walls were cut like stairways, locking concrete to bedrock. In front of the anchorages went concrete pylons to support the approaches.

Construction then moved into the channel. Work began on the pier--or base--for the tower off the Marin shore. Surveys had shown the best site to be a bedrock ledge jutting into the channel twenty feet below water. Strauss decided to construct a cofferdam--a thick-walled box--on top of the ledge, pump the water out, and build "in the dry," a bridge-building technique developed by the ancient Romans. Heavy timber cribs, thirty-two feet high and stacked three deep, were towed to the site, filled with rocks, and sunk, enclosing three sides of an area almost as large as a football field. The fourth side was formed by the cliff. After driving steel posts around the timber, like a fence, and dumping still more rocks inside the posts to make the cofferdam watertight, workers pumped it dry, leaving a muddy floor covered with fish that they scooped up and took home for dinner.

Building the pier off the San Francisco shore, on the other hand, took twenty-two months. With no bedrock ledge available, surveys found the best site to be eleven hundred feet offshore, in water a hundred feet deep. Since no pier--or anything else of this magnitude, for that matter--had been constructed in the open sea, Strauss and his engineering team had to figure out how to do it. Later he called the laying of the south pier the "most difficult engineering feat men have ever tackled."

The builders decided to create a calm lake in the midst of the turbulent channel by building an oval-shaped fender--a steel and concrete barricade the size of a sports stadium, with walls almost twenty-eight feet thick. The walls would stand on rock ledging sixty-five feet below the water and rise fifteen feet above. Inside the fender, safe from strong currents, workmen could build the pier by sinking forms and pouring concrete underwater.

Rather than construct the fender from barges pitching about on the waves, the crew built a 1,125-foot-long-trestle to carry trucks and machinery across the water to the site. A construction feat in itself, the steel trestle stood on pilings sunk in the channel floor. It had just been completed in August 1933 when a ship lost its way in the fog and crashed into it. Strauss rebuilt. In October a Pacific storm rolled into the channel, smashing the new trestle and the first sections of the fender. Strauss rebuilt again. Then, in December, hurricane-force winds twisted the third trestle into a steel pretzel. Calling bridge building a "battle of man against the sea," Strauss ordered still another trestle built--this one anchored in bedrock. The fender itself, he decided, would also stand on bedrock, a hundred feet deep.

This meant that thirty-five feet of the channel floor had to be excavated to bedrock. Aboard the derrick barge Ajax, workmen lowered dynamite charges through hollow pipes, while deep-sea divers connected detonator leads to the charges in pitch-darkness under the water. They could work only in slack water, four times a day, twenty minutes at a time. Still, the water would come "awhipping in and knock you for a loop," as one diver recalled.

After connecting the leads, the divers clambered aboard the Ajax, which moved out of the way before the channel exploded. Then the barge returned, workers lowered more charges, and divers connected more leads. After each explosion the Ajax dropped steel baskets to scoop out loosened rock--more than 1.5 million cubic feet in all.

Once again Strauss began building the fender. Steel frames, 115 feet high, were dropped from the end of the trestle. Divers followed them down and bolted them together underwater. Concrete was then poured into the frames through a tremie, a long, funnel-topped pipe. As fender sections rose fifteen feet above the water, derricks and concrete-mixing trucks, new at the time, moved out onto the top to build the new sections.

Several sections of the fender were left open. Strauss intended to float a caisson--a wooden form four stories high--into the fender, sink it with concrete, and construct the pier on top. On a calm October day in 1934, with

407

everything going according to plan, barges towed the caisson into place. That evening a storm came up, crashing the caisson against the fender.

Strauss was called at home. He hurried to the site and made what he called "a quick decision." He ordered the caisson pulled from the fender. Instead of using it as a base for the pier, he would close the fender, turn it into a giant cofferdam, and build the pier "in the dry." (The caisson was subsequently towed to sea, dynamited, and sunk.)

As soon as the last sections of the fender were in place, rivers of concrete were poured underwater over the entire floor, creating a base six stories high. As the base rose, water was pumped out of the fender, leaving a dry, paved area-- "a giant bathtub," said a workman--in the midst of the channel. Next, in the center of the fender, workmen began constructing the seven-story pier. When it was completed, water flowed through pipes into the fender, which was left in place to protect the pier from ships and from scouring--wearing away by the tides.

With the piers in place San Franciscans watched the towers rise. Each tower consisted of two legs containing a maze of three-and-one-half-foot square hollow cells. The cells would give the towers enough flexibility to bend eighteen inches toward the channel and twenty-two inches toward shore. Made of steel, with walls less than an inch thick, they had been fabricated in sixty-five-ton sections and shipped from Pennsylvania's Bethlehem shops via the Panama Canal. Lower legs were set on top of the piers, held by the steel dowels that stuck out of the concrete. Once they were secure, derricks were bolted between them. The derricks then began hoisting up other sections.

When completed, each tower stood 746 feet above the channel and contained a total of 5,004 cells, with openings connected by 23 miles of ladders. Workmen climbed through the cells, riveting them together. "Finding your way out at quitting time was always a problem," said one, even though they used a twenty-six-page booklet of instruction. "Although I designed this weird labyrinth," Strauss confessed, "I doubt if I could find my way out."

Massive cross bracing, usually found on bridge towers, was used only below the roadway level. Above, four horizontal "portal" struts connected the two legs, giving the towers a light, graceful appearance -- like "majestic doorways," Strauss said. At each strut the number of cells decreased, making the towers seem even higher.

During construction of the Marin tower the bridge crew suffered its first injuries--an outbreak of lead poisoning. Lead-based paint had been applied to cell sections at the fabricating shop to retard corrosion. When riveters bolted cells together using hot steel rods, the paint melted, releasing toxic fumes. Whole shifts were rushed to the hospital. One riveter remembered, "Doctors were diagnosing it as appendicitis--60 men all coming down with appendicitis at the same time." After doctors had hit upon the correct diagnosis, workmen got orders to remove paint before driving the bolts. And the steel fabricators got orders to use nonleaded paint for the San Francisco tower.

The next step, spinning the bridge's two 7,125-ton cables, proved the most complicated. A century before, the bridge builder John A. Roebling had made modern suspension bridges possible by developing a method of spinning--or building--cables of compressed wire right at a building site, designing machinery that would lay cables wire by wire across long spans. Now Roebling's Sons, the firm he had founded, landed the contract for the Golden Gate Bridge--"the biggest cable spinning job in the world." Before work could begin, footbridges, or catwalks, had to be constructed across the channel three feet below the position of the cables.

On August 2, 1935, the Golden Gate itself was closed for the only time in history while a barge moved back and forth, stringing support wires for the footbridges. First, the wires were hoisted over the towers. Pulleys then hauled long trains of redwood planks, bolted together on steel runners, along the wires from the shores to the towers. Derricks lifted other redwood trains to the tower tops to be pulled across the center span, and rope railings were fastened in place. Working on the footbridges, high above the channel, the crew would make sure the cable wires fell correctly into place.

Now cable spinning could begin. Traveling wheels rolled along wire cables the footbridges, stringing 27,572 pencil-thin wires for each cable. On each trip across, the wheels carried 24 wires, which looped around strand shoes--steel anchors--on the opposite shores. When an average of 452 wires had been strung, a strand was complete, and its strand shoes were clamped to steel rods embedded in the anchorages. Sixty-one strands, five feet thick, were compacted by hydraulic sculpting machines into one cable 36 1/16 inches thick. Suspenders--long wire ropes--were then looped over the cables at fifty-foot intervals and clamped into place. From these wires would hang the roadway.

In the fall of 1936 workers began constructing the final section of the bridge. Derricks bolted to the towers at the roadway level started laying long girders ninety feet apart out over the water. Work on the deck progressed from both towers simultaneously, balancing weight on the cables.

Since most bridge fatalities occurred during deck construction, with riveters doing what newspapers termed "a dance of danger" on the narrow steel beams, Strauss had ordered a heavy hemp net strung shore to shore, the first emergency net in bridge building. Within two months nineteen men had fallen into the net. All became members of the "Halfway to Hell Club." The net was not the only safety measure Strauss had insisted on. Determined to beat the accepted odds of one death for every one million dollars in a bridge, he ordered workmen to wear hard hats for the first time on a major construction job and tinted goggles to guard against fog blindness. Steelworkers who horseplayed on the high beams were fired, and anybody who came to work with a hangover had to drink sauerkraut juice to "ward off lightheadedness."

The safety measures paid off. During four years of construction there was only one fatality among 684 workmen. Three months before the completion date, however, ten others lost their lives. The steel roadway frame was in place, and concrete had been poured into wooden forms supported by the floor beams. Standing on a traveling platform hung from the side, a crew was stripping the forms. Suddenly a metal grip snapped, and the ten-ton platform plunged into the net. Pulled from its moorings, the net dropped into the channel, tangling men and debris. Only two workmen managed to jump free and stay afloat until a fishing boat reached them. Another, Tom Casey, jumped upward as the platform gave way and grabbed hold of the girder. He dangled for seven minutes in midair before fellow workers could pull him to safety. "It seemed a hell of a lot longer than that to me," he told reporters. Tragically, safety inspectors had just condemned a similar platform and were walking across the bridge to inspect the platform when it fell.

The calamity marred the construction, but the work continued. At last, with railings in place, the bridge awaited only a coat of paint. Paint in a choice of three colors--proved best to protect the steel from corrosion in the salty, humid air. Black and gray were too somber, said Strauss's assistant Irving Morrow, who had argued earlier for a "rainbow-colored bridge." The bridge should be orange, he insisted--a bright streak against the green Marin hills and blue sky and water. After the first coat of paint had gone on, commented one reporter, no one could imagine the bridge in any other color.

At 6:00 a.m. on May 27, 1937, a chorus of foghorns signaled the start of "Pedestrian Day" on the Golden Gate Bridge. Two hundred thousand people strolled across the bridge, pushing baby carriages and carrying picnic lunches. They gazed out over the city and the ocean, awed by breathtaking new views, and reverently touched the railings, as if to convince themselves the bridge really existed.

The formal dedication came the following day. As President Roosevelt announced the opening to the world, Strauss and San Francisco's mayor Angelo Rossi rode in the first automobile across the bridge. A caravan of cars followed, carrying governors, mayors, politicians, and celebrities from eleven Western states, Canada, and Mexico. Through the channel sailed the U.S. Pacific Fleet, while 450 navy planes flew overhead. San Francisco erupted in a weeklong celebration, with fireworks, singing by Al Jolson, and dancing in the streets. The cheering had barely subsided six months later, when Joseph Strauss, his health broken, resigned as chief engineer of the Golden Gate Bridge. The following spring, one year after the opening, he died of a heart attack. He had built his masterpiece--a bridge both practical and beautiful-- and he had done so for a total cost of $33.7 million. A vital part of the Pacific Coast highway system, the bridge linked San Francisco to northern California and southern California to the Northwest. Traffic was to increase with every year. More than three million vehicles crossed that first year, and nearly forty-one million went across in the year ending June 30, 1986.

Named the "Most Beautiful Steel Bridge" by the American Institute of Steel Construction in 1937, the Golden Gate Bridge has continued to capture the public imagination. To millions of Americans it has come to symbolize the West; to nations rimming the Pacific it is the symbol of America itself.

Over the years the bridge has undergone two major changes: additional steel bracing was placed under the roadway after hurricane-force winds caused gyrations in the deck in 1951, and the concrete roadway was replaced between 1983 and 1985. But any proposals to change the bridge's appearance--by installing high "suicide barriers" or overhead arches carrying electronic traffic signals--have been met with howls of protests from all across the Bay Area.

On its golden anniversary the Golden Gate Bridge remains much the same as when Joseph Strauss built it, and the words he spoke on opening day still apply. "This bridge needs neither praise, eulogy nor encomium," he said. "It speaks for itself."

Who Wrecked Galloping Gertie?

An Excerpt from Chapter 22 - **Bridges and Men**
Joseph Gies

*Despite the fact that no lives were lost, the failure of the Tacoma Narrows
Bridge (Galloping Gertie) has become one of the most notorious bridge failures
in American civil engineering history. Part of this notoriety may be attributed
to the fact that, for the first time, the failure of a major structure as well as its
performance under stress in the few preceding hours were recorded in moving
pictures. However, Galloping Gertie has also become notorious because of the
effect its failure had on design assumptions and practices. In his account of the
investigation following the failure, Joseph Gies provides insight into the way in
which a failure becomes the basis for new understandings in design.*

"Drove across the bridge at about 8:30 as usual to observe the behavior as the
wind during the latter part of the night had been quite severe and was still
blowing moderately. The bridge appeared to be behaving in the customary
manner, the east side span being practically quiet, the main span oscillating in
a four-noded manner and the west span oscillating considerably from the
temporary holddown to the tower. All of these motions, however, were
considerably less than had occurred many times before so I came to the office
at about nine o'clock. . . .

"Yesterday in a conference it had been determined that we would proceed
immediately to streamline the southerly side of the main span. I was to
prepare detailed sketches . . . as quickly as possible. . . . Upon my return to
the office at nine o'clock, I immediately undertook the preparation of these
sketches. At about ten o'clock Mr. Walter Miles called from his office to come
and look at the bridge, that it was about to go. This was the first indication I
had that anything of an unusual or serious nature was occurring.

"I immediately drove with Mr. Miles to the dock, from which we could see the
bridge. The center span was swaying wildly, it being possible first to see the
entire bottom side as it swung into a semi-vertical position and then the entire
roadway.

"It was at once apparent that instead of the cables in the main span rising and
falling together, they were moving in opposite directions, thereby tilting the
deck from side to side. I could observe one car, stationary, some distance east
from the center. It appeared that the center of the span was remaining about
horizontal and the two halves were revolving about a longitudinal axis of the
bridge.

"I then returned to the office, took my own car and went to the bridge, where
all traffic had been stopped and several people were coming off the bridge
from the easterly side span. I walked to Tower No. 5 and out onto the main
span to about the quarter point. The east side span was practically quiet,
there being but a few inches of vertical motion. . . . Then I observed on the
main span that the concrete sidewalk around the stiffeners of the girders was
failing badly. The curbs at the construction joints were also failing. Adjacent
to the girders, it appeared that the concrete was entirely free and the girders
and the concrete were working back and forth continuously three or four

411

inches. The concrete roadway showed no signs of cracking. The main span was rolling wildly. . . . The wind had not moved it a noticeable amount sidewise. The deck, however, was tipping from the horizontal to an angle approaching forty-five degrees.

"Beyond the center of the span at the lamp posts the deck was tilting in an opposite direction. The entire main span appeared to be twisting about a neutral point at the center of the span in somewhat the manner of a corkscrew.

"At Tower No. 5 I met Professor Farquharson, who had his camera set up and was taking pictures. We remained there a few minutes and then decided to return to the east anchorage warning people who were approaching to get off the span. . . .

"The main span was still rolling badly and the east side span was still quiet. . . . At that time it appeared that should the wind die down, the span would perhaps come to rest and I resolved that we would immediately proceed to install a system of cables from the piers to the roadway level. . . . I returned to the administration building. . . . I called Mr. Frincke of the Bethlehem Steel Company and requested him to send us a superintendent and a pusher immediately. . . . I then called the Weather Bureau . . . and was informed that the barometer was rising and in all probability the wind would quiet later in the day. . . .

"I was then informed that a panel of laterals in the center of the span had dropped out and a section of concrete slab had fallen. I immediately went to the south side of the plaza. . . . The bridge was still rolling badly about the center as it had been doing previously. I returned to the toll plaza and from there observed the first section of steel fall out of the center. From then on successive sections towards each tower rapidly fell out. . . . I requested that all persons be cleared from the administration buildings. . . . Shortly thereafter...coinciding with the dropping of the sections of the center span, I observed the side span settle rapidly and was momentarily expecting the towers to come down. I did not observe the exact time that the center section fell out although I was later informed that it was 11:10."

--Clark H. Eldridge, Bridge Engineer
Washington Toll Bridge Authority

Galloping Gertie's tumble into Puget Sound was the most spectacular of all bridge collapses. It was also the most completely observed and minutely recorded. Several engineers were present at the dance of death, and one, Professor F. B. Farquharson, wind expert from the University of Washington, made a movie which became a newsreel classic. If firsthand testimony alone were needed, there would have been no difficulty in uncovering the cause of Gertie's demise.

Certainly, no time was lost in fixing the blame. All the engineers connected with the erection immediately stepped aside and allowed the full spotlight to focus on the designing engineer, Leon S. Moisseiff of New York. In one sense this was fair. Moisseiff's design for the Tacoma Narrows had been accepted by the state and federal authorities, including two boards of consulting engineers,

412

very largely on the basis of his reputation. How had Moisseiff acquired his reputation? By designing or helping to design practically every major longspan bridge of the century. From 1897 to 1914, he had been engineer of design in the Department of Bridges, New York City, for both the important Manhattan Suspension and the Hell Gate Arch. He was designing engineer for the Delaware River (suspension) Bridge at Philadelphia. As Consulting Engineer for the New York Port Authority, he participated in the design of the George Washington and The Triborough bridges. He was Consulting Engineer on the Ambassador at Detroit, the Golden Gate, the San Francisco Bay Bridge, and, most recently, the Bronx-Whitestone in New York. At sixty-eight he was indisputably the most eminent design engineer for suspension bridges in the world.

It was hardly any wonder that nobody questioned Moisseiff's design for the Tacoma Narrows. True, it represented a considerable departure from previous long-span suspensions. But Moisseiff was the recognized authority, and the departure involved in the Tacoma bridge was of a kind that does not readily stir doubts. It was not a move in a new direction, but simply a further advance in the direction in which suspension bridges had been moving steadily ever since the Williamsburg. Ammann, Steinman, Modjeski (chief engineer of the Philadelphia-Camden), Woodruff (design engineer of the Golden Gate), and all the rest had steadily lengthened and narrowed their bridges. They had a logical theoretical basis for their trend. The clumsy, heavily trussed Williamsburg had been built in accord with the primitive and erroneous "elastic" theory of the nineteenth century, now supplanted by the "deflection theory" propounded by Melan, the Austrian engineer who also had contributed significantly to the development of the reinforced-concrete arch. Simultaneously the railroad was replaced by the automobile as live loading--requiring far less stiffening against deflection.

The trusses got steadily smaller, which, of course, represented an economy, and also contributed to the gracefulness of the bridge's appearance. In the early thirties a still further economy, heightening still further the bridge's slender charm, was introduced, first on several small suspension bridges, then on larger ones. This was the plate girder--a solid but very shallow metal strip that took the place of the truss. Plate girders were used by Moisseiff on both the Bronx-Whitestone and the Tacoma Narrows. At first glance, this tendency may seem so obviously dangerous an economy that common sense might have avoided it, but this is by no means the case. In the first place, the purpose of the stiffening truss was not to protect the bridge against the wind, but to keep the roadway from deflecting vertically under live loading--to keep cars from sinking into a trough as they drove over it. The wind itself was not regarded as a hazard. If this in turn sounds rash in view of some of the disasters previously recorded--such as the Tay Bridge--it should be clearly realized that wind-tunnel tests, carried out by the same Professor Farquharson who took the famous film of the collapse, proved that winds even of gale force would not blow down the Tacoma Narrows Bridge. And Professor Farquharson's tests were, as far as they went, perfectly valid--a gale could not have blown down Galloping Gertie. The trouble with the experiments was that they were too limited. Professor Farquharson himself felt concern over their incompleteness. The wind is a complex force, and under certain circumstances its action can have very complex results. The wind tunnel, still a novel device in the 1930s, required considerable further sophistication.

At the same time, the dead weight of bridges varied, quite apart from the design. The George Washington, with its eight-lane deck, weighed a solid 56,000 tons. The Tacoma was dramatically lighter, owing to no quirk on the part of the designer, but simply to the fact that the project called for only a two-lane roadway plus sidewalks. Anticipated traffic was not heavy enough to justify a wider, heavier bridge.

Galloping Gertie, in short, was long, narrow, shallow, and light. All the same--as Moisseiff's experience, Melan's theory, and Professor Farquharson's wind-tunnel tests agreed--such a bridge would stand up. What then was the trouble?

The three-man board of engineers appointed by the Federal Works Agency, which had sponsored the project, consisted of Theodore von Karman of the California Institute of Technology, an expert on wind effects, Glenn B. Woodruff, design engineer of the Golden Gate and Transbay suspension bridges, and Othmar H. Ammann. Their report, a heavy volume dated March 28, 1941, four and a half months after Gertie's fall, attributed the failure to "excessive oscillations caused by wind action." These excessive oscillations, in the opinion of the distinguished engineers, were "made possible by the extraordinary degree of flexibility of the structure and of its relatively small capacity to absorb dynamic forces." They noted significantly: "At the higher wind velocities torsional [twisting] oscillations, when once induced, had the tendency to increase their amplitudes."

Their investigation showed that the initial failure was the slipping of the cable band on the north side of the bridge to which the center ties were connected, a slipping that apparently triggered the torsional oscillations, causing breaking stresses at several points. . . .

While the report was in preparation, two interesting articles appeared in the **Engineering News-Record** of December 5, 1940. One was a description of corrective measures taken to combat oscillations on the Bronx-Whitestone Bridge in New York. The measures consisted of the installation of stays running from the tops of the towers to the roadway. The other article was entitled "Two Recent Bridges Stabilized by Cable Stays," and by it hung an extraordinary experience.

Two years earlier, David Steinman had built the Thousand Island bridge complex, consisting of five separate bridges, including two suspension spans. Like all the bridge projects of the depression-bound 1930s, the job was constricted by a tight budget. One week before the dedication ceremony, at which President Franklin D. Roosevelt and Prime Minister Mackenzie King were to speak, Steinman received a telephone call in New York from his engineer on the spot. It was a third enactment of the drama in which James B. Eads and Theodore Cooper had been earlier protagonists: Steinman's superintendent reported that both the suspension spans at the Thousand Islands were acting oddly--vertical heaving motions with back-and-forth longitudinal motion in a quartering wind.

Steinman took the next train, went straight to the bridge and waited for a quartering wind. It came, and the span Steinman was watching began to

414

waver. Making an on-the-spot analysis, Steinman came up with two physical answers to the problem: first, a pair of inclined stays forming an inverted V at the middle of the span, rigidly connecting each plate girder to the cable above it; second, stays running radially from the ends of the roadway to selected points on the cables. Borrowing some steel hoisting rope, Steinman rushed his stays in time for the dedication, replacing them some months later with specially fabricated steel struts.

As a matter of fact, Steinman himself did not entirely understand what he had done, for the Thousand Islands spans never went into torsional oscillation. Steinman, who had translated Melan's epoch-making exposition of the deflection theory back in 1908 (Moisseiff had applied it to the Manhattan Bridge the following year), took immediate note in 1940 of the motions reported at Tacoma Narrows. Similar though less dramatic disturbances were also occurring on the Bronx-Whitestone, which was long and shallow like the Tacoma, though with a wider (originally four-lane, today a tight six) roadway, and was stiffened with a plate girder. Steinman wrote the engineers at Tacoma offering to help. Tacoma referred him to Moisseiff in New York. Steinman apparently telephoned Moisseiff, and received a polite no-thank-you to his offer.

According to Steinman's biographer, the engineers of the Tacoma Bridge "took over Steinman's idea of mid-stay spans," but copied it incorrectly. Because his photographic slides showed the stays made of wire rope, they used wire-rope stays for the Tacoma design. Steinman asserted:

> They did not know that the wire-rope mid-span stays on the Thousand Islands Bridge were only an emergency, a temporary installation that we replaced a few months later by the permanent stays made of rigid structural-steel angle members. Even so, the mid-span stays at Tacoma, copied from me, although inadequate, were the only thing that kept the bridge from going into destructive oscillations during the four months of its life. On the fatal morning, one of the rope-stays became slack and snapped, whereupon the span went into its dance of death.... If they had let me help them, I could have saved the bridge, as I have saved several other bridges. I could have made the Tacoma span safe for a very small expenditure. But my offer went begging.

Whatever validity there was in Steinman's claim, this style of talk was hardly calculated to endear him to fellow engineers. Steinman already had a reputation as a thorny individualist and a "self-advertiser"--this last as great a sin among engineers as it is among doctors. Without mentioning him by name the three-man board investigating Galloping Gertie dealt rather rudely with Steinman's opinion:

> These bridges [the Thousand Island and Deer Isle] are more nearly comparable in size to some of the early flexible suspension bridges. They give no clue to the possible behavior of a suspension bridge three and one-half times longer and six times heavier.

However, in the Summary of Conclusions the report stated:

415

It was not realized that the aerodynamic forces which have proven disastrous in the past to much lighter and shorter flexible suspension bridges would affect a structure of such magnitude....

The reference to disasters of the past was principally to the suspension bridge Charles Ellet built in 1849 at Wheeling, West Virginia, and which collapsed in 1854 under a moderate wind. Steinman has pointed out in his own books that the eyewitness reports of the fall of Ellet's bridge are strikingly reminiscent of those of the Tacoma Narrows. For that matter, Ellet's span was not so short: 1010 feet.

It should be emphasized that until Steinman's experience at the Thousand Islands, his bridge theory differed in no way from that practiced by Ammann, Moisseiff, and the rest. The Thousand Islands bridges, of 800 and 750 feet, had stiffening girders only 6-1/2 and 6 feet deep--a ratio of 1/125. Steinman always had had an exceptional interest in aerodynamic theory. During the First World War, he had taught the first course in aeronautics in an American university at City College of New York. After reading the Tacoma board's report, Steinman wrote a paper entitled, "Rigidity and Aerodynamic Stability of Suspension Bridges" and submitted it to the American Society of Civil Engineers. It was not immediately accepted, and a fair guess is that it was returned to Steinman for revision because it was thought to reflect unfairly on Moisseiff. If so, Steinman himself changed his views, for in his book, **Bridges and Their Builders**, he defends Moisseiff and places the blame for Galloping Gertie on the whole profession. His paper was accepted in 1943 and published in the Society's **Proceedings** for November. It touched off a tempest. The controversy ran on for several years, marked in 1945 by a revision of Steinman's original article with replies to criticisms.

Summing up the whole bizarre accident, Galloping Gertie tore itself to pieces because of two characteristics:

1. It was a long, narrow, shallow, and therefore very flexible structure standing in a wind-ridden valley;

2. Its stiffening support was a solid girder, which, combined with a solid floor, produced a cross section peculiarly vulnerable to aerodynamic effects.

Neither Steinman nor Ammann nor von Karman nor anyone else knew the full answer to the profound mystery of Galloping Gertie. In a quite real sense it remains unsolved today, for aerodynamics is a still-growing science. "We haven't any more got the wind all figured out than the space people have the sky all figured out," one engineer remarked to this writer.

Leon Moisseiff, who never attempted to pass the Tacoma failure off on the shoulders of colleagues or subordinates, . . . did his best to aid in the investigation . . . His disaster is universally attributed to the state of engineering and aerodynamics in 1940, and so esteemed was he by his colleagues in the American Society of Civil Engineers that a few years after his death a fund was established for the Moisseiff Award, to be given annually for an important paper on structural design.

The Meat Ax

Chapter 36, **The Power Broker**
Robert Caro

Author Robert Caro's biography of Robert Moses, **The Power Broker,** *won the Pulitzer Prize for its fascinating portrayal of this man who was responsible for the evolution of a "can do" philosophy of public works development that permeated the attitudes of civil engineers in the middle half of this century. Other selections from the book have been used to illustrate various aspects of the character of the man who was recognized by the American Society of Civil Engineers as an Honorary Memeber in 1970 (despite the fact that he was not a civil engineer!), to convey the challenge and the excitement of working with a man who has a dream and an overwhelming desire to see that dream fulfilled, and to portray the types of conditions under which the great projects of another era were constructed. However, choosing such selections from the Caro biography without including at least one selection that conveys some idea of the technical accomplishments of this man known as "The Master Builder" would be a gross injustice. Furthermore, the reader is reminded that reducing the depiction of Robert Moses' technical accomplishments to a single selection from a book that describes his involvement in scores of projects can, at best, only hint at the true magnitude of his achievements.*

The following selection from the book--Chapter 36, The Meat Ax--vividly portrays both the incredible complexity and difficulty of major construction projects in modern American cities and the extent to which such projects are dependent on the solution of social, economic and political problems as well as resolution of the technical difficulties.

The tremendous extent of Robert Moses' power, and the extent to which, with that power, he shaped the greatest city in the New World and the great suburbs stretching out from it, is demonstrated by the roads he built during the quarter of a century following World War II.

These were roads like no other roads in history, for these were roads through a city.

Most of the great roads of antiquity--the 1,500-mile Royal Road of Persia, laid across three mountain ranges and lined with artificial oases at which relays of horses were kept shaded and fresh so that the royal couriers of Darius could cross Asia Minor in nine days instead of three months ("There is nothing in the world that could be faster than these couriers," Herodotus exclaimed); the three "silk roads," the longest roads ever built, laid out centuries before the birth of Christ so that caravans emerging from a gate in the Great Wall of China could carry bales of the material all the way to Europe, where it was valued so highly that it was weighed against gold; the post roads with which Genghis Khan tied together the vast Mongol empire; the twenty-nine military highways of Rome which, built by "the greatest men of the republic" ("None but those of the highest rank were even eligible to the office of superintending them") and radiating from Rome to which all roads led, ran with Roman directness (to avoid curves, mountains were cut through at enormous expense, marshes were bridged or simply filled up with solid masses of concrete) to the most remote provinces ("Even seas did not stop their progress, for the roads

417

were built up to the water's edge and then continued upon the opposite shore") to speed the marches of the legions and engines of war which kept Rome mistress of the known earth--were roads through open country. Their builders may have had to contend with mountains and marshes, with the snow of the Alps and the heat of deserts, but they did not have to evict from their homes tens of thousands of protesting voters, demolish those homes, tunnel under or cut across subways and elevated railroads, sewers and water mains and gas mains and telephone and electric conduits and cables, all of which, providing a city with essential services, had to be kept in operation during construction. They did not have to solve these problems in space almost unbearably constricted because to obtain a single extra foot of width would require additional thousands of evictions. A few major roads were built within ancient cities (some of the Roman highways ran right up to the golden milestone in the Forum, for example), but ancient cities did not have subways and gas mains. These were, moreover, cities on a different scale from modern cities -- imperial Rome was one-eighth the size of New York; Athens at the height of its glory was never larger than Yonkers -- so the problem of eviction was on a different scale. And since the traffic for which these roads were designed was different from modern traffic -- not only in volume but in size and speed -- they were constructed on a different scale. The major roads in Rome, the widest paved highways in any ancient city, were, even including their "service roads," the margins to which carriages were restricted to keep the central portion free for infantry and pedestrians, only sixty-five and a half feet wide at their widest point; the highways Moses was proposing to build were two hundred feet wide. A horse-drawn carriage can turn fairly sharply; the monster tractor-trailers of the twentieth century require a turning radius so great that a single interchange connecting one highway to another can cover eighty acres. Not only did these roads of antiquity have no underpasses or overpasses to carry intersecting roads across them--access to these roads was not controlled; they could be entered from any intersecting thoroughfare-- their very dimensions were so much smaller than those of modern highways that they were really comparable not to those highways at all but only to modern streets or avenues. Nor were the roads even of modern times--of the swollen cities of the nineteenth century and the Industrial Revolution. The greatest intracity road development of modern times before Robert Moses was the boulevarding of Paris envisioned by Emperor Napoleon III and carried out by his Prefect of the Seine, the "brawny Alsatian" Georges-Eugene Haussmann, between 1852 and 1870. But the roads of Haussmann, impressive though they were, were nonetheless still roads designed for the carriage rather than the car.

The automobile age created in the twentieth century a need for roads of a new dimension, roads a hundred feet or more across, roads with underpasses and overpasses and with interchanges so immense that to create them hundreds of acres of earth must be covered over with concrete--gigantic roads, not highways but superhighways. But the greatest of these roads--Mussolini's autostrade and Hitler's Autobahnen and the Long Island parkways (which predated autostrade and Autobahnen), Belt Parkway and West Side Highway of Robert Moses--had been built around the edges of cities and between cities. Except for rare instances and short stretches, no superhighways had been built within cities. And even those short stretches of superhighways that had been built within cities had almost invariably followed open paths within them--undeveloped river banks, for example or sparsely populated corridors-- as if their creators had shied away from pushing huge roads through the city's dense fabric. The most noticeable exception had been the Triborough Bridge approach highways Moses had built through Astoria and the East Bronx--but the total length of these highways had been no more than eight miles. Now Moses was proposing to build through the heart of the city more than a hundred miles. No one had dared lay superhighways through a heavily populated modern city on anything like such a scale: lump together all the superhighways in existence in all the cities on earth in 1945, and their mileage would not add up to as many miles as Robert Moses was planning in 1945 to build in one city.

The immensity of the physical difficulties in Moses' path could be grasped only "on the ground," and on the ground they made even engineers accustomed to immense difficulties quail.

One of his proposed superhighways, for example, was the "Cross-Bronx Expressway," a seven-mile-long road that would run straight across that densely populated borough. The Cross-Bronx Expressway would be a huge trench gouged across a city. And it would have to be gouged across the city without disturbing the city's lifelines, the water and gas mains, electric cables and telephone wires, sewers and steam pipes, streets and subways, that supplied hundreds of thousands of residents of the Bronx with services too essential to be interrupted for the long months it would take to build each section of the expressway. General Thomas F. Farrell, builder of World War II's legendary Burma Road, did not fully comprehend what that meant until, now a Moses consultant, he was sent out to look over the proposed route.

Standing on a bluff in Highbridge Park in Manhattan looking across the Harlem River at the Bronx, Farrell saw staring back at him from the top of the bluffs across the river a wall, a wall sixty and eighty and a hundred feet high, a wall of apartment houses. And crossing the river, entering the Bronx, Farrell saw that the wall was seven miles deep. Athwart the route Moses had chosen for his road stood literally hundreds of buildings, close to half of them apartment houses.

And an engineer like Farrell, accustomed to grasping at a glance the essentials of even the largest engineering problems, could see on his first tour of the route that the apartment houses were the least of those problems.

Stepping out of his limousine at a high spot on Jesup Avenue to look out over a half-mile valley to the east, the general saw that apartment houses crammed that valley solid -- a staggering panorama of massed brick and mortar and iron and steel. Looking down at the map Moses had given him, he saw that Moses was preparing to gouge the huge trench of the expressway straight across the valley's heart. But what staggered Farrell most was not what was in the valley but what was on the other side of it, glaring down at him from the high ridge on its far side, a ridge even higher than the one on which he was standing.

On top of that ridge was not only a wall of apartment houses, big, luxurious buildings of a notable sturdiness, but, running along the top of the ridge, a steady stream of automobiles, toy-sized in the distance, for on top of that ridge was the Grand Boulevard and Concourse, the "Park Avenue of the Bronx." For the Concourse--built at the turn of the century in imitation of Haussmann's boulevards with separate, tree-shaded lanes for pedestrians,

bicyclists and horse-drawn carriages--was now a major automobile
thoroughfare. Construction of an expressway would take years, Farrell knew;
the stream atop the ridge could not possibly be dammed for that long: the
Cross-Bronx Expressway could not cross the Grand Concourse at grade. A
glance told the general that carrying the expressway over the Concourse on a
gigantic viaduct was unfeasible; the ascent up from the valley floor would be
almost three hundred feet, far too steep for the big trucks that would be using
the expressway. The expressway would have to avoid the Concourse by diving
beneath it, by diving down through the ridge, tunneling through with
dynamite while not disturbing the apartment houses and road above. And,
from a cross-section map he had been given, Farrell knew what was inside
that ridge--not merely a huge storm sewer and a maze of smaller utility mains,
but another utility somewhat more formidable. What was inside that ridge
was a railroad, the Concourse line of the Independent Subway. Tens of
thousands of persons rode that subway every day; it, too, would have to be
kept in operation. And its triple tracks lay sixty feet below the top of the
ridge; to get beneath them while going through the ridge, the expressway
would have to dive deep indeed. And "deep" in the Bronx, Farrell knew, as all
New York engineers knew, meant Fordham gneiss, a rock that combined
layers of unusual hardness, requiring intensive and prolonged blasting, with
frightening instability that caused slipping of the rocks on even the simplest
engineering jobs. The engineers building the expressway would have to blast
it through the ridge while holding up above it--holding absolutely steady even
while igniting dynamite blasts that would shake a mountain--not only a tangle
of sewers and mains but a boulevard, a subway and a row of apartment houses.
And they would have to hold boulevard, buildings and subway steady while
trying to find a footing for the necessary massive supports in unstable rock.

Because the expressway had to dive under the subway, it couldn't go over the
valley on a viaduct; that would make the dive beneath the subway too steep.
It couldn't cross the valley at grade; that would mean closing the north-south
streets in it that cut across the expressway's path, and among those streets in
the valley were no fewer than five major thoroughfares that couldn't be closed
for long. It would have to burrow across the valley, and that meant building
up those streets while struggling through another maze of mains. And atop
one of those streets, Farrell saw a distant skeleton of steel, the girders and
tracks of the Jerome Avenue elevated rapid transit line, that would have to be
kept running. While building the expressway under Jerome Avenue, Farrell
realized, Moses would somehow have to hold up, for months if not for years,
not only the broad, heavy avenue but the spindly elevated structure above it--
and hold it steady enough for the trains to run along it in safety.

The ridge and valley, in fact, were only a microcosm of the physical difficulties
in the way of the Cross-Bronx Expressway. The path of the great road lay
across 113 streets, avenues and boulevards; sewers and water and utility
mains numbering in the hundreds; one subway and three railroads; five
elevated rapid transit lines, and seven other expressways or parkways, some of
which were being built by Moses simultaneously. All had to be kept in
operation while the expressway ran below or above them. This would be a
difficult enough engineering task if the engineers had sufficient space in which
to work. But on the Cross-Bronx Expressway, there was, Farrell could see,
never going to be enough room. Blasting a tunnel and building a road while
holding up above it a major street that itself is holding up a transit line is
difficult enough. But holding it up when the girders which held up the transit

line turned out to be resting on the spots--seemingly the only spots--capable of holding the weight of the tunnel required the fastening to those girders of "needle beams" of immense strength, beams built with legs stretching out to either side that could be sunk into the next available firm rock to hold the weight. The rock was blasted and chiseled out from under the girders so carefully that the road's designer, Ernie Clark, would recall years later that "we took the stuff out with a teaspoon." In one 466-foot section, the expressway ran under four major avenues and an elevated rapid transit line. Working with girders some of which were a hundred feet long and weighed nineteen tons, the engineers were constantly hemmed in on either side by the foundations of apartment buildings that could not be condemned because the condemnation would add additional millions to the cost and that were in constant danger of being damaged--some of them were damaged--by blasting. Blasting a tunnel under a rapid transit line is difficult enough. Building a viaduct over the street and under the rapid transit line is less difficult--if there is thirty feet, the required clearance for streets and expressway, between the top of the asphalt of the street and the bottom of the steel of the transit line. When there isn't, the room can be created only by lifting the rapid transit line into the air--so delicately that its operation is not disturbed--by jacking it into the air, three-tenths of an inch at a time, with immense hydraulic jacks and holding it solid, until new girders of the right height can be installed, with timbers so huge that one man who lived near the Third Avenue jacking operation said, "I never knew there were trees like that in the world before." Throughout the construction of the massive superhighway, Ernie Clark says, "we were always figuring in inches and tenths of inches." In the face of such difficulties, moving a river five hundred feet, a job required where the expressway crossed the Bronx River, was a feat so insignificant that in the speeches Clark made to the delegations of engineers who came from all over the United States and Europe to hear him describe the expressway's engineering, he hardly bothered to mention it.

If building the huge new highways was tough, tying them together was tougher--for the knots, the interchanges between them, required so much space that even what looked like immense amounts turned out to be insufficient.

So immense was the mass of swirling, intertwined lanes of links between great roads that had to be built between and up the sides of those rocky 170-foot-high cliffs along the Harlem River that the unassuming Clark once ventured to suggest to an engineering convention, in his quiet way, that a new word would have to be invented to describe it: "'interchange' does not adequately describe the construction in this area."

Two great north-south roads--the Major Deegan Expressway and the Harlem River Drive--were being built by Moses along the two banks of the Harlem. They would have to be linked up with the Alexander Hamilton Bridge, which Moses was building 170 feet above them to carry the expressway across the river valley--and both the bridge and the two river-bank highways would have to be linked as well to local streets on both sides of the river, as well as to the old Washington Bridge which crossed the river a few hundred feet to the north of the Hamilton. A total of twenty-two separate ramps and eighteen separate viaduct structures would be required to carry the thirty-one lanes of roadway necessary for the links. Making the rise in the links shallow enough so that huge tractor-trailers could negotiate them would have been simple if there had been sufficient room to work in: just start the climb far enough away so that the rise in grade could be gradual. But there wasn't nearly enough room. Two thousand feet south of the Washington Bridge was another nineteenth-century structure, the Highbridge Aqueduct, and the massive steel and stone piers of both these structures plunged down to the river banks, so the Hamilton Bridge--and its connecting links--would have to be built between those piers, and two thousand feet was a pittance in terms of the size required. On the Manhattan side of the river, moreover, was an existing roadway resembling an ancient Roman aqueduct, rising from the river to a tunnel cut beneath 178th Street as a connection under Manhattan to the George Washington Bridge. That roadway was supported on columns one hundred feet high. Knock out one of those columns and the roadway might collapse. The Cross-Bronx Expressway would have to be fitted between them, and the expressway's width was only five feet less than the space between the columns; there was practically no room for maneuver at all in the placing of those twenty-two ramps with their thirty-one lanes. The grades could not be kept shallow; to keep them from being impossibly steep, they would have to wind around and around each other; visualizing it in his mind's eye, Clark knew that the interchanges with which Moses would be filling the air on both sides of the deep valley would be the largest bowl of concrete "spaghetti" cooked up to that time by any highway builder in history. Some of the strands in the bowl would have to be almost incredibly thin and long. Because of the space limitations, normal-sized columns could not be used; the diameter of some, in fact, could be no more than 78 1/2 inches. And some of these slender columns, needed to support immense weights, would have to be 100 feet high! Radically new column designs would have to be evolved, Moses' engineers saw. The ingredients in the sauce, moreover, would have to be varied, indeed; as it turned out, no two strands of spaghetti curved exactly alike, so that each piece of steel for the dozens of columns involved, for the girders supporting the roadbeds which sat atop the columns, for the beams which formed the floor of those roadbeds, for the brackets which held those beams and girders in place, had to be fabricated individually.

The Cross-Bronx was one of thirteen expressways Robert Moses rammed across New York City. Its seven miles were seven out of 130. The physical problems presented by its construction were by no means unique. Even for the "easiest" of those monster roads, those traversing relatively "open" areas of the city, there were always private homes, small apartment houses--and whole factories--which had to be picked up and moved bodily to new locations. For most of these roads, Moses had to hack paths through jungles of tenements and apartment houses, to slash aqueducts in two and push sewers aside, to lift railroads into the air or shove them underground. For one expressway, the Van Wyck, he had to hold up in the air the busiest stretch of railroad in the world, the switching yard through which thirteen tracks and sidings of the Long Island Rail Road pass over Atlantic Avenue in Jamaica--hold it up and hold it steady enough so that during the seven months it took to slide the huge expressway underneath, the 1,100 train movements which took place daily in that yard could continue uninterrupted.

None of Moses' previous feats of urban construction--immense though they had been--compared with the roads he was planning now; as is demonstrated by the cost. Highways had always cost millions of dollars. In the whole world, only a handful had cost as much as $10,000,000. These new highways would cost $10,000,000 per mile. One mile, the most expensive mile of road ever built, cost $40,000,000. Their total cost would be computed not in tens but hundreds of millions of dollars. The total cost of the roads Robert Moses built within the borders of New York City after World War II was over two billion dollars.

The roads, of course, were not the largest elements in his transportation program. They were, in fact, in one sense only links between the water crossings he was planning to carry their users over or under the water that divided the city into boroughs.

The scale of these crossings made the mind boggle. No suspension bridge anywhere in the world would be as long (or expensive) as the Verrazano-Narrows Bridge; it would be the longest such bridge ever built, its towers so far apart that in designing them allowance had to be made for the curvature of the earth; their tops are one and five-eighths inches further apart than their bases. There would be enough wire in the Verrazano's cables to circle the earth five times around at the equator or to reach halfway to the moon, enough concrete in its anchorages to pave a single-lane highway reaching all the way from New York to Washington, and more steel in its towers--taller than seventy-story skyscrapers--and girders than was used in the construction of the Empire State Building. No underwater vehicular tunnel in the Western Hemisphere--and only one underwater vehicular tunnel anywhere in the world--would be as long as the Brooklyn-Battery Tunnel. The tile used to line it would have tiled 4,500 bathrooms; to ventilate it adequately against the fumes of 60,000 cars and trucks per day, air would have to be driven through huge ducts at the velocity of a Force Twelve hurricane, and the fans which drove that air would consume daily as much electricity as is used daily by a small city. Among such marvels even a huge suspension bridge like the $92,000,000 Throgs Neck--itself an engineering feat that would make most cities proud--would hardly be noticed by New York. Comparisons among public works of different types are difficult. In terms of size, however, Moses' road-building program was certainly comparable to any public works feat in history. In terms of physical difficulty, his program would dwarf them all.

Immense as were the physical obstacles in Moses' path, however, the Coordinator* was equal to them.

A technological system--engineering and construction techniques and equipment--capable of solving those physical problems was already in existence. The methods and machines required to build mammoth highways even within a congested city had never been used to the capacity Moses was planning to use them.

As for the tangle of red tape in his way--every main and cable and sewer relocation, for example, required approval by several city departments--that was sliced through with his customary directness. Moses' aides were under standing orders to go straight to the department head at the first sign of

----

* In 1945, newly elected New York City Mayor William O'Dwyer named Robert Moses "Coordinator of Construction", a position created especially for him, and gave him sweeping powers over construction of all major public works projects of all types.

423

resistance from any underling. Most city agencies closed up tight at five o'clock--or earlier. Working weekends was unheard of. But hours and weekends meant nothing to men who knew that when their boss wanted something done, "he wanted it done--period--he didn't care how it was done." Commissioners were routed out of bed at midnight--and long after midnight-- by their telephone calls. Watching a Broadway play, a commissioner would feel a tap on the shoulder, and, in the flickering darkness of the theater, would see the tall form of Arthur Hodgkiss or Bill Chapin beckoning him peremptorily to the rear of the theater. One refused to leave his seat; he found himself signing forms on his lap in the third row of a darkened theater. And if some commissioner balked at overruling an underling who had refused, say, to O.K. a Chapin-proposed sewer relocation, his secretary would soon be telling him: "Commissioner Moses is on the line--himself!" And if--as almost never happened--some commissioner remained recalcitrant, the next call his secretary would announce would probably be from the Mayor. Frustration might be piled on frustration; Moses faced them all down. After he had whipped into line behind the vast over-all expressway program--after years of effort that can only be guessed at--Mayor, Governor, Legislature, Board of Estimate, City Council, Federal Bureau of Public Roads, State DPW [Department of of Public Works] and an army of city bureaucrats, after all agreements were signed and the bidding for contracts under way, inflation of unforeseen dimensions raised the bids to levels beyond the state's ability to pay its share. Painstakingly, he worked out and obtained legislative and voter approval for a $500,000,000 bond issue which allowed him to get many of the expressways under way and even to finish a few. But costs continued to soar. He had underestimated the city's share so drastically that it could not even assume those minor costs that, by law, neither state nor federal government could assume. For years the expressways lay stalled--until the Federal Interstate Highway Act of 1956 allowed the feds to pick up 90 instead of 50 percent. Working through his banking allies, Moses persuaded Congress to include in the Act--despite the fact that it would circumvent its drafters' original intent of creating a tollfree system--clauses allowing roads linked to toll bridges to be included in the system, thus making his expressways eligible. Then, through a dozen ingenious subterfuges, he persuaded the state to use some of its own highway building funds, freed by the reduction in the share of the costs it was to assume, to pick up some of the city costs. There were other minor--but irritating--inconveniences: wars, for example. The Korean conflict was a source of real irritation. Steel was the precious metal to the highway builder, and the National Production Administration was obstinately insisting that available steel should go first to the war effort. Other cities accepted the situation without protest; Moses fired off telegrams to and pulled strings in Washington. Federal officials believed they would placate him by allocating his highways well over 10 percent of all steel available for civilian use, but they didn't know their man. Moses fired his next shot on the front page of The New York Times, charging that the officials were turning civilian defense efforts into a "monstrous joke" by sabotaging construction of arterials needed "to prepare for bombing evacuation, troop and supply movements." When federal officials tried to counter his charges with facts, Moses termed their statements "gobbledygook," the Times editorialized that roads are "essential in wartime . . . [the federal decision] mustn't be the last word"--and New York's allocation was quickly increased by another 10,000 tons. Next it was copper. Another attack, another victory. Then a strike kept the copper he had been allocated in the warehouses. But he intervened--and the warehouse doors opened.

424

To obtain his precious rights-of-way, Moses dealt with other giant city real estate holders--insurance companies, railroads, banks, the Catholic Church--as if the city were a giant Monopoly board, shuffling properties as casually as if they were playing cards, giving the Catholic Church, for example, space for an addition to a Fordham campus in the Bronx in exchange for an easement in Queens, handing Con Ed* half a square block for a new gas storage tank (complete with guarantees of Board of Estimate easements for the concomitant underground pipeline) in exchange for two hundred feet of right-of-way through a Con Ed open storage area. At Randall's Island luncheons he made himself the broker between a dozen disparate interests, reaping, always, the commission in right-of-way that he wanted. At one location near Fordham Road, for example, the path of the Major Deegan Expressway was blocked by both a housing development being built by the Equitable Life Assurance Society and a 217-foot-tall Con Ed gas storage tank. Negotiations were stalled --until a luncheon. By dessert, in a complicated land exchange, Equitable had been served up even more land for its development, Con Ed had agreed to "rearrange its distribution facilities" to "eliminate the necessity of the tank," and Moses was savoring the taste not only of the necessary right-of-way but of sufficient additional land adjoining it to create a park and playground for the residents of the Equitable development.

Robert Moses didn't merely solve these "physical" problems. He gloried in solving them. A reporter who was permitted to drive around with him on one highway inspection tour saw Moses "mentally readjusting houses as though they were so many toy building blocks." One of the blocks was a three-story factory--Moses turned it around and reset it on the same plot at a different angle. Another was a church--he turned it sideways. Another was an apartment house six stories high, which--with highway officials who had flown in from all over the country watching in awe, most of them expecting the structure to collapse--was inched a hundred yards out of the Van Wyck Expressway right-of-way with the possessions of thirty-five families still inside it. It cost at least as much--and possibly more--to move the building than it would have cost to demolish it, and in later years, Moses was quite frank about why he had decided to move it. "I moved it because everybody said you couldn't do it," he would tell the author. "I'll never do that again, broke a lot of gas mains . . . That was an absolutely crazy stunt, you know." But at the recollection, a broad, genuine grin spread across Moses' face, a grin of achievement and pride. He was overflowing with pride at his construction feats. The reporter painted a picture of a man happy as he played with his toy blocks. When the limousine reached Van Cortlandt Park, the reporter wrote, Moses began chuckling over reminiscences of the attempt by "the bird lovers" to stop him from running the Major Deegan Expressway through a swamp in the park that they had wanted preserved as a bird sanctuary. They had tried to obtain an injunction, he said, "but we just filled in a little faster." During construction of the Brooklyn-Queens Expressway, Moses rented the penthouse floor of the Marguerite Hotel--an old, sedate establishment right next to the expressway's route--and used it as an office. It had two advantages: only a very few people knew of its existence, so he was interrupted by few telephone calls, and he could look down on the construction as he worked. And he spent a lot of time looking down at it, watching the cranes and derricks and earth-

----

* Consolidated Edison, a New York electric utility

425

moving machines that looked like toys far below him moving about in the giant trench being cut through mile after mile of densely packed houses, a big black figure against the sunset in the late afternoon, like a giant gazing down on the giant road he was molding. "And I'll tell you," says one of the men who spent a lot of time at the old hotel with him, "I never saw RM look happier than he did when he was looking down out of that window."

It was not the physical problems that were the most difficult to solve, however, but the political.

A technology for solving the physical problems had been perfected, but not the methods and machinery for the creation of large-scale urban public works in a democratic society; the American system of government almost seemed designed to make such creation as difficult as possible.

It is no coincidence that, as Raymond Moley puts it, "from the pyramids of Egypt, the rebuilding of Rome after Nero's fire, to the creation of the great medieval cathedrals...all great public works have been somehow associated with autocratic power." It was no accident that most of the world's great roads--ancient and modern alike--had been associated with totalitarian regimes, that it took a great Khan to build the great roads of Asia, a Darius to build the Royal Road across Asia Minor, a Hitler and a Mussolini to build the Autobahnen and autostrade of Europe, that during the four hundred years in which Rome was a republic it built relatively few major roads, its broad highways beginning to march across the known earth only after the decrees calling for their construction began to be sent forth from the Capitol by a Caesar rather than a Senate. Whether or not it is true, as Moley claims, that "pure democracy has neither the imagination, nor the energy, nor the disciplined mentality to create major improvements," it is indisputably true that it is far easier for a totalitarian regime to take the probably unpopular decision to allocate a disproportionate share of its resources to such improvements, far easier for it to mobilize the men necessary to plan and build them; the great highways of antiquity awaited the formation of regimes capable of assigning to their construction great masses of men (Rome's were built in large part by the legions who were to tramp along them); at times, the great highways of the modern age seemed to be awaiting some force capable of assigning to their planning the hundreds of engineers, architects and technicians necessary to plan them. And most important, it is far easier for a totalitarian regime to ignore the wishes of its people, for its power does not derive from the people. Under such a regime it is not necessary for masses of people to be persuaded of an improvement's worth; the persuasion of a single mind is sufficient.

This last point has especial significance for the construction of public works in a city. For in a city such construction requires the eviction of people from their homes. Even when the public agrees in theory that a work is needed, no members of the public want to lose their homes for it. People never want their neighborhood disturbed by it. If it is to be built, they inevitably feel, let it be built somewhere else. A totalitarian regime can ignore such feelings, which is why the great city rebuildings of history--not only Haussmann's of Paris but St. Peterburg's by Peter the Great, and Rome's first by Nero and later by Augustus--have almost invariably been carried out by such regimes, the notable exceptions being cases (such as the great London fire of 1666 or the saturation bombings of the German cities in 1944) in which a monumental catastro-

426

phe destroyed so much of a city that it had no choice but to rebuild--and in which the catastrophe had removed from the scene the people who might have objected.

But Moses was not building under a totalitarian regime. Moses was building under a system in which permission to build could be granted only by officials who derived their power from the people. And, in that light, what was most significant about the Cross-Bronx Expressway was not that seven miles of brick and mortar and steel and iron had to be removed from its path but that seven miles of people had to be removed, removed from homes which in a time of terrible housing crisis in New York were simply irreplaceable. "People said that [the route] was so built up that you'd never get the politicians to say okay," Ernie Clark would later recall, and engineers who had built bigger roads even than Ernie Clark agreed. Farrell and Chapin's legendary Burma Road would symbolize to history the epitome of difficulty in construction. But Chapin understood political as well as engineering problems. Years later, he would recall the feeling that had swept over him when he had stood on Jesup Avenue staring down at that valley in the Bronx packed edge to edge with voters' homes. "I said to myself:"The [Burma] Road was tough. But that was nothing compared to this son of a bitch.'" People--the people whose homes stood in the path of all Moses' urban expressways--were the most difficult problem of all.

But Moses solved this problem, too.

Democracy had not solved the problem of building large-scale urban public works, so Moses solved it by ignoring democracy.

Critics who said the Coordinator simply ignored the people in his path oversimplified; he may have wanted to, but political considerations, the considerations that mattered to other public officials, made it impossible for him to do so--at least until after Mayor William O'Dwyer had been safely returned to office in 1949, Moses tried to take the people into account--tried hard. It was he who, to persuade apartment dwellers (hitherto uncompensated for eviction since they did not own the land involved) to move, persuaded the Legislature to provide for their reimbursement: $100 per room and $100 for moving expenses. Finding that they still balked--for middle- and lower-class families in New York, no few hundreds of dollars could compensate for the loss of a comfortable apartment they could afford--he even moved a few apartment houses smaller than the one on the Van Wyck. He moved entire blocks of private homes--263 homes on the Van Wyck Expressway alone-- where there was room to move them. But along most of his routes, there was not room. And, as even his admirer Jacob Lutsky puts it: "He thought about people. But if it came to a project or people, he'd take the project."

He had the power to do so--to ignore or override the procedures democratic government establishes to govern the planning of public works. Was it mostly dictators who had built great urban public works of the past? In road-building in and around New York, he had a dictator's powers. And he used them.

He enjoyed using them--for using them gave him what was his greatest pleasure: the imposition of his will on other people. One evening, he was sitting with Sid Shapiro and several other aides in his limousine parked on a side street in Queens, studying possible locations for the Clearview Expressway. Suddenly there appeared at the end of the street hundreds of citizens bearing torches and a scarecrow effigy labeled, in large letters,

ROBERT MOSES. The aides realized that they had happened upon an anti-expressway torchlight rally. The big black car sat at the end of the street unnoticed in the dusk by anyone in the crowd as the effigy was hoisted to a lamppost and set afire. "I didn't dare look at RM," Shapiro recalls. But to his surprise, his boss threw back his head and roared with laughter. And when someone suggested they drive away, RM said no. He wanted to stay for a while. He didn't want to miss a thing. He sat there all through the speeches comparing him to a "dictator," "a Hitler," "a Stalin." And, says Shapiro, "he laughed and laughed. RM really got a kick out of it."

When he replied to protests about the hardships caused by his road-building programs, he generally replied that succeeding generations would be grateful. It was the end that counted, not the means. "You can't make an omelet without breaking eggs." Once, in a speech, he said:

> You can draw any kind of picture you like on a clean slate
> and indulge your every whim in the wilderness in laying out
> a New Delhi, Canberra or Brasilia, but when you operate in an
> overbuilt metropolis, you have to hack your way with a meat ax.

The metaphor, like most Moses metaphors, was vivid. But it was incomplete. It expressed his philosophy, but it was not philosophy but feelings that dictated Moses' actions. He didn't just feel that he had to swing a meat ax. He <u>loved</u> to swing it.

428

Tapping a Floating Kingdom

An Excerpt from Chapter 14 of **The Earth Changers**
Neil C. Wilson and Frank J. Taylor

*One of the aspects of a civil engineering career that appeals to its most
adventuresome aspirants is the opportunity to work on a construction project in
a foreign country. American engineers and contractors are in great demand in
almost every nation on earth not only because of the need for American
equipment and expertise, but also because of the American engineer's reputation
as a "hands-on," "can-do" technical expert and manager. Deserts, jungles,
mountains, tundras, oceans and icefields have all, at some time or another,
provided the backdrop for scenes in the drama of American civil engineering.*

*This selection describes the adventures of one of the country's premier
construction and engineering companies, the Bechtel Corporation, in the
Middle East during the years after World War II. The observant reader will
note that the American engineer working in a foreign country must learn to cope
with new environments, new politics and new cultures, as well as new technical
challenges. In such encounters, the key to success is often the ability to adapt to
the realities of an unfamiliar work environment, and not the technical expertise
one possesses.*

Afternoon on the Arabian desert found a party of American engineers far from
a water hole. They had a tank on a trailer and they went into camp. Before
long they discovered that they were not alone. Not by several hundred people
and thousands of horses, camels, sheep, and goats. The livestock kept coming
over the low hills. A pavillion arose, evidently the evening quarters of some
important chieftan.

The Americans were aware that Arab tribesmen are jealous about who camps
in their bailiwicks. Politeness suggested paying a call before the sheik and his
armed men paid one of their own. George Colley* and a companion went over.
The companion knew some Arabic. With his best bow, George inquired if he
might offer the sheik a drink from his water tank. For this courteous gesture
George and pal were welcomed and soon found themselves cross-legged on the
rug, sipping coffee. By the size of the armed retinue they knew they had
encountered a powerful chief indeed. During the coffee-sipping and the
silences which went with it, George hazarded the information that he was a
friend of King Ibn Saud.

----

* George Colley Jr. had grown up with the Bechtel Corporation. His father
had been a friend of Warren Bechtel, the founder of the corporation. George
Jr. had been in charge of Bechtel work in the South Pacific immediately
before and during the early years of World War II. He (and his wife, along
with another Bechtel employee and his wife) had been captured by the
Japanese while trying to escape when the Japanese armed forces began
overrunning areas where Bechtel had work underway. They spent almost
four years in a Japanese prison camp.

The sheik's sunbaked countenance turned flintlike. After an interval for word-choosing, "I," pronounced the sheik witheringly, "am an Arab who has always been an Arab. Ibn Saud is an Arab who became an Arab."

Later George solved the riddle of this haughty remark. The man he had met in the desert came of a local lineage that traced back three or four thousand years. Ibn Saud's people, only a thousand or so years ago, had come to this part of Arabia from Damascus. In the sheik's book that made Ibn Saud a Johnny-come-lately. Time, George perceived, is measured by millennia in the East.

Bechtel people had been at work in this part of the globe since 1943. The first job had been a batch of refinery projects for Bahrain Petroleum Company-- Bapco--on the little oil-soaked island of Bahrain just off the coast. An underwater pipeline from the mainland followed. A 50,000 barrel-a-day refinery for Aramco (Arabian-American Oil Company) at Ras Tanura on the mainland was next. Postwar needs shoved this capacity up to 100,000 barrels in '46; 200,000 by '56. By the end of the 1940s, Bechtel had helped build a whole city at Dhahran, Aramco's headquarters. It was only a step in the parade of towns, refineries, ports, and pipelines Bechtel put up for this client. Between 1944 and 1948 they totaled around $200 million worth.

Then came Tapline.

In 1945 the owners of Aramco started planning a pipe that would take the Arabian peninsula at a bound and bring oil to the shore of the Mediterranean. The line would eliminate the twenty-day, 7,000-mile round trip by tankers down the Persian Gulf, through the Indian Ocean to the Red Sea, and through Suez Canal with its toll of 12 to 13 cents a barrel. Thus was organized Trans-Arabian Pipe Line Company--Tapline for short.

Burt E. Hull went in as Tapline's head man. Back in America, Hull had built the Big Inch and Little Inch [pipelines] during the war.

Steve Bechtel[*] and some other figures in American construction were invited to Arabia in 1957 to discuss this pipeline. They came back with the news that International Bechtel Inc., with associates, was going to lay the line from the Abqaiq field to beyond the Trans-Jordan border. A line almost as long as Big Inch from Texas to New York and of twice its capacity. In addition to the Tapline contract, there would be port, transportation, and other facilities built for the Saudi Arabian government. Said Steve when he got home, "In the Middle East program I foresee potentially the biggest development of natural resources ever undertaken by American interests."

When a suitable co-venturing group had been collected, old rivals and partners once more were found on an all-star team: Morrison-Knudsen Company, Inc., of Boise, Idaho; H. C. Price Co., of Bartlesville, Oklahoma; Bob Coynes of Oakland, California; Sverdrup & Parcel of St. Louis; J. H. Pomeroy & Co. of San Francisco, and the Bechtels as sponsor. They were lined up to haul, weld,

----

* Steve Bechtel, one of Warren Bechtel's three sons, had taken over the Bechtel Corporation upon his father's death in 1933.

and heave into place 850 miles of the 1,100-mile champion pipeline of its time. Williams Brothers Corp. of Tulsa was awarded the tough, short portion from Jordan down to Sidon. Burt Hull warned that the chief problems of the desert would be water, transport, and morale.

The portion taken over by Bechtel lay through an exceedingly empty region. A tree was as rare as a three-hump camel. Reconnaissance parties moved out into that windy void. From the Persian Gulf westward, the first hundred miles or so contained heavy sand dunes. The next 750 were wastes of gravel and rock, with occasional wadies where surface water stood or flowed after the rare showers. The route climbed to 2,975 feet near the Jordan border.

Clark Rankin and Ray Hamilton flew from San Francisco* to go over the route and pick out the site for a base camp. From then on, construction men moved back and forth between America and Arabia the way commuters ride the Long Island Railroad. Van W. Rosendahl headed up the two Bechtel international corporations. Clark Rankin was put in charge of project planning. Over came George Colley, then two years out of the Kuching prison camp on Borneo where the Japs had trimmed his stocky figure to 125 pounds. He was senior Bechtel construction officer for some of the international projects. Al Berlander was construction manager. Don Roberts, who later would be project manager for the Trans Mountain pipeline across Canada's Rockies, was International Bechtel's chief engineer until sand and bugs mowed him down. Bob Bowman was general superintendent, Harry Waste project engineer, and Q. Poggi in charge of transport.

Surveyors went out into the field that summer and found the temperatures, the winds, the sandstorms, and the August humidity no worse than Arizona or New Mexico in summer . . .

Back at Dhahran the amenities of civilization, American style, had developed in a wide and handsome way. Aramco's doings in the heart of the Middle East had converted the town into the number one American community between Europe and the Philippines. It was even air-conditioned.

So Dhahran became headquarters for the contractor's principal field officers. Here IBI [International Bechtel Inc.] executive vice-president John Rogers set up shop; he was from the original refinery construction job at Ras Tanura by way of Arizona, where he'd been getting out wartime copper. Rudy Grammater, alumnus of the Canol Pipeline** from the Mackenzie to the Yukon, was IBI administrative chief. Here general superintendent Lou Killian arrived in June, 1947, with nineteen American technicians. All were a picked bunch; it had cost $5,000 a head to select, process, and transport these "Camel Legionnaires." They were men who'd known and worked with each other from the Arctic Circle to South America, and now most of them started

----

* The Bechtel Corporation headquarters offices are in San Francisco.

** The Canol Pipeline was built by Bechtel during World War II. It was 1200 miles long, 1,000 miles from the nearest city, and crossed some of the most rugged and remote wilderness on the globe. Temperatures during the winters were as low as -70 degrees.

sprouting Islam-like whiskers. Lou Killian, although without the assist of a set of whiskers, is now Bechtel executive vice-president in the Middle East.

For equipment the construction forces had the benefit of the outsized equipment pool and experience of Aramco, which had been fighting local conditions for years and licking them the way Montgomery licked Rommel. As Caterpillar's best customer, Aramco stood second only to the United States Army; as a buyer of tires and special-type trucks, it knew no peer. Nothing has ever been seen on American highways like the 165-passenger personnel carriers which Aramco operates to haul its workers about; no trucker, outside of Arabia, has steered such tires as the monsters on which Aramco grinds over the dunes. With such a devotee of Bigness for a client, Bechtel's people could rejoice; and they set about developing some big stuff themselves.

There wasn't any suitable marine terminal on the Persian Gulf. So one was made.

With rangy Lou Killian in charge, camp was set up north of Dhahran on the gulf shore where some protection was afforded by a sandy hook of land that looked like a camel driver's stick. From this resemblance the place bore the name Ras al Mish'ab. The Neutral Zone which divides Saudi Arabia from oily Kuwait, at the head of the gulf, was 14 miles beyond. Lou and his party trucked up from Dhahran and staked out the camp in a howling sandstorm. The tents just did stay up.

"Our tents," recalled an old timer, "were nicely floored with wall-to-wall sand." The occupants cussed sand, ate sand, and breathed sand along with sand flies and plain flies, the latter as thick as currants in a mince pie. Presently the mud-brick buildings of a semipermanent camp began to rise, but sand and insects remained as dependable ingredients of the atmosphere.

Soon a jetty[*] and pier stretched out into the blue water. Mish'ab's first cargo ship arrived two months after camp was laid out. Barges brought ashore tools and the first pipe. Unloading was a problem. Tapline borrowed an idea from the fir loggers in Oregon and Washington. At certain places in those faraway evergreen forests, loggers use overhead cables, or "skyhooks" for hauling timber from steep mountain slopes. So a skyhook was erected 3 miles out into the Persian Gulf. Twenty-four A-frames[**] went up to hold the cable, from which self-propelled cars were hung, each capable of trundling 10 tons of pipe.

Operated in tandem, these high-wire cars made the 3-mile journey in five minutes, and presently they were bringing ashore 1,100 tons a day, swinging it along 80 feet above the water with all the fun of riding a scenic railway.

----

[*] A jetty is a barrier built out from a shore or streambank to protect it from erosion or wave action.

[**] An A-frame is a tower with two slanting legs, similar in appearance to the letter "A".

Ashore, the 31-foot pipe sections were welded into triple-lengths. And soon the Abqaiq-Qatif section was in place, . . . The pipeliners were on their way, Gulf to Mediterranean.

For vertical alignment, the long steel pipeline was supported over the first stretches on crosspieces set on wooden piles. A machine called the "loping camel" drove these at a smart clip. Later concrete supports went in. Care was taken to weld the lengths under atmospheric temperatures as close to the mean as possible; excessively hot hours were avoided. Provision was made for camel-crossings--ramps of sand--every mile or two, unless the pipe ran under a dune or hill.

Winds blew. Sun blazed. . . . Grit storms and rare raindrops hit like machine-gun bullets. Sand dunes shifted and slued. But men who'd built Hoover [Dam], [Grand] Coulee [Dam], Pacific bases, and many "big inches" had been through a good deal of difficulty before, and knew how to hunt for the answers. To keep the truck engines cool, two radiators were used. Everything that moved got double servicing.

Long before the first black goat-hair Arab tent had been raised at Ras al Mish'ab, Ray Hamilton and his logistic experts back home had been figuring out problems to be met. Trucks and trailers too wide and too heavy for any United States highway had been built and tested on the desert of New Mexico under conditions approaching those in Saudi Arabia.

But while Ray had been able to produce trailer trucks hauling 40 tons of pipe in lengths up to 93 feet, he couldn't rid Arabia of truck-driver lonesomeness. Roads were nonexistent until the project made its own. The first convoy to a place called Duwald had been dispatched from Ras Tanura in September 1947, and consisted of three Diamond-T tractor-trailer combinations with loads of gasoline, oil, water, and food. The round trip took eleven days and the trucks had twenty-two flat tires. The road at that time was a winding trail, rocky most of the way and elsewhere deep in powder. The tractors were without cabs and the men learned something about driving through Arabian winds. In one section the trail wound through a dry wash and it took the convoy six hours to make 1,000 feet. Dust pockets became so deep that clouds raised by the trucks sometimes choked out the engines.

In the early stages the men generally didn't see another vehicle outside of their own convoy after the first three or four days out. The convoy system was abandoned with the arrival of super-husky Kenworth equipment, and single trucks with American drivers went through from Ras Tanura to Ras al Mish'ab in nine hours instead of days.

At the height of construction, transportation was up to 10 million ton-miles a month. Trucks were on the go twenty-four hours a day, with regular changes of crews. . . .

The mighty sand-tires, invented by Aramco's engineers to whip the desert--12-ply affairs inflated to only 6 or 12 pounds of air--plunged at the dunes like snowplows, and rode over the obstacles or through them. The vehicles hit bogs and all but sank to their beds; out from camp came the rescue hooks, and sometimes more hooks to rescue the rescue hooks, and the trucks churned on again. Often drivers saw mirages of lovely lakes and turned aside for them.

One time a leader, who'd been fooled often enough, refused to turn aside, but that time it wasn't a mirage, it was honest-to-Allah water. Work for the hooks again.

While Tapline was abuilding, Bechtel used only one spread of men and machines, but it was a gang that was really spread. Portions of it were 150 miles apart.

When a man got into trouble in that world of dune and rock, he used an arm-signal code to the occasional planes that passed. He stood out in plain view with arms straight down at his sides if no help was needed; he waved down in front if he desired only mechanical help; he held arms straight out if he needed food and water; he waved both arms aloft if he wanted medical aid.

The "pleasures" of this alien desert continued to make life interesting for the men from overseas. At Ras al Mish'ab, a Yank was sitting in a tent telling of how, under similar circumstances, a fellow worker had discovered a big viper or blacksnake between his knees. At that point a tentmate remarked, "Don't look now, but there's one between yours." The narrator did look, and went out by backward somersault.

Prior to the coming of Tapline, the only water wells along the right-of-way, other than isolated wells drilled by Aramco exploration crews, were a few ancient cisterns along the caravan routes. These went dry during the summer months. Since water is a first requisite to any construction job, Aramco early in 1957 had begun drilling for Tapline's account. Ultimately twenty-five producing wells were in use--and how they were in use!--and to obtain that water it had been necessary to drill sixty holes. Some holes proved dry, and some were just bad water. The producing wells were from 200 to 1,700 feet deep. Pumps were required at all but one. Under the agreement between Tapline and the Saudi government, Tapline was obligated to furnish water free to all travelers along the pipeline route and, remembers George Colley, "it soon developed that all Bedouins were classified as travelers."

A new water hole was news of first importance to the desert tribes. As soon as the word traveled--and nobody ever did find out how it traveled so fast--large numbers of nomads moved their families, tents, and livestock over great distances to camp near the new supply. At one pump station a check was made twenty-four hours a day for two weeks and it was discovered that an average of 1,642 Arabs, 5,457 camels, and 7,195 sheep, goats, and donkeys a day used the watering trough.

By 1950 the job was going full blast with more than 1,000 Americans and 5,000 Arabs hard at it. Roy McAuliffe went in as project manager, and Roy Middleton was running pipeline construction.

Then things slowed down. A steel shortage was on at home and export permits were held up by the U. S. Government. Pipelaying was throttled down to one kilometer a day, and many of the Camel Legionnaires returned to the U. S. A. But it wasn't time wasted. During this curtailment special attention was given to on-job training of Arabs. . . . Then the steel shortage ended and speed picked up again.

Throughout the crossing of Saudi Arabia the American foremen had fine practice at controlling their tongues. The Arab is a proud man, and cussing him out is the height of unwisdom. Though the hard-hat men from the New World sometimes called the headcloth-topped natives "ragheads," or "rags," it was purely for convenience and good-naturedly meant. To the Arabs, especially the Bedouin, or herdsmen Arabs, the Americans showed respect that steadily increased. The respect became honest admiration when they discovered that these untutored children of the desert were picking up ten or fifty words of English for every word of Arabic the Yanks could master.

The ways of Araby turned out to be endlessly varied, often perplexing, and always fascinating. For one thing, the Arabs themselves proved to be beyond surprising, even if the Americans were not. Lifted from a camel and placed in an airplane for his first ride through the sky, the Arab was calm as a rock. Even if, as happened, the plane hit an air pocket and dropped 1,500 feet, the passenger never batted an eye. It was all up to Allah; why worry? The dusky sons of the wind and sand also turned out to have a lively sense of humor. And though they had no national handicrafts of any kind except rough weaving and tentmaking, they took to machinery with zest.

As sections of the line neared completion, the tribesmen here and there found a new use for it. Where it swung in [the] air across a hollow, they draped their tents over the big pipe, using it for a ridgepole, and their delight knew no bounds when the line was finished and the warmed-up oil flowed through it . For in winter the line was just a nice stove, heating the tent deliciously. . . .

West of Hafar at a spot called Hatin the problem was posed of constructing a hauling road across marine limestone beds and through clay and gypsum country beyond. This was a major undertaking. Forty Americans and several hundred Arabs went at it and continued until the Jordan border was reached. All in all, it became necessary to build a highway 930 miles long from Ras al Mish'ab to an intersection with an existing road in Jordan.

The highway across previously impassable desert is now one of the revolutionary features of the Middle East landscape. For the first time it is possible to make motor round trips between Mediterranean ports, Persia, Kuwait, and the Persian Gulf shore of Saudi Arabia. Over this road trucks now speed fruits and vegetables and other goods from the Mediterranean area to the Persian Gulf markets.

The two construction armies, the Williams forces working eastward from Sidon in Lebanon and the Bechtel spread pushing westward, joined hands in September of 1950. And there was the pipeline. It took 4,900,000 barrels of oil to fill it. In addition there had to be working stocks at intermediate pump stations. The total was 6 million barrels, which is more than the oil pumped daily from all the wells in the United States.

The long hose carries, every day, almost a third of a million barrels of oil across the deserts from the Persian Gulf to the Mediterranean Sea. The 450 miles of above-ground pipe is a happy experiment. It requires little maintenance. Rabbits and gazelles crouch in its shade. The nomad knows that it does his ancient range no harm. In the early months the tribesmen shot eight rifle holes through the strange-looking target, probably because they thought it was still carrying water (water had been used for testing). But

The Trans-Arabian Pipeline

the local authorities put trackers out and caught the marksmen, or a reasonable facsimile thereof; there have been no further pot shots in several years. In Syria, Jordan, and Lebanon the line is underground, for in those parts there is more cultivation. At places the line passes within 100 yards of the embattled Israeli border and inspectors can't approach it without armed escort. Still, while nations clash, the oil goes through. Down at Sidon, where the oil finally flows to the ships, 9 Americans, 12 Europeans, and 212 Lebanese operate the lively terminal and a dozen tankers at a time cluster there for their loads. The storage tanks are on a hill just above the spot where tradition says Jonah landed from the whale.

The size of Tapline is hard to realize. As big around as a large barrel, it would extend, if laid on a United States map, along the Pacific from the Mexican border almost to Canada, and the whole route would be pretty much like a traverse of Death Valley. The line includes 1,067 miles of 30-inch pipe or larger, and six major pump stations. The capacity is 317,000 barrels of oil a day.

436

The Bridge at Mackinac
David B. Steinman

*David B. Steinman, the author of this poem and more than 100 other published poems, was an internationally known bridge engineer. In addition to his works in verse, he wrote two books--***Bridges and Their Builders*** *and* **Miracle Bridge at Mackinac** *(co-authored with John T. Nevill)--and more than 600 technical papers and articles.*

Steinman was the chief engineer for the Mackinac (pronounced Mack-in-naw) Bridge and for hundreds of others including the Sydney Harbor Bridge in Australia and the Florianopolis Bridge in Brazil.

He was born in New York in 1886, played in the shadow of the Brooklyn Bridge, and received bachelor (1909) and Ph.D. (1911) degrees in civil engineering from Columbia University. His Ph.D. thesis was entitled "The Design of the Henry Hudson Memorial Bridge as a Steel Arch." A little over twenty years later he was awarded a contract to design the Henry Hudson Bridge over the Harlem River.

Early in his career he developed an interest in aerodynamic effects on suspension bridges, and eventually he became a world-renowned expert on aerodynamic considerations in bridge design. While teaching civil engineering at the University of Idaho early in his career he wrote Gustav Lindenthal, then one of the most eminent bridge designers in the world, to inquire about the possibilities of employment as a bridge engineer. Lindenthal wrote back advising him to seek another field of endeavor because the price of steel was so exorbitant that the days of bridge design and construction would soon be ending! A few years later Lindenthal invited Steinman to join his firm.

The bridge across the straits at Mackinac was the longest bridge in the world at the time of its construction. It is more than five miles in length (including approaches), with a main span of 3800 feet and a total suspended span of 7400 feet. Steinman considered it one of his major accomplishments.

Dr. Steinman died in 1960.

> In the land of Hiawatha,
>   Where the white man gazed with awe
> At a paradise divided
>   By the straits of Mackinac--
>
> Men are dredging, drilling, blasting,
>   Battling tides around the clock,
> Through the icy depths of water
>   Driving caissons down to rock.
>
> Fleets of freighters bring their cargoes
>   From the forges and the kilns;
> Stone and steel--ten thousand barge loads--
>   From the quarries, mines and mills.

Now the towers, mounting skyward
Reach the heights of airy space.
Hear the rivet hammers singing,
Joining steel in strength and grace.

High above the swirling currents,
Parabolic strands are strung;
From the cables, packed with power,
Wonder spans of steel are hung.

Generations dreamed the crossing;
Doubters shook their head in scorn.
Brave men vowed that they would build it--
From their faith a bridge was born.

There it spans the miles of water,
Speeding millions on their way--
Bridge of vision, hope, and courage,
Portal to a brighter day.

Dr. David B. Steinman, P.E., founder of Steinman Boynton Grondquist & Birdsall

Selected Additional Reading

Civil engineering projects are the subjects of more books than any other aspect of the profession. Because so many civil engineering projects are important from a historic or societal perspective, books about them are more widely available than books on any other aspect of civil engineering. Many of the available books, however, do not provide as much information about the civil engineering aspects of a project as one might expect. The books listed here, a small sample of what one might expect to find in even a small library, are limited primarily to books about projects described in these selections or books which provide information on many different civil engineering projects.

Ames, Charles Edgar. **Pioneering the Union Pacific: A Reappraisal of the Builders of the Railroad** (New York: Appleton-Century-Crofts, 1969) 591 pgs.

Anders, Leslie. **The Ledo Road** (Norman, OK: University of Oklahoma Press, 1965) 255 pgs. (Military Construction in World War II)

Brown, Allen. **Golden Gate** (New York: Doubleday & Co., 1965) 231 pgs.

Cassady, Stephen. **Spanning the Gate** (Mill Valley, CA: Squarebooks, 1986) 133 pgs. (Construction of the Golden Gate Bridge - illustrated with numerous photographs)

Dillon, Richard; Moulin, Thomas; and DeNevi, Don. **High Steel** (Berkeley, CA: Celestial Arts, 1979) 168 pgs. (Construction of the Golden Gate and Oakland Bay Bridges - illustrated with numerous photographs)

DuVal, Jr., Miles P. **And the Mountains Will Move** (New York: Greenwoood Press, 1970) 374 pgs. (Construction of the Panama Canal)

Gies, Joseph. **Bridges and Men** (New York: Doubleday & Co., 1963) 343 pgs.

Gies, Joseph. **Wonders of the Modern World** (New York: Thomas Y. Crowell Co., 1966) 241 pgs.

Grey, Zane. **Boulder Dam** (New York: Simon and Schuster, 1967) 199 pgs. (Fiction)

Grey, Zane. **The U. P. Trail** (New York: Harper & Brothers, 1918) 409 pgs. (Fiction)

Griswold, Wesley. **A Work of Giants** (New York: McGraw-Hill Publishing Co., 1962) 367 pgs. (Construction of the Transcontinental Railroad)

Haase, John. **Big Red** (New York: Harper & Row, 1980) 502 pgs. (Fictional account of the construction of Boulder Dam)

Hayden, Martin P. **The Book of Bridges** (New York: Galahad Books, 1976) 152 pgs.

Hopkins, H. J. **A Span of Bridges** (New York: Praeger Publishing Co., 1970) 288 pgs. (An illustrated history of bridges)

Jackson, Donald C. **Great American Bridges and Dams** (Washington, D. C.: The Preservation Press, 1988) 357 pgs. (State-by-state description with photographs of bridges and dams considered worthy of historic preservation)

Jacobs, David and Neville, Anthony E. **Bridges, Canals and Tunnels** (New York: American Heritage Publishing Co., 1968) 160 pgs.

Jordan, Philip D. **The National Road** (Indianapolis, IN: Bobbs-Merrill Co., 1948) 442 pgs.

Kahrl, William L. **Water and Power** (Berkeley, CA: University of California Press, 1982) 583 pgs. (Conflict over Los Angeles' water supply from the Owens Valley)

Kraus, George **High Road to Promontory** (Palo Alto, CA: American West Publishing Co., 1969) 371 pgs. (Construction of the Transcontinental Railroad)

Lee, W. Storrs. **The Strength to Move a Mountain** (New York: G. P. Putnam's Sons, 1958) 318 pgs. (Construction of the Panama Canal)

Mayer, Lynne R. and Vose, Kenneth E. **Makin' Tracks** (New York: Praeger Publishing Co., 1975) 261 pgs. (Construction of the Transcontinental Railroad)

McCague, James. **Moguls and Iron Men: The Story of the First Transcontinental Railroad** (New York: Harper & Row, 1964) 392 pgs.

McCullough, David G. **The Great Bridge** (New York: Simon and Schuster, 1972) 636 pgs. (Construction of the Brooklyn Bridge)

McCullough, David G. **The Path Between the Seas: The Creation of the Panama Canal** (New York: Simon and Schuster, 1977) 698 pgs.

Morgan, Murray. **The Dam** (New York: The Viking Press, 1954) 162 pgs. (Development of Grand Coulee Dam)

O'Neill, Richard W. **High Steel, Hard Rock and Deep Water** (New York: The Macmillan Co., 1965) 280 pgs.

Outland, C. F. **Man-Made Disaster: The Story of St. Francis Dam** (Glendale, CA: A. H. Clark, 1963)

Payne, Donald G. **The Impossible Dream: The Building of the Panama Canal** (London: Hodder and Stoughton, 1971) 284 pgs.

Plowden, David. **Bridges** (New York: W. W. Norton & Co., 1974) 328 pgs. (An illustrated history of bridges in North America)

Reisner, Marc. **Cadillac Desert: The American West and Its Disappearing Water** (New York: The Viking Press, 1986) 582 pgs.

Romley, David A. **A Crooked Road** (New York: McGraw-Hill Publishing Co., 1976) 252 pgs. (Construction of the Alcan Highway)

Rose, Mark H. **Interstate: Express Highway Politics, 1941-1956** (Lawrence, KS: Regents' Press of Kansas, 1979)

Schodek, Daniel L. **Landmarks in American Civil Engineering** (Cambridge, MA: The MIT Press, 1987) 383 pgs. (An illustrated study of major American civil engineering projects)

Scott, Q. and Miller, H. S. **The Eads Bridge** (Columbia, MO: The University of Missouri Press, 1979) 142 pgs.

Seely, Bruce. **Building the American Highway System** (Philadelphia, PA: Temple University Press, 1987) 315 pgs.

Shank, William H. **Towpaths to Tugboats: A History of American Canal Engineering** (York, PA: American Canal and Transportation Center, 1982)

Shaw, Ronald E. **Erie Water West** (Lexington, KY: The University Press of Kentucky, 1966) 449 pgs. (Construction of the Erie Canal)

Sibert, William L. and Stevens, John F. **The Construction of the Panama Canal** (New York: D. Appleton and Co., 1915)

Steinman, David B. and Nevill, John T. **Miracle Bridge at Mackinac** (Grand Rapids, MI: Eerdmans, 1957) 208 pgs. (Construction of a bridge across the Straits of Mackinac)

Stern, Richard M. **The Big Bridge** (New York: Doubleday & Co., 1982) 324 pgs. (Fiction)

Stevens, Joseph E. **Hoover Dam: An American Adventure** (Norman, OK: The University of Oklahoma Press, 1988) 326 pgs.

Sundborg, George. **Hail Columbia - The 30-Year Struggle for Grand Coulee Dam** (New York: The Macmillan Co., 1954)

Talese, Gay. **The Bridge** (New York: Harper & Row, 1964) 140 pgs. (Construction of the Verazzano Narrows Bridge)

Van Der Zee, John. **The Gate** (New York: Simon and Schuster, 1986) 380 pgs. (Design and construction of the Golden Gate Bridge)

Whitney, Charles S. **Bridges - Their Art, Science and Evolution** (New York: Greenwich House, 1983) 360 pgs. (An illustrated history of the evolution of bridge building)

441

Williams, John Hoyt. **A Great and Shining Road: The Epic Story of the Transcontinental Railroad** (New York: Times Books {A Division of Random House}, 1988) 341 pgs.

Wilson, N. C. and Taylor, F. J. **The Earth Changers** (New York: Doubleday & Co., 1957) 312 pgs. (Descriptions of numerous civil engineering projects constructed by U. S. construction companies)

Woodbury, David O. **Builders for Battle** (New York: E. P. Dutton & Co., 1946) 415 pgs. (Construction in the Pacific during World War II)

Woodbury, David O. **The Colorado Conquest** (New York: Dodd, Mead & Co., 1941) 367 pgs. (Construction of Boulder Dam)

Woodward, Calvin **A History of the St. Louis Bridge** (St. Louis, MO: G. I. Jones & Co., 1881) 391 pgs. (Construction of the Eads Bridge)

The Golden Gate Bridge Under Construction

ETHICS AND PROFESSIONALISM

The readings in this section either describe ethical problems that confront modern civil engineers or discuss aspects of civil engineering practice that are important from a philosophic perspective rather than from a technical or practical perspective. The readings are concerned more with how a civil engineer believes or feels than with what he or she does.

Most of the selections in the other sections of this book were written *about* engineers and civil engineering projects by writers who were not engineers themselves; most of these selections were written by engineers. Careful study of these readings should dispel the notion that civil engineers do not have or need the capacity to understand and appreciate the philosophic, moral and artistic aspects of their work.

Civil engineers face ethical and professional problems different in many respects from the problems encountered in other professions. Foremost among these is the problem that the civil engineer's client is frequently the whole of society rather than a single individual. In such cases knowing and understanding the needs and will of the client is a challenge unlike that faced by any other professional. The competing needs within society and the failure of many individuals to make their opinions and desires known in any public forum makes it difficult for civil engineers to ascertain just what it is that society wishes to have done. Finding a means for developing civil engineering projects that reflect the needs of society as a whole--the powerless as well as the influential and those who cannot or do not voice their opinion as well as those who are outspoken--is one of the most formidable tasks of the modern civil engineer in a democratic society.

The astronomical amounts of money involved in civil engineering projects is the source of another frequently-occurring major ethical problem for civil engineers. Although the overall record of the civil engineering profession is remarkably good considering the vast sums of money they are responsible for spending wisely, the recurring instances of bribery and kick-back schemes involving civil engineers are a reason for concern. Because these schemes often involve appointed or elected public officials they are widely publicized and result in extensive unfavorable publicity for the civil engineering profession.

An important professional issue facing the civil engineering profession is the development of an appropriate level of interest and concern for the humanistic dimension of the profession. The intensity of the scientific, mathematical and technical aspects of the civil engineering curriculum often results in students devoting insufficient attention to the general knowledge aspects of their education. Yet, of all the engineering and scientific professions there is no other that demands the breadth of knowledge, understanding and appreciation of the social sciences, interpersonal relationships and communications, and individual and societal values. Students often miss their best opportunity to acquire knowledge in these areas due to excessive emphasis on their technical courses. Accreditation standards for engineering curricula, increased emphasis by professional societies, and a greater level of societal awareness and concern for social and moral impacts of technical decisions are causing engineering educators to devote more attention to this problem.

Like the other sections of this book, the readings in this section are not selected with a view toward providing a comprehensive description of the ethical and professional problems confronting civil engineers. The intent is to create interest and awareness, not to serve as a tutorial. For that reason these readings should not be used as a formal course in ethics and professionalism. They will, however, serve as a useful adjunct to a formal course based on a more traditional textbook on engineering ethics.

In general, the readings herein are intended to be inspirational--to cause the reader to reflect on the positive aspects of the civil engineering profession and the sense of personal satisfaction that should be inherent in the pursuit of a professional career in civil engineering. A few of the readings do dwell on the negative consequences of a failure to recognize one's responsibility to the public at large. It is important to point out that situations conducive to moral and ethical failure exist within the profession in order to emphasize the need for developing an ability to think about philosophic issues and for forming proper moral values early in one's career.

Concrete and Kafka: A Personal Overture

Chapter 1, The Civilized Engineer
Samuel C. Florman

Samuel Florman's retrospect speaks volumes to those engineers who share his educational experiences (whether in whole or in part), but it is particularly touching to those he describes in the penultimate sentence of the penultimate paragraph--those who "muse about the meaning of life." If there is one selection in this anthology which should be read and pondered by every person considering a career in civil engineering, this is the one.

"I became an engineer."

Thus begins John Hersey's novel, **A Single Pebble**, in which the protagonist travels to pre-revolutionary China seeking a site for a dam along the Yangtze River. As he encounters a civilization little changed since the Middle Ages, the young man finds his faith in technology giving way to awe and self-doubt.

I, too, became an engineer and have spent a number of years thinking about, as well as practicing, this much misunderstood profession, albeit in less dramatic settings than the chasms of the Yangtze, and with less discouraging conclusions than Mr. Hersey's.

How does one decide to become an engineer? I made the decision in 1942 during my senior year at the Fieldston School, a sylvan campus in the Riverdale section of the Bronx, forty-five minutes by subway and footpath from where I lived in Manhattan. The idea had occurred to me earlier-- especially during several visits to the 1939 World's Fair--but I was far from being the stereotypical engineer. I did not, for example, build radios, assemble models, or fiddle with car engines. Living in the city, I had no access to cars, and when some mechanical device failed in our apartment, my parents called on the building superintendent. Like my fellow students at Fieldston, I read a lot of books and wrote a lot of papers. My favorite subject by far was English particularly a senior seminar in which we reviewed great Western literature from Aeschylus to James Joyce. Nevertheless, I did my best work in mathematics, and was gently urged by several of my teachers to consider a career in science.

There were no "two cultures" in those days and I can recall no division between students of different sorts of talents, rather mutual respect and a shared appreciation of achievement. If this sounds idyllic, well it was. Not that we lived in a state of constant elation--we were teenagers, after all--but academically the place was heaven. We knew we would follow many different career paths: the world seemed incredibly open and full of possibilities--in the arts, the sciences, and the professions. Business, however, we regarded with a scorn compounded of intellectual elitism and post-1930s radicalism. Ironically, the fathers who paid out not inconsiderable tuition were mostly hard-working small businessmen.

Of the acceptable career alternatives, science, medicine, and engineering were considered more or less on a par with law, journalism, and the arts. Excellence is what counted; our class had an abundance of it and our expectations were

445

high. We were not surprised in later years when the most accomplished student in the class studied physics at Harvard, got his doctorate there, and ended up at Los Alamos, any more than we were when the president of the student council became a nationally syndicated newspaper columnist, ofttimes called a pundit. We were a class full of potential pundits.

Although I wanted my life's work to be creative and stimulating, I was not totally oblivious to money. A part of my Depression-bred consciousness was concerned about some day being able to support a family. For all the appeal of mathematics and physics, it wasn't clear to me how one made a living in those fields. This was even more true of writing and the arts. Business, as I have said, was out of the question, and as for medicine, needles made me queasy. So I chose engineering. Engineers, from the little I knew, studied science and used their brains. They also got jobs and earned salaries. And, after a fashion, they were cultural heroes. The newsreels that I saw every weekend between two movies at Lowes 83rd Street often featured the dedication of a new TVA dam or some other impressive public work. There was much cutting of ribbons and drinking of toasts, each event celebrating a counterattack against rural dust bowls or urban slums. And when the movies themselves depicted engineers--usually in the B film, to be sure--they were stalwart men in high-laced boots engaged in heroic endeavors such as building railroads or prospecting for oil. Intellectually challenging, financially sensible, and withal a touch of romance and adventure--engineering seemed like an ideal calling.

I had never heard it suggested that engineers were lower-middle-class, eccentric, or uncultivated--today I believe the epithet is "nerd"--nor did it occur to me that anybody held such opinions. The only sour note was sounded by an uncle who observed that instead of wanting to be an engineer I should aim to be someone who hires engineers, thus implying that I was about to join an exploitable sub-class. The remark enraged my father, who had no clear idea of what engineers did but was proud to have a son who was going to enter a profession.

When it came time to select a college, I naturally thought about M.I.T. [Massachusetts Institute of Technology]. Two of my engineering-bound classmates went to that august institution and never regretted their choice. But there was something about the huge labyrinth of laboratories that made my spirit sink, and still does in spite of all the good things I know about the place. Instead, I chose Dartmouth College, whose beautiful New Hampshire campus captured my heart.

I had only the vaguest idea of how one went about getting an engineering education. According to the Dartmouth catalog it seemed that I would "go to college," earn a Bachelor of Arts degree while majoring in the sciences, and then pursue an engineering degree in graduate school. This is how General Sylvanus Thayer thought it ought to be when, in 1867, he gave Dartmouth $40,000 for the purpose of establishing the Thayer School of Engineering. As Superintendent of the U.S. Military Academy from 1817 to 1833, the general had overseen the development of that institution into a distinguished school of applied science, and in his later years he decided to endow a graduate school of engineering at a liberal arts college. He believed that before embarking upon professional training one ought to become "a gentleman." The Thayer School's two-year program originally was designed to follow a full four-year undergraduate education, but in 1893 a five-year program was devised combining the senior year of college with the first year of engineering school. That program endures to this day.

In most of the nation, however, engineering education evolved along different lines. The technical institutes, and later the land grant colleges, developed four-year programs that carried students directly from high school into engineering studies. This effectively did away with the concept of a liberally educated engineer, although the accrediting arm of the profession eventually required that an engineering curriculum have a minimum 12 1/2 percent liberal arts component.

Of all this I was blissfully unaware as I arrived in Hanover, New Hampshire, in July of 1942. (A year-round program had been instituted because of the war.) I embarked on a typical course of study: English literature and French, sociology and economics, psychology and political science. As a pre-engineering student I also took mathematics, physics, and chemistry, and two other subjects--then required but long since discarded--graphics and surveying. I will not argue that these courses deserved to maintain their place in the curricula of higher education, but I recall vividly the delights of T-square, triangle, and india ink,* and the thrill of carrying a transit** through the autumn woods. These sorties into the tangible world, combined with the abstract fancies of mathematics and the sciences, reinforced my conviction that I was headed toward the best of all possible careers.

I was barely into my sophomore year when, almost imperceptibly, I began to undergo a metamorphosis. As if under a spell, I became increasingly absorbed in my technical, pre-professional studies. Looking back, I find it difficult to explain what happened, although since many of the hundred thousand-plus Americans who enter engineering each year go through the same experience, how extraordinary can it be? All I know is that the liberal arts began to pale and seem trivial, even annoying. Mathematical formulas took on the quality of fun-filled games, and the physical world became an enchanted kingdom whose every secret seemed worth exploring. Also I began to think of courses in terms of how they would help me become a better engineer, more thoroughly grounded in the sciences, more perceptive and quick-witted, and--let us face facts--more desirable to some future employer. Despite the educational advantages I had enjoyed in high school, notwithstanding the proclaimed policy of the liberal arts college I was attending, I came down with a bad case of vocationalism. I lost interest in becoming an educated person--the "gentleman" envisioned by Sylvanus Thayer. I wanted to become an engineer.

Could an inspiring humanities professor have prevented this from happening? I like to think so. Surely the situation was not helped by a freshman English course devoted mostly to the painstaking dissection of **Lord Jim**, nor by the introductory social science courses which were informative but deadly dull. More exciting teachers and better-planned classes might have made the difference, but it is common knowledge that when one is embarked on an affair of the heart, the most prudent counsel, even skillfully presented, often

----

* Drafting paraphernalia.

** An instrument used in surveying

447

falls on deaf ears. And there can be no doubt but that my feeling about engineering was not altogether different from falling in love.

As it happened, my most exciting professors were mathematicians. I recall winning a prize in a mathematics competition--second prize to be exact--and being invited along with the other winners to dinner at the home of the department head. After dining, we sat in the living room sipping brandy and listening to recordings of Mozart sonatas. Although at the time my musical taste ran more to Glenn Miller and Artie Shaw, I found the experience extremely agreeable. I associated my euphoria with the delights of mathematics, not giving adequate credit, I now believe, to Mozart and the winning of prizes, to say nothing of brandy.

Along with my commitment to mathematics and science, I developed a taste for extracurricular activities that I can only characterize as anti-intellectual. I and my pre-engineering fellows spent our leisure hours attending movies and sporting events. Occasionally we hitchhiked to Smith College and looked for girls. When lectures, concerts, and plays were offered on campus, it seemed natural that they be attended by other students, those increasingly strange young men who had decided to major in history, philosophy, or literature. One of my dormitory mates was enrolled in a special course with Robert Frost, who was at the time poet-in-residence at the college. Several times this friend invited me to join him for an evening of readings and discussion with the noted poet, but I was always too busy writing up my laboratory experiments, or else committed to a party at some local tavern. Today I cannot believe -- simply cannot believe--that I never even saw Robert Frost, much less spent an evening with him when I had the chance.

Shortly after entering college, I had enlisted in the Navy V-12 program on campus, and at the end of each term there was a period of uncertainty while we recruits waited to hear what the government had planned for us. After a year, we were called to active duty, but this merely meant putting on a uniform and learning how to march in formation. Those of us who were heading for engineering were encouraged to continue our studies. After twenty-four months of nonstop schooling, I accumulated three years worth of credits and was ready to enter the professional phase of my education.

I had by this time resolved to become a "civil" engineer. The term was coined in mid-eighteenth-century England by John Smeaton, builder of the Eddystone lighthouse, who wanted to demonstrate that his work had no military implication (which is ironic in view of the fact that most military engineers subsequently have been trained in civil engineering). Civil engineers design and construct buildings, dams, and bridges; towers, docks and tunnels--structures of all sorts. Civil engineering also encompasses highways, railroads, and airports, along with water supply and sewage disposal. In short, civil engineering is basic and of the earth, historically--along with mining--the root of all engineering. In the eighteenth century the development of the steam engine led to a new specialty called mechanical engineering, and each major technological advance has brought with it a fresh division of the profession: electrical, chemical, aeronautical, petroleum, computer, and so forth. I make this digression into the self-evident only because so many otherwise well-informed people keep asking me what it is that engineers do. Every technological product has to be designed and its fabrication overseen, and that is what engineers do. They occupy the vast middle spectrum between theoretical scientists and sub-professional technicians.

Buildings are usually planned by architects, but engineers design the structural and mechanical components within them, and civil engineers often oversee the actual construction process. These overseers are sometimes called construction engineers, and this is what I have become -- more a business manager, I suppose, than a creative spirit, more a master builder than a man of science, yet still a member of the engineering family.

I have long forgotten most of the theorems that I learned in engineering school, but I recall vividly the nature, the "feel" of that learning. Like all engineers, I took basic courses in electricity, fluid mechanics, metallurgy, and thermodynamics (the study of heat and energy, particularly the workings of internal combustion engines, air conditioners, and the like). As a civil engineer, I took a series of courses in "structures," learning how to design beams, walls, slabs, and trusses. Then there were the more specialized studies: highways, water supply, and sanitation. In all of this there was a good amount of "hands-on" work. We poured concrete, cured it, and tested it to failure, analyzed the behavior of water in pipes and over weirs, and experimented with a variety of motors and generators. Occasionally we ventured out into the field, visiting construction jobs and sewage treatment plants, or--a great favorite--measuring the flow of a river while perched above it in a tiny hand-operated cable car. The theoretical work was difficult--some of it exceedingly so--but the physical *doing* made it seem worthwhile.

Nowadays engineering education is much more "scientific" than it used to be. In addition to the subjects that were taught in the 1940s, a contemporary curriculum will include computing and information processing, probability and statistics, systems, optimization, and control theory, even system dynamics (policy design and analysis based on feedback principles and computer simulation). Much of the so-called "shop work" has fallen by the wayside, relegated to students who take two-year technician courses or four-year engineering technology programs. The change came about in the 1950s, particularly in the aftershock of Sputnik. Also, the growth of new disciplines has meant that there is simply more material to learn and so less time for knocking about in overalls or muddy boots. This has been inevitable, appropriate, and a darned shame.

In the spring of 1945, the Navy decided to call in my debt, so to speak, and I was ordered to officers training school. After a few weeks of shooting guns, large and small, and studying semaphore code and naval etiquette, I was commissioned an ensign in the Civil Engineer Corps and sent to serve with the Seabees. I arrived in the Philippines just as the war was ending and went with the 29th Construction Battalion to occupy Truk, an atoll in the Caroline Islands that had been bypassed during the westward counteroffensive in the Pacific.

It was a pleasant enough life for a young would-be engineer. During the day we worked on construction projects, repaving the airstrip with fresh coral dredged from the sea, erecting Quonset huts, and building roads and a water supply system. In the evenings we drank beer, played cards, and talked-- mostly about our work, baseball, and girls. Also--and I could not remember this happening since early childhood--I found myself with long periods of spare time. As the tropical sun sank into the sea behind implausible palm trees, it was impossible not to become introspective.

Among tales of self-discovery and inner change, life on a desert island has an honored place. It helps, of course, if the island has a supply of books. Our battalion had a library of sorts, stocked mostly with the chunky, squarish paperbacks that were printed especially for the armed forces. To help pass the time, I started to read again. At first it was purely recreational: the mysteries of Erle Stanley Gardner, the historical novels of Thomas Costain and Samuel Shellabarger, the best-sellers of Lloyd Douglas and Irving Stone, and the outlandish burlesques of Max Shulman and H. Allen Smith. I had forgotten how much fun books could be, even ordinary unprepossessing ones. In addition to contemporary reprints, there were in our island library a number of Modern Library classics, and it was to these that I turned next. One evening I started to read **Crime and Punishment**. It was as if I had stepped through a looking glass and found myself back in my high school English seminar. Here once again were supercharged words and ideas and people and passions and questions of justice and the meaning of life. It was like high school and yet different because, though I was only four years older, I felt forty years more mature. On a Pacific island, thousands of miles from what we call civilization, the decades condense and urge for meaning wells up in the young wanderer. At least this is what happened to me as I read **Crime and Punishment**, then **Madame Bovary, The Scarlet Letter, Pride and Prejudice, Fathers and Sons, Tess of the D'Urbervilles,** and other great works I cannot now recall.

I did not discuss with my fellow officers the books that I read and the thoughts that these books engendered. During the day, as the bulldozers roared and dust clouds rose, I immersed myself in the work as if nothing had changed. And in the evenings I idly chatted and joked like one of the guys. But I began to wonder why it was that we engineers, as a group, seemed to live in a world so far removed, intellectually and emotionally, from the ferment I was rediscovering in literature.

The only officer in the battalion who was not an engineer was the chaplain. One night he joined several of us in a game of cards, and between hands the conversation droned on in the usual way, which, as I have said, meant anecdotes about our work, the current baseball season, and girls. The chaplain tried to interject some thoughts about the United Nations and prospects for international order. We did not respond. He then tried other topics: the morality of nuclear weapons, the ethical responsibility of war criminals, the role of religion in the post-war world, and so forth, but each time we returned to our trivia. Finally, the chaplain slammed his cards down on the table, looked upward, and said in a loud voice, "Dear Lord, I know that I am unworthy, I confess that I have sinned, but why did you have to abandon me on this island with nobody for company but these boring engineers?"

I cannot say that from that moment on my life was totally changed, but it was an epiphany of sorts, a moment that I have never forgotten. I do not, needless to say, advocate wars or even universal military service. But there is much to be said for a forced interruption in a professional career. Every young engineer (or doctor, or physicist, or lawyer) would benefit, I am convinced, from a year on a desert island. The island should be well stocked with books and ideally should include among its inhabitants at least one outspoken chaplain.

When I returned to New York in the summer of 1946 I intended to look for a job. I had, after all, received a degree from Dartmouth and was entitled to call myself an engineer. (It was not until a few years later that I took the examinations necessary to acquire a state professional engineer's license, but that document is not a prerequisite for practice except in special circumstances, and most American engineers never bother to get one.) Many of my friends, however, were returning to school to resume educations that had been interrupted by the war, and this tended to make me feel that I was still properly a student. Besides, I was only twenty-one years old, the GI Bill of Rights offered free tuition, and the hunger for books that had been aroused during my stay in the Pacific was far from satisfied. I thought things over for a few weeks, then one September morning took a bus uptown to Columbia University and impetuously signed up for a master's program in English literature. My parents reacted to this as if I were manifesting some form of battle fatigue--without having been in a battle--and they were only partly placated by my assurances that I still planned to make my way as an engineer. The people at Columbia were likewise bemused. With only two semesters of English to my credit-- freshman English at that--I was surely one of the least qualified candidates ever to knock at their door. But my intentions appeared honorable, and the mood of the day favored giving veterans a break.

I enrolled in four courses: American Literature Since 1870, taught by Lionel Trilling; Modern Drama, taught by Joseph Wood Krutch; The History of the English Novel, taught by a professor whose name I do not recall; and The Romantic Movement, taught by a professor whose name I do recall but will not record since he read his lectures in a monotone and almost managed to make Wordsworth, Shelley, and Keats seem lifeless. Aside from this one bad choice--for which I compensated by monitoring an undergraduate course in the Age of Reason--I found the classes totally absorbing. I remember with delight not only the lectures and the books we read, but also the arguments and small storms that constantly erupted. When Trilling announced in class one day that Henry James was a far greater novelist than John Steinbeck, and that anybody who didn't think so ought to reconsider his plans to teach English for a livelihood (most of my classmates were budding academics), the booing was loud and raucous. On the other hand, when Trilling entered the classroom after having testified in court on behalf of Edmund Wilson's **Memoirs of Hecate County**--which, strange as it may seem today, was almost proscribed for being pornographic--he was greeted with a standing ovation. When Krutch argued on behalf of trusting one's own judgment instead of kowtowing to so-called literary authorities, he was challenged by a student who said individual taste was suspect since "even monkeys know what they like"; to which Krutch replied, after a moment's thought, "I guess I prefer an honest monkey to a dishonest student." On another occasion, Krutch pointed to a political demonstration that was taking place on campus within view of the classroom window. "Who here," he asked, "believes that there is a radical solution to the discontents of mankind?" There then ensued a political debate followed by a discussion of the tragic view of life. This was totally different from anything I had ever experienced in an engineering classroom.

The entire campus seemed to seethe with excitement and I hardly knew where to look first. When Jacques Barzun was scheduled to lecture to the freshman Contemporary Civilization course, I wangled my way into the hall. A friend of mine recommended the music appreciation classes of the noted composer Douglas Moore, so when I was in the mood--which was often--I dropped in. It was a small class where I couldn't pass unnoticed, but Mr. Moore was a gentle man who said I was welcome. To this day, I remember how he analyzed with us the final movement of Mozart's 41st Symphony, the four themes intertwining in heavenly combinations. Paul Henry Lang, a well-known musicologist and critic, also was on the faculty and for a while I added one of his classes to my itinerant schedule. One day he confessed that he had begun to write a book on the aesthetic theory of opera only to give it up in dismay. Opera is beautiful and soul-stirring, he said, but its formal rules are undefinable. It simply is what it is. This is the way I felt about my experience at Columbia. I was no closer to Truth than I had ever been, but the overall experience was as thrilling as a Verdi duet.

In the evenings, after I had finished the required reading in Dickens, Ibsen, and Faulkner, I turned hungrily to the classics of earlier times, Homer, Shakespeare, Dante, and then, for a nightcap, read The Partisan Review, where I expected to learn what the heavy thinkers of the day were up to. On social outings, I started frequenting Carnegie Hall instead of the hotel ballrooms and Fifty-second Street jazz joints that had previously been my favorites. About the only habit that was not affected by my galloping intellectualism was the way I approached The New York Times on Sunday: I still started with the sports section--as I do to this day.

In order to qualify for a master's degree, I was required to write a thesis, and here I wanted to pick a topic that was all-embracing, not some hidden facet of an obscure author, but something "hot" that would reveal to me the essence of the contemporary cultural scene. With some trepidation, but determined to get advice from the highest possible source, I took the problem to Lionel Trilling. He was not enchanted by my concept, saying that it lacked subtlety, but then, on second thought, the question seemed to amuse him. After a moment's reflection he suggested Franz Kafka. Kafka was culturally hot, he said, and getting hotter. This idea presented two problems: first, I had never read Kafka and had scarcely heard of him; second, and more serious, my selected area of interest--to which my thesis had to relate--was American literature. Well, said Trilling, why not American criticism of Franz Kafka? Why not, indeed? It was an inspired choice, for not only did I immerse myself in the anguished fantasies of this quintessential twentieth-century writer, but in tracking down his reviewers I came face to face with Marxism, Freudianism, existentialism, and all the other isms of the day. This excursion into the world of literary criticism showed me that intellectuals can be as petty, inconsistent, and plain foolish as anybody else, a fact that the awe-struck student of literature--and that is what I was--tends to overlook.

I had anticipated that the thesis would be a chore, but it turned out to be hard work, which is different and better. I enjoyed the research, delighted in the reading, and much to my surprise, relished the writing. I don't suppose that anybody has read that opus, or ever will, other than the kindly assistant professor whose job it was to do so, but a more rewarding task I have never undertaken. There were times when I found myself beginning to envy the men and women whose life's work is reading, researching, writing, and teaching, who traffic in words and ideas instead of goods.

452

Yet when the academic year ended, I said farewell to the campus on Morningside Heights the way one leaves an enchanting foreign land to return home. It never occurred to me that I might stay, become a scholar, and change the direction of my life. For one thing, it was high time to start earning money. Also, I was committed to engineering, I was trained, and deep down I believed that I was better suited to building than to reading and writing. In fact, deep down I had concluded that it was the main business of humankind to build, to be technologically creative. Literature, I felt in my bourgeois heart of hearts, was commentary upon life, while engineering was the stuff of life itself. What was the meaning of all those great novels except to explore the ways in which people work, trade, farm, war, earn, spend, cooperate, and compete? (Also love and hate and create art, to be sure, but only within the framework of an economically viable society.) What was the bitter joke with which Kafka wrestled during his short, sad life except that the state of grace that ever eludes the artist is found in the daily round of ordinary affairs?

As much as I was enchanted with literature and the life of the mind, I had come to resent the condescending attitude that some intellectuals evinced toward "materialism" and the technological impulse. It was all very well for Plato to say that thinking is better than doing, but it seemed ironic when one considered that the glory of Athens rested upon the marvels of Greek technology. (My resentment of Plato and his intellectual descendants was counterbalanced by my affection for Homer, whose literary world was rich with the feel of metals, woods, and fabrics, and whose robust characters took delight in buildings and ships, in objects designed, manufactured, used, given, admired, and savored.)

Shortly after leaving Columbia, I got my first honest-to-goodness job. As aide and lackey to a housebuilder on Long Island, I could not consider the work professionally exalted. But when I started to measure and calculate, to set up my transit and wave directions to the backhoe operator, I knew I was in a place where I belonged. Eventually my career took me into an office, and today I visit construction sites less often than I would like. But when the cranes lift steel beams to dizzying heights, or when the wet concrete flows into huge wooden forms as the laborers yell and bang their shovels against the chutes, I say to myself, as did Robert Louis Stevenson when he came upon a scene of railroad construction on the western American plains, "If it be romance, if it be contrast, if it be heroism we require, what was Troy to this?"

I have had no reason to regret the career choice I made, somewhat cavalierly, in high school. If life has not been one grand adventure interspersed with ribbon-cutting ceremonies like those shown on the screen at Loews 83rd Street, well, it is common knowledge that life isn't the movies. Given real choices in the real world, I think that engineering ranks high in providing psychic rewards. Although I have a special affection for civil engineering, I recognize the appeal of the other branches of the profession, and vicariously enjoy the work of my more "scientific" fellows, for example, those on the frontiers of electronics.

Yet for all the satisfactions inherent in my work, I thank the fates that I had my time with the humanities and that I do not, like so many engineers in mid-career, lament lost opportunities in academe. Mark Van Doren overstated the case, but not by much, when he wrote of the happiness that comes to the student of the liberal arts:

> That happiness consists in the possession of his own powers, and in the sense that he has done all he could to avoid the bewilderment of one who suspects he has missed the main thing. There is no happiness like this.

No one person can be all things at once. It is unrealistic to expect engineers to be artists or poets or literary scholars, and vice versa. Civilization flourished because specialists rely upon one another to perform particular tasks. By nature and inclination, engineers are different from poets and artists. Having said that, I still think it regrettable that so many engineers arrive at the gates of their profession without having experienced much of what the world has to offer.

Occasionally I try to sell my personal brand of salvation to young engineers, but I do not delude myself about the extent of my success. If I try to speak of my own experience, some smart whippersnapper will tell me what I know only too well, that it was something of a fluke. Not every engineer can go to a private high school and a liberal arts college, then spend a year on an island in the Pacific followed by another year studying the humanities at government expense.

Yet who can predict the future? When the present frenzy over computers and technical training has run its course, the inevitable counterreaction will set in. Engineers, being bright, curious, and pragmatic, may well conceive new patterns for their education and their careers. When older engineers get together they invariably agree that immediately after graduating from college they wished they had taken more technical courses. Ten years later, advancing along career paths, they wished they had learned more about business and economics. Ten years again, in their forties, thinking about the nature of leadership and musing about the meaning of life, they regretted not having studied literature, history, and philosophy. This pattern has become something of a clich , confirmed by studies and polls.

Can it ever change? Of course. At least, "Hope springs eternal in the human breast." The line is by Alexander Pope, who, as I learned at Columbia in 1947, was an eighteenth-century English poet with a penchant for quotable couplets. The sentiment, however, is very much in tune with the engineering temperament.

The Man Who Bought Route 128

Chapter 3, **Getting Sued and Other Tales of the Engineering Life**
Richard L. Meehan

Richard Meehan's recounting of the corruption of Thomas Worcester and its subsequent adverse consequences describes one of the greatest occupational hazards facing civil engineers. Because much of the work of civil engineers revolves around projects involving large sums of money and because the process of securing contracts for completion of these projects seems, at times, to be totally irrational and carried out with total disregard for the engineer's professional ability, the temptations to circumvent the process by exerting extraordinary influence, by monetary means if necessary, are sometimes overwhelming.

Unfortunately--as if succumbing to these temptations is not, in itself, bad enough--circumventing the system to get a contract is usually not the end of the nefarious undertaking, but only the beginning. As Thomas Worcester learned, the demands for money and favors seldom end when the contract has been secured. Honest and competent performance of the engineering work once the contract has been secured (which is often all the engineer ever wanted from the outset) is usually not an important consideration to others who participate in such schemes, and the engineer often finds his or her ability to carry out professional responsibilities has been compromised.

Cases such as the Thomas Worcester case need to be studied carefully by anyone who is considering a career as a civil engineering consultant.

By all accounts I have heard, Thomas Worcester was a gentleman and a fine engineer, a worthy successor to his father, who some years ago had established the Boston consulting engineering firm of Thomas A. Worcester, Inc. Indeed, most New Englanders are well acquainted with the defunct firm's work without necessarily being aware of it, for Worcester's green steel bridges and turnpike cuts and fills are a functional if inconspicuous part of the Massachusetts landscape. Worcester himself was a Harvard man, gray suited and dignified, hardly distinguishable from the investment bankers and lawyers who walk with brisk dignity on Federal and Milk streets. He was a proper Bostonian, what my family would describe, not without some admiration, as a Yankee. And a successful one at that, for after the war, when that particular breed had weakened and many old Boston businesses were on the decline, the Worcester engineering firm was booming, growing in the late 1940's from a few dozen to over two hundred employees.

The success of Worcester's firm during those lean years was based in part on the widely recognized quality of its work. Worcester's resume included $100 million of successful wartime government contracts, and one of his new Route 128 bridges had won a national prize for its design. The firm's success was less obviously but undeniably attributable to a man named Frank Norton, whom Worcester hired in the late 1940's. Norton's calling card described him as assistant to the president, and the assistance he provided concerned that most delicate aspect of the consulting engineering business, which those of us in it frequently describe to our clients as "business development" but discuss among ourselves as "getting jobs." As I well know, having been at it some ten years, it is a constant source of worry, for consulting work is subject to the worst feasts and famines and is highly competitive besides. How often we find ourselves having hired and trained a large and fine staff to complete projects in hand, only then to find the work dry up so we are faced with the prospect of laying off those people we have trained and to whom we have become attached. Or worse, watching the firm's debts pile up while we maintain our staff, anxiously hoping for the arrival of a big new contract. And of course, beyond the goal of corporate stability, there are those little prickles of ambition, the firm of one hundred employees instead of only ten, the invitations to be keynote speaker at conferences or to testify at important hearings, pricey schools like Milton and Beaver Country Day for the children, the teak sailboat at Marblehead.

According to well-established canons of professional behavior, civil engineering consultants, like architects and lawyers, are supposed to be selected for public work on the basis of quality and reputation, through a gentlemanly process of discussion and negotiation, not by grubby competitive bidding. After all, the cheapest fee is not likely to yield the best design. Yet it is sad but true that in the absence of bidding, the selection process sometimes boils down to a matter of personal influence, contacts, favors, or even worse.

So when in the late 1940's a highly political official named William F. Callahan recaptured the post of Massachusetts state director of public works, traditionally known as a most powerful position because of the broad authority of the director in awarding contracts, it became essential that firms like Worcester's establish contacts with Callahan. And Frank Norton was the link, for as Worcester would later recall in court, Norton was "certain he could get the jobs for me because he knew Callahan." Norton was hired at a good salary and given a generous expense account. There are, of course, certain expenses that go along with such a key position as assistant to the president--dinner with clients and cronies at Jimmy's Harborside, an occasional trip to Florida to work out details.

And just as he had promised, Frank Norton began to deliver the jobs in the form of contracts from the Massachusetts Department of Public Works. One was a real plum: the design of the first eighteen miles of Boston's new peripheral highway, Route 128, which included forty-four bridges.

Meanwhile Frank Norton's expenses were as impressive as the contracts he brought in. During one period when Norton delivered $2,750,000 in new engineering work, his expenses amounted to $275,000, or exactly 10 percent of the contract amount. True, sales costs of 10 percent are not entirely out of line in most businesses, including consulting engineering. But what was troublesome was that Norton needed the money in cash, vaguely promising but never providing an account of its disposition.

As owners of small firms well know, getting hundreds of thousands of dollars in cash out of a company is difficult unless the cash is recorded as going to some legitimate subcontractor, supplier, or landlord or unless it is paid out as taxable salary or dividends to the company officers. The company check registers are, after all, subject to audit by IRS tax agents, who are not likely to accept, without further explanation, some $275,000 in checks written to petty cash or to someone's cousin for interior decorating services. Like so many others who have found themselves in this awkward position, Thomas Worcester began to "borrow" a little money from the firm here and there, to tide things over. There was, of course, no going back, and it was not long before the distraught engineer was writing corporate checks to companies and employees who did not exist, then cashing them himself, delivering the bundles of tens and twenties to Norton. Did Worcester, we ask now, really know what was happening to this money? From what he later told me, I would assume he did; but Norton was good enough not to trouble Worcester. a gentleman, with unsavory details. He just delivered the jobs, as they had agreed.

If Thomas Worcester will emerge from the end of this story ennobled by tragedy, Norton is but a minor player, dead now twenty-seven years. But as I reflect on the fragments of court testimony that I have read and on the story that Worcester himself told me one rainy afternoon in his office, I am compelled to pause for a moment, for I feel an odd sympathy for Francis C. Norton, whose name is, I suppose, appearing here in print for the last time ever. In these righteous, post-Watergate days, we are much inclined to moralize, perhaps superficially, on the conduct of men like Frank Norton. But perhaps the Nortons of the world smile at our moral naivet . Perhaps they say to themselves, "You, hypocrites all, are comfortable with your lying and thievery, because they conform to the standards of your peers. But we see you in ways you would never dare to see yourselves. For we are the boundary mediators, ambassadors, Indian scouts. We alone have the courage to slip through that perilous no-man's land between your petty clans. It is a lonely calling, but not without honor."

There is a detail of Norton's demise that touches me. There had been a minor stroke, a chronic heart condition. Frank was not in good health. The pressures were slowly bringing him down, affecting his arteries. A fragment of court testimony harshly illuminates the beginning of the end. The scene was a friend's house, a Judge Charles Flynn's, a few days after Christmas 1951. A children's Christmas party went on in the background, and Flynn poured Norton a stiff drink. Norton had hired Flynn as a lawyer to represent the Worcester firm. The IRS had begun to breathe down Worcester's neck.

"I feel better," said Norton, sipping the drink, his second.

"It's going badly," said Flynn.

"For who?" asked Norton.

"You know for who. It's going badly for the Worcester Company. And for you."

"For me?"

"Yes."

"Well, I'll take care of that," said Norton, rising to leave.

Who among us does not know how sad New Year's Day can be when matters go badly! I imagine Norton sitting at his dining room table, that first day of 1952, staring at his tax returns. It was going badly, and it was up to Frank to take care of it. And he did. Mrs. Norton found him there later that day, slumped over the papers, dead of a heart attack.

If Thomas Worcester had felt some discomfort at the act of passing money to Norton, he must have found it that much more distressing to wrap up the bills in an A&P grocery bag and pay a call to the Norton household. For with or without Frank, certain scheduled payments were due nonetheless; without Frank around and until some other arrangements could be made, it had been decided that Mrs. Norton would have to handle matters. As Worcester had expected, there was another guest at the Norton home when he arrived one evening on his distasteful errand, and although he never saw the man--he had stepped out of the room for a moment when Worcester arrived--Worcester testified that Mrs. Norton had explained that the overcoat hanging in the hall, into which she put the packet of money, belonged to William F. Callahan.

As time passed, the fat public contracts on which the Worcester firm was now becoming dependent began to bring other obligations, other favors for Worcester to perform. There were the out-of-office politicians like Ed Rowe. Somebody in high Democratic party circles had decided that Ed Rowe should be a Republican candidate for governor, a spoiler, to split the Republican election effort; to further this, Worcester hired Rowe for $20,000 to assist in representing the firm. And certain political lawyers, like Henry Santuososso, who had to call Worcester's office one day to ask what the firm's records showed his $10,000 legal consultation was supposed to have been about. A certain city councilor, who was paid $50 a week for seventy-seven weeks for doing nothing. A firm called Public Relations, Inc., hired for $29,000 to "stand by."

One day, an associate would later recall, Worcester just buried his face in his hands. "It's got to be stopped if we're going to keep business going . . .it's almost impossible to do business today without somebody looking for something," said the unhappy engineer.

If we're going to keep business going. The responsibility for enterprise, that moral imperative accepted so eagerly by those engaged in business, entirely overlooked by those who are not, with the result that this fundamental and recurrent rationale is almost always overlooked by both sides in any debate of the ethics of shortcutting the legal and bureaucratic system.

Those readers who are unaccustomed to the ways of big cities in the East should recognize that what I have described so far was, and for all I know still is, the normal manner of doing public business in Boston, just as it is and probably always will be in Bangkok and Honolulu and Accra. This was no secret. As recently as two years before Norton's employment by Worcester, James Michael Curley had taken office for the fourth time as Boston's mayor. On election day, Curley was under indictment for his acts as president of an organization known as the Engineers Group, Inc., a business promotion enterprise started by an ex-convict friend of Curley's to peddle war contracts to engineers. The mayor's election-winning slogan was "Curley gets things done"; but whatever the mayor elect got done for the first five months of his mayoral term had to be done from Danbury prison where he was simultaneously serving a jail term for mail fraud. This was not the first occasion Curley had done time while in public office; indeed the veteran politician had capitalized politically on an earlier jail sentence--he had taken a civil service exam to help a friend--and his authorized biography, **The Purple Shamrock**, which Curley gave as a gift to his admirers, indicated that "there wasn't a contract awarded that did not have a cut for Curley."

This kind of corruption was condemned by proper Yankee Boston, but the continued political success of Curleys and Callahans proved that the majority of voters romanticized their behavior, many considering the shakedowns of businesses like Worcester's as a kind of retroactive tax on past injustices perpetrated by the Boston brahmins on poor immigrants. Who couldn't remember a sweet Irish aunt or grandmother who had scrubbed the floors of fine houses on Beacon Hill? And wasn't that Yankee moralizing really just sheer hypocrisy? "They all paid him--the banks and the utilities and the rest," one of Curley's underlings would recall when Curley had passed from the scene. "But who were the ones who did the paying? All of them, old Yankees." Not only was Yankee justice, encoded in those abstract and unintelligible leather-bound tomes in which the evil spirit of Cromwell still lurked, basically slanted to favor the established rule over the people, but worse yet, even the Yankees themselves cheated when it suited their purposes.

If James Michael Curley remains in public memory as the last of the flamboyant Boston bosses, observers of the Massachusetts scene agreed that the state's real power broker during the postwar years was its elusively quiet but iron-willed director of public works, later chairman of the State Turnpike Authority, William F. Callahan. Callahan, into whose overcoat pocket Thomas[*] Worcester said he had seen his money disappear, was Boston's Robert Moses the man who really got things done, whose political power was such that he never lost a vote in the state legislature. The son of a shoe factory worker, Callahan had earlier lost his public works job during a shakeup initiated by the Republican Governor Saltonstall in the 1930's but recovered his position when the Democrats came back to power in the late 1940's.

Like most other men in our suburban Newton neighborhood, Bill Callahan had worked hard. He had built an organization that had accomplished visible works. "There is a certain pride in building bridges, building roads," he said. The Callahans lived in a fine white house; one of my secret schoolboy shortcuts went through a grove of forsythia at the back of their ample lot. He was a balding, jowly Irishman. A cancer operation left him without a larynx, but with great difficulty and characteristic determination he learned to speak through a hole in his chest. "In 1952 I had my throat cut by a doctor. I've had it cut in other ways many times since," he said. The Callahans had lost a son, Bill Jr., in World War II. He was inclined to spoil his daughter, Jane. She married a Johnny Kelley. Callahan hired a firm called Highway Traffic Engineers, Inc. to work on the Massachusetts Turnpike on the condition that half the firm's stock go to Johnny and that he be paid a salary of $20,000 a year. Kelley's principal contribution to the firm seems to have been attendance at the annual shareholder's meeting. The Kelleys lived in Wellesley, which is even nicer than Newton.

The several reformers who had set out to topple Callahan from his position discovered a crafty and determined opponent. One crime commission lawyer recalled the Callahan style: "He was able to see to it that business was placed in the hands of persons from all walks of life, sometimes on a very thin basis as far as return services were concerned, and he was able to combine the natural

----

* Robert Moses, a contemporary of Callahan's, was Coordinator of Construction for New York City.

loyalty which arises out of a financial relationship of that kind with a particular skill in appealing to the variety of persons who were the subject of his benefaction. His sources of communication were remarkable. He was able to find out what every man cared about, and somehow, directly, or indirectly, to appeal to those desires."

Snooping investigators were never physically threatened but they were audited, received strange calls and callers, fed anonymous leads and tips that later proved false, which led to blind alleys, and which made them appear foolish to the public or within their profession. "The shots that were fired by this remarkable man were of a subtle and perceptive kind and created an atmosphere of embarrassment, confusion, and uncertainty which tended to sap one's resolve," said one prosecutor.

And yet William Callahan was said to live a clean personal life. He did not drink or smoke or even play golf. He was a man who loved work, politics, and family. His interest ran to power, and he made a clean distinction between power and knowledge. Of those critics, especially academic critics, who opposed his high-handed methods, he said, "You've got to have critics. But usually they're people who haven't accomplished anything in their lives. I call them grocery-store philosophers, pen pushers."

One of the critics Callahan most definitely had in mind was a lean, lawyerish-looking young MIT professor of civil engineering, A. Shleffer Lang, who in my junior year there had allied himself with some Harvard Law School people to oppose one of Callahan's pet projects, the extension of the Massachusetts Turnpike into central Boston. One can imagine and even sympathize with the aesthetic objection to ramming a huge, reinforced-concrete, traffic-bearing structure right into the heart of Richardson's and Olmsted's Boston. There is, it has always seemed to me, a fundamental issue of quality, of taste, to be considered in these matters. But it is a characteristic of our times that these public questions cannot be debated in terms of quality; perhaps such terms carry excessively aristocratic associations. It becomes necessary instead to demonstrate that the offensively crude is fiscally irresponsible or that it constitutes a health hazard. So in opposing the project, the professors used the same arguments of financial irresponsibility used by the Saltonstall commission that had fired Callahan back in the 1930s. In the spring of my junior year, their anti-turnpike campaign, which had culminated in a series of public advertisements opposing the project, successfully frightened Wall Street investment bankers from purchasing $175 million in bonds for the proposed extension.

If Callahan was furious with the interference of these "pen-pushing" Harvard and MIT professors, by the end of that same year he had encountered an even more determined enemy. This was Judge Charles E. Wyzanski of the U.S. District Court of Massachusetts, who had taken on the 1960 tax fraud case of *United States v. Thomas Worcester* after a previous judge, William McCarthy, had been too busy to try the case in the last three years before his retirement. Wyzanski's starchy disapproval of Massachusetts politicians and their methods was well known. The judge was fond of comparing, for didactic purposes in his prep school commencement speeches, the righteousness of Massachusetts's earliest governor, John Winthrop, with the depravity of its recent governor, Curley. Presented with a tax-evasion conviction of consulting engineer Thomas Worcester, Wyzanski seized this opportunity to "sound a clarion," as he described his own actions, to expose the corruption that was rotting what he imagined had once been a commonwealth of honorable men. He accordingly offered Worcester a deal: in exchange for suspension of the eighteen-month jail sentence that the engineer had received for evading taxes on $180,000 of illegitimate payoffs, Worcester agreed to testify with complete candor on the disposition of the "Worcester bounty," as the judge called the payoffs and favors that the engineering firm had so liberally dispersed in return for public contracts. Wyzanski's purpose was nothing less than exposing a statewide "network of corruption," which included public officials at all levels of the state, along with the other institutions--banks, insurance firms, the press, and the bar--extending even to members of the judiciary, for he believed that all of them were, with greater or lesser degrees of enthusiasm, participants in the system of bribery and extortion. Worcester, whom Wyzanski said had "completely persuaded me of his decency," agreed to testify in order to avoid jail, and for several months in the fall of 1960, Wyzanski, in an unusual series of hearings, used Worcester's testimony as a starting point for a major expos of the corruption rooted in the Massachusetts system of public works, and most specifically in that perennial soft spot, the award of engineering and insurance service contracts by Callahan's Turnpike Authority. For three months, Wyzanski subpoenaed and exposed state officials, including Callahan himself, to the questioning of both Worcester's attorneys and the U.S. attorney, Elliot Richardson.

Predictably, Wyzanski's efforts were harshly criticized in some sectors of the press, perhaps even more so in the streets and bars, after Sunday mass and over corned beef dinners in Boston's predominantly Irish neighborhoods. The ethnicity of the name was unclear--was he a Catholic or a Jew?--but the affiliations were soon enough revealed. Wyzanski had attended the prestigious Exeter Academy, then gone on to Harvard. He was a member of Boston's Signet and Tavern clubs, was prone to sprinkle his college and prep school commencement speeches with references to Pericles and Locke and Wordsworth, to reminisce about his Exeter friends, boys with names like Francis Plimpton and Dudley Orr and John Cowles. Whatever he was, Catholic or Jew, one thing was clear enough about this Wyzanski he was trying to pass himself off as a Yankee.

Indeed they were all Harvard men: Worcester, Wyzanski, and Richardson. Elliott Richardson had been a legislative assistant to the same Governor Saltonstall who in 1939 had dismissed Callahan from his public works position for squandering public funds. Richardson's career and affiliations had been a carbon copy of Wyzanski's; both had graduated from the same school, clerked for the same justice, Learned Hand, and been associated for many years with the oldest Yankee law firm in the city. Were these not grounds to think that Wyzanski was Richardson's mentor?

Surely this highly publicized hearing, initiated by Richardson and blessed by Wyzanski, had a larger purpose than adjudication of the trivial question of whether Thomas Worcester's probation should be revoked. Clearly it was an attempt to try the political system of the state, an assault on the institutions of family and church, a campaign to undercut the personal loyalty that protected people from exploitation by the traditionally powerful Yankee establishment. Wasn't this only an extension of the communist subversion of American ideals, which--as Senator McCarthy and others had shown in the early 1950's--was infecting institutions like Harvard and MIT? And who could

461

ignore the fact that it was an election year, that Bobby Kennedy was already searching Cambridge for candidates for high-level federal posts? Wasn't Wyzanski, who had served with John Kennedy as a Harvard overseer, really just creating a little election year publicity for himself, perhaps bucking for a Kennedy appointment? Could it not be said that behind that Harvard mask of righteousness there lurked more in the way of personal ambitions than Judge Wyzanski would care to admit?

If men of action like Bill Callahan cut corners to get things moving, took care of their families and friends, spread the work around to the small people, what of it? Consider that Callahan had been accused, in which his public relations man called "Mr Wyzanski's Nazi trial," of pushing turnpike insurance business toward friends among the fifty-four insurance brokers who were members of the state legislature. But why shouldn't he? Callahan pointed out how when the Republicans were in office, the big, old-line insurance companies got the business. "The little fellows didn't get a look. Somebody has to get the insurance business. And it don't cost you any more whether you insure through John Smith or Mike McCarty." The difference was that Mike McCarty, barely getting along with six kids, would remember the favor when it came time to vote. John Smith--or Thomas Worcester--offered money, not loyalty.

That same fall I had finally reached the relatively safe anchorage of my senior year at MIT. It was a presidential election year, and for those of us from Boston a year of paradoxical choices. Presidential candidate Richard Nixon, coming from that land of pinko lotus-eaters, California, was a staunch anticommunist. John Kennedy, grandson of a Boston mayor said to have been even more corrupt than Curley, was a liberal who sat with Judge Wyzanski on Harvard's Board of Overseers. And yet it was Bill Callahan, not Charles Wyzanski, who was invited to Kennedy's private birthday party; Callahan sold most of the fund-raising advertisements on the souvenir menu. Neither presidential candidate measured up to my standards of honor. Friday evenings, at my proletarian hangout, the Paradise Cafe, I defiantly played the pinball machine instead of watching the televised debate. Why vote, I would say to my parents, when the choice is between two phonies like that?

It was the year in which I was too cynical to participate in the new dance craze, the twist, but earnest enough to criticize our unfair play toward Castro's new Cuba. Like most of my classmates, I was resigned to compulsory military service after graduation, but there would be plenty of good jobs waiting after that. It was 1960, and progress was still an important word. The first commercial nuclear power plant had just been licensed. A couple of young engineers were starting a new company called Teledyne. The laser, the felt-tip pen, and Enovid were perfected that year, and the space race would go on. Few things were as bright as the future of technology.

With the exception of one structural design course, MIT seemed easy to me that fall. I wrote a FORTRAN computer program that aimed to beat Fats, the bookmaker at the Paradise (it did not); studied the strength of the building materials of the future, foam core aluminum panels and fiberglass-reinforced plastic: broke asphalt cylinders in the basement of Building One. For the fun of it, I took two humanities courses. One was taught by a wickedly funny visitor, John Hawkes, whose novels were just coming out as New Directions paperbacks. The other, a course in moral philosophy, was taught by Yale philosopher John Rawls. I did not understand much of the scholarly Rawls, about whom a book has recently been published, **Understanding Rawls**. Rawls's spare, bespectacled appearance and scholarly manner were just what one would expect of that endangered species, the moral philosopher. Most of the specimens, one gathered from our assigned readings, were to be found in English academic preserves, tut-tutting each other's "fallacies" at croquet or over tea. I soon read enough of G.E. Moore and A. J. Ayer to realize that they would never presume to replace the authority of the church fathers, whom I had chosen to dismiss a few years earlier, thereby creating a gap in my moral sensibility as troublesome as a missing tooth. It seemed to me that the principal concerns of modern ethical philosophers were linguistic; they wanted to define precisely the meaning of the word good in "Bill Callahan is a good man" or should in "Thomas Worcester should have demanded an account of Norton's expenses." I gathered that one school of philosophical thought had gone so far as to declare that such moral statements had no meaning whatsoever; *should* and *good* in their view were no more than animal grunts of approval. If this were true, then of course these Harvard lawyers and Boston Irish pols were no more than two clans of peeled apes, screeching at each other across the Charles River, and Thomas Worcester's only wrong was the practical mistake of trying to operate across tribal boundaries. Or perhaps, in accordance with Immanual Kant's basic and to me more sensible rule, a man would be right provided only that he acted as if he willed his actions to be universal laws. In that case, wouldn't Callahan and Wyzanski, each of whom would be content presumably to live in a world of their own values, both be logically "right"? Then only Worcester, whose respectability did not match his conduct, would be wrong. Could it be that the existence of diverse hardy ethnic subgroups, each with its own "operational code," provided a benevolent ecological stability to Massachusetts, a political state which might be preferable to mass anomie, to spells of weak leadership broken by periodic revolutionary violence? Was honesty to be expected in a true leader? "Our rulers will find a considerable dose of falsehood and deceit necessary for the good of their subjects," said Socrates. It was all very confusing, from an ethical standpoint. Now in leafing through my twenty-year-old text for Rawls's moral philosophy course, Sellars and Hospers's **Readings in Ethical Theory,** seven hundred pages sturdily bound, I am pleased to find that I underlined the following statement, which still best expresses my retrospective findings on the practical applications of moral philosophy: "It is therefore very easy when we investigate these matters philosophically to go wrong because we do not have before us at the time a genuine ethical experience, and at the very moment when we have such an experience we are too much concerned with it as a practical issue to philosophize about it."

Spring term of 1961 was my last at MIT, and as the putty-colored limestone buildings of the institute began to warm in the April sun and the grass of the quadrangle grew thick and green, I began to feel a sense of lightheartedness that I had not known since high school summer vacations. For me the struggle for grade-point averages was over; the last lion, "Structural Analysis and Design," was replaced by a lamb, "Social and Political Factors in Engineering." Said to be a gut course, it was taught by the same Professor Lang who earlier had tangled with William Callahan on the turnpike extension bond issue. Lang's aim was to expose us to some of the nontechnical problems that engineers are famous for not recognizing. I had been following the Worcester case in the newspapers, and when I suggested that as a course project I undertake a study of the scandal, Lang told me that would be an excellent idea. I accordingly called Worcester's attorney, Calvin Bartlett, and explained my interest in the case. Bartlett invited me to his office and gave me a large cardboard box full of court transcripts. When I asked him a question about Worcester's attitude, Bartlett suggested that I ask Worcester himself, picked up the phone, and called the engineer. Would Worcester talk to me? The answer was apparently yes. Bartlett asked me what day would be convenient for me the following week and set up an appointment.

So it happened one rainy spring afternoon that I took the MTA downtown to visit Thomas Worcester. I was nervous and wore what seemed to me to be an appropriately funereal charcoal suit and dark blue tie. It was a blustery day and by the time I found the correct street address, my London Fog raincoat and suit pants were soaked.

In the cramped lobby of one of the older buildings in Boston's financial district, I found one of those glass-encased, black felt building directories on which press-on letters indicated that what remained of the Worcester firm occupied the fifth floor. A fifty-foot ascent in a creaking elevator and a short walk down a linoleum corridor brought me to a heavy oak and glass door. Inside a male secretary greeted me suspiciously, then escorted me through a series of empty offices with dusty drafting tables to Worcester's office. Worcester greeted me like a gentleman, and we sat talking for an hour or so. My recollection is of a tall, stooped man with a starched white shirt, who seemed tired but glad to tell me the story, much as I have told it here. It seemed to me then as it does now that Worcester, of all of them, on a purely logical basis had committed the most grievous wrong. And yet Worcester that afternoon had a certain stoic way about him, a certain dignity that elevated him, the one man who had actually been convicted, above all of the others, including the righteous Wyzanski and the determined Callahan. It was, I suppose as I look back on it now, residual traces of Catholicism in me that made Worcester's judgments carry a certain high authority, for only he among the many sinners had done penance. I believed, moreover, that Worcester's trial had yielded a good, that Massachusetts politics would never be the same.

Only once that afternoon did that cool Yankee seem bitter, and that anger was directed at his fellow engineers, who had put him out of the Boston Society of Civil Engineers for his conduct. "They denounced me and yet they all conduct their business the same way," he said.

After he talked, Worcester accompanied me to the door, we shook hands, and I walked out into the thickening gloom. Worcester and the male secretary had been the only people on the fifth floor of the building. Walking back to Park Street, past the slate stones of Granary Burying Ground glistening black in the rain, the realization came to me that Worcester's only reason for coming to the office that day may have been to talk to me.

464

Loyalty, or Why Engineering Is Sometimes Like Baseball

Chapter 9, **The Civilized Engineer**
Samuel C. Florman

Dealing with conflicts between one's professional obligations and one's personal conscience is not a matter for which most civil engineers are thoroughly prepared by their formal education. There was a time, as Florman indicates, when such conflicts were almost non-existent. Today, however, the diversity of viewpoints within American society, the complexity of many scientific and technological problems, and heightened awareness of moral and ethical issues on the part of a better-educated public make the possibility of such conflicts very real. If for no other reason, Mr. Florman's long experience as a practicing civil engineer would be reason enough for his views on resolving such conflicts to be given serious consideration by aspiring engineers.

"What should I do if I am assigned work that I think is not in the best interest of society?" The young man who asked me this question was an engineering student soon to graduate and embark on a career. He was frowning and looked deeply troubled. Even in this time of conservative self-interest, I have met a number of young engineers with similar misgivings.

When I was in school we didn't worry about such matters. We had concerns aplenty, but they were of a different kind: What sort of work did we want to do? What were we good at? How could we best achieve success, whatever that was? As for *loyalty*, it never occurred to us that conflicts could arise. We assumed that an engineer was loyal to superiors, colleagues, and clients, and that service to the community came naturally as the result of a job well done.

Today, in the wake of the environmental crisis and widespread doubts about technology, such a view smacks of insensitivity. If engineering projects can turn out to be harmful--to the environment, say, or to public safety--then it appears possible that one's professional work might run counter to one's obligation to society. Which gives rise to the question put to me by the frowning student.

. . . This young man was thinking intently about his own personal future and his own very personal feelings. After a moment's thought, I assured him that he was unlikely to encounter the moral crisis he feared. I could say this with some confidence because every engineer I have ever met has been satisfied that his work contributes to the communal well-being, though admittedly I had never given much thought to why this should be so. As I pondered the question, reflecting on the way most people manage to blend personal morality with loyalty and commitment, I found myself thinking--of all things--about the trade of Eddie Stanky.

In the late forties Eddie Stanky played second base for the then Brooklyn Dodgers. He lacked outstanding talent but was known as a great competitor. He was feisty and inventive. He was particularly adroit in reaching first base by managing to be hit with pitched balls. One day I was horrified to learn that this dirty player, this bad sport, had been traded to the team I rooted for, the New York Giants. But then a strange thing happened. Once he was playing for my team he seemed morally reformed. We Giant fans no longer called him

465

a bad sport or a dirty player. He was spunky little Eddie. He was clever little Eddie. He was resourceful little Eddie. That was one of my earliest lessons in the relativity of ethics.

I told my questioner about this experience. It may seem frivolous to compare such an incident to the moral dilemmas of engineers, but I think the analogy is worth considering. Is it not true, at work or play, that we instinctively root for the home team? Of course, we do not condone real knavery; but we tend to view our own cause in a favorable light. An engineer who works for Exxon feels differently about offshore drilling than he would if he were employed as town engineer for a seaside resort. This has less to do with good and evil than it does with where one stands at a given moment. The petroleum engineer feels that he is called upon, both by his company and society as a whole, to provide an adequate supply of oil at a reasonable price. The town engineer is responsible for the well-being of the people who rely upon him to keep their community safe, clean, and prosperous.

Obviously an engineer's attitudes are in some measure formed by assignment. Not every project need be exactly to his taste, . . . but on the job he will likely find his fellow workers to be, by and large, decent people with whom he will want to join in common cause. From the outside, Exxon may appear to be a giant, heartless corporation obsessed with making profits. From the inside, Exxon doubtless is perceived as an association of earnest individuals working hard on constructive projects. "Groupthink" is a well-documented phenomenon from which none of us is immune.

"But isn't this bad?" the troubled student asked. "Not necessarily," I replied. Nobody wants a professional to pledge blind allegiance to an employer or a client, but since the *sine qua non* of accomplishment is effort, society is clearly well served if most people approach their work with zest--even if the zest is founded in something as apparently trivial as team spirit.

The student and I did not have time to pursue the discussion further. Later, however, I wondered if my answer had not rung hollow. After all, the classic response in such an encounter has always been "To thine own self be true." Yet I could not persuade myself that this was what one ought to stress to a young person preparing for a first engineering job.

In addition to enthusiasm, the good work that assures communal well-being requires order and cooperation. Any group of people undertaking a complex task must establish an organizational structure. Without assigning responsibility--without discipline--large-scale technological enterprise is unthinkable. Accordingly, a young engineer will have to temper any personal misgivings, such as they are, with an awareness of the group's needs for order. This may sound slightly unsavory, but only until it is considered in depth. Our daily lives teach us to live with disappointment. Our particular point of view cannot always, or even usually, prevail. In politics we learn to accept this and proudly call it democracy. We do what we can to influence the tide of events without being excessively disruptive. It is called working within the system.

Indeed, the system seems to work best when different groups, each energetically pursuing its own goals, clash and are obliged to resolve their differences. The term "healthy competition" has become a clich , but it expresses a profound truth. Competition *is* healthy--ecologically, politically, and, of course, technologically. Our society's strength stems in no small measure from its multiplicity of often conflicting institutions: corporations, labor unions, bureaucracies, universities, trade associations, charities, foundations, political parties, and public interest organizations (which often more accurately could be termed "private interest organizations"). In this respect the body social reminds me of the human body, in which health is maintained by a variety of defense mechanisms: antiseptic fluids, white blood cells, the lymphatic system, and an incredibly complex chemical immune system.

Society's interests are better served, I believe, by the resolution of conflict between organizations than by the disaffection of individuals within organizations. Let oil companies search for oil--not recklessly, without care and common sense--but enthusiastically, without the kind of inner dissension that results in paralysis. Let the rest of us establish the limits beyond which we do not want the oil companies to go. And let the same process apply wherever technological progress impinges on other values that society holds dear--particularly public safety and environmental quality.

When engineers are loyal, each to his own organization the system works and the public is served. But more is involved than efficiency. The quality of communal life depends upon the trust and respect that prevails among families, friends, and co-workers. An abstract devotion to "the good of humanity" is no substitute for devotion to real human beings. Out of fruitful personal relationships comes the decency that sets the moral tone for the good society. A sociologist who interviewed hundreds of workers once told me that what Americans love best about their jobs is the social milieu in which they spend their days--the relationships they establish with their colleagues. The Japanese have developed a sense of group harmony far beyond what we Westerners can, or care to, cultivate. But even in our individualistic society, mutual loyalty and good faith in the workplace is a key to personal happiness, and productivity as well.

Unfortunately, since the 1950s the concept of loyalty has been tarnished by too many loyalty oaths and abusive demands for conformism, just as the concept of patriotism has been sullied to the point where the term is on the verge of becoming pejorative. The problem is as old as humanity: how to balance the rights of the individual with the needs of the group.

I do not advocate unthinking obedience to the group, obedience of a sort that was confronted and discredited at the Nuremberg trial of war criminals. And I do not say that whistle-blowing is never justified. It is, occasionally, not only justified but necessary and heroic. However, one cannot endorse--except in the most exceptional circumstances--the betrayal of one's companions for some "greater good." Ayatollah Ruhollah Khomeini urges students to spy on their teachers and classmates and to secretly tell security forces about those whose dedication to Islamic values is in doubt. This is whistle-blowing of the worst kind, the other end of the spectrum from Nazi-type loyalty. Between the distasteful extremes of betrayal and servility each individual must make his own way. Happily, the need for engineers to make such choices is, to my best knowledge, exceedingly rare.

If I were able to resume my discussion with the troubled young engineer, I would urge him to embark upon his career with great enthusiasm and not be overly apprehensive that his work might harm society. Naturally, he must follow his own star, but to a certain extent he has done this already by choosing to be an engineer, a professional who must work with others, rather than, for example, choosing to be an artist, who can work alone. There are still personal decisions to be made. If he is averse to armaments or nuclear power, by all means let him steer clear of these fields. If he is an ardent environmentalist, let him seek work in that area. But once engaged in an engineering task, let him put his whole heart into it. He can maintain his values and sense of self without indulging his ego at every turn. If he wants to serve society, let him do good work.

What if, in spite of the enormous odds against it, he does encounter base practices and experiences a crisis of conscience? Loyalty to the group will require that he come down on the side of legality and prudence. No assemblage of engineers is well served by deception. And if the very worst should happen--if he becomes involved in a situation that cannot be honorably resolved within his organization--then, speaking as a member of society, I would rather count on the righteous wrath of an engineer whose loyalty has been betrayed than on the pique of an engineer who was from the start a suspicious malcontent.

It is a bittersweet paradox that new-fangled technology, because it depends so much upon group effort, should summon forth old-fashioned morality. The American philosopher Josiah Royce, writing at the beginning of this century, spoke of the need for "loyalty to loyalty." The idea is worth reviving in our time.

A New Tradition: Art in Engineering

An Excerpt from Chapter 1 - The Tower and the Bridge
David P. Billington

Design handbooks, books on technical aspects of design and books describing analytical techniques for design are common in all the civil engineering disciplines, but nowhere are they more common than in the field of structural design. Most such books provide almost no information on the rationale for a particular design approach. David Billington, Professor of Civil Engineering at Princeton University, has produced a book which is a rarity in the engineering profession. It provides a systematic examination of the philosophy and rationale for structural design and an exploration of its historical roots.

In this selection, the introductory chapter to his book, he spells out the significance of the three primary objectives of structural engineering--efficiency, economy and elegance--and describes their role in producing engineering structures that are works of art. Professor Billington also explores in this chapter the relationship between science and engineering, the relationship between structural engineering and architecture, and the relationship between the engineering of structures and the engineering of machines. His exploration of these relationships defines the unique role of the structural engineer.

A New Art Form

While automation prospers, our roads, bridges, and urban civil works rot. Children control computers while adults weave between potholes. The higher that high technology sails the worse seem our earthbound services for water, transportation, and shelter. Yet civilization is civil works and insofar as these deteriorate so does society, our high technology notwithstanding. We forget that technology is as much structures as it is machines, and that these structures symbolize our common life as much as machines stand for our private freedoms. Technology is frequently equated only with machines, those objects that save labor, multiply power, and increase mobility. In reality, machines are only one half of technology, the dynamic half, and structures are the other, static, half--objects that create a water supply, permit transportation, and provide shelter.

This book is devoted to the idea that structures, the forgotten half of modern technology, provide a key to the revival of public life. The noted historian Raymond Sontag titled his book on the period between the two world wars **A Broken World**, and his pivotal chapter called "The Artist in a Broken World" characterized the persistent hopes of the time by "the vision of mending the broken world through a union of art and technology." He had in mind groups like the ill-fated German Bauhaus, but he and all other historians missed the fact that such a union had for a long time already existed. It was a tradition without a name, confused sometimes with architecture and other times with applied science, even on occasion misnamed machine art. It is the art of the structural engineer and it appears most clearly in bridges, tall buildings, and long-span roofs.

This new tradition arose with the Industrial Revolution and its new material, industrialized iron, which in turn brought forth new utilities such as the railroad. These events led directly to the creation of a new class of people, the modern engineers trained in special schools which themselves came into being only after the Industrial Revolution had made them a necessity.

Such developments are well known and almost everyone agrees that they have radically changed Western civilization over the past two hundred years. What is not so well known is that these developments led to a new type of art-- entirely the work of engineers and of the engineering imagination. My major objective in this book is to define the new art form and to show that since the late eighteenth century some engineers have consciously practiced this art, that it is parallel to and fully independent from architecture, and that numerous engineering artists are creating such works in the contemporary world of the late twentieth century. It is a movement awaiting a vocabulary.

The Ideals of Structural Art

Although structural art is emphatically modern, it cannot be labeled as just another movement in modern art. For one thing, its forms and its ideals have changed little since they were first expressed by Thomas Telford in 1812. It is not accidental that these ideals emerged in societies that were struggling with the consequences not only of industrial revolutions but also of democratic ones. The tradition of structural art is a democratic one.

In our own age when democratic ideals are continually being challenged by the claims of totalitarian societies, whether fascist or communist, the works of structural art provide evidence that the common life flourishes best when the goals of freedom and discipline are held in balance. The disciplines of structural art are efficiency and economy, and its freedom lies in the potential it offers the individual designer for the expression of a personal style motivated by the conscious aesthetic search for engineering elegance. These three leading ideals of structural art--efficiency, economy, and elegance--which I shall illustrate throughout this book, can be briefly described at the outset.

First, because of the great cost of the new industrialized iron, the engineers of the nineteenth century had to find ways to use it as efficiently as possible. For example, in their bridges, they had to find forms that would carry heavier loads--the locomotive--than ever before with a minimum amount of metal. Thus, from the beginning of the new iron age, the first discipline put on the engineer was to use as few natural resources as possible. At the same time, these engineers were called upon to build larger and larger structures--longer- span bridges, higher towers, and wider-spanning roofs--all with less material. They struggled to find the limits of structure, to make new forms that would be light and would show off their lightness. They began to stretch iron, then steel, then reinforced concrete, just as medieval designers had stretched stone into the skeletal Gothic cathedral.

After conservation of natural resources, there arose the ideal of conservation of public resources. In Britain, which was the center for early structural art, public works were under the scrutiny of Parliament, and private works were usually under the control of shareholders and industrialists. The engineer had, therefore, always to work under the discipline of economy consistent with usefulness. What the growing general public demanded was more utility for less money. Thus arose the ideal of conservation of public resources. The great structures we shall describe here came into being only because their

470

designers learned how to build them for less money. Moreover, working with political and business leaders was a continuing and intrinsic part of the activity of these artists. They created not alone in a laboratory or a garret but under the harsh economic stimulus of the construction site.

Curiously enough, whenever public officials or industrialists decided deliberately to build monuments where cost would be secondary to prestige this art form did not flourish. Economy has always been a prerequisite to creativity in structural art. Again and again we shall find that the best designers matured under the discipline of extreme economy. At times, when approaching the limits of structure late in their careers, they might encounter unforeseen difficulties which increased costs. But their ideas and their styles developed under competitive cost controls. Economy is a spur, not an obstacle, to creativity in structural art.

Minimal materials and costs may be necessary, but they are not, of course, sufficient. Too many ugly structures result from minimal design to support any simple formula connecting efficiency and economy to elegance. Rather, a third ideal must control the final design: the conscious aesthetic motivation of the engineer. A major goal of this book is to show the freedom that engineers actually have to express a personal style without compromising the disciplines of efficiency and economy. Beginning with Telford's 1812 essay on bridges, modern structural artists have been conscious of, and have written about, the aesthetic ideals that guided their works. Thus, this tradition of structural art took shape verbally as it did visually. The elements of the new art form were, then, efficiency (minimum materials), economy (minimum cost), and elegance (maximum aesthetic expression). These elements underlie modern civilized life.

Civilization requires civic or city life, and city life forms around civil works: for water, transportation, and shelter. The quality of the public city life depends, therefore, on the quality of such civil works as aqueducts, bridges, towers, terminals, and meeting halls: their efficiency of design, their economy of construction, and the visual appeal of their completed forms. At their best, these civil works function reliably, cost the public as little as possible, and, when sensitively designed, become works of art. But the modern world is filled with examples of works that are faulty, excessively costly, and often ponderously ugly.

Such need not be the case. If the general public and the engineers themselves see the extent and the potential of structural art, then public works in the late twentieth century can, more than ever, be efficient, economical, and elegant.

The History of Structural Art

I shall demonstrate the potential of structural art through its history, and have divided the book into two parts to reflect the two major historical periods. The first part of the book traces the history of structural art up to the completion of the Eiffel Tower, the last great work of iron, and the second describes the developments springing from the use of steel and concrete and concludes with a series of the late-twentieth-century works. The historical narrative begins in Britain toward the end of the eighteenth century. Here we can see how the rise of new forms is connected directly to the use of new materials in solving the transport problems posed by industrialization. The

471

transportation networks--canals, roads, and railways--accelerated the pace of technological developments, leading to urbanization and further industrial change. As cities grew more crowded, office buildings became higher, and train terminals of longer span and bridges of truly immense proportions began to be economically feasible.

The second period of structural art begins in the 1880s, when steel prices dropped and reinforced concrete was developed. Engineers soon began to explore new forms with these materials, so that even before the cataclysm of 1914, a bewildering variety of structures arose at a dizzying pace. But the maturity of new forms in steel and concrete came only afterward, when Western civilization careened from one world war to another through boom, inflation, and depression. During this period, movements in art and architecture proclaimed solutions to city decay, focusing on the menace or promise of technology.

The best known of these movements was the German Bauhaus, whose aim was to "avoid mankind's enslavement by the machine" by integrating architecture and machine production, and by getting the artist away from art for art's sake and the businessman away from business as an end in itself. The new architect, in the words of the Bauhaus founder, Walter Gropius, would be "a coordinating organizer, whose business is to resolve all formal, technical, sociological, and commercial problems" and whose work leads from building to streets, to cities, and "eventually into the wider field of regional and national planning." The Bauhaus and other such movements barely recognized the tradition of structural art. For example, in a classic work defining the Bauhaus, Gropius included forty-five illustrations, not one of which shows any work of structural art. Furthermore, in describing the comprehensive education given to the new architect, Gropius noted that there were no courses offered in steel or concrete construction. Although Gropius and others stimulated new thinking about technology and design, they did it from the perspective of architecture rather than structure. Indeed, the great influence of such architects on post-World War II ideas about building has tended to obscure the tradition of structural art. In addition to the common confusion between structural art and architecture, there arose a misconception about the relationship of structure to science and to machine art. Therefore, I must say something about what this new engineering art is not, before showing historically what it is.

Engineering and Science

The confusion of structural art with science assumes that engineering, being applied science, merely puts into practice the ideas and discoveries of the scientist; the engineer is merely the technician, following orders from above. This idea is a common twentieth-century fallacy. It was articulated, for example, by Vannevar Bush, wartime director of the Office of Scientific Research and Development, in his influential report to President Truman which led to the establishment of the National Science Foundation. Bush summarized his ideas vigorously:

> Basic research leads to new knowledge. It provides
> scientific capital. It creates the fund from which the
> practical applications of knowledge must be drawn. New
> products and new processes do not appear full-grown.

> They are founded on new principles and new conceptions,
> which in turn are painstakingly developed by research in
> the purest realms of science.

> Today it is truer than ever that basic research is the
> pacemaker of technological progress. In the nineteenth
> century, Yankee mechanical ingenuity building largely
> upon the basic discoveries of European scientists, could
> greatly advance the technical arts.

Not only is Bush's history of Yankee ingenuity inaccurate, but so is his general belief that "basic research is the pacemaker of technological progress." In a 1973 conference, leading historians of technology presented papers on the subject "The Interaction of Science and Technology in the Industrial Age." The conference summarized the wide variety of studies by then completed and "overwhelmingly, the group agreed in disagreeing with the conventional view (of Bush) that technology was applied science."

There is a fundamental difference between science and technology. Engineering or technology is the making of things that did not previously exist, whereas science is the discovering of things that have long existed. Technological results are forms that exist only because people want to make them, whereas scientific results are formulations of what exists independently of human intentions. Technology deals with the artificial, science with the natural.

Science and technology are best viewed as parallel activities, each one at times drawing on the resources of the other, but more often developing independently. An example of this independence is the fact that of the vast number of technological inventions made since World War II for the military, only about 0.3 percent can be traced to scientific discoveries; the remainder developed independently, from design stimuli within the technological community itself. A leading British scholar recently concluded that there is "very little indication of any clear or close links between basic scientific research and the great mass of technical developments." Having considered a wide variety of case studies, ranging from chemistry in Britain to structures in the United States, he observed that "science seems to accumulate mainly on the basis of past science, and technology primarily on the basis of past technology." In our present context, it is essential that we make the distinction between science and technology, so that we can focus on the true sources of engineering originality.

From the fundamental difference mentioned earlier flow a number of other crucial differences. Science works always to achieve general theories that unify knowledge. Every specific natural event, to be scientifically satisfying, must ultimately be related to a general formulation. Engineering, in contrast, works always to create specific objects within a category of type. Each design, to be technologically satisfying, must be unique and relate only to the special theory appropriate to its category. It is this uniqueness that makes structural art possible. Were engineering works merely the reflections of general scientific discoveries, they would lose their meaning as expressions of the style of individual designers. The fact that these works need not--indeed, in some cases should not--be based on general theories is apparent from concrete studies in the history of technology. I give here two illustrations.

Robert Maillart, the Swiss bridge designer, developed in 1923 a limited theory for one of his arched bridge types which violated in principle the general mathematical theory of structures and thereby infuriated many Swiss academics between the wars. But Maillart's limited theory worked well for that special type of form. Within that category type, Maillart's theory was useful and had the virtue of great simplicity; he developed the theory to suit the form, not the form to suit the theory. In the United States, by contrast, some of our best engineers understood the general theory well, but not understanding Maillart's specific ideas, they failed to see how new designs could arise. They were trapped in a view of an engineering analysis which was so complex that it obscured new design possibilities. Today the undue reliance on complex computer analyses can have the same limiting effect on design.

A second, even more dramatic example occurred with suspension bridge design at the same time. A new and more general theory of analysis became fashionable in the 1920s. Imbued with the idea that more general theories would automatically give more complete insight into bridge performance, all leading designers of the period used that theory, which obscured rather than clarified understanding and helped cause the defective design for a series of major bridges in the 1930s and the Tacoma Narrows Bridge collapse of 1940.

Such examples show how this new perspective on engineering design as an activity independent of basic science suggests a new type of research, basic to a design profession, where historical, humanistic study is as important as the development of scientific analyses.

Structures and Machines

Related to the fallacy that technology is applied science is the fallacy that technology involves only machines. This one-sided view dominated Jacques Ellul's frequently cited **Technological Society**, allowing him to portray the modern world as both mechanistic and demonic, without personality, without art, and without hope. Crucial to Ellul's argument was his insistence on defining technology (or, in French, *la technique*) as "the one best way," the super-rational means by which one inevitably arrives at the single optimum solution to each problem. There is no possibility, in this view, for individuals to express their own personalities except, as Ellul puts it, by adding useless decoration to the machine. Only by compromising function or adding cost, two sides of the same thing, could the engineer inject any art. Ellul argued that this art was merely symbolic of a machine age and did not at all reflect the efficiency of the "one best way."

But technology is not just machines. There are two sides to technology: structures--the static, local, and permanent works--and machines--the dynamic, universal, and transitory ones. The Eiffel Tower, Seattle's covered stadium (the Kingdome), and the Brooklyn Bridge are structures; they were designed to resist loads with minimum movement and to stand as long as their societies stand. By contrast, elevators, air conditioners, and cars are machines; they only work when they move and are continually replaced as they wear out or are made obsolete by newer models. Technology has always meant both structures and machines; they are its two sides.

The civilized world requires both sides of technology. Structures stand for continuity, tradition, and protection of society; machines for change, mobility, and risk. There is a constant tension between these two types of objects--between the extremes of a frozen society where structure dominates and a frantic society dominated by machinery. Yet structures must be built by machines and most only can be built because of machines. Modern city buildings would be almost useless without elevators, and very few bridges would ever have been built without the pressure of railroads and automobiles. In the same way, machines require structure to hold them together and would be useless without structures in which or on which to operate.

As intimately connected as they are, structures and machines must function differently, they come into being by different social means, and they symbolize two distinctly different types of designs. Structures must not move perceptibly, are custom-built for one specific locale, and are typically designed by one individual. Machines, on the other hand, only work when they move, are made to be used widely, and are in the late twentieth century typically designed by teams of engineers. General statements about technology are frequently meaningless unless this basic distinction is first made.

In addition to the two types of objects, technology can be thought of as including two types of systems: networks and processes, which are extensions of structures and machines respectively. Networks--such as canals, roadways, railways, electric lines, and airways--are immovable conduits distributing things. The network is a distributor not a convertor. Processes, on the other hand, are systems that change the state of things--such as internal combustion, oil refining, water treatment, and electric power generation. These are dynamic systems, characterized by change and related intimately to machines such as engines, pumps, reactors, and turbines. Networks are static systems characterized by their permanence, and depend for their operation upon such structures as aqueducts, bridges, dams, airports, power plants, and transmission towers.

I shall consider only structures, but it should be clear that they lose meaning if we forget their complementary relationship to machines. The Eiffel Tower is mainly lost to the general public without its elevators; the Kingdome would be useless without electric lights and air conditioning; and the Brooklyn Bridge was built by the use of all kinds of machinery and serves today as a major route for cars.

Structures and Architecture

The modern world tends to classify towers, stadiums, and even bridges as architecture. This represents yet another, albeit more subtle, fallacy similar to the confusion of technology with applied science and with machines. Here even the word is a problem because "architect" does come from the Greek word meaning chief technician. But, beginning with the Industrial Revolution, structure has become an art form separate from architecture. The visible forms of the Eiffel Tower, the Kingdome, and the Brooklyn Bridge result directly from technological ideas and from the experience and imagination of individual structural engineers. Sometimes the engineers have worked with architects just as with mechanical or electrical engineers, but the forms have come from structural engineering ideas.

Structural designers give the form to objects that are of relatively large scale and of single use, and these designers see forms as the means of controlling the forces of nature to be resisted. Architectural designers, on the other hand, give form to objects that are of relatively small scale and of complex human use, and these designers see forms as the means of controlling the spaces to be used by people. The prototypical engineering form--the public bridge--requires no architect. The prototypical architectural form--the private house--requires no engineer. We have seen that scientists and engineers develop their ideas in parallel and sometimes with much mutual discussion; and that engineers of structure must rely on engineers of machinery just to get their works built. Similarly, structural engineers and architects learn from each other and sometimes collaborate fruitfully, especially when as with tall buildings, large scale goes together with complex use. But the two types of designers act predominately in different spheres.

The works of structural art have sprung from the imagination of engineers who have, for the most part, come from a new type of school--the polytechnical school, unheard of prior to the late eighteenth century. Engineers organized new professional societies, worked with new materials, and stimulated political thinkers to devise new images of future society. Their schools developed curricula that decidedly cut whatever bond had previously existed between those who made architectural forms and those who began to make--out of industrialized metal and later from reinforced concrete--the new engineering forms by which we everywhere recognize the modern world. For these forms the ideas inherited from the masonry world of antiquity no longer applied; new ideas were essential in order to build with the new materials. But as these new ideas broke so radically with conventional taste, they were rejected by the cultural establishment. This is, of course, a classic problem in the history of art: new forms often offend the academics. In this case, it was beaux-arts against structural arts. The skeletal metal of the nineteenth century offended most architects and cultural leaders. New buildings and city bridges suffered from valiant attempts to cover up or contort their structure into some reflection of stone form. In the twentieth century, the use of reinforced concrete led to similar attempts. Although some people were able to see the potential for lightness and new forms, most architects tried gamely to make concrete look like stone or, later on, like the emerging abstractions of modern art. There was a deep sense that engineering alone was insufficient.

The conservative, plodding, hip-booted technicians might be, as the architect Le Corbusier said, "healthy and virile, active and useful, balanced and happy in their work, but only the architect, by his arrangement of forms, realizes an order which is a pure creation of his spirit . . . it is then that we experience the sense of beauty." The belief that the happy engineer, like the noble savage, gives us useful things but only the architect can make them into art is one that ignores the centrality of aesthetics to the structural artist. True, the engineering structure is only one part of the design of such architectural works as a private house, a school, or a hospital; but in towers, bridges, free-spanning roofs, and many types of industrial buildings, aesthetic considerations provide important criteria for the engineer's design. The best of such engineering works are examples of structural art, and they have appeared with enough frequency to justify the identification of structural art as a mature tradition with a unique character. That character has three dimensions.

The Three Dimensions of Structure

Its first dimension is a scientific one. Each working structure or machine must perform in accordance with the laws of nature. In this sense, then, technology becomes part of the natural world. Methods of analysis useful to scientists for explaining natural phenomena are often useful to engineers for describing the behavior of their artificial creations. It is this similarity of method that helps to feed the fallacy that engineering is applied science. But scientists seek to discover pre-existing form and explain its behavior by inventing formulas, whereas engineers want to invent forms, using pre-existing formulas to check their designs. Because the forms studied by scientists are so different from those of engineers, the methods of analysis will differ; yet, because both sets of forms exist in the natural world, both must obey the same natural laws. This scientific dimension is measured by efficiency.

Technological forms live also in the social world. Their forms are shaped by the patterns of politics and economics as well as by the laws of nature. The second dimension of technology is a social one. In the past or in primitive places of the present, completed structures and machines might, in their most elementary forms, be merely the products of a single person; in the civilized modern world, however, these technological forms are the products of a society. The public must support them, either through public taxation or through private commerce. Economy measures the social dimension of structure.

Technological objects visually dominate our industrial, urban landscape. They are among the most powerful symbols of the modern age. Structures and machines define our environment. The locomotive of the nineteenth century has given way to the automobile and airplane of the twentieth. Large-scale complexes that include structures and machines become major public issues. Power plants, weapons systems, refineries, river works--all have come to symbolize the promises and problems of industrial civilization.

The Golden Gate, the George Washington, and the Verrazano bridges carry on the traditions set by the Brooklyn Bridge. The Chicago Hancock and Sears towers, and the New York Woolworth, Empire State, and World Trade Center towers--all bring the promise of the Eiffel Tower into the utility of city office and apartment buildings. The Astrodome, the Kingdome, and the Superdome carry into the late twentieth century the vision of huge permanently covered meeting spaces first dramatized by the 1851 Crystal Palace in London and the 1889 Gallery of Machines in Paris.

Nearly every American knows something about these immense twentieth-century structures, and modern cities repeatedly publicize themselves by visual reference to these works. As Montgomery Schuyler, the first American critic of structure, wrote in the nineteenth century for the opening of the Brooklyn Bridge, "it so happens that the work which is likely to be our most durable monument, and convey some knowledge of us to the most remote posterity, is a work of bare utility; not a shrine, not a fortress, not a palace but a bridge. This is in itself characteristic of our time." So it is that the third dimension of technology is symbolic, and it is, of course, this dimension that opens up the possibility for the new engineering to be structural art. Although

477

there can be no measure for a symbolic dimension, we recognize a symbol by its elegance and its expressive power.

There are three types of designers who work with forms in space: the engineer, the architect, and the sculptor. In making a form, each designer must consider the three dimensions or criteria we have discussed. The first, or scientific criterion, essentially comes down to making structures with a minimum of materials and yet with enough resistance to loads and environment so that they will last. This efficiency-endurance analysis is arbitrated by the concern for safety. The second, or social criterion, comprises mainly analyses of costs as compared to the usefulness of the forms by society. Such cost-benefit analyses are set in the context of politics. Finally, the third criterion, the symbolic, consists of studies in appearance, along with a consideration of how elegance can be achieved within the constraints set by the scientific and social criteria. This is the aesthetic-ethical basis upon which the individual designer builds his work.

For the structural designer the scientific criterion is primary (as is the social criterion for the architect and the symbolic criterion for the sculptor). Yet the structural designer must balance the primary criterion with the other two. It is true that all structural art springs from the central ideal of artificial forms controlling natural forces. Structural forms will, however, never get built if they do not gain some social acceptance. The will of the designer is never enough. Finally, the designer must think aesthetically for structural form to become structural art. All of the leading artists of structure thought about the appearance of their designs. These engineers consciously made aesthetic choices to arrive at their final designs. Their writings about aesthetics show that they did not base design only on the scientific and social criteria of efficiency and economy. Within those two constraints, they found the freedom to invent form. It was precisely the austere discipline of minimizing materials and costs that gave them the license to create new images that could be built and endure.

Being Human

Chapter 1, **To Engineer Is Human**
Henry Petroski

Dr. Henry Petroski, a professor of civil engineering at Duke University, wrote **To Engineer Is Human** *to put forth his ideas concerning the role of failure in successful design. He also makes the point in the book that the process of engineering design is a process common to many human endeavors. In this selection from the book he lays the groundwork for his theory regarding the failure of engineered structures.*

Shortly after the Kansas City Hyatt Regency Hotel skywalks collapsed in 1981, one of my neighbors asked me how such a thing could happen. He wondered, did engineers not even know enough to build so simple a structure as an elevated walkway? He also recited to me the Tacoma Narrows Bridge collapse, the American Airlines DC-10 crash in Chicago, and other famous failures, throwing in a few things he had heard about hypothetical nuclear power plant accidents that were sure to exceed Three Mile Island in radiation release, as if to present an open-and-shut case that engineers did not quite have the world of their making under control.

I told my neighbor that predicting the strength and behavior of engineering structures is not always so simple and well-defined an undertaking as it might at first seem, but I do not think that I changed his mind about anything with my abstract generalizations and vague apologies. As I left him tending his vegetable garden and continued my walk toward home, I admitted to myself that I had not answered his question because I had not conveyed to him what engineering is. Without doing that I could not hope to explain what could go wrong with the products of engineering. In the years since the Hyatt Regency disaster I have thought a great deal about how I might explain the next technological embarrassment to an inquiring layman, and I have looked for examples not in the esoteric but in the commonplace. But I have also learned that collections of examples, no matter how vivid, no more make an explanation than do piles of beams and girders make a bridge.

Engineering has as its principal object not the given world but the world that engineers themselves create. And that world does not have the constancy of a honeycomb's design, changeless through countless generations of honeybees, for human structures involve constant and rapid evolution. It is not simply that we like change for the sake of change, though some may say that is reason enough. It is that human tastes, resources, and ambitions do not stay constant. We humans like our structures to be as fashionable as our art; we like extravagance when we are well off, and we grudgingly economize when times are not so good. And we like bigger, taller, longer things in ways that honeybees do not or cannot. All of these extra-engineering considerations make the task of the engineer perhaps more exciting and certainly less routine than that of an insect. But this constant change also introduces many more aspects to the design and analysis of engineering structures than there are in the structures of unimproved nature, and constant change means that there are many more ways in which something can go wrong.

Engineering is a human endeavor and thus it is subject to error. Some engineering errors are merely annoying, as when a new concrete building develops cracks that blemish it as it settles; some errors seem humanly unforgivable, as when a bridge collapses and causes the death of those who had taken its soundness for granted. Each age has had its share of technological annoyances and structural disasters, and one would think engineers might have learned by now from their mistakes how to avoid them. But recent years have seen some of the most costly structural accidents in terms of human life, misery, and anxiety, so that the record presents a confusing image of technological advancement that may cause some to ask, "Where is our progress?"

Any popular list of technological horror stories usually comprises the latest examples of accidents, failures, and flawed products. This catalog changes constantly as new disasters displace the old, but almost any list is representative of how varied the list itself can be. In 1979, when accidents seemed to be occurring left and right, anyone could rattle off a number of technological embarrassments that were fresh in everyone's mind, and there was no need to refer to old examples like the Tacoma Narrows Bridge to make the point. It seemed technology was running amok, and editorial pages across the country were anticipating the damage that might occur as the orbiting eighty-five-ton Skylab made its unplanned reentry. Many of the same newspapers also carried the cartoonist Tony Auth's solution to the problem. His cartoon shows the falling Skylab striking a flying DC-10, itself loaded with Ford Pintos fitted with Firestone 500 tires, with the entire wreckage falling on Three Mile Island, where the fire would be extinguished with asbestos hair dryers.

While such a variety may be unique to our times, the failure of the products of engineering is not. Almost four thousand years ago a number of Babylonian legal decisions were collected in what has come to be known as the Code of Hammurabi, after the sixth ruler of the First Dynasty of Babylon. There among nearly three hundred ancient cuneiform inscriptions governing matters like the status of women and drinking-house regulations are several that relate directly to the construction of dwellings and the responsibility for their safety:

> If a builder build a house for a man and do not make its construction firm, and the house which he has built collapse and cause the death of the owner of the house, that builder shall be put to death.
> If it cause the death of the son of the owner of the house they shall put to death a son of that builder.
> If it cause the death of a slave of the owner of the house, he shall give to the owner of the house a slave of equal value.
> If it destroy property, he shall restore whatever it destroyed, and because he did not make the house which he built firm and it collapsed, he shall rebuild the house which collapsed from his own property.
> If a builder build a house for a man and do not make its construction meet the requirements and a wall fall in, that builder shall strengthen the wall at his own expense.

This is a far cry from what happened in the wake of the collapse of the Hyatt Regency walkways, subsequently found to be far weaker than the Kansas City Building Code required. Amid a tangle of expert opinions, $3 billion in lawsuits were filed in the months after the collapse of the skywalks. Persons in the hotel the night of the accident were later offered $1,000 to sign on the dotted line, waiving all subsequent claims against the builder, the hotel, or anyone else they might have sued. And today opinions as to guilt or innocence in the Hyatt accident remain far from unanimous. After twenty months of investigation, the U.S. attorney and the Jackson County, Missouri, prosecutor jointly announced that they had found no evidence that a crime had been committed in connection with the accident. The attorney general of Missouri saw it differently, however, and he charged the engineers with "gross negligence." The engineers involved stand to lose their professional licenses but not their lives, but the verdict is still not in as I write three years after the accident.*

The Kansas City tragedy was front-page news because it represented the largest loss of life from a building collapse in the history of the United States. The fact that it was news attests to the fact that countless buildings and structures, many with designs no less unique or daring than that of the hotel, are unremarkably safe. Estimates of the probability that a particular reinforced concrete or steel building in a technologically advanced country like the United States or England will fail in a given year range from one in a million to one in a hundred trillion, and the probability of death from a structural failure is approximately one in ten million per year. This is equivalent to a total of about twenty-five deaths per year in the United States, so that 114 persons killed in one accident in Kansas City was indeed news.

Automobile accidents claim on the order of fifty thousand American lives per year, but so many of these fatalities occur one or two at a time that they fail to create a sensational impact on the public. It seems to be only over holiday weekends, when the cumulative number of individual auto deaths reaches into the hundreds, that we acknowledge the severity of this chronic risk in our society. Otherwise, if an auto accident makes the front page or the evening news it is generally because an unusually large number of people or a person of note is involved. While there may be an exception if the dog is famous, the old saying that "dog bites man" is not news but that "man bites dog" is, applies.

We are both fascinated by and uncomfortable with the unfamiliar. When it was a relatively new technology, many people eschewed air travel for fear of a crash. Even now, when aviation relies on a well-established technology, many adults who do not think twice about the risks of driving an automobile are apprehensive about flying. They tell each other old jokes about white-knuckle air travelers, but younger generations who have come to use the airplane as naturally as their parents used the railroad and the automobile do not get the joke. Theirs is the rational attitude, for air travel is safe, the 1979 DC-10 crash in Chicago notwithstanding. Two years after that accident, the Federal Aviation Administration was able to announce that in the period covering 1980 and 1981, domestic airlines operated without a single fatal accident involving a

----

* Two engineers did, in fact, subsequently lose their professional licenses as a result of the Hyatt Regency failure.

large passenger jet. During the period of record, over half a billion passengers flew on ten million flights. Experience has proven that the risks of technology are very controllable.

However, as wars make clear, government administrations value their fiscal and political health as well as the lives of their citizens, and sometimes these objectives can be in conflict. The risks that engineering structures pose to human life and environments pose to society often conflict with the risks to the economy that striving for absolute and perfect safety would bring. We all know and daily make the trade-offs between our own lives and our pocketbooks, such as when we drive economy-sized automobiles that are incontrovertibly less safe than heavier-built ones. The introduction of seat belts, impact-absorbing bumpers, and emission-control devices have contributed to reducing risks, but gains like these have been achieved at a price to the consumer. Further improvements will take more time to perfect and will add still more to the price of a car, as the development of the air bag system has demonstrated. Thus there is a constant tension between manufacturers and consumer advocates to produce safe cars at reasonable prices.

So it is with engineering and public safety. All bridges and buildings could be built ten times as strong as they presently are, but at a tremendous increase in cost, whether financed by taxes or private investment. And, it would be argued, why ten times stronger? Since so few bridges and buildings collapse now, surely ten times stronger would be structural overkill. Such ultraconservatism would strain our economy and make our built environment so bulky and massive that architecture and style as we know them would have to undergo radical change. No, it would be argued, ten times is too much stronger. How about five? But five might also arguably be considered too strong, and a haggling over numbers representing no change from the present specifications and those representing five- or a thousand-percent improvement in strength might go on for as long as Zeno imagined it would take him to get from here to there. But less-developed countries may not have the luxury to argue about risk or debate paradoxes, and thus their buildings and boilers can be expected to collapse and explode with what appears to us to be uncommon frequency.

Callous though it may seem, the effects of structural reliability can be measured not only in terms of cost in human lives but also in material terms. This was done in a recent study conducted by the National Bureau of Standards with the assistance of Battelle Columbus Laboratories. The study found that fracture, which included such diverse phenomena as the breaking of eyeglasses, the cracking of highway pavement, the collapse of bridges, and the breakdown of machinery, costs well over $100 billion annually, not only for actual but also for anticipated replacement of broken parts and for structural insurance against parts breaking in the first place. Primarily associated with the transportation and construction industries, many of these expenses arise through the prevention of fracture by overdesign (making things heavier than otherwise necessary) and maintenance (watching for cracks to develop), and through the capital equipment investment costs involved in keeping spare parts on hand in anticipation of failures. The 1983 report further concludes that the costs associated with fracture could be reduced by one half by our better utilizing available technology and by improved techniques of fracture control expected from future research and development.

Recent studies of the condition of our infrastructure--the water supply and sewer systems, and the networks of highways and bridges that we by and large take for granted--conclude that it has been so sorely neglected in many areas of the country that it would take billions upon billions of dollars to put things back in shape. (Some estimates put the total bill as high as $3 trillion.) This condition resulted in part from maintenance being put off to save money during years when energy and personnel costs were taking ever-larger slices of municipal budget pies. Some water pipes in large cities like New York are one hundred or more years old, and they were neither designed nor expected to last forever. Ideally, such pipes should be replaced on an ongoing basis to keep the whole water supply system in a reasonably sound condition, so that sudden water main breaks occur very infrequently. Such breaks can have staggering consequences, as when a main installed in 1915 broke in 1983 in downtown Manhattan and flooded an underground power station, causing a fire. The failure of six transformers interrupted electrical service for several days. These happened to be the same days of the year that the thousand buyers from across the country visited New York's garment district to purchase the next season's lines. The area covered by the blackout just happened to be the blocks containing the showrooms of the clothing industry, so that there was mayhem where there would ordinarily have been only madness. Financial losses due to disrupted business were put in the millions.

In order to understand how engineers endeavor to insure against such structural, mechanical, and systems failures, and thereby also to understand how mistakes can be made and accidents with far-reaching consequences can occur, it is necessary to understand, at least partly, the nature of engineering design. It is the process of design, in which diverse parts of the "given-world" of the scientist and the "made-world" of the engineer are reformed and assembled into something the likes of which Nature had not dreamed, that divorces engineering from science and marries it to art. While the practice of engineering may involve as much technical experience as the poet brings to the blank page, the painter to the empty canvas, or the composer to the silent keyboard, the understanding and appreciation of the process and products of engineering are no less accessible than a poem, a painting, or a piece of music. Indeed, just as we all have experienced the rudiments of artistic creativity in the childhood masterpieces our parents were so proud of, so we have all experienced the essence of structural engineering in our learning to balance first our bodies and later our blocks in ever more ambitious positions. We have learned to endure the most boring of cocktail parties without the social accident of either our bodies or our glasses succumbing to the force of gravity, having long ago learned to crawl, sit up, and toddle among our tottering towers of blocks. If we could remember those early efforts of ours to raise ourselves up among the towers of legs of our parents and their friends, then we can begin to appreciate the task and the achievements of engineers, whether they be called builders in Babylon or scientists in Los Alamos. For all of their efforts are to one end: to make something stand that has not stood before, to reassemble Nature into something new, and above all to obviate failure in the effort.

Because man is fallible, so are his constructions, however. Thus the history of structural engineering, indeed the history of engineering in general, may be told in its failures as well as in its triumphs. Success may be grand, but disappointment can often teach us more. It is for this reason that hardly a history can be written that does not include the classic blunders, which more often than not signal new beginnings and new triumphs. The Code of Hammurabi may have encouraged sound construction of reproducible dwellings, but it could not have encouraged the evolution of the house, not to mention the skyscraper and the bridge, for what builder would have found incentive in the code to build what he believed to be a better but untried house? This is not to say that engineers should be given license to experiment with abandon, but rather to recognize that human nature appears to want to go beyond the past, in building as in art, and that engineering is a human endeavor.

When I was a student of engineering I came to fear the responsibility that I imagined might befall me after graduation. How, I wondered, could I ever be perfectly sure that something I might design would not break or collapse and kill a number of people? I knew my understanding of my textbooks was less than total, my homework was seldom without some sort of error, and my grades were not straight A's. This disturbed me for some time, and I wondered why my classmates, both the A and C students, were not immobilized by the same phobia. The topic never came to the surface of our conversations, however, and I avoided confronting the issue by going to graduate school instead of taking an engineering job right away. Since then I have come to realize that my concern was not unique among engineering students, and indeed many if not all students have experienced self-doubts about success and fears of failure. The medical student worries about losing a patient, the lawyer about losing a crucial case. But if we all were to retreat with our phobias from our respective jobs and professions, we could cause exactly what we wish to avoid. It is thus that we practice whatever we do with as much assiduousness as we can command, and we hope for the best. The rarity of structural failures attests to the fact that engineering at least, even at its most daring, is not inclined to take undue risks.

The question, then, should not only be why do structural accidents occur but also why not more of them? Statistics show the headline-grabbing failure to be as rare as its newsworthiness suggests it to be, but to understand why the risk of structural failure is not absolutely zero, we must understand the unique engineering problem of designing what has not existed before. By understanding this we will come to appreciate not only why the probability of failure is so low but also how difficult it might be to make it lower. While it is theoretically possible to make the number representing risk as close to zero as desired, human nature in its collective and individual manifestations seems to work against achieving such a risk-free society.

Hired Scapegoats

Chapter 5, **Blaming Technology**
Samuel Florman

Civil engineers, particularly those employed by government agencies, are truly public servants; it is their responsibility as professionals to use their knowledge and experience to provide projects and services in accordance with the demands of society. Sometimes the demands of society are simple and unmistakable. Frequently, however, the demands are complex, garbled, and even conflicting. Furthermore, societal desires are not unvarying; from time to time society as a whole changes its mind about what is desirable. Civil engineers, particularly those involved in developing projects that may take decades from conception to completion, should not be surprised to find themselves castigated for doing what they have been told to do.

Few professions have been subjected to the kind of criticism that was heaped on civil engineers in the decade from 1965 to 1975. From 1945 until 1965 civil engineers were authentic heroes. They planned, designed and constructed an interstate transportation system that surpassed any previously conceived network of roads; they built water resource development projects that provided safe and plentiful water supplies for homes and industries, unparalleled access to water-based recreation, a measure of relief from fear of flooding in both rural and urban areas, abundant water for agricultural needs in the arid west and a water transportation network that provided low-cost routes for important bulk commodities; they built the airports that made air transportation as common as the horse and buggy had been fifty years earlier and the launch facilities that made it possible to send Americans into space; and they planned, designed and constructed the skyscrapers, mass transit systems, water supply and wastewater treatment systems, shopping centers, boulevards, and other facilities that made it possible for the United States to change from an essentially rural to an essentially urban nation. Then, in the mid-1960s civil engineers suddenly found themselves being damned for the same actions that had brought them so much praise in the previous two decades. Societal values were in the process of changing and civil engineers were caught in the transition.

Any professional can experience a "change in heart" on the part of a client. What is distinctive about the plight of the civil engineer is that the client is not a single individual. In theory at least, the client is society as a whole--to the extent that we can discern the desire of society as a whole. What happened to the civil engineer in the last half of the 1960s and early 1970s is that there was no unanimity of opinion on what was necessary, much less desirable, insofar as the work of the civil engineering profession. Large numbers of people believed that "business as usual" was appropriate for civil engineers, but a large number also believed that terrible mistakes were being made--that the pursuit of "business as usual" would result in a disaster.

The civil engineer in this situation is torn between the requirement to be a faithful servant of the public--a fundamental tenet of the civil engineering ethic --and the responsibility to exercise professional judgment which might conflict with public desires. It was a dilemma familiar to politicians--whether to vote one's conscience or one's constituency--but unusual for the engineer who, by and large, had been educated to believe that no one would question "professional judgments."

485

In this selection, Samuel Florman, a civil engineer, argues that one of the agencies involved in civil engineering work, the U.S. Army Corps of Engineers, was unjustly criticized during the late 1960s for responding to the directives of duly elected public officials. He suggests that the agency may, in fact, be serving a useful purpose in providing social stability during a time of changing values and perceptions.

In 1971, about a year after the first celebration of Earth Day, General Frank Koisch, Director of Civil Works, U.S. Army Corps of Engineers, was invited to address a meeting of college newspaper editors in Washington. His announced topic was "Does the Corps Give a Dam?" A more immediate question was, Would the general be allowed to speak? From the moment Koisch strode into the hall, resplendent in military uniform, the audience of young journalists was in an uproar. For almost half an hour a battle raged between the program committee, pleading for silence, and a group of hostile activists, determined that this enemy of the people should not be heard. It was finally agreed that, no matter how repugnant his views, an invited guest should be permitted to speak. Order was restored, and General Koisch stated his case, which was that the corps did give a damn. There is no evidence that he convinced anybody present. As he spoke, a young woman circulated through the audience handing out bumper stickers that read "Dam the Corps -- Not Our Rivers." When the general had concluded his remarks, it was discovered that his hat had been pilfered and that someone had carved obscenities on his leather briefcase.

The incident was a sign of the changing times. The heroic image the U.S. Army Corps of Engineers had enjoyed in an earlier age vanished in the aftermath of Vietnam and the environmental crisis. Its dam-building, dredging, draining, and other works, which once seemed so marvelous, are regarded increasingly with revulsion. Although never without its few vocal critics (Harold L. Ickes complained that is was "above the law"; Justice William O. Douglas labeled it "public enemy number one"), the corps could not have been prepared for the virulent hostility directed against it during the 1970s. Books with such titles as **The River Killers** and **Dams and Other Disasters** chronicled the corps' alleged predations and called for its dissolution. Magazine articles-- from "Dam Outrage" (<u>The Atlantic</u>, April 1970) to "Flooding America in Order to Save It" (<u>New Times</u>, November 1976)--characterized it as "a giant bulldozer out of control, burying villages, disfiguring the landscape." The other media contributed to the swelling expression of public outrage.

Even politicians, who once treated the corps with deference, joined the attack. Congressman Stewart Udall compared the corps to "a giant water-loving dinosaur with less brain per pound of flesh than any other vertebrate." Senator William Proxmire awarded the corps his 1976 "Golden Fleece of the Year" for "the worst record of mismanagement and cost growth in the entire government."

Upon reflection, there is nothing at all surprising about this development. A more fitting villain for this nation in this era could hardly be imagined. The U.S. Army Corps of Engineers appears to embrace in one entity the three segments of American society that evoke our most intense protest: the military, the bureaucracy, and the environment-savaging technocracy.

But appearances, as we continually say and repeatedly forget, can be deceptive. In the case of the Corps of Engineers, the publicly accepted image happens to be completely at variance with the facts. This does not move me to mount a defense on behalf of the corps, which has, after all, not suffered anything more cruel than a bad press. But I think the matter deserves examination because it exemplifies a combination of public misunderstanding and frenzy that seems to be a recurrent feature of our national behavior. It also demonstrates that even where engineers are nominally in control of important enterprises, their work is responsive to the needs and desires of the community in which they live.

What are the facts about the Corps of Engineers, and how do they differ from the image?

That part of the corps which builds dams and dredges harbors, and attends to other civic works--the Civil Works Directorate--is simply not, by any reasonable definition, a part of the military establishment. Although technically a branch of the Army, this organization is, in fact, an agency of the United States government. The directorate's 40 offices across the nation are staffed by 32,000 civilian engineers, technicians, and other civil servants. A mere 300 Army officers nominally oversee the activities of this huge organization, and their involvement is circumscribed by the fact that their service in the directorate is limited to three-year tours of duty. More important, the power to authorize the study of a corps project, initiate it, and appropriate the money for it is held, not by any arm of the military, but by the Public Works Committees of the Congress and the Public Works Subcommittees of the Appropriations Committees. The Secretary of the Army rarely interferes in these matters. Even the Budget Bureau and the White House think twice before getting involved. The Corps of Engineers is an agency through which Congress studies, evaluates, and executes public works projects, particularly in the area of water resources development.

Why, then, the anachronism of keeping this institution as a branch of the Army? Why not establish it as a government agency known simply as the Department of Engineering, into which other federal engineering organizations could also be integrated, including the Bureau of Reclamation of the Department of the Interior, which provides irrigation facilities for the 17 Western states? Someday it may come to pass. Common sense favors the idea; tradition, however, opposes it.

Engineering, for almost all of recorded history, was closely linked to the military. Fortifications and weapons were major engineering concerns. Transport and water supply came within the province of military planning. The term civil engineer did not even exist until the mid-eighteenth century, when it was coined by the famous English engineer John Smeaton, builder of the Eddystone lighthouse, in an attempt to differentiate his work from that of the military. The United States Military Academy at West Point was established in 1802 as an engineering school, and for several decades was the main source of engineers for the nation. When Congress embarked on mapping the unexplored West, and developing harbors, canals, and other massive public works, it quite naturally got into the habit of delegating projects to the Army Corps of Engineers. A tradition so entwined with the history of our land is not quickly cast aside. For all practical purposes, however, the corps has almost nothing to do with the military.

What of the corps' reputation as an arrogant, unresponsive bureaucracy? Here again the facts belie the myth. The *sine qua non* for an unresponsive bureaucracy is an established, independent, and relatively invulnerable fiefdom. The Civil Works Directorate has nothing of the sort. Each year the Public Works Appropriation Act provides funds for the corps' civil-works program on a project-by-project basis. No other major federal agency has its work funded in this way. Critics of the corps say that the annual appropriation for each project serves to obscure the long-term cost of these projects. At the same time, however, it also makes corps activities supremely sensitive to every wish of Congress. The appropriation of funds for the corps is, in fact, the major "porkbarrel" legislation of each Congressional session, and it reflects unerringly the mood and the shifting power relationships in the Senate and House. Projects are started, stopped, expedited, and delayed, and the action is parcelled out in each of the 50 states according to agreements arrived at in the labyrinths of the Capitol.

Doubtless some members of the corps have learned their way about in those labyrinths and have proved themselves adept at such bureaucratic tricks as juggling cost-benefit ratios and rationalizing tremendous cost overruns. But it is clear that they have no real power, being dependent at 12-month intervals on Congressional whim. Far from being an intransigent bureaucracy, the corps appears to have evolved as an instrument exquisitely tuned to work the will of the people.

All right, critics of the corps might concede, but which people? Corps projects traditionally come into being when some local citizens' group gains the political support of a Congressman and the technical approval of the local corps district engineer. Typically, the local group is a Chamber of Commerce or some other representative of monied interests. Yet even if many projects are conceived in greed, and sponsored under slightly unsavory circumstances, the entire local community often benefits from increased employment and prospering business climate. Some studies have sought to demonstrate that other types of federal programs would be more effective in aiding local communities, and that is probably true. But pending the evolution of such programs, the Corps of Engineers serves as a conduit for Congressional revenue-sharing. The gains of sponsoring entrepreneurs, ill-gotten as they may be, are not in the long run a major consideration. As A. Den Doolard, a Dutch author, has written of the contractors who work on building the dikes in Holland: "Profit is merely the bait that destiny has offered to these calculators."

The destiny of America, as perceived until recently by the vast majority of its people, has been to grow economically and to develop its water resources to this end. Wilderness areas have been flooded, rural families uprooted, archaeological sites inundated, and important caves damaged, not because these were objectives of the Corps of Engineers, but because commercial development was mandated by the citizenry. As the values of people change, and as Congress reflects such change, the activities of the corps will change automatically. Indeed, changes are occurring at this time. Two college professors, writing in Public Administration Review, claim that such changes demonstrate how aging, entrenched bureaucracies are capable of innovative and progressive response to new conditions. But this misses the point about the corps, which is that, because of the unique year-by-year, project-by-project funding of its activities, it is not a comfortable, unresponsive bureaucracy.

If the corps is neither a true branch of the military nor an entrenched bureaucracy, does it not at least stand condemned for its technocratic destruction of the environment? The issue is complex, but again the corps must be found not guilty.

If, by doing away with our inland waterway system, we could have back our wild rivers, how would we then transport the one-sixth of all intercity cargo that is presently water-borne? By truck and railroad, of course. Comparing barges moving slowly upstream with roaring trailer trucks and freight trains, I, for one, cannot see where this would be much of an environmental improvement. The corps has been accused of being in cahoots with the barge industry; this may be true, but since such complaints usually come from railroad and trucking lobbyists, they do not excite the environmentalist in me.

Another corps activity that has troubled environmentalists is in the area of water supply. During those years in which we have been blessed with adequate rain and snow, the people who see fit to excoriate the builders of dams and reservoirs have not had to worry about drought. But as soon as precipitation drops below acceptable limits, we are haunted by visions of failed crops and incipient dust bowls. Water shortages are "natural" occurrences, I suppose, but some modification of the environment in an effort to avert such disasters seems to me to be morally justified, even in a society sensitive to ecological concerns. Surely the majority of our citizens have supported and continue to support public works projects whose purpose is to assure an adequate supply of water.

On the other hand, in the area of flood control, things have been done which are difficult to defend. By damming and leveeing, and permitting commercial and residential development of the flood plains, the corps has restricted rivers to artificial channels, where they flow more swiftly and become potentially more dangerous than they ever were. Now belatedly, the corps is stressing "non-structural" methods of flood control. It is only proper that it be held accountable for its past errors in this field. But in a nation where people persist in living on cliffs which are crumbling into the sea, and build houses atop major earthquake faults, there is little likelihood that even the most far-seeing engineers could have prevented the rush onto the low-lying plains. It is a lot easier to hold back torrents of water than it is to stand in the way of land-hungry Americans.

Perhaps the most serious problems caused by the corps result from its indiscriminate filling, dredging, and draining of our wetlands. But it is hard to place the blame for these acts entirely, or even mainly, on corps personnel. Until recently the importance of wetlands in the ecological scheme of things was not understood. Estuaries and swamps, we have only lately learned, moderate our climate, provide natural pollution control, and play a vital role in the life-cycle of a multitude of marine organisms and other animal life. If these facts were not sufficiently known to biologists, meteorologists, agronomists, and other environmental specialists in our society, how can we expect the corps' civil engineers to have been uniquely prescient? Considering the general lack of knowledge in these matters, was it such an inexcusable manifestation of hubris for the corps to have tinkered with nature? Hardly--unless we condemn all humanity for its dreams, dating from earliest times, of draining malarial swamps and "reclaiming" what appeared to be fetid marshland. Goethe's Faust, remember, found his final salvation in a land-reclamation project.

In light of the new scientific knowledge, Congress in 1972 gave the corps responsibility for protecting all the wetlands in the nation. When the corps defined its responsibility as limited to its traditional province, the *navigable* waters, environmental groups brought suit to establish the corps' authority over all waters. In a situation not without irony, the courts agreed with the environmentalists that the corps should have total control. Taking its new responsibility with all earnestness, the corps in 1976 stopped the Deltona Corporation of Florida from turning 2,000 acres of mangrove swamp into a housing development, even though Deltona had already sold the land to prospective home-builders. Commercial developers all over the nation were aghast. "The decision was a shock," said the president of Deltona. "I still can't get over it. The corps--they've been like us. They're engineers, our kind of people."

Obviously, the president of the Deltona Corporation, like a lot of people in this country, has no understanding of what engineering is all about. The profession is dedicated to performing works "for the general benefit of mankind" or "for the good of humanity," to quote from two definitions of long standing. But this does not mean that engineers take unto themselves the right to define what such benefit or good might be, or, even less, that they are committed to a real estate developer's ideas on the subject. Society establishes its own goals, and engineers, like jurists, educators, politicians, and the rest of the body social, work toward achieving these goals. When the nation wanted to fill in its wetlands and tame its rivers for the sake of commerce, the engineer did the job. If the nation has become more sensitive to environmental considerations, then so, by definition, has the engineer. Engineering is not anti-environment. Environmentalism itself is a branch of engineering.

This is not to say that engineers are automatons without conscience or conviction; they are philosophically an integral part of the community. This is a thorny issue to which we will have to return in a discussion of engineering ethics in chapter 15. Engineers are called upon to be guided by conscience at the same time that they are urged to serve the will of the public. As far as the U.S. Army Corps of Engineers is concerned, there is ample evidence to demonstrate that it is responsive to the desires of "the people."

Members of the corps are action-oriented, to be sure, but they are no more devoted to building than they are to protecting wetlands. In fact, in the Mississippi delta they are experimenting with ways of creating new wetlands. A certain amount of footdragging is inevitable among those long associated with particular undertakings. But the brightest, most alert, and most ambitious members of the corps will see to it that the times do not pass them by. Careers are not made by defying the will of the electorate.

Some of the corps' new ecological awareness will be discounted as cynical lip service. When the chief of engineers jovially passes out big buttons saying "The Corps Cares," this proves only that a public relations department is at work. But wanting a reputation for caring can clearly be considered a step in the right direction.

490

In response to the changing national mood, the corps is constantly reevaluating and deauthorizing many of its projects; at the same time it is moving ahead with some projects to which there is much public opposition. Even after new standards of caution and sensitivity are applied, there remain areas of disagreement which are essentially a matter of taste. What sort of a landscape do we want? What mix of wilderness and factories, parks and highways, suburbs and cities will do? It is said that there is no disputing taste, but in fact there is nothing more important to dispute. John Dewey put it this way:

> The formation of a cultivated and effectively operative good judgment or taste with respect to what is esthetically admirable, intellectually acceptable and morally approvable is the supreme task set to human beings by the incidents of experience.

This challenge is one that the citizens of a free society must recognize and accept. We cannot avoid it by pretending that our fate is in the hands of organizations such as the Corps of Engineers.

There is no way in which we can recapture the wild continent that once was. To regret this, I believe, is an elitist conceit. But we can stop any new project at any time. All we have to do is convince ourselves, and then our Congress, that this is what we want to do. The Carter administration, when it first came into office, selected 59 water-resource development projects as "high-priority projects for reevaluation" and then recommended a halt to 19 of them. It discovered that objections came, not from engineers, but from ordinary citizens and their Congressmen, who chose continuing commercial development of their home areas over environmental concerns and budgetary restraint. The experience of the Reagan administration was similar.

We say that we oppose the corps for being militaristic, bureaucratic, and anti-environmental, but upon inspection these reasons are seen to be invalid or feeble. We actually oppose the corps because it so unerringly shows us what we are -- or what we just were.

The concept of scapegoat has come down to us from biblical myth. Perhaps, like so many mysterious phenomena of mass psychology, a combination of confused and misdirected blame is, in some way that we cannot see, a vital element in maintaining social stability. This must be the hope of those who wish not to be overly discouraged by recent public behavior toward the U.S. Army Corps of Engineers.

God Would Have Done It in the First Place . . .
George Fisher

The mood of the time in the decade between 1965 and 1975 was one which saw engineers in general subjected to severe criticism for their failure to demonstrate sufficient understanding of non-technical dimensions of the problems they were trying to solve. No group of engineers was criticized more vehemently than the civil engineers whose efforts in building highways, airports, dams and other public works projects were viewed as incompatible with the trend toward living in harmony with the natural environment. And among the civil engineers the U.S. Army Corps of Engineers was singled out for the most intense criticism, at least in part because as a branch of the U.S. Army, engaged in an unpopular conflict in Vietnam, it was a convenient target for those seeking to register their disapproval of that activity.

George Fisher, editorial cartoonist for the <u>Arkansas Gazette,</u> was an early critic of the Corps' water resources development programs in the state of Arkansas. His cartoons never demonstrated the irrational hostility that characterized the work of some Corps critics, but his sharp jabs at what he considered the Corps' "Keep Busy" approach to problem-solving were very popular with Corps opponents.

"God would have done it in the first place if He'd had the money."

492

An Engineer Is More Than a Man With a Diploma

An excerpt from **The Dam**
Robert Byrne

In his novel, **The Dam,** *civil engineer Robert Byrne explores some of the intellectual and ethical problems encountered by both experienced and inexperienced engineers. The fictional situation which forms the nucleus of the story--a young, inexperienced engineer bolstered by results obtained with a computer model questioning the judgment and experience of his employer--is more dramatic in the telling here than one would expect such encounters to be in the typical engineering office, but it is certainly not unrealistic. One of the most troublesome experiences for any engineer is the determination that serious errors have been made in work which has already been accepted as the basis for design or construction decisions. Finding such an error, determining that it is significant and communicating that finding to other engineers is almost always at least mildly traumatic, if for no other reason simply because it causes one to face up to the conflict between the very human desire to avoid conflict and the ethical responsibility to put foremost the safety and welfare of society.*

This problem arises so frequently because, unlike many other professions, the engineering profession has developed what amounts to an unwritten principle that engineering analyses and decisions should be subjected to one or more peer reviews. Many engineering offices and agencies assign young engineers to such work both as a means of familiarizing them with work currently underway and as a form of on-the-job education--a way of bridging the gap between textbook exercises with specified assumptions and "right" answers and the complexities of real-world engineering problems in which all of the assumptions come from the knowledge and experience of the engineer and for which there are no "answers in the back of the book."

In **The Dam,** *as in his other books with engineering themes, Mr. Byrne doesn't just tell a story; he also tells his readers about engineers and the engineering profession.*

The Los Angeles Headquarters of Roshek, Bolen & Benedetz occupied three floors of the 500 Tishman Tower on Wilshire Boulevard. Most of the senior company officers were on the top floor, where also were located the advance planning and project development departments. On the middle floor were sections specializing in highways, structures, and tunnels. The lower floor was devoted to computer services, hydroelectric, nuclear, and mining. There were over a hundred employees on each floor, more than half of them engineers, who worked at desks or drafting tables in a central area surrounded by a ring of offices. Departments working on petrochemical developments, pipelines, ocean facilities, and foundations were in other buildings in the Los Angeles area, and one of the proposals being evaluated on the upper floor was the construction of an office tower that would consolidate all operations under one roof.

On Tuesday morning, May 26, Phil Kramer was at work an hour early, sitting at a terminal feeding his remodeled dam failure program into a computer. The revision was the result of five evenings of collaboration with Janet. She knew nothing about dams, but she knew how to forge a chain of logic and she knew how to ask questions that made him alter some of his numerical assumptions. It was she who suggested that the original mathematical model was too small and too simple to be applicable to Sierra Canyon. The model had to be expanded to accommodate the sheer size of the structure and the greater-than-average volume of data the instrumentation provided. The work they had done together had resulted in a program that was no longer appropriate for an "average" dam. It was tailor-made for Sierra Canyon.

When he had completed the required preliminary operations, Phil typed "List Gallery D meter points." A touch of the Execute key brought a column of two dozen code letters onto the screen. By touching the four Cursor control keys in a certain sequence, he brought the green indicator line to the top of the column. Phil opened a copy of the latest inspector's report from Sierra Canyon. The readings had been taken three weeks earlier, when the surface of the reservoir was still five feet below the lip of the spillway at the crest of the dam. With his left hand holding the report open and his right working the keyboard, he entered a number after each of the letters on the screen. When the column was completed, another appeared.

Thirty minutes later, all of the available data had been fed into the system. Phil instructed the computer to make estimates of the dam's condition under his "best case" assumptions. Four minutes later, columns of figures appeared relating to ten-thousand-cubic-yard blocks of the dam. Since there were ninety million cubic yards of material in the dam, there were nine thousand coded blocks, but the instructions were such that only those with an above-normal seepage, pressure, settlement, or shift were identified. Twenty blocks came on the screen under the heading "Exceed Values Predicted in Design." Five were labeled "Critical." "Conduct Visual Inspection," the computer suggested. Touching another sequence of keys brought to the screen code letters for the dam cross-sections containing the critical blocks.

Phil pursed his lips and shook his head, wondering if he should junk the program and start from scratch. Apparently it was even more skewed than before. First he wanted to see just how far off it was. He asked the computer to calculate the "worst case." This time forty-seven blocks appeared under the "Exceed Values" heading, and twelve were called "Critical." The characters faded and were replaced with the command "Take Immediate Action." Phil asked for successive displays of the critical cross-sections. As the triangular images came on the screen, dotted lines moved from right to left indicating the plane of maximum weakness--in each case it was at the lowest elevation, apparently the interface between the embankment and the foundation rock.

A new message appeared on the screen: "Garbage coming out? Don't cry. Recheck garbage going in."

It was one of the phrases Phil had included in the program to relieve the tedium.

Janet greeted him cheerfully when he phoned her at work. "How is everything over at Colossal Engineering?"

"Wonderful. According to the giant brain, our finest dam is dissolving in forty-seven places at once. I'm calling to advise you to dump any shares of Colossal that might be in your portfolio."

"You know I can't afford a portfolio. I keep my shares in a drawer."

"Janet, the results are worse than before. I don't think anything's wrong with the logic. My initial assumptions, the arbitrary coefficients, must be too pessimistic."

"I can't help you there. Until you dropped this dam on me, the biggest thing I ever analyzed was a silicon chip. Why don't you explain it to old what's-his-name--Roshek? He could probably spot the flaw in a second."

Phil laughed. "You must want to get me killed. That guy scares me to death. You should see the way he swings through here in the morning on his crutches. You can hear him coming down the hall from the elevators like something out of a monster movie. It's funny, the way everybody stops talking and starts working. That old man, I swear, can look around the room and just with a glance knock a man right off his stool."

"You've got to talk to somebody. You can't just sit there worrying about it."

"Should I call the senior partners together and tell them that according to my calculations Sierra Canyon Dam is on its way to Sacramento? They would fall on the floor laughing. They would say, 'Gee, those must be swell calculations.' I'm fresh out of college. I'm not supposed to act as if I know anything. My eyes and ears are supposed to be open and my trap shut."

"You're too bashful. You've got an ingenious program and you should feel more confident about it. Don't you have a boss you could talk to?"

"Two. One doesn't know enough about computers and the other doesn't know enough about dams. I suppose I could go to Herman Bolen, who interviewed me before I was hired. A pretty nice guy, I think, a little on the pompous side."

"Talk to Bolen. If the dam collapses tomorrow, you don't want to have to say that you knew it was going to but were too embarrassed to mention it. Oh-oh, I've got to hang up--here comes my supervisor looking at her watch. You think you've got a movie monster at *your* place. . . ."

Phil spent the rest of the morning as well as his lunch break trying to summon up enough courage to ask Bolen's secretary to make an appointment. Twice he put his hand on his desk phone and withdrew it in the face of dreadful visions. "Are you crazy?" Bolen might rage at him. "I have better things to do than talk to children about their hallucinations. Do you seriously think I give a good goddamn about your old school project?" No, Phil corrected himself, Bolen wasn't the type to rage. He was more likely to belittle him with paternalism. "Your little computer program is very nice. You should look at it again in a few years when you've had some experience. Now if you'll excuse me, I'm expecting the Chancellor of West Germany." Another possibility was that Bolen would fire him on the spot for wasting valuable computer time and for not devoting a hundred percent of his attention to his assigned work. He certainly didn't want to risk losing his job.

Phil was part of a four-man team designing a rock-fill dam for an agricultural development in Brazil. Most of his time was spent double-checking the drawings and computations of others, which was educational and to which he

495

didn't object. He was sure that if he applied himself and avoided serious mistakes he would be given more responsibility and original work to do. Already there was a possibility that he would be asked to accompany the team leader on a trip to the jobsite later in the year. A junket to Brazil! There would have been no chance of that had he taken his father's advice and joined one of Wichita's small consulting firms. He would be stuck laying out sewer connections for tract homes and hoping for a driveway design job as a change of pace. He was glad he had decided to take a crack at a big firm in California. As the days went by, the danger receded that he would make a fool of himself and have to slink back home in disgrace. Quit worrying, he scolded himself. You've got a good brain, a decent education, and a willingness to work. Bolen is the type who puts a high value on things like that. You're not nearly as tongue-tied as you used to be. So *what* if you look too young to be taken seriously? Time would correct that condition.

He puts his hand on the phone again. Bolen was nothing to fear. He might be impressed with a new employee who looked beyond his immediate assignment, and would think of him when a promotion was to be made. Phil thought of the various pieces of advice his father had given him on how to succeed in the business world. Present yourself as a person who solves problems rather than causes them, but if you have a problem you can't solve, take it to someone who can, and make sure your facts are right. Phil frowned. His facts were three weeks old. He'd better call the dam and get the latest meter readings. Another thing his father believed was that the higher a person is on the ladder the easier he is to deal with. Bolen was certainly high enough: the second rung from the top.

Phil picked up the phone and put in a call to Sierra Canyon.

Herman Bolen had the second largest and second best equipped office in the company. There was all-wool carpeting, a private bathroom, and floor-to-ceiling walnut paneling--the real McCoy, not plastic or veneer. The left side of his desk resembled the instrument panel of his private plane. At the touch of a button he could summon his secretary, ring fifty phones around the world, get a weather forecast or a stock quote, rotate the louvers outside the windows, light a cigar, heat a cup of coffee, or manipulate numbers in every manner known to mathematics.

He didn't mind being number two when number one was Theodore Roshek. Roshek was a brilliant engineer with an inhuman capacity for work; he deserved his prestige and his larger share of the profits. Herman Bolen wasn't envious at all. He was, in fact, grateful to the older man. If Roshek hadn't taken a chance on him years ago, he would probably still be lost in the Bureau of Reclamation labyrinth, a federal drudge nobody ever heard of. As it was, thanks to Roshek, his own hard work, and a run of luck in the form of illnesses that had struck down rivals inside the firm, he was now enjoying considerable power and prestige. He was making more than he ever dreamed he would--more than a hundred thousand dollars after taxes in the previous year. He had played a role in some of the twentieth century's most notable engineering achievements, as could be gathered from the framed photographs and artist's renderings on the walls: Mangla Dam, the Manapouri Power Scheme, the Iraqi Integrated Refineries, the Alaska Pipeline, the Sinai Canal, Sierra Canyon Dam.

496

He worked well with Roshek. Roshek could turn on the charm when he had to, but his normal manner was harsh and cutting. Bolen's was soft and fatherly. He smoothed the feathers that Roshek ruffled. Not that life was perfect. Bolen mourned the retreat of his hair and the advance of his waistline. His body, pear-shaped and high in fat content, was gaining weight relentlessly--the current rate, according to his desk-top calculator, was approximately 0.897 pounds per month. *Reading* about exercising and dieting was, obviously, not enough. He touched a button. Instantly appearing on a small glass screen was the time to a hundredth of a second in twelve time zones. In Los Angeles it was 5:06.34 p.m. Time to call it a day.

There was a light knock on his door, followed by the gray head of his secretary. "Mr. Kramer is here to see you," she said.

"Who?"

"Mr. Kramer, the young man from downstairs. He asked early this afternoon for an appointment. Tall, with the reddish hair? The hydro design section? Computers?"

"Oh, oh, oh, yes. Send him in."

For some preposterous reason he thought she meant that Jack Kramer, the old-time tennis pro, was in the outer office. He had once seen Kramer play a very fine match against Pancho Segura . . . good grief, over thirty years ago! *Phil* Kramer was the lad who had just come aboard. Bolen had interviewed him himself, recruited him, recommended that he be hired. Likable young fellow, and well mannered. Presentable. Bright future if he applied himself. Just the kind of raw material that Bolen, Roshek & Benedetz was looking for. *Bolen*, Roshek & Benedetz? *Roshek*, Bolen & Benedetz . . .

Kramer was thanking him for his time with a trace of awkwardness as he sat down on the edge of a chair. "You said that if I ever had any trouble I should feel free to come to you."

Bolen smiled in a friendly fashion. The boy was nervous and had to be put at ease. "I said that and I meant it. I know how hard it is to come right out of college into a huge organization. It's a kind of cultural shock. The shock of the real world, eh? When I graduated from St. Norbert's, I enlisted in the Seabees! There was a shock for you!" He chuckled at the memory, mirth that wasn't shared by his young visitor, who sat staring at him, frowning. Bolen joined his hands and leaned forward, lowering his voice. "Now, then, what seems to be the problem? We are both engineers. If you can state the problem in specific terms, we'll solve it."

"Well, Mr. Bolen, yes, there is a problem. I think that one of the firm's structures . . . that is, according to some computer modeling I've been doing . . . Sir, I think that Sierra Canyon Dam is, or *could be*, and I might be completely wrong, probably am completely wrong, and I'm hoping you can show that I'm off base in thinking that the dam is . . . well, is . . . I hate to go over people's heads, but I thought before mentioning this to anybody I would talk it over with somebody who . . ."

497

"Mr. Kramer, just lay out the problem in an orderly manner. Sierra Canyon Dam is what?"

Phil composed his thoughts before beginning again. "In graduate school I worked out a computer program to analyze the performance of embankment dams with the goal of being able to detect conditions that might precede . . . um, failures. It's a mathematical model built up of data from ten dams on pore pressure, settlement rates, seismic response, seepage under various hydrostatic loads, and so on. There's a built-in comparison with conditions that prevail when dams fail, which I got from studying Baldwin Hills and Teton."

"I remember reading about it on your resume. We chatted briefly about it during our interview. Nice piece of work for a student. Imaginative. But as for its *practical* value . . ." What was he working up to?

"I use a three-dimensional matrix that has been very productive in the chemical process field. Not just the *amount* of pressure, seepage, settlement, and movement in different parts of the embankment, but their relationship to each other, how each one affects the others, and, most important, the rate of change of the values as the reservoir rises."

Bolen nodded and tried to adopt an expression that would convey both sympathy and a slight impatience. "I like the general concept, but there are too many unknowns to get the solution you want. Baldwin Hills and Teton dams weren't well monitored before they failed. The trouble with your approach would lie, it seems to me, in assigning meaningful numerical values to such things as relationships and rates of change."

"I made a lot of assumptions."

"Ah . . ."

"Mr. Bolen, on my own time I've been trying my program on Sierra Canyon. What happens is that . . ." He paused.

"The dam, according to the model, is not . . . is not doing too well."

Bolen shook his head and smiled faintly. "Now, really, Mr. Kramer . . ."

"It sounds ridiculous, I know, and that's how it struck me at first. When I made this appointment to see you, I was intending to ask your advice on revising the program. But this afternoon I began to wonder if maybe I'm not onto something."

"Oh?" Bolen was beginning to regard the young engineer in a new and less flattering light. He was clever, but there was something immature about him. He seemed to lack a sense of propriety and proportion. Bolen couldn't help thinking back to the beginnings of his own career and how impossible it would have been for him to approach a senior officer with such a wild tale. Times change and not always for the better.

"I had been using readings from three weeks ago, when the lake was five feet from the top. This afternoon I used values from last Friday, May 22, when the water was eleven inches deep going over the spillway. The computer showed that . . . that . . ."

"That the dam is failing."

Phil exhaled. "All the way from the maximum transverse section to the right abutment."

While Bolen searched for a remark that was suitably sarcastic without being contemptuous, he asked how Phil got Friday's figures.

"I called the dam," Phil said.

"You *what*?"

"I called the dam and talked to the man in charge of maintenance and inspection. A Mr. Jeffers. I wanted to have all the facts I could when I saw you. The lake is higher right now than it's ever been--eleven inches deep going over the spillway."

"You called Jeffers? And told him you were with R. B. & B.?"

"Yes, sir. I asked him if there was excess seepage in Gallery D. He said the inspector hadn't mentioned anything. I asked him about the meters that weren't registering and--"

"I hope to God you didn't tell him that the dam was failing."

"No, sir. I was surprised to learn from him that in the earthquake five years ago--"

"I've heard enough." Bolen raised his voice slightly and lifted his hand for silence. He could be firm and decisive when he had to be. "This is a serious matter. Something must be done and I'm not sure what."

"Well, the spillway gates could be opened to start lowering the reservoir and a special inspection could be made of--"

"I don't mean the dam," Bolen said, practically shouting: "I mean *you*! I don't know what should be done about *you*." Surprised at his own vehemence, he lowered his voice to a whisper. He didn't want Charlene sticking her head in again. "I hired you and so I feel a special responsibility. The failure lies with you, not with the dam."

"I was only trying to--"

"You are behaving like a partner in the firm without putting in the required twenty years of hard work. The only failure we need concern ourselves with is a loss of perspective. Have you ever *seen* Sierra Canyon Dam? Have you ever worked on the design or construction of a dam of any kind, even in the summer months? I thought not." Bolen studied the young man, seated across from him, whose eyes were round in shock and whose cheeks were turning red. He couldn't help feeling sympathy for him. There was something affecting about his naivete. He was sincere. He was open. He probably expected to be complimented for his efforts instead of bawled out. Kramer had talent and could become an asset if brought along properly. Bolen adopted his well-practiced soothing manner and said, "I want you to attend to the duties for which you were hired. Don't use the computers for anything not authorized by Mr. Filippi or myself. Don't mention what you've done to anybody or, I can assure you, you'll be the butt of jokes for years to come. Years from now, you and I can have a private laugh about it. Above all, don't call the site again. Leave the dam in the hands of those of us who have lived with it since the day it was conceived. You have a lot to learn, Mr. Kramer, before you can think about telling us how to run the company. All right? Agreed?"

Kramer gestured with his hands, then let them fall helplessly to his lap. "The readouts scared me," he said in a soft voice. Now he was having trouble meeting Bolen's eyes. "I still think an inspector should be sent down to Gallery D. The readings there are high by anybody's standards."

"You have the courage of your convictions, I'll give you that, even if they are wrong." Bolen waved vaguely toward the door to indicate that the meeting was over. He watched Kramer struggle to his feet like a man who had been flogged. The poor kid was obviously one of those people who couldn't hide their emotions, a trick he'd have to learn if he wanted to go very far in the engineering profession. Bolen stopped him at the door with a final comment designed to cheer him up. "I won't mention this to Mr. Roshek. He would take a dim view of one of his own employees, especially one without experience, raising doubts about a dam that happens to be one of his special favorites. This will just be between the two of us."

Kramer nodded and closed the door behind him.

Thirty minutes later, after checking the Sierra Canyon report himself, Bolen touched a button on his console. A phone rang five hundred miles away in an underground powerhouse.

"Jeffers."

"Herman Bolen. I was afraid you'd be gone for the day."

"Hello, Herman! Hey, we work day and night up here in the mountains. Not like you city slickers."

"I'll trade places with you anytime. You breathe this air for a while."

"It's a deal. You ought to pay us a visit, at least, to see the water spilling over the top. It's quite a show. Makes Niagara look like a leaky faucet. I shot a roll of film today; I'll send you some prints if they turn out."

"Do that. Larry, did you get a call from our Mr. Kramer this afternoon?"

"Yeah, what was that about? He seemed all exited, especially when he found out that a lot of the instruments haven't worked since the quake."

"You volunteered that?"

500

"I sort of mentioned it in passing. I figured a man in the company would already know. I called Roshek to find out what was up, and damned if the girl didn't put my call through to him in Washington! I didn't mean to bother him there. He didn't know Kramer or what he was up to . . ."

Bolen had been doodling on a scratch pad. The lead broke when he heard about the call to Roshek. Now he would have to try to reach the old man himself so he wouldn't think something was being hidden from him. "Thanks for getting him riled up," Bolen said. "He's difficult enough when he's happy."

Jeffers laughed. "Sorry about that. Did he call you?"

"No, but I'm sure he will. Kramer is a young engineer we just hired . . . green, right out of college. I prefer a man with a little experience, myself. We gave him a research job to do, checking some hydro stuff in our files. We didn't intend for him to start phoning around the country. God knows what kind of bill he ran up!" He chuckled to give the impression that the affair was trivial. "But in looking at some of the figures he rounded up, I see that drainage in Gallery D is a little on the high side. Wouldn't you say?"

"Up from last year, maybe, but not much. We're going to have plenty of water all year to run through the turbines, if that's what you're worried about. The reservoir isn't leaking away."

"My thought is that the dam is under maximum stress for the first time in years. A lot can be learned that would help in some other designs we're working on. I want you to do me a favor, Larry. Make a visual check of Gallery D. Personally."

Jeffers moaned. "Jesus, Herman, do you know what a drag it is to go down there? Two hundred steps! Like climbing down the stairwell of a fifteen-story building. Duncan was down there last Friday, anyway. . . ."

"I'm sure Duncan is a very fine inspector, but he can't bring your wealth of experience to the task. Go yourself, Larry, and report back to me. I don't mean a written report, just phone and describe what you see. You might as well record the instrument readings while you're at it to make sure Duncan is doing his job."

"You mean tonight? I had three Chinese engineers in here today with some guy from the State Department and . . . I'm bushed."

"Yes, I'm afraid I mean tonight. If there is anything down below that needs attention it should be tended to right away, the reservoir elevation being what it is. If by some miracle our young Mr. Kramer has hit on something, we don't want him coming around later saying 'I told you so.'"

Jeffers sighed. "I'll check it out later. First I'm going to have some dinner and read the paper. I'll phone you tomorrow if I see anything unusual."

"Phone me in any case. I want direct confirmation."

"Okay, boss, whatever you say. Hello to the wife."

501

The twisting, twenty-mile-long groove that the Sierra Canyon River has worn
through the foothills northeast of Sacramento is too narrow for most of its
length for more than a county road and a loose string of cabins and cottages.
Twelve miles upstream from the mouth of the canyon, the valley widens
enough to accommodate the tree-lined streets of Sutterton. Before the dam,
Sutterton was a quietly aging village with a population of less than a thousand.
Named after John Sutter, whose sawmill a few dozen miles away was the
scene of the gold strike in 1848, the town flourished and floundered under
successive waves of prospectors, miners, loggers, railroad builders, and
pensioners. By the 1930s it had subsided into little more than a point of
departure for fishermen and hunters.

Architecturally the town offers little. There is a gargoyle on the Catholic
church. The wooden tower atop City Hall is a curious example of carpenter
Gothic, and the cracked bell that hangs therein was shipped around the Horn
from a Belgian foundry, now defunct. Three buildings with foundations dating
from the 1870s are California State Historical Landmarks, which means that
their owners can be strangled in red tape should they try to upgrade them.
What is now the Wagon Wheel Saloon began as a brothel where, according to a
widely believed local legend that is almost certainly false, Mark Twain and
Bret Harte once knocked each other's teeth out.

In the 1960s the town was assaulted by a new wave of invaders: geologists,
surveyors, soil analysts, hydrographers, and civil engineers looking for the best
possible site for a dam of record-breaking size. Close on their heels, as
enabling legislation was enacted, permits obtained, and court challenges
beaten back, came representatives of the Corps of Engineers, the Bureau of
Reclamation, the Bureau of Land Management, the Department of
Agriculture, the Forestry Service, the California Division of Water Resources,
the State Department of Fish and Game, the California Division of Highways,
and thirty-seven other local, county, regional, state, and federal agencies that
claimed an interest in or jurisdiction over one part of the project or another.

Preparation of plans and specifications and supervision of construction were
assigned by the owners of the project, the Combined Water Districts, to its
engineering consultant, Roshek, Bolen & Benedetz, Inc. A year before the
design of the dam was completed and final authorizations received from the
regulatory agencies, R. B. & B. awarded two preliminary contracts that had to
be started early if they were to be completed on time: the driving of a
diversion tunnel to carry the river around the site and the excavation of a
cavern in which the powerhouse would be built.

The fifteen-foot-diameter diversion tunnel entered the mountainside at river
level and emerged four thousand feet downstream. Workmen experienced in
underground drilling, blasting, and rock removal started from both ends and
worked toward each other. Temporary barriers--cofferdams--of earth were
built to keep the river away from the tunnel portals while work was under
way. Diverting the river into the finished tunnel was a well-publicized event
that was witnessed by hundreds of people from the overlooks and by
thousands on Sacramento television. The feat was accomplished in
September, when the river flow was seven thousand cubic feet per second, a
tenth of what it reached during the annual spring flood. At a signal from a

flagman, a fleet of thirty trucks and bulldozers dumped load after load of rock into the river, building the banks on each side toward each other until the channel was pinched off. The water rose quickly, but before it could overtop the barrier and wash it away, it found the opening of the diversion tunnel. Cheers were heard in the canyon when the water first entered the tunnel and again when it emerged from the downstream portal.

In the year following the award of the two-hundred-million-dollar contract for construction of the dam, the population of Sutterton doubled, and it doubled again in the second year. The newcomers were specialists in such things as heavy-equipment operation and maintenance, concrete production and placing, steel erection, and earthmoving. They were part of a nationwide fraternity of men whose skills and temperaments led them from one big construction job to another and whose children were used to being strangers in suddenly overcrowded small-town schools.

After the river was diverted, scrapers, power shovels, loaders, and bulldozers stripped away the topsoil along the axis of the dam from one side of the canyon to the other. Exposing bedrock required excavating a trench two thousand feet long, five hundred feet wide, and a hundred and fifty feet deep. Cracks in the foundation rock were sealed by pumping grout under pressure into hundred-foot-deep drilled holes. A solid concrete core block eighty feet high and a hundred and fifty feet wide was built along the bottom of the trench. Inside the core block were drainage and inspection tunnels, access to which was by stairs leading downward from the underground powerhouse.

When foundation work was finished, the dam took shape rapidly. Fifty scrapers and trucks shuttled twenty hours a day between the site and nearby quarries and borrow pits. Impervious clay was placed in the center while earth and rock, in precisely specified zones, were placed on either side. The material was spread into foot-deep layers by bulldozers and graders and packed down by rollers.

For nearly four years the residents of Sutterton were jolted by explosions, vibrated by passing trucks, and coated with dust. Few complained. The dam was putting the town on the map, and people were crowding in with money practically falling out of their pockets. New gas stations, car lots, realty offices, souvenir shops, and trailer parks sprang up like weeds. The highway south of town eventually was lined with every fast-food franchise known to man, plus one that was invented on the spot: Dorothy's Damburgers.

A popular form of entertainment was watching construction operations from the overlooks on the hillsides, one of which was equipped with bleachers, loudspeakers, and chemical toilets. According to a statement broadcast every hour whether anyone was listening or not, the dam required placement of ninety million cubic yards of material, enough to duplicate the Pyramid of Cheops thirty times over or to fill two cereal bowls for every human being in the world.

"Although Sierra Canyon is not a concrete dam," the voice intoned, "enough concrete is required for the core block, the powerhouse foundations, the spillway, the intake and outlet works, and a highway across the top--a million cubic yards in all--to build a sidewalk from San Francisco to New York and back with enough left over to continue it to the vicinity of Milpitas. The lake that will form behind the dam will at its maximum elevation have an area equal to eighteen thousand seven hundred and seventy-five football fields.

"The chimney-like structure you see under construction directly beneath you is located just upstream from the upstream toe of the dam and will be eight hundred and forty-five feet high, with its top twenty feet above the surface of the lake. It is the ventilation-intake tower, which among other things will provide a means of emergency ingress and egress for powerhouse personnel. Inside will be a massive vertical pipe leading to the powerhouse turbines. Water will be admitted through remotely controlled gates at ten elevations.

"The Combined Water Districts hope you enjoy your visit and ask you not to throw garbage over the guardrails. Keep children and pets under control at all times. This message will repeat in fifty-five minutes or upon the insertion of a quarter."

Sidewalk superintendents in the bleachers who took their duties seriously became familiar with a certain blue pickup truck. The driver was the chief designer of the dam and the project's resident engineer, Theodore Roshek, who had taken it upon himself to make sure the contractor followed every line of fine print in the specifications. Construction crews learned that it was useless to try to cut even the smallest corner, because Roshek would wave his canes, turn red, and threaten to shut the job down. Was a small mistake already buried under the fast-growing embankment? Dig it up. Was faulty concrete poured and set? Break it out and pour it again. He was constantly on the move, either in his pickup or on foot, despite the discomfort he felt when walking over rough ground. Each week he spent three days in Los Angeles tending to the affairs of his consulting firm and four days at the dam. During those four days, it was agreed by the project's seventeen hundred and sixty workmen, he succeeded in making life miserable for everybody.

When the dam was completed, a platform draped with bunting was set up in front of City Hall and a dedication ceremony was held featuring oratory, six massed high-school bands, barbecued chicken, German potato salad, and Popsicles. Several of the speakers mentioned Roshek. The contractor's project manager said that the engineer's nit-picking and hair-splitting, his policy of never giving the contractor the benefit of the doubt, and his refusal to negotiate even trivial points had resulted in an overall loss of four million dollars for his company over the life of the job. The audience laughed. Contractors were always claiming to lose money, even when they were driving to their private planes in their Cadillacs. The laughter changed to applause when he added that the result of "that s.o.b.'s meanness" was the best-built dam in the history of the world.

The Mayor of Sutterton, an overweight, perspiring man with a penetrating voice and the largest hardware store within fifty miles, read from a script written by his wife, who had once studied journalism at Chico State. Before describing how Sutterton intended to march bravely forward into a future garnished by economic benefits, he thanked "a great engineer for his unstinting efforts. Theodore Roshek leaves behind more than a dam that is justly famous. He leaves behind more than a legacy of dedication and personal integrity. No, my friends, he leaves behind a good deal more. Those of you who have come to know him as I do realize that he navigates on what my grandfather used to call 'gimpy' legs. His labors over the past four years

504

haven't done those gimpy legs any good. In fact, he told me this morning at breakfast that they are a hell of a lot worse than when he arrived amongst us, you should pardon my French. Now maybe you can see what he is leaving behind. Invested in the great dam that rises behind me like the very aspirations of civilization itself are not just the best years of Theodore Roshek's life, not just the essence of his audacious and unquenchable genius, but a big chunk of his health as well."

*   *   *

Phil crossed the carpeting and sat down tentatively on a leather chair in front of Roshek's massive mahogany desk. Good God, Phil thought when he saw the expression on the old man's face, he looks as if he's going to spring at my throat! What's he so mad about--that I didn't bring a cap I could twist in my hands while he bawls me out? If he tries to slap me the way Sister Mary Carmelita did in grade school for shooting spitballs, he's going to have a fight on his hands.

"You're Kramer? You look intelligent. Why don't you act it?"

"I beg your pardon?"

"Tell me if I've got this straight. You have no practical experience except for a few summers' work with the highway department in Kansas or some goddam place. You've never been involved in the design or construction of a dam. You know nothing about the subject except what you've read in books, which is worse than nothing. You've come up with a cockamamie computer model that makes you think you can sit in an office five hundred miles from a dam you've never seen and understand it better than men with lifetimes of experience who are sitting right on top of it. Is that right?"

Phil stared, speechless. Was Roshek kidding? Was he trying to be funny? "No," he managed to say, "that's not right at all."

"It isn't? What's not right about it?"

Phil crossed and uncrossed his legs. "I don't know where to begin. You've put the worst possible interpretation--"

"Furthermore, you had the audacity to call the chief maintenance engineer on the site and lead him to believe that his headquarters people think something is wrong with his dam."

"I did not! I called him only to get some current meter readings. He may have thought I sounded excited on the phone, but he wasn't looking at what my computer program was telling me."

"He doesn't *need* to look at what your computer program is telling you, and neither do the managers of ten thousand other dams. He doesn't *need* intimations from a greenhorn that he isn't doing his job properly. This company doesn't need and *I* don't need an employee casting doubts about one of our structures. We depend on two things for success: the ability to provide technical services of the highest professional quality, and the *reputation* for being able to provide them. Ruin our reputation and we are out of business,

505

it's as simple as that. What you have been doing amounts to a whispering campaign against your own employer."

Phil waited until he was sure it was his turn to speak. Keeping his voice calm, he tried to make his point. "Mr. Roshek, I no doubt deserve some criticism for taking too much initiative. But the data clearly indicate, to me anyway, that some sort of investigation is called for."

"You're not an engineer. I hope you realize that. Not yet. Not by a long shot."

"I'm not a licensed engineer, that's true. In California you need five years' professional experience after graduation before you can apply. I have a doctorate in--"

"An engineer is more than a man with a diploma and five years' experience. True engineering can't even be taught in school, because it involves a man's personality, his willingness to consider every last detail, the way he respects the materials he works with, his sense of history and the future, and his integrity."

For Christ's sake, Phil thought, he's not listening to a word I say! He's using me to practice some sort of goddamned commencement address.

"Most important," Roshek went on, "is maturity. A sense of proportion. Judgment. A doctor wouldn't say to a patient, 'You probably don't have cancer. Then again, you might have. We'll know when we get the lab results. In the meantime, don't worry about it.' See how stupid that sounds? What you've been doing is along that line. Jeffers called me in Washington to find out what was going on."

"He did? I didn't mean to get everybody so upset. I didn't realize that a phone call would--"

"You didn't realize that you could have caused a panic? What if word leaked out that we were worried about the safety of the nation's highest dam? A secretary overhears part of your conversation, a switchboard operator listens in, rumors start flying, newspapers pick it up, politicians demand an investigation, environmentalists charge a cover-up . . . I've seen it happen. Over nothing. Because a prematurely smart college student gets excited over drivel in a computer."

Phil felt his cheeks turning red. He knew he should simply let Roshek's tirade run its course without risking disaster by fighting back, but he felt he should put up some sort of defense. Silence would imply that he agreed with Roshek's distortions. Besides, he was getting mad. There was nothing in life he hated more than to be accused of something he didn't do.

"Sir, I didn't say to Mr. Jeffers that I thought the dam was failing. I showed my surprise that so many of the meters were out of order and that the seepage was so high. I did tell Mr. Bolen, in private, that my program indicated something was wrong. Maybe I should have discussed it with him first before calling the dam, but I wanted to make sure I had the latest figures." Phil let his voice trail off because his words were being ignored. Roshek's eyes were

506

wandering around the room resting on the photographs of his projects, and he was reciting their names like a litany.

"Sinai, Maracaibo, San Luis, Alyeska. These tremendous developments are as sound as the day they were built. " He gestured toward a glass display case in which was a realistic scale model of an earth and rock dam complete with tiny trees on the abutment slopes and a center stripe on the road across its crest. "Sierra Canyon Dam. Recognize it? Probably not, since your knowledge is confined to textbook abstractions. Not one of the structures this company has had a hand in designing has ever had the slightest question raised about its safety. Not one has suffered a failure of any kind. Engineered structures fail, yes, usually because foundation conditions aren't properly assessed. It was Karl Terzaghi, the father of modern soil mechanics, who said that when Mother Nature designed the crust of the earth, she didn't follow the specifications of the American Society of Testing and Materials."

I'm a dummy audience, that's all I am. Phil said to himself, wondering how much more he could take before boiling over.

"The structures I've designed will be in use two hundred years from now, if the civilizations of the future want them. Durability like that is a result of skill, hard work, intuition. and uncompromising insistence on top-quality work every step of the way. When a design philosophy like that is brought to bear, structures don't fail."

Did Roshek really believe that? Phil wondered. That if a skilled engineer did his best nothing could go wrong? The proposition was absurd on its face.

Roshek gazed with a kind of rapture at the display case. "Sierra Canyon Dam, about the supposed inadequacies of which you have developed such a lunatic obsession, has a design life of three hundred years. It is an engineering landmark, and not because of its height or its cost-benefit ratio."

"I don't have an obsession," Phil said quietly, "except possibly for a computer programmer I met recently."

"It represents an unprecedented effort to insure safety, from the thoroughness of the geophysical investigations right through to the ongoing system of inspection and maintenance. I insisted on the most extensive network of sensors ever implanted in a dam."

"Half of those sensors don't work anymore."

"There is a matter of justice involved, too. Of percentages. Of fairness." Roshek pushed his swivel chair away from his desk and looked at his legs. "I was struck by polio two years before the vaccine was developed that would have saved me. That's enough bad luck and injustice for one life." He looked up and seemed momentarily flustered by having become more personal than he had intended. Recovering, he glowered at Phil as if he were to blame for the indiscretion. "I've spent too much time on this already. I need to tell you just one thing. Stop concerning yourself with the dam. Is that clear? If you want to fool around with schoolboy computer models, use your own computers and your own time. That's all. You can go." He began assembling papers from his desk and placing them in his attache case.

Phil had gradually slumped down in his chair while watching the older man's bizarre performance. If he understood the last remarks correctly, Roshek was saying that his structures could not fail because his legs already had, that there was a limit to the bad luck that could happen to any one man. This was the world-famous engineer, the paragon of logic and objectivity?

"Well?" Roshek said, eyeing him. "I said you can go."

Phil straightened up but did not rise. "Mr. Roshek," he said, "you've been very unfair. I thought you'd give me a minute at least to explain my actions. In my defense I could point out that--"

Roshek cut him off. "What do you mean, 'in your defense?' This is not a trial. I pay your salary and so I can tell you to do whatever I want you to do. I'm telling you to drop Sierra Canyon Dam. Your ignorance of it would fill Chavez Ravine. You've caused enough trouble and I want an end to it. Yesterday, as you may or may not know, I was sworn in as president of the American Society of Civil Engineers. You should read the Code of Ethics. Point number two is that engineers should perform services only in their areas of competence."

"I was president of the student chapter in college." Phil said half to himself, "and I know the Code of Ethics, too. Point number one is that engineers should put the safety, health, and welfare of the public above everything else."

"What? What did you say?"

Phil stood up, his cheeks hot and his heart pumping. In a louder voice he said, "I don't deserve to be shouted at and treated like a child," amazed that he was saying anything at all. "In the past few weeks I've been closer to that dam than you. It's true that extensive foundation borings were made before construction, but the earthquake five years ago might have changed everything. It's not true that none of your structures has ever been questioned--the dam leaked so badly after the quake two million dollars had to be spent to plug up the cracks. The reservoir was filled this spring fifty percent faster than you yourself recommended five years ago that it should be."

Roshek was so astounded by the outburst that he couldn't find his voice. His mouth opened and closed and his eyebrows rose high on his forehead.

Phil tore a sheet of paper from a notebook and dropped it on the desk. "Here are the latest seepage figures from Gallery D. In every case they are higher than Theodore Roshek said they should be when he wrote the original specifications. Somebody should go down there right now and take a look because next week may be too late."

Roshek crumpled the sheet into a ball and hurled it against the wall. He found his voice, and it was loud. "I don't need you to tell me how to look after a dam. I didn't order you in here to listen to your sophomoric opinions! Your opinions are more irrelevant now than they were before, because you are fired! Get out! If you are at your desk in the morning, I'll have you arrested for trespassing!"

Phil tried to slam the door on the way out but hydraulic hinges made it impossible.

508

"Say You Gave at the Office . . ."

An Excerpt from **A Heartbeat Away**
Richard M. Cohen and Jules Witcover

Never in the history of American civil engineering has the profession been subjected to the criticism and notoriety that followed in the wake of the investigation and resignation of Vice President Spiro T. Agnew.

Beginning in the summer of 1973, when the nation was captivated by the Congressional hearings concerning the Nixon administration's involvement in the so-called "Watergate scandal," rumors began to circulate about Mr. Agnew's possible involvement in a scandal that would render him politically incapable of assuming the presidency if President Nixon were forced to resign. As summer passed into autumn and President Nixon's situation became increasingly precarious, the rumors about Vice President Agnew's predicament also became more clamorous. Finally, on October 10, 1973, Mr. Agnew resigned the office of Vice President of the United States and appeared before a federal judge in Baltimore to plead <u>nolo contendere</u> [no contest] to a pending charge of income tax evasion.

The income tax evasion charge was the result of an investigation of a kickback scheme involving Mr. Agnew and various civil engineers in Maryland. According to the U. S. Justice Department, Mr. Agnew, while he was County Executive of Baltimore County and subsequently Governor of Maryland, received and spent funds demanded from contractors and consultants who had received contracts from county and state agencies under Mr. Agnew's supervision. Furthermore, contended the Justice Department attorneys, Mr. Agnew had failed to report the receipt of these funds and pay Federal income tax on them.

In his book, **Go Quietly or Else** *. . ., Mr. Agnew contends that he did nothing different than his predecessors in Maryland and that the investigation leading to his resignation was a politically motivated witch hunt. He points out that none of the engineers allegedly involved in the schemes were ever convicted of any illegality connected with the case.*

The fact is, however, that this investigation did result in the identification of illegal and unethical practices on the part of some civil engineers in the State of Maryland. As a result of the investigation and the publicity it generated, the integrity and honesty of the civil engineering profession was subjected to the kind of scrutiny that discredits, to some extent, the entire profession.

This selection describes the events and circumstances leading up to Mr. Agnew's resignation.

On January 15, 1973, five days before the second inauguration of President Richard M. Nixon and Vice President Spiro T. Agnew, the eyes of the nation's political community were focused on a courtroom in Washington, D. C. There, four men arrested in the break-in of the Democratic National Committee headquarters at the Watergate complex--Bernard Barker, Virgilio Gonzalez, Eugenio Martinez, and Frank Sturgis--pleaded guilty to conspiracy. Their pleas were to have been the final act in an embarrassment that the

Nixon-Agnew administration, in the full flush of 1972's landslide victory, hoped to put behind it. On that same Monday, however, another scene was unfolding in nearby Baltimore that was destined to inject an entirely new element into the equation of Watergate--the question of presidential succession in time of national crisis.

The unwilling catalyst in this unforeseen development was an old friend and business associate of Agnew named Lester Matz. During the morning, Matz arrived in the lobby of downtown Baltimore's Mercantile Building for an appointment with a lawyer. Matz was extremely nervous. He walked into the elevator, turned, punched number 18, and watched the elevator's polished metal doors close in front of him. As the car rose to the law offices of Venable, Baetjer and Howard, the glistening doors reflected the image of an athletic-looking man, tanned the year round by the sun of Saint Croix, where he and his friend Ted Agnew had bought condominiums, and trim from the ski slopes of Aspen, where he maintained an apartment.

At the law firm's reception desk the dapper Matz asked for Joseph H. H. Kaplan, whom he had never met. . . .

*　　*　　*

Matz handed Kaplan a subpoena *duces tecum* issued in the name of a special federal grand jury that had recently been impaneled in Baltimore. It called for the engineering firm of Matz, Childs and Associates, Inc., to produce certain corporate records. Kaplan was not surprised that the firm had received a subpoena. Every lawyer in town had heard by then that George Beall, the U. S. attorney for Maryland, was investigating kickbacks in Baltimore County, with the objective of indicting Dale Anderson, the county's Democratic executive and political boss. Reaching for a yellow legal pad, Kaplan jotted notes as the extremely agitated Matz told his story.

The records, Matz admitted at once, would indicate that his engineering firm had been generating cash for the purpose of kicking back 5 per cent of its fees for county public-works projects to Baltimore County politicians. It was an old familiar story, and Kaplan had a ready strategy to deal with it. The government, he explained to Matz, was not interested in making a case against Matz or his partner, John Childs. What the government wanted was information to use against higher-ups. Kaplan's advice was orthodox and blunt: tell the government everything. Withhold nothing. Then the U. S. attorney, as was the custom, would offer Matz and Childs immunity from prosecution. All they had to do was be absolutely candid and agree to testify as government witnesses at any subsequent trial.

Matz was hesitant. "Do I have to tell them everything I know?" he asked.

Yes, Kaplan said, he did. If the prosecutors discovered that he had withheld any information, the immunity grant would be voided and he would be prosecuted with a vengeance.

In that case, Matz replied uneasily, he could not cooperate.

Why couldn't he? the perplexed Kaplan demanded.

510

"Because," Matz blurted out, "I have been paying off the Vice President."

For the rest of that morning, as Joe Kaplan sat stunned, Lester Matz related a story that before long would jolt a nation nearly inured to shock after the many months of revelations in the Watergate case. It was--unlike the Watergate chronicle of arrogance, excess and stupidity in high places that defied imagination--a story of old-fashioned graft and greed practiced by the one man regarded more than most as the epitome of righteousness in American politics. Systematically since 1962, the year Agnew became executive of Baltimore County, Matz had been making cash payments to him in return for county public-works contracts. The arrangement had continued after Agnew became governor of Maryland in 1967, except from that point on the *quid pro quo* involved state contracts. In fact, Matz had made payments to Agnew even after his election as Vice President. On one occasion, he had visited Agnew in an office he then occupied in the basement of the White House, and there handed him an envelope containing about $10,000 in cash.

Matz, having concluded his story, asked to use a phone. Kaplan led him to one in the outer office, and from there Matz made two calls to close associates of Agnew. He was in trouble and something had to be done. Agnew must somehow stop this blasted investigation. But the Vice President was very busy. It was a period of intense activity for him. In just five days he would be inaugurated for his second term and would then leave almost immediately for the Far East as a presidential emissary extraordinary.

After Matz left the office, Kaplan remained in his chair. The lunch hour came and went and still the young lawyer sat there, tumbling Matz's admissions over and over in his head. . . .

*   *   *

All over Baltimore that third week in January, other consulting engineers and architects rushed to the offices of the city's available legal talent, advised by some cold and blunt-talking assistant U. S. attorneys that they should retain counsel "familiar with the federal criminal code." The first subpoenas of nearly one thousand were being served, all bearing the name of assistant U. S. Attorney Russell T. Baker, Jr., and all issued in the name of a special federal grand jury.

George Beall, the thirty-five-year-old U. S. attorney for Maryland, was papering the Baltimore metropolitan area with subpoenas, attempting, or so it was rumored, to prove at the bar of justice what was already no secret in the state of Maryland: Baltimore County public officials were receiving kickbacks.

*   *   *

On January 4, Baker [had] sent out his first wave of subpoenas--twenty seven to the firms doing the most business with Baltimore County and the twenty-eighth to the county government itself. Then he waited, hoping . . . that he would soon discover kickbacks in the county's construction industry. But Baker would have been in for a long wait had the subpoenas all gone to construction firms. No unusual cash flow was found in any of them. It so happened, however, that one of the twenty-seven top firms receiving government business was an architectural company. This was in itself unusual, but in

511

addition, there in the books of Gaudreau, Inc., a largely family-owned company headed by Paul Gaudreau and his brothers, Robert Browne's* IRS agents found the cash they were looking for. Almost by accident, the prosecutors who had plunged into what amounted to a fishing expedition in the wrong industry suddenly stumbled on a key to another world of corruption--smaller but still numerous contracts for architectural and engineering work that were let not by public, competitive bidding but by private negotiation with public officials.

The agents promptly reported their findings to Browne, who checked again with his counterparts in New Jersey. The New Jersey IRS was not surprised; engineers and architects were the soft touches of the construction industry and therefore the easiest to extort. Unlike the ruggedly independent builders or any of the building tradesmen, the designers were invariably college-educated men with little or no inclination for a brawl. In addition, they were generally the proprietors of firms with many employees and unlike the builders could not pick up a crew at the union hall, so the loss of a contract could mean layoffs and economic ruin.

On January 11, Baker . . . authorized a new wave of subpoenas, this one to wash on all those engineers and architects who dealt directly with Baltimore County officeholders to get county business. The next day, a Friday, an IRS agent served a subpoena on the consulting engineering firm of Matz, Childs and Associates, demanding that the corporation produce its books. Neither John Childs nor Lester Matz was at work that day, so it was not until the next Monday that Matz, the president of the firm, went to see his lawyer, Kaplan.

*   *   *

Within days, two of the engineers subpoenaed by Baker signed aboard.** . . . [W]hen the witnesses were asked to provide the names of other engineers who might also be kicking back, the same names surfaced: Matz, Childs and Associates; and Jerome B. Wolff. The name of Wolff would haunt the investigation from the winter day in late January when his name was first mentioned until the muggy summer day in July when his testimony sealed the case against Agnew. Already the two engineers who had joined Gaudreau in cooperating had supplied deeply incriminating information about Wolff.

Among his fellow engineers, the brilliant Wolff was hailed as a genius, and it was at least true that he was nationally recognized for his expertise in the new and complex field of environmental planning. He was both lawyer and civil engineer, able to provide legal and scientific advice to a client concerned about the environmental impact of a particular project. Short, compact, and robust

----

* Robert Browne was chief of intelligence for the Internal Revenue Service [IRS] in Baltimore. He and Beall had consulted with their counterparts in New Jersey for advice on the most fruitful means for conducting such an investigation because the U. S. attorney for New Jersey had successfully completed several similar probes over the past eight years.

** That is, they had agreed to cooperate with the government by providing evidence and serving as witnesses in the prosecution of others involved in the schemes.

at the age of fifty-nine, Wolff was justly proud of his professional standing, honored that other engineers turned to him when they had a seemingly insoluble problem. He considered himself to be above all a scientist; certainly, he was no mere hack engineer lusting for a municipal sewer contract. His hobby, fittingly, was astronomy; his compulsion, ominously for Ted Agnew, was recordkeeping.

To others who were not engineers, Wolff was far better known for a different reason--his long and enduring relationship with Ted Agnew. Thumb out, Wolff had hitched a ride with him from Towson* to Annapolis and finally to Washington. In Washington, he held the title of vice-presidential adviser for science and technology until 1970, when he returned to private business. In Annapolis, where he was the Agnew-appointed chairman of the State Roads Commission, Wolff performed a pivotal task--he advised the governor on the awarding of state engineering contracts.

In the winter of 1973, though, what the prosecutors were hearing about Wolff had nothing to do with Agnew. Jerry Wolff was just another engineer, president of Greiner Environmental Systems, Inc., a subsidiary of one of Maryland's largest engineering firms, the J. E. Greiner Co. Wolff's firm, the prosecutors were hearing, was one of those that was kicking back a portion of its fees. Routinely, another subpoena was authorized by Baker. When Wolff received it, he took two actions to protect himself: he called a lawyer--and he sent a message to Agnew.

*   *   *

Nothing in Joe Kaplan's thirty-six years had prepared him for what Matz told him that day. . . . Matz declared that he would rather go to jail than add to the woes of an already troubled nation by implicating the Vice President. Friendship apart, the best thing for him to do for his country was to keep his mouth shut, he said. But there was friendship, too, and an association that went back more than ten years. Matz had courted Agnew in 1961, and had donated $500 to his 1962 campaign for county executive; both Matz and Childs hoped that with a friend in the county seat of Towson their firm would receive the public-works contracts that up to then had been denied them. They were right; their friend rewarded them with the long-sought contracts. All he asked was 5 per cent of their fees, an expense both men thought was only fair.

Years before, Matz had learned his lesson. As he told the story, his engineering firm and two others had formed a consortium to negotiate a public-works contract in nearby Anne Arundel County. Matz, trained at the Johns Hopkins University and proud of his talents, worked hard for that contract, drafting his plans meticulously and presenting them with the preparation of a doctoral candidate taking his oral examinations. He was rewarded with a note telling him that the contract had been let to one of the other firms in the consortium. Matz henceforth would often reach for his wallet before bothering to reach for his slide rule.

----

* Towson is the county seat of Baltimore County.

Now Matz's firm prospered. He rented an apartment in Aspen, Colorado, to indulge his love of skiing and purchased a condominium on Saint Croix in the Virgin Islands, near the one bought by Agnew. He became a man of wealth and substance, a first vice president of the Chazic Amuno synagogue, a *mensch* who honored the Jewish tradition of charity by sharing his wealth with his less fortunate coreligionists in Israel. In 1970, his charities cost him $90,000. The year before, a different tradition had cost him at least $10,000, for it was in 1969 that he had placed that sum in an envelope and handed it to the Vice President of the United States.

*   *   *

[The prosecutors], . . . reviewing the evidence already assembled, decided that the next thing to do was to bring to terms Lester Matz, his partner John Childs, and Jerry Wolff, none of whom were turning out to be very cooperative.

In fact, it was their very lack of cooperation that first aroused the suspicions of the prosecutors. Originally, they had been just three more engineers caught in the net thrown over Baltimore County's engineering industry. More and more, though, the prosecutors were hearing their names. They offered all three immunity from prosecution in return for their cooperation, but when they were summoned before the grand jury, all three invoked their Fifth Amendment rights against self-incrimination. It was a most peculiar performance. By invoking their right against incriminating themselves, all three were in the prosecutors' eyes admitting that they had something to hide. Whatever it was, the prosecutors were determined to find out. Already, there were hints that the firm of Matz, Childs and Associates was worth investigating. Its rapid success in the highly competitive engineering field was reason enough for suspicion.

Of the approximately two hundred engineering firms in Maryland, Matz, Childs was by far one of the most prosperous. Founded in 1955 with only a small nucleus of engineers, the company eighteen years later had more than three hundred and fifty employees and was ranked the ninety-second largest design firm in the nation. In the last decade, more than half of its contracts-- 231 in all--were awarded by various government agencies, notably those headquartered in Towson. From what the prosecutors had already learned, it seemed doubtful that Matz, Childs and Associates owed its success only to the quality of its work. In Baltimore County, success did not come with hard work alone.

In addition to all that, the corporation's books, now in the hands of Browne's IRS agents, were turning informer. Matz, like Gaudreau before him, stood by helplessly as the seemingly meaningless entries made by anonymous accountants fell into a pattern that told a story. Once again, the IRS agents were finding indications that cash was being generated. What made them suspicious was the way in which the Matz, Childs firm rewarded some of its key employees with bonuses. Nothing can look so suspicious to an IRS agent as a pattern of bonuses. A cooperating employee, an IRS agent knows, can give the bonus back to the firm in cash almost immediately, retaining just enough to pay his income tax on it. In the case, of Matz, Childs and Associates, the cash accumulated in this fashion was kept in a wall safe. And the books of Matz, Childs and Associates tantalized the IRS agents for another reason. The firm frequently used the services of a consultant already under investigation as a suspected bag man and money launderer.* The agents suspected that the only service this particular consultant provided the firm was a kickback of his fee (less, of course, his percentage).

Taken together, the pattern of bonuses and the steady use of the consultant was a suspicious bundle of information, but it was not hard enough evidence. In order to breathe life into the cold ledgers someone would have to talk. Matz and Childs would not; they had already been before the grand jury and taken the Fifth Amendment. So Baker turned logically enough to the employees themselves, the ones who had been receiving bonuses. On April 10, he subpoenaed six of the highest ranking employees of the firm and directed them to appear before the grand jury on Thursday, April 12. Then Baker set his trap.

Baker naturally expected that the Matz, Childs employees, like the firm's owners, would invoke the Fifth Amendment before the grand jury. Therefore, he decided to take them before a judge and confer "use immunity" on them. A relatively new legal device enacted by the Nixon administration, use immunity enabled a prosecutor to compel a witness to answer any question put to him before the grand jury, in exchange for which the witness was granted immunity so that anything he admitted to under the line of questioning could not be used against him in any prosecution for a crime. It was a *quid pro quo* that honored the witness's Fifth Amendment right while also compelling him to testify. The witness could still be prosecuted on evidence the government had learned independently. In the case of the Matz, Childs employees, use immunity was a perfect solution. Beall's staff had no interest in prosecuting them. All they wanted was information with which to build a case against Matz and Childs, who in turn might prove instrumental in the investigation of . . . [Baltimore County Executive Dale] Anderson.

Joseph H. H. Kaplan had been expecting the prosecutors to subpoena the Matz, Childs employees, who he was now also representing. He advised them to invoke the Fifth Amendment, forcing the government to fish elsewhere for its information. So confident was Kaplan of his defensive parry that he planned to be in Delaware later in the day, attending to the problems of a different client. Kaplan had to break that appointment. While he stood helplessly by, the confused and panicked Matz, Childs employees were whisked before a judge, use-immunized, and escorted by a triumphant Baker into the grand jury room. There, one by one, they confirmed the suspicions of the IRS agents. They had been kicking back their bonuses.

Suddenly both Lester Matz and John Childs were in very bad trouble--tax trouble, to be precise. Their firm had been deducting the bonuses as wages--a legitimate business expense--but now there was abundant evidence that the bonuses were not salaries at all. The employees themselves were not sure exactly what one would call them. Bribes? Illegal campaign contributions?

----

    * Bag man in a term used to describe an individual who collects money for someone else, usually to make it difficult to prove exactly who the recipient is. A money launderer is a middle man who manipulates money to make it difficult to trace.

Cash for the apartment of a mistress? Whatever it was, it was not deductible.

Word of what had happened in the grand jury room was soon flashed to Washington. Matz, who all along had been receiving pep talks from Agnew [via an intermediary] . . . was no longer confident. Wolff, too, was beginning to feel pressed, knowing that only a steadfast Matz stood between him and total incrimination. . . . Wolff sat down and wrote directly to the Vice President in Washington. Choosing his words meticulously in case the letter was read by a vice-presidential aide, Wolff composed what one of the prosecutors later called a "carefully worded scream for help" that left little doubt that it was in Agnew's best interest that the investigation be halted. Having mailed the letter, Wolff destroyed his only copy, reciting it later from memory.

*   *   *

Kaplan, meanwhile, was doing some probing of his own. He was in frequent touch with Baker, often discussing some of the Matz, Childs corporate records the government had seized months before and the firm wanted returned. On Friday, May 18, Kaplan made another of these routine calls to Baker, giving him no hint that he was about to blurt out the secret that had been weighing upon him since January. Baker reminded Kaplan that his clients were certain to be indicted. Kaplan didn't flinch. Both Matz and Childs were aware of that possibility, he replied, and in fact were reconciled to it. They had always been prepared to cooperate with the government, but they had nothing the government would consider of value. Anyway, what they *did* have to say, the government would not be willing to hear.

Baker reacted as if Kaplan had slandered him. What did Kaplan mean by that? Both Matz and Childs had been offered immunity months ago, and it was Kaplan who had rejected the idea. Kaplan cut right in as if Baker had not said a thing. . . . As for Anderson, his clients had no information to offer.

Then, without the slightest fanfare, Kaplan dropped the bombshell. The only person his clients were uniquely in a position to incriminate was--the Vice President of the United States! The government would undoubtedly not be interested in that sort of information.

Baker was outraged at this suggestion that the Justice Department would retreat in the face of an investigation of the Vice President. . . . The Baltimore U. S. attorney's office, Baker sternly lectured Kaplan, was non-partisan. It was prepared to investigate and prosecute all federal crimes. Period. . . .

*   *   *

Kaplan . . . had conferred with his clients . . . and they were both concerned about the "national implications" of the information they possessed. They were worried lest the downfall of Agnew, added to the national trauma of Watergate, prove to be more than the nation could endure. This was no mere political figure they were talking about. The subject of the discussion was the Vice President.

516

Well, Baker replied, both Matz and Childs would have to have faith in the American criminal-justice system. They were in no position to worry about what effect the investigation would have on the government's ability to function. If they refused to cooperate, the government would have only one recourse--to make a case against them, haul them before a judge, immunize them, and compel them to talk about Agnew before a grand jury.

* * *

Matz and Childs, with an eye on the events in Washington, had cause to worry about the effect their testimony would have on an already troubled nation. But the two engineers had more mundane concerns.

They were afraid they would look like turncoats, men who had squealed on their friends. They requested, and received, assurances that they would be given "use immunity" so it would appear they had been compelled to testify. Another worry was that Wolff might not cooperate, leaving them to face Agnew on their own. That would be an uneven confrontation, a swearing contest between two admittedly corrupt engineers and the Vice President of the United States. Lastly, they feared that if they admitted all their transactions they would implicate politicians with whom they had nothing but honest dealings but who might not have reported legitimate campaign contributions they had made. If the politicians disowned the contributions, again Matz and Childs would look like liars. Over the weeks, Baker met one objection after the other and disposed of them.

Finally, Kaplan played what appeared to be his trump card. His clients, he said, were prepared to be indicted. They had discussed the possibility with their families and were ready, stoically, to go to jail. . . .

There was little cheer . . . in the Matz household--a ranch-style house not far from Jerry Wolff's in the Baltimore County area of Stevenson. His attorney's word notwithstanding, Lester Matz was far from reconciled to an indictment. He was not, after all, a criminal. What bank had he held up? What woman had he raped? He had done what was necessary to do. He had not created the system. In Kaplan's presence Matz had repeatedly said, "The United States of America versus Lester Matz," and shuddered. It was a travesty. He had fought for his country in the war. He loved his country. Now it was threatening him with jail.

* * *

Lester Matz and John Childs met during the 1950s when they both worked as municipal engineers for Baltimore. They formed their own firm in 1955, establishing their offices in an old building at 2129 North Charles Street. Matz, a gregarious fellow, went out to seek clients. He found them in Baltimore County, to which thousands of Baltimoreans were fleeing. All over the county, fortunes were being built as farms made way for housing tracts and sewers and roads were constructed--a heady sight, a Comstock Lode of opportunities for an engineer. One of the major developers was Bud Hammerman, whose father, Sam, had founded the S. I. Hammerman Organization, a real-estate conglomerate. Another was Wolff, then an engineer in private practice. Before long, the three--Matz, Hammerman, and Wolff--established a business relationship.

517

The firm of Matz, Childs began to prosper. As a new and politically unconnected firm, however, it received none of the county's public-works contracts. Despite repeated attempts to break into the favored circle of firms that did, Matz, Childs and Associates remained outsiders, watching with mounting chagrin as the contracts flowed to their better-connected rivals. Still, Matz was not idle. By 1960, he had befriended the chairman of the county zoning board of appeals--Spiro Agnew. (Within two years, Matz and Agnew became involved in certain transactions with a man who will be referred to in this book as The Close Associate. Matz made the man's identity known to the prosecutors, but it was not publicly disclosed because the man was not cooperating with the investigation.) When Agnew announced that he would run for county executive, Matz and Childs threw in with him, donating $500 to what appeared then to be a doomed cause. The two engineers genuinely admired Agnew, and of course also hoped that his victory would bring them the contracts they believed they deserved.

Over the next four years, Matz, Agnew, and The Close Associate became even friendlier--visiting in each other's homes and celebrating milestone family occasions together, such as the bar mitzvah of a Matz son. There was also a business relationship. Shortly after Agnew's election, The Close Associate told Matz that the two of them figured to make a lot of money. The comment, though cryptic, was not lost on Matz, and a short time later, they met with Agnew. The new county executive told Matz that he had great confidence in The Close Associate. Matz unscrambled this message to mean that he was supposed to work through The Close Associate. That was fine with him.

Not long after, The Close Associate asked Matz to prepare a chart listing how much money the engineers receiving county contracts could be expected to kick back. Matz calculated the likely profits on certain jobs, concluded that a 5-per cent kickback was not unreasonable, gave a copy of the chart to The Close Associate, and took the original to Agnew. The county executive thanked him for his work.

The chart then became a manual by which kickbacks in Baltimore County and to Spiro Agnew were determined. When he turned over the copy of the chart to The Close Associate, Matz was told that he would be expected to pay 5 per cent on engineering contracts and 2 1/2 per cent on surveying contracts. This arrangement, in which both parties would benefit, was soon implemented.

Whenever Matz learned which contracts the county was about to let, he would contact The Close Associate and tell him which ones he wanted. Matz usually delivered the money to The Close Associate in his office, handing him a plain white envelope containing the cash. He paid in installments, generally, making each payment when the county sent him an installment for the work performed. And when the size of the cash payments increased and Matz and Childs found themselves in a cash bind, they began to generate cash by having key employees kick back bonuses.

At mid-point in Agnew's Baltimore County administration, Matz complained to The Close Associate that he was not getting enough county work. They all met together at Agnew's home, and Agnew promised to contact the appropriate county officials and order them to step up the flow of contracts to Matz, Childs.

It was no surprise, then, that in 1966, Matz and Childs were enthusiastic supporters of Agnew's gubernatorial campaign. Their faith in the man's abilities--and his financial value to them--was undiminished. With Jerry Wolfe as chairman of the state roads commission and Agnew in the governor's mansion, Matz, Childs and Associates soon began to enjoy a steady flow of state contracts. By then, however, circumstances had made Matz reluctant to continue paying through The Close Associate, for he suspected that the intermediary was skimming money off the top and taking all the credit for the cash he handed over. Matz went to Annapolis for a face-to-face talk with the new governor. In Agnew's ornate office with its majestic fireplace, Matz proceeded to denigrate The Close Associate, warning that he lacked discretion and would sooner or later get them in trouble.

Matz had a proposition. Instead of paying through The Close Associate, why not deliver the cash to Agnew directly? He would put the money in a savings account from which Agnew could draw after he returned to the practice of law. The savings-account money, Matz continued, could perhaps be accounted for later in the form of legal fees. Agnew liked that idea especially.

Subsequently, Matz reconsidered the savings-account scheme and decided that it involved keeping too many records. He did not, however, reconsider his determination to make his payments personally, and from that time forth he dealt directly with Agnew. The contracts kept coming, increasing substantially as the Agnew administration matured. On one occasion, Matz recalled, he was asked by Wolff if he was taking care of his "obligations." Matz replied that he was taking care of them directly.

All through 1967, Matz, Childs and Associates continued to share in the largess of the Agnew administration. So large were the contracts that Matz and Childs had to defer their payments to Agnew until they received their fees from the state. The fees began to arrive in the summer of 1968, and Matz, now far behind in his obligations to Agnew, was determined quickly to catch up, lest he be suspected of welshing. By July, 1968, his payments totaled about $20,000. With the fees that would soon be in the mail from the state, Matz figured he would owe $30,000 more. He showed his calculations to Childs, who agreed on the sum.

The firm, however, was in a fix. Matz and Childs felt they could not safely generate $30,000 in cash. So Matz turned to a former client who generally dealt in large sums of cash and arranged a "loan." In a complicated transaction, Matz, Childs and Associates loaned this former client $30,000, transferring the funds by corporate check; the client agreed to deliver $30,000 in cash to Matz. On the books of the client's firm the loan was recorded as being repaid in installments of $1700--a sum Matz and Childs thought they could safely manage in cash. When they received a "loan" installment, they simply transferred $1700 in cash to their "debtor."

The friend was able to produce $20,000 of the total almost immediately. Matz showed the cash to Childs, then he stuffed it in a manila envelope and drove to the State House in Annapolis. Taking the elevator to the second floor, he passed through the governor's reception room, with its oversized portraits of past Maryland governors going back to the Lords Baltimore, father and son, to the governor's office. There, Matz handed the envelope to Agnew, thanked him for the state contracts, and left. It was the last payment Matz made while

Agnew was governor of Maryland. Within the month, Agnew was chosen by Richard M. Nixon to be his running mate.

(Matz's payments to Agnew were not always for his personal use. In 1967, Agnew asked Matz for a $5000 donation to Nelson Rockefeller's presidential campaign. Cash or check? Matz asked. A check would be fine, Agnew said, and Matz put one in the mail. When the campaign temporarily collapsed with Rockefeller's declaration that he would not be a candidate, Matz's check was returned to him. By the time Rockefeller changed his mind again and re-entered the race at the end of April, 1968, Agnew was well on his way into Nixon's fold.)

By 1969, Agnew had been promoted by Richard Nixon and by the American people out of Maryland and down the Baltimore-Washington Parkway to the seat of national power. What he gained in stature, however, he lost in the authority to grant contracts. There were virtually none at his disposal. Nevertheless, Matz felt that he owed Agnew money for Maryland contracts received under the old Agnew administration. On a piece of yellow paper, he calculated the sum he thought was due Agnew and called the Vice President's office for an appointment. Matz took the yellow paper and an envelope containing $10,000 in cash and went to see Agnew in his office in the basemen of the White House. The engineer showed Agnew his calculations, reviewed them with him, and handed the Vice President the envelope. Agnew took it and put it in a desk drawer. Matz then told Agnew he might "owe" him more money as the contracts negotiated during Agnew's Maryland administration continued to generate fees. Agnew told Matz to call his secretary when the next payment was ready and tell her he had more "information" for the Vice President.

On his return to Towson, Matz told Childs about his White House transaction with Agnew. This was no longer something he could be casual about, and he admitted to Childs that he was shaken. He had just paid off the Vice President of the United States in the White House. Matz told one other person about the payoff--Jerry Wolff, then vice presidential assistant for science and technology.

From there on, Matz's common sense conflicted with his sense of obligation. Since Agnew was no longer in a position to award contracts, the pace of the payments diminished, though Matz did make one to him for $2500 in return for a federal contract awarded in 1971 to a subsidiary of Matz, Childs. Then about a year later, in the spring of 1972, Matz was contacted by The Close Associate, who pressed hard for a $10,000 contribution. Matz complained to Agnew himself.

"Say you gave at the office," the Vice President told him.

*   *   *

In the fantasy world of every criminal investigator is the discovery of the perfect witness--the man or woman with total recall and a personal diary to back it up. The Baltimoreans . . . were greatly in need of such a witness. They had Lester Matz who . . . came in, edgy and obsequious, and spilled out his story. It was substantially the same account that Joseph Kaplan . . . had conveyed earlier [to Baker on the telephone], including the payoff in the Vice

President's office in the spring of 1971, and it was plenty damaging to Agnew. But what the prosecutors needed now was solid documentation.

Beyond their wildest expectations, they began to get it . . . On July 10, Wolff . . . provided the deeply incriminating details. Tense, agitated, punctuating his story with overly loud, nervous laughter at things that were not funny, he offered this account:

In April 1966, Wolff was approached by The Close Associate of Baltimore County Executive Agnew. The Close Associate asked Wolff for money in return for county contracts that Agnew had arranged for him to receive. Wolff paid The Close Associate $1250 in cash, another payment of indeterminate amount to another Agnew associate for "legal fees," and one or two other payments.

Later in the same year, when Agnew ran for governor, Wolff gave him a cash contribution of $1000 and also worked in the campaign. If he were elected, Agnew suggested, Wolff might be made chairman-director of the state roads commission. Governor Agnew made good on the promise and Wolff took office on March 1, 1967. One of his chief tasks was to monitor every consulting engineering and construction contract in the state. It was a commanding position; under state law, for all practical purposes he controlled the selection of engineers and architects on every road commission contract, subject only to Agnew's approval.

Shortly afterward, Agnew's old friend Bud Hammerman approached Wolff. Agnew had instructed him to ask Wolff to join in an arrangement whereby Wolff would notify Hammerman which engineering firms were in line for state contracts, so that Hammerman could contact them for cash payoffs. Agnew had advised Hammerman that this conformed with a long-standing Maryland system under which engineers made large "cash contributions" in return for government contracts. Both he and Hammerman would be burdened with substantial financial demands from the political community, Agnew said, and it was only fair that those benefiting from the contracts bear their share of the burden. Wolff agreed to participate in the scheme and proposed that he, Agnew, and Hammerman split the proceeds in three even slices. Agnew balked; at first he told Hammerman he didn't see why Wolff should receive anything at all, but he agreed as long as he--Agnew--got his *half* of the pie. Hammerman went back to Wolff and they agreed to split the remaining half between them.

For the next eighteen to twenty months, Wolff told Hammerman which engineers were in line for state contracts, and Hammerman kept him informed of which engineers were paying off. In time, the contracting community came to know that Hammerman was the man to see, and in time there was no need for Hammerman to make a hard pitch. The engineers were expected to make "political contributions," almost always in cash, and even when there was no campaign to contribute to. Wolff told Hammerman the kickbacks should average 3 to 5 per cent of the contract, but Hammerman took any reasonable amount--sometimes in a lump sum, sometimes in installments. When Hammerman got the name of an engineer with a new contract, he would call, "congratulate" him, and arrange for a meeting at which the payoff was made. Hammerman would keep his 25 per cent, give

Wolff his 25 percent, and put Agnew's 50 per cent in a safe-deposit box until Agnew called for it.

At first Wolff kept his share at home, then he transferred it to two and later to three safe-deposit boxes, two in Baltimore and one in Washington. He spent most of the money on ordinary personal expenses over the next four years, but he used a small portion of it for kickbacks to other public officials in return for contracts given to two consulting firms in which he retained an interest-- kickback money to pay kickbacks!

Wolff was a highly qualified engineer, and so it did not seem out of the ordinary that he would make recommendations to the governor on who should do government work, and that Agnew generally would concur in his selections. As a basic premise, Wolff insisted that the firm chosen be competent to do the job, and Agnew and Hammerman on occasion would suggest to Wolff that a particular company ought to receive special consideration. Sometimes Wolff was asked to "recommend" a firm that was not kicking in--in order to create a pattern of general fairness and to avoid large and transparent deviations from that pattern, such as giving any one company a disproportionate share of the available contracts. But Wolff was so clearly the czar in making contract awards that some of the engineers and architects who were not kicking in took to wearing buttons that said, "Who's Afraid of Jerry Wolff?" And in pressuring engineers, Hammerman was so heavy-handed that one of them once went to Agnew and complained. Agnew, apparently fearful that the engineer would make his complaint public, gave orders that the engineer's firm receive some work.

Sometime after Agnew's election as Vice President in November 1968, but before his inauguration, Agnew asked Wolff to draw up a list of the contracts awarded during his term as governor to Green Associates, Inc., a Maryland engineering company. Wolff discussed the list with the firm's president, Allen Green, revised it somewhat, and turned it over to Agnew. The clear inference to be drawn from this exercise was that Green had been paying off and Agnew planned to use the list to persuade him to continue.

The details of Wolff's story were damaging enough. But what made his willingness to cooperate with the government even more important was the supporting material he brought with him. He was, in the description of one of the prosecutors, "a pack rat, a guy whose nature is just to keep a lot of documents . . . who had kept an incredible amount of paper contemporaneous with events and had destroyed none of it." He also kept diaries, small pocket-sized day-timers in which he wrote painstakingly detailed notes in tiny script of what had gone on each day. (An example, in blue ink, read: "Paid to _____ , $9024, Bay Bridge.") He had, the prosecutors found out to their joy, a little book for almost every month of every one of the last ten years, with only a few gaps--none of them in the years 1966-68, when Agnew was governor. The inks in the small diaries were chemically tested and other tests of authenticity were made. A fifty-two page report by the Scientific Services Division of the Treasury Department's Bureau of Alcohol, Tobacco, and Firearms established that the inks Wolff used, of various colors, were on the market at the times the notations were said to have been made.

Wolff was able to produce the list he prepared of Green's business with Agnew and other documents that, in one of the prosecutor's words, "screamed authenticity because of the way they were. They were little handwritten pencil things on a piece of notepad paper . . . some scruffy little thing, which is all dog-eared from its age." He also had lists prepared in the summer of 1968, in advance of Agnew's campaign for the Vice Presidency, of all the engineers whom he understood to be paying Agnew, on the assumption Agnew would want to have them to raise campaign funds. One was a list of the top ten engineering firms that had received government contracts under the Agnew administration, in order of how much they received. He even had a code of pluses and circles by which he marked those who were paying through Hammerman and those who were paying directly to Agnew. It was from this list that the prosecutors first identified the seven firms that were making kickbacks through Hammerman and the two or three that were paying Agnew directly. Wolff's detailed contemporaneous documents, with dates and amounts of contracts and kickbacks he received through Hammerman, were all turned over to Beall and retained as corroboration for presentation to the grand jury and, eventually, for use in trial.

*   *   *

Long before Governor Agnew's tenure, Green had become accustomed to making cash payments to public officials in return for various state and local contracts. It was seldom necessary that an express agreement be talked out; everybody involved knew the system, which functioned on a tacit understanding that engineers who paid got contracts and those who didn't got none, or at least few, and not the most lucrative ones.

Green first knew Agnew in mid-1963, when Agnew was Baltimore County executive. When he ran for governor in 1966, Green gave him between $6000 and $10,000 for his campaign, and after his inauguration, Green met him several times in the governor's offices in Baltimore and Annapolis. At one of those meetings, Agnew began to complain about the heavy financial costs of being governor. As leader of the Maryland Republican Party, he said, he needed money for his own political organization as well as funds to help Republican candidates around the state. Not only that, but he had to adopt and maintain a life-style far beyond his means. As county executive, he had served at a financial sacrifice, given the low salary of $22,500. The governor's newly raised salary of $25,000 was only the barest improvement; it still wasn't enough to handle the new load. Throughout Agnew's tenure, this theme recurred in his conversations with Green.

As one who had been around the state capitol for some time, Green did not need the message spelled out. He told Agnew he recognized that the governor had financial burdens and wasn't a wealthy man. His own firm had done well on public contracts and probably would continue to do so; there was no reason why he couldn't help with periodic cash payments. Agnew said he would appreciate such assistance very much.

In the past Green had paid public officials up to an average of 1 per cent of the fees he received on public engineering contracts, both in campaign contributions and in straight kickbacks. On this basis, he calculated that he could make six payments a year to Agnew in amounts of $2000, $2500, or $3000 each, depending on how much cash he had at the time--always cash, to prevent anyone tracing the payoffs on the company books. Six times a year he would ask for an appointment with the governor to deliver the money.

On the first occasion, Green handed Agnew an envelope containing between $2000 and $3000 in cash, saying as he did that he knew of the governor's financial bind and wanted to help. Agnew took the envelope and placed it either in his desk drawer or in his coat pocket, and thanked Green. Over the next two years. as Green made regular payoffs, they gradually said less and less about them to each other; Green would just hand over the envelope and Agnew would take it.

The two men usually discussed state business at such meetings, and Green nearly always would take the opportunity to bring up the subject of special interest to him--state road contracts. He would tell Governor Agnew which road and bridge contracts his company was interested in; sometimes Agnew would promise him a contract, sometimes tell him it had been committed to somebody else. In all this, the payoffs were never mentioned, but they continued on a regular basis.

In each of the two years Agnew was governor, 1967 and 1968, Green paid him $11,000. In the same period, his firm received about ten contracts from the state roads commission with fees of between $3 and $4 million. Occasionally, Wolff asked Green whether he was taking care of his "obligations" with respect to the state contracts he was getting. Green said that he was.

As Wolff had testified in his own account of his arrangement with Agnew, he came to Green shortly after Agnew was elected Vice President and showed him the list he had prepared at Agnew's request of the contracts Green had gotten from the state roads commission under Agnew. Green concluded that the purpose of the list was to assess what he owed Agnew. Wolff and Green discussed the matter and bargained it out, Green arguing that some of the contracts on the list had been awarded during the previous administration of Governor Tawes. True, Wolff said, but the Agnew administration could have canceled some of them or switched portions of them to other firms. Green prepared a revised list and gave it to Wolff.

Only once during the two years that Agnew was governor did he ever expressly mention any connection between the payments Green made and the state work he received. That occasion came just before Agnew's inauguration as Vice President, when Green made a payoff in Agnew's Baltimore office. Agnew, referring to the list, noted that Green's firm had received a lot of work from the state roads commission: he was glad matters had worked out that way, Agnew said. Then came the poverty plea again. During the two years in the governor's mansion, he still hadn't been able to improve his finances and although his salary as Vice President ($43,000 plus $10,000 for expenses, raised to $62,500 in 1970) would be much higher than his salary as governor, the social and other demands of lofty national office would put even greater pressure on his personal funds. So he hoped Green could continue the help he had been giving him, and he in turn hoped he could help Green get federal contracts. Green told the Vice President-elect he was willing to continue the payments, but he wasn't certain he could produce such large amounts as he had in 1967 and 1968. Contracts already awarded to his firm in Maryland would generate some income over the next several years, and so he could make the payments for a while; and he hoped his firm's federal contracts would indeed increase as a result of what the new Vice President could do for him. But there was one thing that did worry him, he told Agnew. The new state Democratic administration might take credit for and possibly demand payments on the basis of contracts that had been awarded to his firm by Agnew. The Vice President-elect told him he didn't believe that would happen.

Thus it was that Green continued to pay Agnew off personally, delivering $2000 three or four times a year either to the Office of the Vice President in the Old Executive Office Building, or to Agnew's apartment in the Sheraton Park Hotel. As in the past, cash was always in a plain envelope, and Green and Agnew were alone when it was handed over.

The first time he did it, Green felt particularly uncomfortable. Making a payoff in the very office of the second-ranking official in the government of the United States, with the Seal of the Vice President on the wall behind Agnew's desk, was bad enough; but Green was concerned that his conversation with the Vice President might be overheard or even taped. So when he handed over the envelope he told Agnew the money was part of a continuing and unfulfilled commitment in "political contributions." As he said it, he raised his eyes to the ceiling, silently conveying to Agnew the reason he was saying something the Vice President patently knew was not so.

The last payment Green made was during the Christmas season in 1972, after the U. S. attorney's office in Baltimore had begun to look into corruption in Baltimore County. All told, in addition to the $22,000 Green paid Agnew when he was governor, in 1969 and 1970 he paid the Vice President $8000 a year-- four payments of $2000 each; and in 1971 and 1972, $6000 a year--three payments of $2000 each. That brought the total Agnew received from Green over the six years to $50,000.

The Green story was explosive in its detail, and in the portrait it painted of Agnew: a blatantly greedy public official who somehow justified his demands for graft on the social and political obligations placed on him as he climbed the political ladder.

Green himself had been matter-of-fact about the relationship he had enjoyed with Agnew. He told the prosecutors, as he had told others, that he long ago concluded that payoffs were an integral part of the engineering scene in Maryland and some other states. He had plunged into this with his eyes open, waiting for the day when his firm would reach the size at which it could no longer be ignored--payoffs or no payoffs.

*　　*　　*

At precisely three minutes after two o'clock [on October 10, 1973], Spiro T. Agnew, Vice President of the United States, strode briskly into the courtroom. He was dressed impeccably, as always, in a perfectly pressed blue suit and blue-and-tan striped tie, his graying hair slicked neatly back off his tanned but now thin and tight-lipped face. There were murmurs as he made his unannounced entrance.

*　　*　　*

Now Judge Hoffman walked in. All rose, then sat again as the clerk announced his arrival and concluded with the traditional "God save the United States and this honorable court." . . .

Hoffman . . . asked, " . . . what plea . . . does the defendant wish to enter in connection with the charge as stated in the criminal information?"

"On behalf of the defendant, your honor," [his attorney] replied, "we lodge a plea of *nolo contendere*."

Hoffman turned to [Agnew]. "Mr. Agnew, is that correct and is that your plea?"

"That is my plea, your honor," Agnew said, in a firm voice.

"I am required to advise you, Mr. Agnew," Judge Hoffman said, "that a plea of *nolo contendere* is, insofar as this criminal proceeding is concerned today, the full equivalent to a plea of guilty . . . Do you thoroughly understand the consequences of a plea of *nolo contendere?*"

"I do, your honor," Agnew--struggling lawyer who became Vice President, voice of the law-and-order society--replied.

*   *   *

"Thank you, Mr. Agnew," the judge said, "and now you may take your seat. Subject to further proceedings, I will accept your plea of *nolo contendere*."

. . . [his attorney] got up. "At two-oh-five p.m.," he told the court, "there was delivered into the office of the secretary of state a letter, subscribed by the Vice President, in which he resigns his office."

*   *   *   *

With that, Spiro T. Agnew, private citizen, sat down.

"Naysaying Nabobs"

Civil engineering firms that depend on public works projects for all or most of their activity and revenue are easy targets for unscrupulous and greedy individuals who have managed to work themselves into positions where they can control decisions concerning which firms will receive such work. The temptation to yield to demands for illegal kickbacks or to "pad" the cost of a project to generate a false record that conceals thefts of public monies is almost overwhelming when the owners of a firm are faced with the possibility of work being withheld, especially when failure to receive the work would mean financial disaster for the firm. Small firms without a diverse clientele are particularly susceptible to this type of corrupt influence. Often the civil engineers involved in these schemes receive nothing for themselves other than the work which makes it possible for the firm to continue to operate.

Despite the fact that the American Society of Civil Engineers has consistently warned its members about the necessity for refusing to yield to this form of economic blackmail, cases continue to occur.

Probably the most celebrated such case was the case involving Spiro Agnew, Vice President of the United States--a case which led to his resignation in disgrace. The following selection consists of a portion of an editorial criticizing civil engineers for their participation in such schemes, the response of a former ASCE official to the editorial, and the reaction of an ASCE member to the response. The first two parts of the selection are reproduced here just as they appeared in the "Professionally Speaking" section of the February, 1974, issue of Civil Engineering magazine. The third part is the entirety of a letter to the editor of Civil Engineering. An abbreviated form of the letter was published in the magazine several months later.

In the aftermath of Vice President Agnew's resignation last October, civil engineers--and their insistence that competitive bidding for engineering projects is not in the public interest--have come in for their fair share of lumps from the nation's press. Typical of the current press pillorying of civil engineers is an editorial which appeared in the October 30, 1973 issue of the *Gainesville* (Fla.) *Sun*, entitled "Naysaying Nabobs." ASCE Executive Director Emeritus William H. Wisely, a resident of Gainesville where he is associated with the University of Florida, chose to refute the allegations made in the editorial. Portions of the editorial--and the Wisely reply--are here reprinted.

"Now that Spiro Agnew has been side-lined as the country's most prominent crook, let's take a look at his fellow nattering nabobs of negativism--the civil engineers who made it all possible.

"It began fairly early, sometime after 1962 when Agnew became Baltimore County's executive officer. For beginners, engineer Lester Matz fell into the habit of kicking in 5% of his government fee work to Spiro Agnew's friends. In 1968, after Agnew became Governor of Maryland, Matz decided to express his appreciation in person--the first $20,000 tucked into a manila envelope.

527

"Engineer Allen Green deserves a medal for Agnew fidelity. All the time that Agnew was Governor and Vice President--six years total--he ceremoniously handed over cash in $2,000 and $3,000 lots six to nine times a year. The final payment was made right in the Vice President's office last December.

"Over time, things were pretty well organized. An Agnew boyhood friend named I. H. Hammerman was the go-between. In 1967, with Agnew barely installed as Governor, Hammerman contacted eight engineering firms and promised to deliver state business in return for cold cash.

"And how many of these eight true-blue, educated and refined, top-notch professional engineers told Agnew's bagman to go to hell? One.

"Such is the evidence which confronted Spiro Agnew if he had chosen to fight in court. Instead, as history books will tell future generations, he resigned the vice presidency and accepted a fine and probation for income tax evasion.

"Our point is that Spiro Agnew was corrupt, but there also were corrupters in abundance, nice guys all. Invite those civil engineers to a Richard Nixon prayer breakfast and you couldn't tell 'em from the preacher of the day.

"The problem is nationwide. The *Wall Street Journal* recently rounded up a clutch of contemporary examples in Pennsylvania and New Jersey, with Louisiana an especial hot spot. In Louisiana, 10% is considered a normal "finders fee" for public works.

"Engineers are a sanctimonious lot and apt to scream foul at the obvious conclusion that engineers are crooks. Not all, of course, but enough--as witness the green outstretched to enrich Florida Governor Claude Kirk's thinly disguised slush fund called a Governor's Club.

"Our files on that 1966-70 episode would choke Jonah's pet whale. Consultants who poured $450,000 into Kirk's slush fund got 70% of the $12.6 million state consultant fees.

"That's when the *Gainesville Sun*, and a lot of other papers, came out for competitive bidding.

"The engineers squawked like Mary Poppins with a punctured umbrella. They declared they were 'professionals' and bidding was against their 'ethics.' But they fell back too hard on their ethical code, and a handy-dandy anti-trust suit has removed that weak reed.

"To ward off anticipated blows, we acknowledge the honesty of a great many engineers. But the engineers have not cleaned their own nest; the corrupt pattern is indisputable; and the public welfare demands remedy.

"The engineers might consult with their old buddy, ward-heeler Claude Kirk. He said that if anything waddles like a duck, quacks like a duck, and associates with ducks, then it must be a duck.

"And that defines the civil engineer's problem."

*　　*　　*

To the Editor
The Gainesville Sun

"Your October 30 references to the civil engineering profession under the caption 'Naysaying Nabobs' must be responded, much as I dislike to react to the sarcasm, bias and shallow logic of your statement.

"You can be sure that the civil engineering profession is well aware of the problem it faces in providing the public facilities and services that are so readily taken for granted--water supply, sewerage, highways, waterways, railroads, flood control, irrigation, power, pollution control, pipelines, airports, tunnels, bridges, buildings and other constructed works. Unfortunately, these projects must usually be carried on in the public sector, where the major decisions are made by officials who are elected or appointed by the political process. Since public works projects involve huge expenditures of money, they have since the time of (Roman Emperor) Hadrian and before been beset by graft and corruption of political origin.

"The American Society of Civil Engineers, with 68,000 members, recognized this problem long ago in its code of ethics, which includes these principles:

> It shall be considered unprofessional and inconsistent with honorable and dignified conduct and contrary to the public interest for any member of the American Society of Civil Engineers:
>
> 8. To exert undue influence or to offer, solicit or accept compensation for the purpose of affecting negotiations for an engineering engagement.
>
> 9. To act in any manner derogatory to the honor, integrity or dignity of the engineering profession.

"Regrettably, the punitive authority of the Society in ethical matters is limited to expulsion from membership of those found guilty by 'due process' of violations. Also, the Society has no resources for in-depth investigations of allegations of misconduct, and often must rely on the courts and public prosecution agencies for evidence. Nevertheless, the record will show that no other engineering society in the entire world has equalled the efforts of ASCE to enforce its code of ethics since its first ethical proceeding in 1899.

"At this time there are six cases before the Society involving alleged violations relating to political contributions or loans for the purpose of securing engineering engagements. Members of the Society in Maryland and in Florida are among those confronted with these charges. I can assure you that any of these members found guilty of ethical misconduct will not escape punitive action under any 'immunity' rule.

"Professional licensing of engineers is done under separate laws in the 50 states, administered by registration boards that are not under the control of the profession. ASCE has long encouraged the adoption and enforcement of ethical standards by these state boards. Loss of license for serious professional misconduct would be very much in the public interest, and this instrument will hopefully find greater application as a result of recent events.

"Your admission that all civil engineers are not corrupt is appreciated, but some numbers are in order. The largest segment of the profession, about 40%, are employees of federal, state and local government. About 30% are employed in industry, construction and education. Another 30% are working in private practice consulting firms, but the vast majority of these are also employees. Less than 5% of all civil engineers are principal partners or stockholders in consulting firms who may negotiate professional service contracts. Again, the vast majority of these consultants are scrupulously honest men of high professional caliber. Although I have no supporting data, I firmly believe that the civil engineering profession will compare quite favorably with other professions with respect to the prevalence of 'shysters' among its practitioners.

"In a response to one of your editorials some months ago I pointed out that there is no connection between influence buying through political contributions and the reluctance of engineers to make their professional services available on the basis of competitive bids. Competitive bidding will not prevent political favor-buying and kickbacks in one form or another. The practice has been extended to suppliers of products and services to all levels of government.

"These evils were not invented by civil engineers. They originated in a political system which permits and encourages dishonest people and double-dealers to divert funds from public works projects to personal or political use.

"I respectfully suggest that you direct your attacks not upon the unwilling victims of this vicious system, but instead against those who have created it and who perpetuate it."

Sincerely,

William H. Wisely
(ASCE) Executive
Director Emeritus

*   *   *

February 28, 1974

Mr. Ned Godfrey, Jr.
Editor, Civil Engineering
345 East 47th Street
New York, NY 10017

Dear Ned:

I read Director Emeritus Wisely's letter (responding to the *Gainesville Sun's* editorial attacking civil engineers) in Professionally Speaking, February, 1974, and found myself nodding my head in agreement--until I reached the last two paragraphs. If Mr. Wisely had ended his letter by replacing these two paragraphs with a statement to the effect that ASCE would continue its efforts to enforce its code of ethics and redouble its efforts to encourage state registration boards to enforce ethical standards, he would have done the civil engineering profession a real service. As it is, I fear he may have undone much of the good which his letter might have done by choosing to blame ". . . a political system which permits and encourages dishonest people . . . to divert funds from public works projects . . ."; by implying (to me at least) that civil engineers are the "unwilling victims of this vicious system"; and by suggesting that the newspaper's attacks would be more properly directed against "those who have created it (the system) and who perpetuate it" than against the unwilling victims (civil engineers?).

Blaming the "vicious political system" for all sorts of related and unrelated evils is in vogue these days, so one can hardly blame Mr. Wisely for jumping on the bandwagon. After all, its much easier to let a nameless and faceless system take your lumps than to bear them yourself. Similarly, we seem to have an oversupply these days of people who are anxious to tell us who's doorstep needs cleaning next--pointing to the other fellow's stoop while their own is crumbling beneath their feet. Despite the fact that blaming the "system" and finger-pointing are so common now (or maybe because of that fact) people, I think, are thoroughly disgusted with the use of these tactics-- whether or not the purpose of their use is evasion of responsibility. Thus the harm done by Mr. Wisely's language in those two paragraphs is not so much that it implies an absence of guilt on the part of the civil engineering profession, but instead that it puts us in the corner with all those who, whatever their true degree of guilt, refuse to recognize their wrongdoing and persist in attempting to blame others. There are already too many unsavory characters in that corner, and I, for one, don't care to join them.

The reality of the situation is that the civil engineering profession, squirm though it may, cannot escape at least a portion of the blame for the ugly practices that have been exposed in Maryland, Florida and elsewhere. The involvement undoubtedly extends far beyond those few engineers who have been convicted or who have plead guilty. Surely there are others, maybe many others, whose participation in these despicable schemes has not been exposed. Undoubtedly there are many, many more engineers who knew of the existence of the schemes and failed to take their knowledge to the proper authorities. Under our code of ethics they too share the guilt. So, while I agree with Mr. Wisely's statement that the vast majority of civil engineers are scrupulously honest, I believe the problem is much more pervasive than indicated by the small number of criminal convictions obtained in the various cases.

Mr. Wisely says that "the civil engineering profession will compare quite favorably with other professions with respect to the prevalence of 'shysters' among its practitioners." This may be quite true, but it begs the question. There is no other profession quite so dependent on the public trust for its survival and well-being. It is true that other professions are damaged by revelations of illegal and unethical practices by their members, but by and large these professions are more dependent on the trust and confidence of individual citizens because they deal primarily with individuals. We, on the other hand, are almost exclusively the agents--either directly or indirectly--of society as a whole. Events which erode public confidence in our profession strike at the very essence of our professional existence. Furthermore, because of the the scope and magnitude of civil engineering projects and because of the vast sums of public monies required to plan, design, construct and operate them, the "occasions of sin" are much greater than for most other professions. And every time an engineer succumbs to the temptation and is subsequently exposed, every single taxpayer who has been taxed to pay for a tainted project feels that the misspent funds came right out of his pocket--and rightfully so. Corruption in engineering doesn't eat away at public trust an individual at a time; it washes public trust away in swirling floods of thousands of citizens at a time.

Because we are so dependent on the public trust we cannot afford to take the chance that we need do nothing more than we've done in the past to enforce our code of ethics. Whether our record of enforcement my exceed that of any other engineering society in the world, as Mr. Wisely claims, is immaterial. We have not done enough. Mr. Wisely says "the Society (ASCE) has no resources for in-depth investigation of allegations of misconduct . . ." That is not correct. We have the resources; we have simply chosen not to use them for that purpose. Whether we should do so in the future is a question that I think should be submitted to the entire membership.

In the final sentence of his letter Mr. Wisely implies that the engineers involved in the Maryland scandals are "unwilling victims." I disagree. The true unwilling victims are the honest engineers who refused to participate in the illegal schemes and suffered the economic consequences of that refusal and the thousands of honest engineers who have never been involved in illegal practices but are nevertheless forced to share in the blackened professional reputation resulting from the dishonest activities of a few scoundrels.

Sincerely,

Augustine J. Fredrich, M. ASCE

The Bridge

A Chapter from **The Monkey's Wrench**
Primo Levi

*Primo Levi, educated as a chemist, was one of Italy's most gifted modern
writers until his suicide in 1988. **The Monkey's Wrench** is one of his most
entertaining works. The novel consists of stories told to one another by
Libertine Faussone, a skilled Italian construction worker, and the narrator who
is known to the reader only by his occupation as a chemist. Each of the stories
deals with some aspect of the story teller's work experience. This particular
selection tells, in a extraordinarily readable way, about a worker's view of a
bridge construction project, the engineer in charge of the construction and the
engineers who designed the bridge. Faussone is the story teller.*

"On the other hand, when they offered me a job in India, I wasn't all that
inclined. Not that I knew much about India. You know how easy it is to get a
mistaken notion of a country, and since the world is big, and it's all made up of
different countries, and practically speaking, you can't visit all of them, you
end up with a head full of crazy ideas about countries, maybe even including
your own. All I knew about India I can tell you in a few words: they have too
many babies; they starve to death because it's against their religion to eat
cows; they killed Gandhi because he was too good; the country's bigger than
Europe and they speak God knows how many languages, and so, for want of
anything better, they settled on English; and then there's the story of Mowgli
the Frog that, when I was a kid, I thought was real. Oh, I was forgetting the
Kamasutra business and the hundred and thirty-seven ways of making love, or
maybe it's two hundred and thirty-seven, I don't remember exactly any more,
I read it once in a magazine while I was waiting to get my hair cut.

"In other words, I would almost rather have stayed in Turin. I was in Via
Lagrange in those days, living with those two aunts of mine; sometimes
instead of going to a pensione* I visit them, because they make a fuss over me,
cook special dishes, in the morning they get up without a sound so as not to
wake me, and they go to the early Mass and buy me fresh rolls still hot from
the oven. They have only one fault: they want me to get married, and that in
itself wouldn't be so bad, but they're kind of heavy-handed about it, and they
keep introducing me to girls who aren't exactly my type. I've never figured
out where the old ladies find them: maybe in convent schools. They're all
alike: they seem made of wax, when you talk to them they don't dare even
look you in the face, they make me terribly embarrassed, I don't know where
to begin, and I get as tongue-tied as they are. So it may happen that, other
times, when I come to Turin, I don't even get in touch with my aunts, and I go
straight to the pensione; also to keep from disturbing them.

"Like I was saying, that was a time when I was kind of tired of roving around,
and in spite of this mania of my aunts' I would gladly have stayed put; but at
the office, they poured it on, they know my weak spot, and they know how to
twist me around their finger: it was such an important job, and if I didn't go

----

* A type of boarding house or hotel in Europe and Latin America.

they couldn't think of anyone else to send. What with one thing and another, they telephoned me every day, and besides, like I said before, I can't keep the engine idling, and I can take the city only for a short while, so the fact is that in late February I began to think it's better to wear out shoes than sheets, and at the beginning of March I was at Fiumicino[*] climbing into a Boeing, all yellow, Air Pakistan.

"The trip was a laugh from beginning to end: I mean I was the only serious traveler. Half were German and Italian tourists, all keyed up from the start at the idea of going to see Indian dance because they thought it was belly dancing, whereas I actually saw it and it's very prim: they dance just with the eyes and the fingers. The other half, on the contrary, were Pakistani workers going home from Germany, with their wives and little kids, and they were happy, too, because they were on their way home and it was their vacation. There were also some women workers; in fact, in the seat next to mine there was a girl in a purple sari. A sari is that dress they wear without sleeves, without any front or back. This girl, I was saying, she was a beauty; I don't know how to describe her: she looked like she was transparent, and with a kind of glow inside, and she had eyes that talked. Too bad she talked only with her eyes; I mean she only spoke Indian and some German. And I've never wanted to learn German; otherwise, I would gladly have struck up a conversation, and I swear it would have been livelier than my conversations with those girls my aunts dig up; no offense, but they're all as flat as if Saint Joseph had run his plane over them. Well, enough about that. Besides, I don't know if this happens with you, too; but with me, the more foreign a girl is, the more she appeals to me, because there's the curiosity.

*   *   *

"You know how it is when they're about to land, the engines slow down a little; the plane tilts forward, and it seems like a big, tired bird; then it moves down lower and lower, until you can see the lights of the field, and when the flaps are raised and the ailerons come out, the whole plane shakes, and it's as if the air had become bumpy. It was the same this time, too, but it was a troublesome landing. Obviously the tower wasn't giving permission, so we began circling; and whether there was turbulence or whether the pilot was green, or there was something wrong, the plane rattled like it was riding over the teeth of a saw; and through the window you could see the wings flapping like a bird's, or like they were on hinges; and this went on for about twenty minutes. Not that I was worried, because I know this happens sometimes, but it came back to me later when I saw what happened to the bridge. Anyway, we landed somehow or other, the engines died, and they opened the exits. Well, when they opened them, instead of air, it seemed like warm water poured into the cabin, with a special smell, which is actually the smell you smell everywhere in India: a heavy odor, a mixture of incense, cinnamon, sweat, and rot. I didn't have much time to waste, I collected my suitcase and ran to catch the little Dakota that was to take me to the work site, and luckily it was almost dark because one look at the plane was enough to scare you; and then, when it took off, even if you couldn't see it, it scared you still worse, but by that time there was nothing to be done, and anyway it was a short trip. It

----

[*] Fiumicino is the location of Rome's international airport

was like those cars in old comic movies; but I saw that the others were all calm, so I kept calm, too.

"I was calm, and pleased, because I was almost there, and because it was a matter of doing a job that suited me. I still haven't told you: it was a big job, rigging a suspension bridge, and I've always thought that bridges are the most beautiful work there is, because you're sure they'll never do anybody harm; in fact, they do good, because roads pass over bridges, and without roads we would still be like savages. In other words, bridges are sort of the opposite of boundaries, and boundaries are where wars start. Well, that's how I thought about bridges, and actually I still think like that; but after I worked on that bridge in India, I began to think I would also have liked to study more. If I had studied, I would likely have become an engineer; but if I was an engineer, the last thing I'd do would be to design a bridge, and the last bridge I'd design would be a suspension bridge."

I pointed out to Faussone that what he was saying seemed somewhat contradictory to me, and he agreed that it was, but before passing judgment I should wait till the end of his story: it often happens that a thing can be good in general and bad in particular, and that is exactly how it was in this case.

"The Dakota landed in a way I'd never seen before, and I've done a lot of flying. When it was in sight of the field, the pilot came down till he was grazing the strip, but instead of slowing down, he gunned the engines to the maximum, making the devil's own racket. He flew over the whole field at an altitude of two or three meters, he zoomed up just over the sheds, made a low turn, and then landed, bouncing three or four times like when you skip a flat stone over a stretch of water. They explained to me that it was to chase off the vultures, and in fact I had seen them, while the plane was coming down, in the beam of the searchlights; but I hadn't realized what they were. They looked like huddled-up old women, but later I wasn't amazed, because in India a thing always looks like something else. In any case, it's not that they were scared off: they shifted a little, hopping with half-opened wings, not even taking flight, and as soon as the plane stopped, they gathered all around it like they were waiting for something, and every now and then one of them would take a quick peck at the next one. They're really ugly animals.

"But there's no point in me telling you about India, there'd be no end to it, and you may even have been there . . . . You haven't? Anyhow, these are all things you can read in books; but how you draw the cables of a suspension bridge isn't in any book, or at least not the impression it makes on you. So we arrived at the work site's airport, which was only a field of packed earth, and they sent us to sleep in the dormitory. It wasn't all that uncomfortable, except that it was hot. But this heat business is another thing I don't have to go into. Just assume it was hot all the time, day and night, and down there you sweat so much that, excuse the expression, you don't have to go to the bathroom. I mean, all through this story it was hot as damnation, and I won't keep repeating that or I'd be wasting time.

"The next morning I went to the site manager to introduce myself. He was an Indian engineer, and we spoke English, and we understood each other fine, because the Indians, if you ask me, speak English better than the English, or at least it's more clear. The English just don't have an inkling, they talk fast and chew all their words, and if you don't understand, they seem surprised,

535

but they don't make any effort. Well, he explained the job to me, and first off he gave me like a little veil to put under my helmet, because down there they have malaria, and in fact at the windows of the buildings there were mosquito nets. I saw that the Indian workers on the site didn't wear the veil, and I asked him, and he replied that they all had malaria already.

"The engineer was very worried. I mean, in his place I would have been worried: but he, even if he was, it didn't show. He spoke very calmly and told me that they had hired me to draw the supporting cables of the suspension bridge, that the bulk of the work was done, namely they had already deepened the bed of the river in five places, where the five piers were to be made; and it had been a lousy job, because the river carries a lot of sand, even when it's low, and so the excavations kept filling up as fast as they were dug. And then they sunk the caissons and sent miners down inside, to dig out the rock, and two of the men had drowned, but in the end they sank the caissons, filled them with gravel and cement, and the dirty part of the job was over, finally. Listening to him, I began to get worried myself, because he mentioned the two dead men casually, like it was something natural, and I began to get the idea that this was a place where you'd better not count on other people being careful, and you'd better be double-careful yourself.

"I was telling you in that engineer's place I wouldn't have been quite so calm. About two hours earlier they had phoned him to say that something incredible was happening, namely, that now the piers were finished, a flood was on its way and the river was flowing in a different direction. He told me this like another person would have said the roast was burned. He really must've had slow reactions. Another Indian arrived with a turban and a jeep, and the engineer said very politely that we'd get together again some other time and now he was very sorry. But I realized he was going to have a look, and I asked to go with him. He made a face I didn't understand, but he said yes. I don't know; maybe because he had respect, maybe because you never turn down an offer of advice, or maybe it was just out of politeness. He was very polite, but the kind of man who lets things take their course. He also had imagination: as we were riding in the jeep--and I won't bother to tell you about the road-- instead of thinking about the flood, he told me how they had managed to throw walkways across the river (he called them catwalks, but I don't think any intelligent cat would have walked over them; I'll tell you about them later, anyway). Another man would have taken a boat or would have fired a harpoon like the ones they use for whales; but this guy sent for all the kids of the village there and offered a prize of ten rupees for the one who could fly a kite over to the opposite side. One kid did it, the engineer paid him the reward (and he wasn't throwing his money around, because that's about three dollars), then he had a thicker rope knotted to the kite string, and so on, until they got to the steel cables of the catwalk. He had just finished telling this story when we reached the bridge, and the sight took his breath away, too.

"Here at home we don't think much about the power of rivers. At that point, the river was seven hundred meters wide, and it made a bend. To me it didn't seem all that smart to put the bridge just there, but apparently it couldn't be avoided, because an important railway line had to go over it. You could see the five piers in the midst of the current, and further on, the approach piers, shorter and shorter, to connect with the plain. On the five big piers, the support towers were already in place, fifty meters high; and between two of the piers there was already a service structure, laid on its side, a light, temporary bridge, in other words, for the final span to be set on. We were on the right bank, which was buttressed by a concrete embankment, nice and strong, but there the river had vanished: during the night it had begun to eat at the left bank, where there was an identical buttress, and early that morning the river had broken through it.

"Around us there were maybe a hundred Indian workers, and they didn't bat an eye. They were calmly looking at the river, sitting on their heels, the way they do. I wouldn't last five minutes, and I don't know how they manage; obviously they were taught as children. When they saw the engineer, they stood up for a minute and greeted him, putting their hands to their chest, this way, folded, like they were going to pray. They made a little bow, then sat down again. We were too low to see the situation clearly, so we climbed the ladder of the tower on the bank, and then we could really take in the whole show.

"Below us, I told you, there was no water any more, just some black mud that was already beginning to steam and stink under the sun, and in it was a big mess of uprooted trees, planks, empty drums, and dead animals. The water was all running against the left bank, like it wanted to carry it off, and in fact, while we were standing there in a daze, watching, not knowing what to do or what to say, we saw a chunk of the embankment come loose, about ten meters long, and slam against one of the piers, bounce back and then swirl down on the current, like it was wood, not concrete. The water had already carried off a good part of the left bank; it had penetrated the breach and was flooding the fields on the other side: it had made a round lake more than a hundred meters across, and more water, like a mean animal bent on doing harm, kept pouring into it, spinning because of the thrust behind it, and spreading out before our eyes.

"The stream brought down all sorts of things: not only flotsam, but what seemed floating islands. Obviously, upstream the river passed through a wood, because trees were coming down with their leaves and roots, and even whole hunks of shore, and you couldn't figure out how they kept afloat, with grass on them, earth, trees still standing or else on their side, patches of landscape, in other words. They traveled at top speed; sometimes they slipped between the piers and sped away on the other side; sometimes they bumped against the bases and broke into two or three pieces. You could see that the piers were really solid, because a kind of tangle had formed against the bases: planks, branches, trunks. And you could see the force of the water, piling against it and unable to dislodge them; and it made a strange racket, like thunder, but underground.

"I tell you, I was glad he was the engineer; but if I had been in his place, I think I would have made myself a little busier. Not that much could be done there and then, but I had the impression that, if he obeyed his instincts, he would sit on his heels, too, like his workers, and stay there watching till God knows when. It seemed rude for me to give him advice, me who had just arrived, and him the engineer; but it was plain as the nose on your face that he didn't know which way to turn, and he paced up and down the bank without opening his mouth, and, in other words, he was spinning his wheels, so I plucked up my nerve and told him that in my opinion it would be a good thing to send for some stones, some boulders, the biggest they could get, and throw them down on the left bank, but sort of quick, because while we were standing there talking, the river had carried off, at one whack, two more slabs of the embankment, and the whirlpool in the lake had started whirling faster than ever. We started to get into the jeep, and at that very moment we saw a mass of trees coming down, with earth and branches, the whole thing, and I'm not exaggerating, big as a house, and it was rolling like a ball. It stuck into the span where there was the service structure and bent it like a straw and pulled it down into the water. There really wasn't much to be done; the engineer told the workers to go home, and we also went back to the camp to telephone for the stone; but along the way the engineer told me, still very calm, that all around that area there was nothing but fields, black earth, and mud, and if I wanted a stone the size of a walnut, I would have to travel at least a hundred miles to find one; like stones were a craving of mine, the kind women get when they're expecting. In other words, he was a polite character, but strange; he seemed to be playing rather than working, and he got on my nerves.

"He started telephoning somebody or other; I think it was a government office. He talked Indian, and I couldn't understand any of it, but it sounded like first came the operator, then the secretary's secretary, then the real secretary, and the man he wanted never came, and in the end they were cut off, a bit like home, in other words; but he didn't lose his temper and began again from the beginning. Between secretaries he told me, however, that in his opinion there wouldn't be anything useful for me to do at the site there for some days. I could stay if I liked, but he advised me to take the train and go to Calcutta, and so I went. I didn't clearly understand if he was giving me this advice to be polite or to get rid of me. In any case, I didn't profit much. To tell you the truth, he warned me not even to try to find a hotel room; he gave me the address of a private house, and I was to go there because they were friends of his, and I would be comfortable with them, also from the hygienic point of view.

"I won't tell you about Calcutta. I was there five days, a waste of time. There are more than five million inhabitants and terrible poverty, and you see it right off. Imagine, the minute I came out of the station, and it was evening, I saw a family going to bed, and they were going to sleep inside a length of cement pipe, a new one, the kind they use for sewers, four meters long, one meter diameter. There was the Papa, the Mama, and three kids. They had put a little lamp in the pipe, and two pieces of cloth, one at one end, one at the other; and they were lucky because most people slept on the sidewalk, wherever.

"In Calcutta everything is cheap, but I didn't dare buy anything, or even go to the movies, because of the filth and the infections. . . . I was on tenterhooks, and every day I telephoned to the site, but either the engineer was out or he wouldn't come to the phone. Then I caught him on the fifth day and he told me I might as well come back, the river was down, and the work could be started; and so I went.

"I reported to the engineer, who still looked like his mind was somewhere else; and I found him in the midst of the yard where the sheds were, with about fifty men around him, and it seemed like he was waiting for me. He said hello in their way, with his hands to his chest, and then he introduced me to the gang: 'This is Mr. Peraldo, your Italian foreman.' They all bowed, with their hands pressed together, and I stood there like a dope. I thought he had forgotten my name, because you know how foreigners always have trouble with names, and to me, for that matter, it seemed like all Indians were named Singh, and I figured the same thing must've happened to him. I told him I was Faussone, not Peraldo, and he gave me this angelic smile and said, 'Sorry, but you know, all you Europeans look alike.' In other words, it gradually came out that this engineer, whose name was Chaitania, screwed things up not only on the job but also when it came to names. And this Mr. Peraldo wasn't somebody he had dreamed up: Peraldo really existed; he was a master mason from Biella, who, by coincidence, was also supposed to arrive that morning, and he was in charge of anchoring the cables of the bridge. And he actually did arrive a little later, and I was glad, because meeting somebody from your own country is always a pleasure. How the engineer managed to take me for him and say we looked alike is one of life's mysteries, because I'm tall and thin and Peraldo was dumpy; I was about thirty, and he would never see fifty again; he had a little Charlie Chaplin mustache, and even then the only hair I had on my head was this little patch back here. In other words, if we looked anything alike, it was only the way we bent our elbows, because he also enjoyed eating and drinking well, and down there it wasn't all that easy.

"Meeting a mason from Biella in such an out-of-the-way place didn't surprise me all that much, because if a man roams the world, wherever he goes he finds a Neapolitan who makes pizza and a Biellese who makes walls. Once I met a Biellese in Holland on a job, and he said that God made the world, except for Holland, which was made by the Dutch, but the Dutch, for the dikes, had called in masons from Biella, because nobody has yet invented a machine that will build walls. And it seemed a nice saying to me, even if nowadays it isn't all that true any more. This Peraldo I was lucky to run into, because he had roamed the world much worse than me, and he knew a thing or two, even if he didn't talk much; and also because, I don't know how he managed, in the dormitory he had a good supply of Nebiolo, and every now and then he would offer me a drink. He would offer me a little, not much, because he too wasn't exactly openhanded and he didn't want to make a dent in his capital. But he was right, too, because the job dragged on and on, and here I really have to say it's the same the world over; jobs finished by the deadline in the specifications are something I haven't seen much of.

"He took me to see the tunnels for the anchorages: because the cables of that bridge, you understand, would be under big tension, and then the usual anchorages aren't enough. The cables had to be fixed in a block of concrete, made like a wedge and set in an inclined tunnel dug out of the cliff. There were four tunnels, two for each cable. But what tunnels! They were like caverns. I'd never seen anything of the kind before: eighty meters long, ten meters wide at the mouth and fifteen at the back, with a rake of thirty degrees....No, no, don't make that face, because you're going to write these things down afterward, and I wouldn't want mistakes to get printed, or at least--excuse me for saying--not through my fault."

I promised Faussone I would be very careful to follow his indications, and under no circumstances would I yield to the professional temptation to invent, embellish, and expand; and therefore I would add nothing to his report, though I might pare away a little, as the sculptor does when he carves the form from the block. And Faussone declared himself in agreement. So, carving from the great block of details he supplied me, in some disorder, I

539

discerned the emerging form of a long, slender bridge, supported by five towers made of boxes of steel, and hung from four festoons of steel cable. Each festoon was 170 meters long, and each of the two cables consisted of a monstrous braid of eleven thousand individual strands, each five millimeters in diameter.

"I already told you the other evening how for me every job is like a first love; but this time I caught on right away that the love was a real involvement, the kind that if you end up in one piece you can consider yourself lucky. Before beginning, I spent a week, like I was in school, taking lessons from the engineers. There were six of them, five Indians and one from the company: four hours every morning with a notebook for taking notes, and then all afternoon studying them, because it was really like the work of a spider, only spiders are born already knowing their job, and besides, if they fall, they don't have far to fall and they don't do themselves much harm because they have a built-in lifeline. For that matter, after this job I'm telling you about, every time I see a spider in his web I remember my eleven thousand strands, I mean twenty-two thousand, because there were two cables; and I feel like I'm sort of a relative of his, specially when the wind's blowing.

"Then it was my turn to teach the lesson to my men. They were genuine Indians, not like those Alaskans I told you about before. At first, I have to confess I didn't have much confidence, seeing them sitting around me on their heels, or some with their legs folded and their knees sticking out, like the statues on their churches I saw in Calcutta. They would stare at me, never ask any questions. But then, a bit at a time, I dealt with them one by one, and I saw they hadn't missed a word; and if you ask me, they're more intelligent than we are, or maybe it's because they were afraid of losing their job, because down there they don't pull their punches. They're men like us, after all, even if they have a turban and don't wear shoes and every morning, no matter what, they spend two hours praying. They also have their problems; there was one with a sixteen-year-old son who was already shooting dice and his father was worried because the boy always lost; another one had a sick wife; and another had seven children, but he said he didn't agree with the government and didn't want the operation, because he and his wife liked kids, and he also showed me their picture. They were really beautiful, and his wife was beautiful, too. All Indian girls are beautiful, but Peraldo, who had been in India for quite a while, explained to me that there was nothing doing with them. He also said that it's different in the city, but there are certain diseases in circulation that it's best to steer clear of. In other words, to conclude: I've never done the fasting I did that time in India. But back to the job.

"I told you about the catwalks, and about the trick with the kite to draw the first strand. Obviously, they couldn't fly twenty-two thousand kites. To draw the cables of a suspension bridge into place there's a special system: you set up a winch, and six or seven meters above each catwalk you install an endless cable, like one of those old belts, stretched between two pulleys, one on each bank. To the endless cable you attach an idle pulley, with four grooves; inside each groove you pass a loop of the single strand, which comes from a big spool. And then you start the pulleys and you draw the idle pulley from one bank to the other. That way, with one trip you draw eight wires. The workmen, apart from those who set up the loops and those who remove them, stand on the catwalk, two men every fifty meters, to make sure that the strands don't overlap. But saying it is one thing, and doing it is another.

540

"Luckily Indians take orders easily. Because you have to remember that on those catwalks it isn't like taking a stroll down Via Roma. First, they're tilted, because they have the same angle that the cable will have afterward. Second, a puff of wind is enough to make them dance, but I'll be telling you about the wind later. Third, since they have to be light and not offer any resistance to the wind, the flooring is made of wire mesh, so it's best not to look at your feet, because if you look down you see the water of the river below, mud-colored, and some little things moving in it; and from up there they might seem like little fish, but actually it's the backs of crocodiles. But I told you, in India one thing always looks like some other thing. Peraldo told me there aren't so many now, but the few that there are all come to where a bridge is being erected because they eat the garbage from the mess hall, and because they're waiting for someone to fall in. India's a fine country, but it doesn't have likable animals. Even the mosquitoes--apart from the fact that they give you malaria, and besides a topee* you always have to wear a veil like ladies in olden times--they're beasts this long, and if you're not careful they give you a bite that takes away a scrap of flesh. And I was told there are also butterflies that come at night to suck your blood while you're asleep, but I never actually saw them, and as far as sleeping goes, I slept fine.

"The trick in that job of drawing wires is that the wires all have to have the same tension, and with a length like that it isn't easy. We worked two shifts, six hours each, from dawn to sunset, but then we had to organize a special team that worked at night, before the sun came up, because during the day there are always some wires exposed to the sun: they heat up and expand, while the others remain in the shade. So you have to do the calibrating before dawn, when all the wires have the same temperature. And this calibration, it was always up to me to do it.

"We went on like that for sixty days, with the idle pulley always going back and forth; and the cobweb grew, nice and taut and symmetrical; and you could already get an idea of the shape the bridge would have afterward. It was hot. I told you that; in fact, I told you I wouldn't tell you again, but it really was hot. When the sun went down it was some relief, also because I could go back to the dormitory and drink a glass and exchange a few words with Peraldo. Peraldo had started out as an unskilled laborer, then he had become a bricklayer, then a cement mixer. He had been just about everywhere, including four years in the Congo, building a dam; and he had a lot to tell, but if I start telling other people's stories besides my own there'll be no end to it.

"When the cable part of the job was finished, from the distance you could see the two cables stretching from one bank to the other with their four festoons, nice and light, just like the strands of a spiderweb. But when you looked at them close up, they were a pair of fearsome bundles, seventy centimeters thick; and we compacted them with a special machine, like a press, ring-shaped, that travels along the cable and squeezes it with a hundred-ton force. But I didn't have anything to do with this: it was an American machine; they sent it down there with its own American specialist, who snooted everybody, didn't speak to a soul, and didn't let anyone near him. Obviously he was afraid they would steal his secret.

----

* A pith helmet

541

"At this point it seemed the dirty work was done. We pulled up the vertical suspension cables in a few days, we fished them up with tackle from the pontoons down below, and it was like catching eels, but these eels weighed about a ton and a half each. And finally it was time to lay the deck; and nobody would have guessed, but this is exactly where the adventure began. I must tell you that, after the disaster of that flash flood I told you about, they didn't say a word, but they followed my advice: while I was in Calcutta they brought in an army of trucks all loaded with big rocks, and when the water subsided, they reinforced the embankments nicely. But you know about the singed cat that, afterward, was afraid of cold water: all during the rigging, from the top of my catwalk I kept an eye on the water, and I also made the engineer give me a field telephone, because I was thinking that if there was another flood, it would be best to keep ahead of it. But I never thought the danger would come from a different direction, and to judge by events, nobody was thinking of that, and the designers hadn't thought of it either.

"I never saw those designers face to face; I don't even know where they were from. But I've met others, plenty of them; and I know they come in different species. There is the elephant-designer, the kind that is always on the safe side; he doesn't care about elegance or about economy: he just doesn't want trouble, and where one would be enough, he puts four; and as a rule he's an older man, and if you think about it, you realize it's a sad thing. Then there's the stingy type, on the other hand: you'd think he had to pay for every bolt out of his own pocket. There's the parrot, who doesn't work out the plans himself but copies them, like in school, and he doesn't realize people are laughing behind his back. There's also the snail: the bureaucratic type, I mean. He moves very slowly, and the minute you touch him, he draws back and hides in his shell, which is the rule book. And, if you'll excuse the expression, I'd call him the nitwit designer. And, finally, there's the butterfly, and I actually think the men who designed this bridge belonged to that category: they're the most dangerous type, because they are young and daring and they fool you. If you mention money and safety to them, they look at you like you were spit, and all they think about is making something new and beautiful, never considering that when a work is planned carefully it comes out beautiful automatically. Excuse me for letting myself go like this, but when a man puts his whole heart into a job, and then it ends like this bridge I'm telling you about, well, it makes you feel bad. You feel bad for lots of reasons: because you've wasted all that time, because afterward there's always a big stink with lawyers and courts and all that stuff, because even if you personally are out of it, you always feel a little responsible. But most of all, seeing a piece of work like that come down, and seeing the way it came down, one bit at a time, like it was in agony, like it was struggling, was enough to make your heart ache like when a person dies.

"And also like when a person dies, and afterward everybody says they saw it coming, from the way he was breathing, from the way he looked around, after the disaster they all had to speak their piece, even the Indian who didn't want the operation: how it was obvious, the suspensions were weak, the steel had blowholes the size of beans. The welders said the riggers didn't know their job, the crane operators said the welders didn't know theirs; and they all took it out on the engineer, and they gossiped and said he was asleep on his feet and he goofed off and he didn't know how to organize the work. And maybe they were all of them right, at least partly, or maybe none of them, because again it's sort of like with people. I've seen it lots of times: a tower, for

542

example, checked and rechecked and tested, and it looks like it ought to stand for centuries, and it begins to teeter after a month; and another one you wouldn't give five cents for, and it never shows a crack. And if you leave it to the technical experts, the ones appointed by the court, then you're in trouble: three of them will come and they'll hand down three different opinions. I never saw an expert who was any good. Obviously, if a person dies, or a construction comes apart, there has to be a reason, but that doesn't mean there was only one, or if there was, that it's possible to discover it. But I'm getting ahead of myself.

"I told you that all during this job it was hot, every day: a damp heat it's hard to get used to; but toward the end I did get used to it. Well, when the job was finished, and the painters were already crawling up more or less everywhere, and they looked like flies on a spiderweb, I realized that all of a sudden the heat had stopped: the sun was up, but instead of the usual heat, your sweat dried on you and you felt cool. I was on the bridge, too, halfway across the first span, and besides the cool I noticed two other things that froze me in my tracks, like a hound when it points: I felt the bridge vibrating beneath my feet, and I heard something like music, but you couldn't tell what direction it was coming from: music, I mean a deep sound, distant, like when they're trying out the organ in church, because when I was a kid I used to go to church. And I realized it all came from the wind. It was the first wind I had felt since I was in India, and it wasn't a great wind, but it was steady, like the wind you feel when you drive the car slowly, and hold your hand out of the window. I felt nervous--I don't know why--and I started walking toward the bank. Maybe it's an effect of our line of work: we don't like things that vibrate underfoot. I reached the abutment, I turned around, and I felt my hair stand on end, hair by hair, and all together, like each hair was waking up and wanted to run off. Because from where I was, you could see the whole length of the bridge in profile, and something was happening that you wouldn't believe. It was like, in that breath of wind, the bridge was waking up, too. Yes, like somebody who's heard a noise and wakes up, gives himself a shake, and gets ready to jump out of bed: the deck was wagging to left and right, and then it began to move vertically, too: you could see ripples running from my end to the one opposite, like when you shake a slack rope. But they weren't vibrations any more; they were waves a meter or two high, because I saw one of the painters who had dropped his things and was running toward me, and one minute I could see him, and the next, not. Like a boat at sea when the waves are high.

"Everybody ran off the bridge; even the Indians moved a bit sharper than usual; and there was a great yelling and great confusion. Nobody knew what to do. The suspension cables had also started moving. You know how it goes in moments like that: one man says one thing, and another says something else. But after a few minutes you could see that the bridge hadn't actually stopped shaking, but the waves had become sort of stabilized. They ran and bounced from one end to the other, always with the same rhythm. I don't know who gave the order, or maybe someone just took the initiative; but I saw one of the camp tractors start over the deck of the bridge, dragging two three-inch cables. Maybe the idea was to stretch them on a diagonal, to stop the swaying. Certainly, the driver was really brave, or really irresponsible, because I don't honestly believe that those two cables, even if they had managed to anchor them, would have held together a structure like that. Imagine: the deck was eight meters wide and a meter and a half high, so just figure out how many tons that meant. Anyway, it was too late to do anything, because from that moment on, everything happened in a rush. Maybe the wind got stronger; I couldn't say, but by ten o'clock the vertical waves were four or five meters high, and you could feel the earth shaking, and hear the racket of the vertical cables loosening and tightening. The tractor driver saw things were taking a bad turn; he left the tractor on the spot and ran to the bank. And it was a good thing he did, because just afterward the deck began to twist like it was made of rubber; the tractor swerved left and right, and at a certain point it jumped the railing, or maybe smashed through it, and ended up in the river.

"One after the other, we heard what sounded like shots from a cannon. I counted: there were six of them. It was the vertical suspensions snapping: they snapped neatly, at the level of the track, and the stumps, in the backlash, flew up toward the sky. At the same time, the deck also began to bend; then it came apart and fell into the river in pieces. Some pieces, however, still hung from the girders, like rags.

"Then everything stopped. Everything was still, like after an air raid, and I don't know what I looked like, but a man next to me was all shaking and his face was greenish, though he was one of those Indians with the turban and with dark skin. In the final analysis, two spans of the deck had collapsed, almost whole, and a dozen of the vertical suspensions. On the other hand, the main cables were in place. Nothing moved. It was like a photograph, except the river kept on flowing as if nothing had happened. And yet the wind hadn't dropped; in fact, it was stronger than before. It was like somebody had wanted to do all that damage and afterward was satisfied. I had a stupid thought: I read in some book that, in ancient times, when they started building a bridge they killed a man, and they put him in the foundations; and later they killed an animal, instead, and then the bridge wouldn't fall down. But, like I said, it was a stupid thought.

"I left then. The big cables had held, after all, and so my work didn't need redoing. I learned that afterward they started arguing about the why and the wherefore, and they couldn't agree and are still arguing. As far as I'm concerned, when I saw the deck beginning to slam up and down, I immediately thought of that landing in Calcutta, and the wings of the Boeing that flapped like a bird's and gave me a nasty moment, even if I've flown lots of times. But I don't know: surely the wind had something to do with it, and I'm told they're rebuilding the bridge now, but with some vents in the deck so the wind won't meet too much resistance.

"No, I never worked on any suspension bridges after that. I left without saying goodbye to anybody, except Peraldo. It was a bad business. It was like when you're fond of a girl, and she drops you overnight, and you don't know why, and you suffer, not only because you've lost the girl, but also your self-confidence. Well, pass me the bottle and we'll have another drink. I'm paying tonight anyway. Yes, I came back to Turin, and I almost got myself into trouble with one of the girls picked out by those aunts I was telling you about at the beginning, because my morale was low and I didn't put up any resistance: but that's another story. Eventually, I got over it."

544

An Engineer Looks at His Profession

An Excerpt from **Years of Adventure, 1874-1920**
Herbert Hoover

*Although he was educated (at Stanford University) as a geologist, Herbert
Hoover spent the early years of his professional career working as a mining
engineer. His reflection on his engineering career could just as well have been
written by a civil engineer of that era. His career as a mining engineer occurred
principally in the years around the turn of the century during which he was
primarily engaged in consulting work in this country and abroad. His
reminiscences suggest a degree of physical hardship rarely encountered by
young engineers now, but his observations concerning the challenges and
rewards of an engineering career are as timely today as they were when he
experienced them almost a century ago.*

*The extent to which Hoover enjoyed his relatively short engineering career is
evident from the pleasure he takes in writing about it, even after a long and
distinguished public service career that included assignments as an
administrator of domestic and international emergency relief, appointment as
Secretary of Commerce under Presidents Coolidge and Harding, and election to
the office of President of the United States.*

I cannot leave my profession without some general comment upon it. Within
my lifetime it had been transformed from a trade to a profession. It was the
American universities that took engineering away from the rule-of-thumb
surveyors, mechanics and Cornish foremen, and lifted it into the realm of
application of science, wider learning in the humanities with higher ethics of a
profession ranking with law, medicine and the clergy. And our American
profession had brought a transformation in another direction through the
inclusion of administrative work as part of the engineer's job.

The European universities did not acknowledge engineering as a profession
until long after America had done so. I took part in one of the debates at
Oxford as to whether engineering should be included in its instruction. The
major argument put forward by our side was the need of University setting and
its cultural influences on the profession. We ventured to assert that not until
Oxford and Cambridge recognized engineering as a profession equal to others
would engineering secure its due quota of the best English brains, because
able young men would always seek the professions held in the highest public
esteem. I cited the fact that while various special technical colleges had
been existent in England for a long time, yet there were more than a thousand
American engineers of all breeds in the British Empire, occupying top
positions.

Soon after the Oxford discussions, I returned to America. At my ship's table
sat an English lady of great cultivation and a happy mind, who contributed
much to the evanescent conversation on government, national customs,
literature, art, industry, and whatnot. We were coming up New York harbor
at the final farewell breakfast when she turned to me and said:

> "I hope you will forgive my dreadful curiosity, but I should
> like awfully to know--what is your profession?"

I replied that I was an engineer. She emitted an involuntary exclamation and [said]:

"Why, I thought you were a gentleman!"

Hundreds of times students and parents have consulted me upon engineering compared with the other professions. My comment usually is:

"Its training deals with the exact sciences. That sort of exactness makes for truth and conscience. It might be good for the world if more men had that sort of mental start in life, even if they did not pursue the profession. But he who would enter these precincts as a lifework must have a test taken of his imaginative faculties, for engineering without imagination sinks to a trade. And those who would enter here must for years abandon their white collars except for Sunday."

In . . . [this] profession, those who follow the gods of engineering to that success marked by an office of one's own in a large city must be prepared to live for years on the outside borders of civilization, where beds are hard, where cold bites and heat burns, where dress-up clothes are a new pair of overalls, where there is little home life--not for weeks but for years--where often they must perform the menial labor necessary to keep soul and body together. Other branches of the profession mean years on the lower rungs of the ladder--shops, works, and powerhouses--where again white collars are not a part of the engineer uniform. But the engineer learns through work with his own hands not only the mind of the worker, but the multitude of true gentlemen among them. On the other hand, men who love a fight with nature, who like to build and see their building grow, men who do not hold themselves above manual labor, men who have the moral courage to do these things soundly someday will be able to move to town, wear white collars every day, and send out the youngsters to the lower rungs and the frontiers of industry.

It is a great profession. There is the fascination of watching a figment of the imagination emerge through the aid of science to a plan on paper. Then it moves to realization in stone or metal or energy. Then it brings jobs and homes to men. Then it elevates the standards of living and adds to the comforts of life. That is the engineer's high privilege.

The great liability of the engineer compared to men of other professions is that his works are out in the open where all can see them. His acts, step by step, are in hard substance. He cannot bury his mistakes in the grave like doctors. He cannot argue them into thin air or blame the judge like the lawyers. He cannot, like the architects, cover his failures with trees and vines. He cannot, like the politicians, screen his shortcomings by blaming his opponents and hope that the people will forget. The engineer simply cannot deny that he did it. If his works do not work, he is damned. That is the phantasmagoria that haunts his nights and dogs his days. He comes from the job at the end of the day resolved to calculate it again. He wakes in the night in a cold sweat and puts something on paper that looks silly in the morning.

All day he shivers at the thought of the bugs which will inevitably appear to jolt its smooth consummation.

On the other hand, unlike the doctor, his is not a life among the weak. Unlike the soldier, destruction is not his purpose. Unlike the lawyer, quarrels are not his daily bread. To the engineer falls the job of clothing the bare bones of science with life, comfort, and hope. No doubt, as years go by, people forget which engineer did it, even if they ever knew. Or some politician puts his name on it. Or they credit it to some promoter who used other people's money with which to finance it. But the engineer himself looks back at the unending stream of goodness which flows from his successes with satisfactions that few professionals may know. And the verdict of his fellow professionals is all the accolade he wants.

President Herbert Hoover (center) on an inspection trip of Boulder Dam in 1932 Construction Superintendent Frank Crowe is at the far right. Others pictured are (left to right) Bureau of Reclamation Chief Engineer R. F. Walter, Mrs. Hoover, Mrs. Wilbur, Secretary of Interior R. I. Wilbur and E. O. Wattis, one of the Directors of Six Companies, Inc., the general contractor for the project.

The Mighty Task Is Done

Joseph Strauss

Upon the completion of the Golden Gate Bridge in San Francisco, Joseph Strauss, who served as chief engineer for the project, expressed his feelings in the following verse. Engineers such as Strauss and David Steinman express in their verses the feeling that many civil engineers have for the projects on which they are employed. Those who contend that engineers build only for the sake of building, caring little for utility or beauty, can not have been exposed to the depth of feeling revealed by these builder/poets.

At last the mighty task is done
Resplendent in the western sun,
The bridge looms mountain high;
Its Titan piers grip ocean floor,
Its great steel arms link shore to shore,
Its towers pierce the sky.

On its broad decks in rightful pride,
The world in swift parade shall ride,
Throughout all time to be;
Beneath, fleet ships from every port,
Vast land-locked Bay, historic fort,
And dwarfing all--the sea.

Launched 'midst a thousand hopes and fears,
Damned by a thousand hostile seers,
Yet ne'er its course was stayed;
But ask of those who met the foe,
Who stood alone when faith was low,
Ask them the price they paid.

Ask of the steel, each strut and wire,
Ask of the searching, purging fire
That marked their natal hour;
Ask of the mind, the hand, the heart,
Ask of each single stalwart part
What gave it force and power.

An honored cause and nobly fought,
And that which they so bravely wrought
Now glorifies their deed;
No selfish urge shall stain its life,
Nor envy, greed, intrigue, nor strife,
Nor false, ignoble creed.

High overhead its lights shall gleam,
Far, far below life's restless stream
Unceasingly shall flow;
For this was spun its lithe fine form
To fear not war, nor time, nor storm,
For Fate had meant it so.

From Slide Rule to Computer:
Forgetting How It Used to Be Done

Chapter 15, **To Engineer Is Human**
Henry Petroski

Nothing has changed the practice of civil engineering more dramatically than the advent of the electronic computer. In this brief essay Dr. Petroski identifies some of the most important questions arising from the use of the computer. At the time the book was written the personal computer had not become as ubiquitous as it now is, so the dimensions of the problems discussed herein were not as widespread as they now are. The problems Dr. Petroski attributes to the use of "large computers" are now inherent in the use of software for the personal computer, since the "pc's" are now capable of handling analyses that only five years ago were restricted to the largest computers then in existence.

This article deals primarily with the potential problems encountered by civil engineers who use computers in structural engineering--only one of the many disciplines within the civil engineering profession. Similar problems exist for the water resources engineers, the geotechnical engineers, the environmental engineers and all the other specialty disciplines.

Ever since computers became widely available for use in engineering the "old timers" have wondered what will happen when there's no one around who remembers how to do the calculations "by hand." At first that concern was based as much as anything on a fear that a power failure might make computers unusable just when they might be needed for an emergency. Now it has become clear that that fear is the least of our problems; far more important is the fear that there will some day be no one around who really understands or "has a feel for" the essence of a design or problem-solving approach.

It is already apparent that some engineers are more proficient in manipulating the intricacies of a particular software package than in assessing the validity of results obtained through the use of the package. As the use of software becomes more and more an essential element of engineering education, it becomes increasingly likely that less and less time will be available for inculcating students with a true understanding of the problems being solved through the use of the software. How that tradeoff (between proficiency in the use of analysis tools and understanding of the analysis process) is resolved is one of the most important problems facing today's engineering educators.

Twenty-five years ago, the undisputed symbol of engineering was the slide rule. Engineering students, who at the time were almost all males, carried the "slip sticks" in scabbard-like cases hanging from their belts, and older engineers wore small working models as tie clips that in a pinch could be used for calculations. When I became an engineering student myself, one of my most important decisions was which slide rule to purchase. Not only was $20 a big investment in 1959, but also I was choosing an instrument that I was told I would use for the rest of my professional life; I was advised along with all the other freshmen to get right at the start a good slide rule with all the scales I would ever need. After much comparative shopping, I chose a popular Keuffel & Esser model known as the Log Log Duplex Decitrig, and for a long time it was my most prized possession. Many of my fellow students also chose K & E rules, and the company was selling them at the rate of twenty thousand per month in the 1950's.

A slide rule was indispensible for doing homework and taking tests, for all our teachers assumed that every engineering student had a slide rule and knew how to use it. If we had not learned in high school, then we quickly studied the manual folded into the box. What our engineering instructors were interested in teaching us was not all the grand things that our various models of rules could do, but their common limitations. They told us about significant digits, for most engineering instruments then had analog dials and scales from which one had to estimate numbers between the finest divisions in much the same way we have to estimate sixteenths of an inch on a yard stick or tenths of a millimeter on a meter stick. The scales on the slide rule have the same limitations, and we were expected to know that we could only report answers accurate to three significant digits from our rules, unless we were on the extreme left of the scale where finer subdivisions existed.

We often had these things inculcated in us by trial and error. If the answer to a test question required us to multiply, say, 0.346 by 0.16892 and we reported the result as 0.05844632 we would be marked for an error in significant digits, for the result of a calculation could not have a greater accuracy than the least accurately known input number. (When older engineers write 0.346 it is implied to be known only to three digits after the decimal point, otherwise it would have been written as 0.3460 or 0.34600 or to whatever decimal place the number is known.) Since no one could ever read as many digits as those in 0.05844632 from his slide rule, the closest he would be expected to get would be 0.0585. (The extra digits were a dead giveaway that the student had forgotten his slide rule and had done the multiplication longhand on some scrap of paper and, worse yet, had forgotten the significance of significant digits.) We also learned how to estimate the order of magnitude of our answers, for the slide rule could not supply the decimal point to the product of 0.346 and 0.16892, and we had to develop a feel for the fact that the answer was about 0.06 rather than 0.6 or 0.006. These requirements on our judgment made us realize two important things about engineering: first, answers are approximations and should only be reported as accurately as the input is known, and, second, magnitudes come from a feel for the problem and do not come automatically from machines or calculating contrivances.

As I progressed through engineering school with my slide rule in the early 1960s, electronic technology was being developed that was to change engineering teaching and practice. But it was not then widely known, and as late as 1967 Keuffel & Esser commissioned a study of the future that resulted in predictions of domed cities and three-dimensional television in the year 2067--but that did not predict the demise of the slide rule within five years.

In 1968, an article entitled "An Electronic Digital Slide Rule" appeared in The Electronic Engineer. It could dare to prophesy, "If this hand-size calculator ever becomes commercial, the conventional slide rule will become a museum piece." In the article the authors, two General Electric engineers, described a prototype that they had built with some off-the-shelf digital integrated circuits. Their "feasibility model" looked like an electric blanket control and, 1-1/2 x 5 x 7 inches, it resembled a novel in size. Yet their marvel could give four-digit answers to any four-digit multiplicands, and it could also divide and calculate square roots, exponentials, and logarithms. It had, however, one shortcoming, and the engineers made the concession that, "Since it has no decimal points, you must figure out your decimals as with a regular slide rule." As far as cost was concerned, that of course would depend upon the cost of the components, but there remained one big obstacle in 1968: "Only the digital readout still poses a problem, since at present there are no low-cost miniature devices available. But there is no question that this last barrier will soon be overcome."

They were right, of course, and within a few years Texas Instruments had developed the first truly compact, handheld, pocket-sized calculator using an electronic chip. Texas Instruments started manufacturing pocket calculators in 1972, but they were still expensive in 1973, costing about ten times as much as a top-of-the-line slide rule. However, price breakthroughs came the next year, and Commodore was advertising its model SR-1400, a "37-key advanced math, true scientific calculator" that could do everything my Log Log Duplex Decitrig could do--and more. If one knew input to ten significant digits, then this calculator could handle it.

I was teaching at the University of Texas at Austin at the time of this great calculator revolution, and there were some engineering students whose daddies did not have to wait for the pocket calculator to become competitive in price with the slide rule. We faculty were thus faced with the question of whether students with electronic slide rules had an unfair advantage on quizzes and examination over those with the traditional slip sticks, for the modern electronic device was a lot quicker and could add and subtract-- something a slide rule could only do with logarithms. The faculty members generally were unfamiliar with all the features of the calculators that were still priced out of their reach, and there seemed to be many pros and cons and endless discussions over the issue of whether an electronic slide rule was equivalent to a wooden one. The question soon became moot, however, as prices plunged and just about anyone who could afford a conventional slide rule could afford an electronic model. By 1976 Keuffel & Esser was selling calculators made by Texas Instruments faster than traditional slide rules, which by then made up only five percent of K & E's sales, and the company consigned to the Smithsonian Institution the machine it once used to carve the scales into its wooden slide rules.

By the mid-1970s calculator manufacturers were making fifty million units a year, and soon just about everyone, including engineers who went through school in the old days, had a calculator. But no older engineer that I know discarded or consigned his slide rule to any museum. At most the old slip stick was put in the desk drawer, ready for use during power failures or other emergencies. A study conducted by the Futures Group in the early 1980s found that most engineers in senior management positions continued to keep slide rules close at hand and still used them "because they are more comfortable." But the always-growing younger generations naturally feel just the opposite. In 1981 I asked a class of sophomore engineering students how many used a slide rule, and I got the expected answer--none. (Some did own slide rules, perhaps because their engineer fathers bought one for the freshman to take away to engineering school. And K & E was selling out its remaining stock of 2,300 at the rate of only two hundred per month in 1981.) I did not ask my class how many used a calculator, for that would be like asking how many use a telephone. And I did not ask how many used a computer, for that was by then a requirement in the engineering curriculum. The trend is clearly that eventually no engineer will own or use a traditional slide rule, but that practicing engineers of all generations will use--and misuse--computers.

Engineering faculty members, like just about everyone else, got so distracted by the new electronic technology during the 1970's that more substantial issues than price, convenience, and speed of computation did not come to the fore. The vast majority of faculty members did not ask where all those digits the calculators could display were going to come from or go to; they did not ask if the students were going to continue to appreciate the approximate nature of engineering answers, and they did not ask whether students would lose their feel for the decimal point if the calculator handled it all the time. Now, a decade after the calculator displaced the slide rule, we are beginning to ask these questions, but we are asking them not about the calculator but about the personal computer. And the reason these questions are being asked is that the assimilation of the calculator and the computer is virtually complete with the newer generations of engineers now leaving school, and the bad effects are beginning to surface. Some structural failures have been attributed to the use and misuse of the computer, and not only by recent graduates, and there is a real concern that its growing power and use will lead to other failures.

The computer enables engineers to make more calculations more quickly than was conceivable with either the slide rule or the calculator, hence the computer can be programmed to attack problems in structural analysis that would never have been attempted in the pre-computer days. If one wished to design a complicated structure of many parts, for example, one might first have made educated guesses about the sizes of the various members and then calculated the stresses in them. If these stresses were too high, then the design had to be beefed up where it was over-stressed; if some calculated stresses were too low, then those understressed parts of the structure could be made smaller, thus saving weight and money. However, each revision of one part of the structure could affect the stresses in every other part. If that were the case, the entire stress analysis would have to be repeated. Clearly, in the days of manual calculation with a slide rule--wooden or electronic--such a process would be limited by the sheer time it would consume, and structures would be generally overdesigned from the start and built that way. Furthermore, excessively complex structures were eschewed by designers because the original sizing of members might be too difficult to even guess at, and calculations required to assure the safety of the structure were simply not reasonable to perform. Hence engineers generally stuck with designing structures that they understood well enough from the very start of the design process.

Now, the computer not only can perform millions of simple, repetitive calculations automatically in reasonable amounts of time but also can be used to analyze structures that engineers of the slide rule era found too complex. The computer can be used to analyze these structures through special software packages, claimed to be quite versatile by their developers, and the computer can be instructed to calculate the sizes of the various components of the structure so that it has minimum weight since the maximum stresses are acting in every part of it. That is called optimization. But should there be an oversimplification or an outright error in translating the designer's structural concept to the numerical model that will be analyzed through the automatic and unthinking calculations of the computer, then the results of the computer analysis might have very little relation to reality. And since the engineer himself presumably has no feel for the structure he is designing, he is not likely to notice anything suspicious about any numbers the computer produces for the design.

The electronic brain is sometimes promoted from computer or clerk at least to assistant engineer in the design office. Computer-aided design (known by its curiously uncomplimentary acronym CAD) is touted by many a computer manufacturer and many a computer scientist-engineer as the way of the future. But thus far the computer has been as much an agent of unsafe design as it has been a super brain that can tackle problems heretofore too complicated for the pencil-and-paper calculations of a human engineer. The illusion of its power over complexity has led to more and more of a dependence on the computer to solve problems eschewed by engineers with a more realistic sense of their own limitations than the computer can have of its own.

What is commonly overlooked in using the computer is the fact that the central goal of design is still to obviate failure, and thus it is critical to identify exactly *how* a structure may fail. The computer cannot do this by itself, although there are attempts to incorporate artificial intelligence into the machine to make it an "expert system," and one might dream that the ultimate in CAD is to have the computer learn from the experience contained in files of failures (stored in computers). However, until such a far-fetched notion becomes reality, the engineer who employs the computer in design must still ask the crucial questions: Will this improperly welded pipe break if an earthquake hits the nuclear reactor plant? Will this automobile body crumple in this manner when it strikes a wall at ten miles per hour? Will any one of the tens of thousands of metal rods supporting this roof break under heavy snow and cause it to fall into the crowded arena?

One *can* ask of the computer model questions such as these. Whether or not they *are* asked can depend on the same human judgment that dismissed the question of fatigue in the Comets* and that apparently did not check the effects of the design change on the Hyatt Regency walkways.** Even if one thinks of the critical questions and can phrase them so that the computer model is capable of producing answers to them, there may have to be a human decision made as to how exhaustive one can be in one's interrogation of the computer. While the computer works very quickly as a file clerk, it cannot work very quickly when it is asked to analyze certain engineering problems. One of the most important problems in design is the behavior of metal under loads that deform structural components permanently. While it takes only seconds to put a bar of ductile steel in a testing machine and pull the bar until

----

* The de Havilland Comet airplanes, three of which failed in flight during the early 1950s before anyone thought to investigate the possibility of a systematic failure--fatigue failure--that could have resulted in the eventual destruction of each of these airplanes.

** The Kansas City Hyatt Regency walkways which failed in 1981 and resulted in 114 deaths and more than 200 injuries. The failure was subsequently determined to be due to a design change that had not been thoroughly evaluated.

it stretches out and breaks like a piece of taffy, simulating such an elementary physical test on the largest computer can take hours.

There can be miles of pipes in a typical nuclear reactor plant, and it could take some of the largest and fastest computers a full day of nonstop calculation to determine how wide and how long a crack in one ten-foot segment of the piping would grow under the force of escaping water and steam. The results of such a calculation are important not only to establish how large a leak might develop in the pipe but also to determine whether or not the pipe might break completely under the conditions postulated (by the human engineer). Since it could take years of nonstop computing and millions of dollars to examine every conceivable location, size, and type of crack in every conceivable piece of pipe, the human engineer must make a judgment just as in the old days as to which is the most likely situation to occur and which is the most likely way in which the pipe can fail. The computer does not work with ideas but with numbers, and it can only solve a single problem at a time. The pipe it looks at must have a specified diameter, a specified crack, a specified strength, and a specified load applied to it. Furthermore, the computer model of the cracked pipe must have a specified idea as to how a crack grows as the postulated accident progresses. All these specifications are made by human beings, and thus the results of the computer are only as conclusive about the safety of the system as the questions asked are the critical ones.

The computer is both blessing and curse for it makes possible calculations once beyond the reach of human endurance while at the same time also making them virtually beyond the hope of human verification. Contemporaneous explanations of what was going on during the accident at Three Mile Island were as changeable as weather forecasts, and even as the accident was in progress, computer models of the plant were being examined to try to figure it out.

Unfortunately, nuclear plants and other complex structures cannot be designed without the aid of computers and the complex programs that work the problems assigned them. This leads to not a little confusion when an error is discovered, usually by serendipity, in a program that had long since been used to establish the safety of a plant operating at full power. The analysis of the many piping systems in nuclear plants seems to be especially prone to gremlins, and one computer program used for calculating the stresses in pipes was reportedly using the wrong value for pi, the ratio of the circumference to the diameter of a circle that even a high school geometry student like my daughter will proudly recite to more decimal places than the computer stores. Another incident with a piping program occurred several years ago when an incorrect sign was discovered in one of the instructions to the computer. Stresses that should have been added were subtracted by the computer, thus leading it to report values that were lower than they would have been during an earthquake. Since the computer results had been employed to declare several nuclear plants earthquake-proof, all those plants had to be rechecked with the corrected computer program. This took months to do and several of the plants were threatened with being shut down by the Nuclear Regulatory Commission if they could not demonstrate their safety within a reasonable amount of time.

Even if a computer program is not in error, it can be improperly employed. The two and a half acres of roof covering the Hartford Civic Center collapsed under snow and ice in January 1978, only hours after several thousand fans had filed out following a basketball game. The roof was of a space-frame design, which means that it was supported by a three-dimensional arrangement of metal rods interconnected into a regular pattern of triangles and squares. Most of the rods were thirty feet long, and as many as eight rods had to be connected together at their ends. The lengthy calculations required to ensure that no single rod would have to carry more load than it could handle might have kept earlier engineers from attempting such a structure or, if they were to have designed it, they might have beefed it up to the point where it was overly safe or to where its own weight made it prohibitively expensive to build. The computer can be used to calculate virtually all the possibilities, which, so long as calculations are not made for rods that stretch or bend permanently, is not nearly so time consuming as the calculation for a cracked pipe, and engineers can gain an unwarranted confidence in the validity of the resulting numbers. But the numbers actually represent the solution to the problem of the space-frame model in the computer and not that of the actual one under ice and snow. In particular, the computer model could have understated the weight on the roof or oversimplified the means by which the rods are interconnected. The means of connection is a detail of the design that is much more difficult to incorporate into a computer model than the lengths and strengths of the rods, yet it is precisely the detail that can transmit critical forces to the physical rods and cause them to bend out of shape.

In reanalyzing the Hartford Civic Center's structure after the collapse, investigators found that the principal cause of failure was inadequate bracing in the thirty-foot-long bars comprising the top of the space truss. These bars were being bent, and the one most severely bent relative to its strength folded under the exceptional load of snow and ice. When one bar bent, it could no longer function as it was designed to, and its share of the roof load was shifted to adjacent bars. Thus a chain reaction was set up and the entire frame quickly collapsed. The computer provided the answer to the question of how the accident happened because it was asked the right question explicitly and was provided with a model that could answer that question. Apparently, the original designers were so confident of their own oversimplified computer model (and that they had asked of it the proper questions) that when workmen questioned the large sag noticed in the new roof they were assured that it was behaving as it was supposed to.

Because the computer can make so many calculations so quickly, there is a tendency now to use it to design structures in which *every* part is of minimum weight and strength, thereby producing the most economical structure. This degree of optimization was not practical to expect when hand calculations were the norm, and designers generally settled for an admittedly overdesigned and thus a somewhat extravagant, if probably extra-safe, structure. However, by making every part as light and as highly stressed as possible, within applicable building code and factor of safety requirements, there is little room for error--in the computer's calculations, in the parts manufacturers' products, or in the construction workers' execution of the design. Thus computer-optimized structures may be marginally or least-safe designs, as the Hartford Civic Center roof proved to be.

The Electric Power Research Institute has been sponsoring a program to test the ability of structural analysis computer software to predict the behavior of large transmission towers, whose design poses problems not unlike a three-dimensional space-frame roof. A full-size giant tower has been constructed at the Transmission Line Mechanical Research Facility in Haslet, Texas, and the actual structure can be subjected to carefully controlled loads as the reaction of its various members is recorded. The results of such real-world tests were compared with computer predictions of the tower's behavior, and the computer software did not fare too well. Computer predictions of structural behavior were within only sixty percent of the actual measured values only ninety-five percent of the time, while designers using the software generally expect an accuracy of at least twenty percent ninety-five percent of the time. Clearly, a tower designed with such uncertain software could be as unpredictable as the Hartford Civic Center roof. It is only the factor of safety that is applied to transmission towers that appears to have prevented any number of them from collapsing across the countryside.

In the absence of these disturbing tests, the success of towers designed by computer might have been used to argue that the factor of safety should be lowered. Conservative opposition to lowering a factor of safety would be hard to maintain for structures that had been experiencing no failures, and time, if nothing else, would wear down the opponents. But a lower factor of safety would invariably lead to a failure, which in turn would lead to the realization that the computer software was not analyzing the structure as accurately as was thought. But it would have been learning a lesson the hard way.

Thus, while the computer can be an almost indispensible partner in the design process, it can also be a source of overconfidence on the part of its human bosses. When used to crunch numbers for large but not especially innovative designs, the computer is not likely to mislead the experienced designer because he knows, from his and others' experience with similar structures, what questions to ask. If such structures have failed he will be particularly alert to the possibility of similar modes of failure in his structure. However, when the computer is relied upon for the design of innovative structures for which there is little experience of success, let alone failure, then it is as likely, perhaps more likely, for the computer to be mistaken as it was for a human engineer in the days of the slide rule. And as more complex structures are designed because it is believed that the computer can do what man cannot, then there is indeed an increased likelihood that structures will fail, for the further we stray from experience the less likely we are to think of all the right questions.

It is not only large computers that are cause for concern, and some critics have expressed the fear that a greater danger lies in the growing use of microcomputers. Since these machines and a plethora of software for them are so readily available and so inexpensive, there is concern that engineers will take on jobs that are at best on the fringes of their expertise. And being inexperienced in an area, they are less likely to be critical of a computer-generated design that would make no sense to an older engineer who would have developed a feel for the structure through the many calculations he had performed on his slide rule.

In his keynote address on the structural design process before the Twelfth Congress of the International Association for Bridge and Structural Engineering held in Vancouver in 1984, James G. MacGregor, chairman of the Canadian Concrete Code Committee, expressed concern about the role of computers in structural design practice because "changes have occurred so rapidly that the profession has yet to assess and allow for the implications of these changes." He went on to discuss the creation of the software that will be used for design:

> Because structural analysis and detailing programs are complex, the profession as a whole will use programs written by a few. These few will come from the ranks of the structural "analysts" ...and not from the structural "designers." Generally speaking, their design and construction-site experience and background will tend to be limited. It is difficult to envision a mechanism for ensuring that the products of such a person will display the experience and intuition of a competent designer.

> In the design office the reduction in computation time will free the engineer to spend more time in creative thought--or it will allow him to complete more work with less creative thought than today. Because the computer analysis is available it will be used. Because the answers are so precise there is a tendency to believe them implicitly. The increased volume of numerical work can become a substitute for assessing the true structural action of the building as a whole. Thus, the use of computers in design must be policed by knowledgeable and experienced designers who can rapidly evaluate the value of an answer and the practicality of a detail. More than ever before, the challenge to the profession and to educators is to develop designers who will be able to stand up to and reject or modify the results of a computer aided analysis and design.

The American Society of Civil Engineers considered the problem of "computer-extended expertise" such an important issue that it made it the subject of its 1984 Mead Prize competition for the best paper on the topic "Should the Computer be Registered?" The title is an allusion to the requirement that engineers be registered by state boards before they can be in charge of the design of structures whose failure could endanger life. Professional engineering licenses come only after a minimum period of engineering work with lesser responsibility and after passing a comprehensive examination in the area of one's expertise. Computers, while really no more than elaborate electronic slide rules and computation pads, enable anyone, professional engineer or not, to come up with a design for anything from a building to a sewer system that looks mighty impressive to the untrained eye. The announcement for the Mead Prize summarized the issue succinctly:

> Civil engineers have turned to the computer for increased speed, accuracy and productivity. However, do engineers run the risk of compromising the safety and welfare of the public? Many have predicted that the engineering failures of the future will be attributed to the use or misuse of computers. Is it becoming easy to take on design work outside of the engineer's area of expertise simply

because a software package is available? How can civil
engineers guarantee the accuracy of the computer program
and that the engineer is qualified to use it properly?

By throwing such questions out to its Associate Members, those generally
young in experience if not in age and the only ones eligible to compete for the
Mead Prize, the ASCE at the same time acknowledged and called to the
attention of future professional engineers one of the most significant
developments in the history of structural engineering.

"He Made You Feel You Were a Part of Something Big"

Excerpts from Chapter 13, The Power Broker
Robert Caro

Although not educated as an engineer, Robert Moses was responsible for some of the most magnificent and controversial public works projects in the history of mankind. He held dozens of positions in the City and State of New York during a career of public service that spanned more than fifty years. In these various positions he was responsible for the planning, design, construction, operation and maintenance of parks and parkways, bridges and highways, powerplants, causeways, housing projects, beaches and bathhouses, stadiums and coliseums, tunnels, playgrounds, fairgrounds and other projects of almost every conceivable type. Perhaps no other man in history was personally involved in as many different projects or in as many different types of projects.

Author Robert Caro's biography of Moses, **The Power Broker,** *won the Pulitzer Prize for its fascinating portrayal of the man, his rise to power, and how he used his personal power to advance projects he felt were needed, regardless of opposition from the public, from organized citizens groups and even from elected and appointed government officials.*

This selection from **The Power Broker** *depicts an experience encountered by many engineers (although few engineers encounter men like Robert Moses): the experience of being involved in a massive project with severe time constraints and seemingly overwhelming problems.*

Despite its outcome, the battle over the bathhouses[*] filled Moses with a sense less of triumph than of dread, of a need for haste so frantic that it was almost desperation. Time, he was afraid, was running out on his dream--and it was running out fast.

The battle had shown him how little chance he would have of completing his dream without the unflinching support of Alfred E. Smith in the Governor's chair. He knew that he could never hope to receive anything near the full amount he still needed from the Legislature, and the public itself might well waver in its support for his parks if it knew their full cost. He had Smith as Governor now. But he wasn't going to have him for long. Al Smith, he knew, was going to be a candidate for President in 1928, and that meant that he could not, under state law, be a candidate for Governor. On January 1, 1929, there was going to be a new Governor. He probably would be a Republican. Even if he were a Democrat, he would not be Al Smith. January 1, 1929, was less than two years away. And two years was far too short a time for completion of a dream that encompassed three long roads and a score of huge parks. His only hope, he knew, was to complete enough of the roads and parks before Smith left office so that the public could see how great they were going to be and would demand that they all be completed.

----

[*] The struggle to secure funding from the New York legislature for the bathhouses under construction at Jones Beach in 1927.

So Moses drove himself and he drove his men. Turning the upstairs bedrooms of August Belmont's mansion into offices,[*] Moses filled the offices with engineers and architects. He himself worked downstairs.

"Belmont had had this tremendous dining room, and there was this huge table there, and Moses made it his conference table," recalls William J. Junkamen, an attorney hired to assist commission counsel Raymond McNulty. "The conferences would start at nine o'clock in the morning and sometimes you couldn't leave until after midnight. And you just worked like hell all day long. Supper was a matter of dashing out and getting a bite to eat as quick as you could." Insisting that the engineers in charge of a project prepare a schedule showing the date on which each of its phases would be completed, Moses would move up the deadlines--by days, by weeks, sometimes by months -- until the engineers felt it was absolutely impossible to meet them. And then he insisted that they be met. "The time was never long enough," Gilmore Clarke says. "But that was all the time you were going to get, and he let you know it." James J. Flynn, captain of the State Police troop assigned to the commission, recalls, "If he wanted a job done, he wanted it done. Period. And he wouldn't take any alibis if it wasn't done; he didn't want alibis, he used to say."

Moses knew exactly what he wanted each of his men to do, and he was impatient when they had difficulty grasping that fact. William H. Latham, a young engineer from the Massachusetts Institute of Technology, was also assigned to observe and report on meetings of the Nassau County Board of Supervisors. "I came back from the first meeting and in giving him my report, I generalized," Latham recalls. "He corrected me, but I guess I didn't get it, and the second time, I started to generalize again. Well, he didn't let me get very far. His palm came down on that table of Belmont's and he jumped up and he paced around that table and he told me he wanted facts, no assumptions. All he wanted from me was what had happened at that meeting. He'd draw his own conclusions. And he let me know right there and then that if I couldn't do it that way, he'd have to find another engineer."

And, Latham adds, Moses wasn't "really angry" at him on that occasion. Latham, an athlete and outdoorsman, is a tall rangy man with huge shoulders and an easy, friendly grin. But the grin fades when he says, "I won't talk about what he's like when he's really angry."

Another technique was the silent treatment. "All of a sudden," a staffer recalls, "you just wouldn't be called in to sit at the conferences any more. He wouldn't talk to you if he passed you in the hall. And then, one day, you'd just be gone."

The men who stayed didn't resent Moses' methods. "If he drove other men hard," says Junkamen, "he drove himself harder."

He was supervising a dozen nonpark projects for Smith, of course, and he was constantly commuting to Albany, a four-hour train trip away. . . . So that he could get in a full day's work in Albany . . . , Moses left his New York

----

* Moses had purchased the Belmont estate for the purpose of creating offices for the Long Island State Park Commission.

apartment at 6 a.m. to catch the early train to the capital and avoid a car trip made tortuous by the lack of a through road. When the session broke up, usually well after the last train back had left at midnight, Moses would ask Smith for the use of a state car and chauffeur. He had a lot of work to do back on Long Island, he'd say. If he drove back at once, he'd be able to start in the morning.

But when his car pulled up in front of the Belmont Mansion, Moses never seemed tired. Charging into Arthur Howland's office, he would slam the door behind him and listen to the chief engineer's problems. Then, one by one, he would call in the other top commission officials and, behind the closed door, listen to theirs. Then the door would fly open and he would charge out into the dining room and sit down at the huge table and begin solving the problems. Junkamen recalls: "He might have been working in New York and not arrive out at Belmont until three o'clock in the afternoon, but then he would work from three until ten o'clock or midnight. Another day, I might get there at nine o'clock in the morning and he'd already be there working at the big table. Then I'd go out on some errand and wouldn't get back until supper time, and he'd still be sitting there. And he'd still be sitting there at midnight." Saturdays were no different. "Hours didn't mean anything to him," Latham says. "Days of the week didn't mean anything to him. You worked when there was work to be done, that was all." And there was always work to be done. Since none of the staffers dared leave the mansion until Moses left, they longed for him to get a telephone call from [Moses' wife] Mary, which usually got him away from the big table and back to his house in Babylon. "Sometimes," Junkamen recalls, "Ray McNulty would slip out and call her and put her up to calling." One Saturday night, repeated telephone calls from Mary failed to bring Bob home, and at ten o'clock she showed up at the mansion. Striding into the conference room, she "just took him by the ear in a very nice but firm way" and pulled him to his feet, Junkamen recalls. "He just laughed and went along--and then we could all go home."

Going home did not, however, necessarily mean a cessation of work for Moses. Giving Howland a key to his Babylon house on Thompson Avenue, he told the chief engineer to stop by every morning on his way to the mansion. Almost invariably, when Howland arrived at about 7:30 a.m., there would be waiting for him on the flat-topped bottom post of the banister a large manila envelope crammed with notes, memos and handwritten letters ready to be typed and mailed, an envelopeful of testimony to what Moses had been doing during the night while his men slept. He tried to keep Sundays free for his family, teaching Jane and Barbara to swim or sail, taking them for picnics, telling them stories. But often the girls would notice that their father had disappeared. When that happened, they knew he was out on the big screened porch on the side of the house, scribbling furiously on a yellow legal pad. And most of the family picnics were held in one of his parks. "Hell," one of the men working under him remembers, "any Sunday at all you could expect to look around and see the boss and his wife and kids and he'd be making notes as he walked along if he saw something he didn't like. So you figured if he was working, why shouldn't you be?"

There were other reasons, too, why the men who stayed didn't resent the driving.

561

"It was exciting working for Moses," one of the commission staffers recalls. "He made you feel you were a part of something big. It was almost like a war. It was you fighting for the people against those rich estate owners and those reactionary legislators. And it was exciting just being around him. He was dynamic, a big guy with a booming laugh. He dominated that scene in the mansion. He would sit there with people running back and forth around him and would be banging his hand down on that big table and giving orders--and when he gave orders, things happened! Howland would go hurrying out of the room and twenty-five draftsmen would hurry to their tables and start drawing or surveyors would jump into their cars and head out on the road."

And, men recalled, incongruous as it might seem to use the word "fun" in connection with unremitting work, it was, nevertheless, fun to work for Bob Moses. "There was a very informal atmosphere in the mansion," one engineer says. "Everyone worked in their shirt sleeves--my recollection of Moses is of him sitting with his tie pulled down and over to one side, sleeves rolled up-- and there was always a lot of joking going on." And the joking, the engineer says, went both ways. "You weren't afraid to kid him."

Of all the reasons why Moses' men didn't resent his driving, the one most frequently mentioned in their reminiscences is that he brought out the best in them. One of Junkamen's duties was to work with Long Island villages on zoning restrictions on the land adjoining the Southern State Parkway. "Mr. Moses wanted five hundred feet on either side of the parkway zoned in the highest residential classifications," Junkamen recalls. "And he didn't want any water tanks or other unsightly structures near them, either. Zoning was a relatively new thing. Hardly a village on Long Island had a zoning ordinance. I had to draw up a lot of what we wanted myself. Then Mr. Moses would have to speak to officials in the municipality and get them to draft over-all zoning ordinances in which our stuff could be incorporated. I'd have to work with the local counsel every step of the way. And then, after the ordinances were adopted, I'd have to work with the local zoning board when people tried to break the restrictions. And I found that at every step of the way, Mr. Moses had ideas that started me thinking along whole new lines."

Constantly, Moses was encouraging his architects and engineers to use their imagination, to make their designs different from and better than any similar designs done before. When the men designing a drawbridge for the Jones Beach causeway submitted their first design, he said, "I know you can do drawbridges. Can you do beautiful drawbridges?" In their second design, the bridge operators' quarters were no longer the standard ugly shacks but turrets* faced with stone worked with the silhouettes of sailing ships. Gilmore Clarke thought he had surpassed himself in the design of bridges for the Bronx River Parkway. But he found Moses had some new standards. He wanted variety, Moses told him. Not only were the bridges to be designed to harmonize with the landscape and not only were they to be stone-faced, but every bridge on every parkway on Long Island--all one hundred of them--was going to be different from every other bridge.

----

* A landscape architect employed by Moses

Engineers assigned to design guard rails and light poles for the parkways expended tremendous effort making the standard iron poles graceful. But iron wouldn't blend in with a rustic setting, Moses said. Guard rails and light poles would have to be made of wood. But it had been proven that no form of wooden guard rails would resist the impact of a speeding car, the engineers said. Moses sent them back to ponder the problem again, and this time one thought of drilling holes in wooden rails and inserting strong steel cable--and now all the rails on all the parkways could be wood.

"Mr. Moses was no lawyer, but he had a great knowledge and grasp of the law," Junkamen would say. "He was not an engineer, but he had a great knowledge of engineering. He knew politics, he knew statesmanship--he was an altogether brilliant man. If you were working with him, you just had to learn from him--if only through osmosis." One of the commission's engineers rhapsodizes: "I don't think there was a man who came into daily contact with him who wasn't inspired to do better work than he had thought he was capable of doing."

If the political difficulties involved in creating a park on Jones Beach were enormous, the physical difficulties were of a size to match. Building on a barrier beach proved to be a very different proposition from building on the mainland. Commission engineers found themselves faced with a succession of problems engineers never encountered on mainland jobs. Cleveland Rodgers, Moses' first biographer, was to write that sometimes it seemed as if Nature herself had "joined forces with the skeptics and obstructionists who had fought Moses all the way from Albany to the beach." But Moses refused to let Nature stand in his way.

None of Moses' engineers had expected work on the barrier beach to continue in winter because drifting ice packs that kept boats off the Great South Bay for days at a time could maroon anyone caught on the strand. But with Smith's time as Governor running out, winters could not be wasted. Moses told the engineers taking surveys for the causeway to cache emergency supplies of food in a shack on the beach and keep working. Even on mornings when wind was whipping the bay into waves and ice was coating the piers, Sid Shapiro' led his hip-booted surveyors into boats for the trip across.

One day, while they were on the strand, ice packs closed the bay. It stayed closed for ten days. All the cached food ran out except pancake batter, and the surveyors lived on pancakes. For the rest of his life, Sid Shapiro would never be able to stomach another pancake. But when the ice cleared and the surveyors returned to Babylon, Shapiro could tell Moses, who was standing on the dock waiting for a report, that ten days' more work had been completed.

The completed surveys contained the worst of news. The mean level of the existing barrier beach, Moses was told, was only two feet above mean sea level. During storms, the ocean rose six, seven or even eight feet and covered the

----

   * Shapiro was a civil engineer who worked with Robert Moses for more than 50 years.

strand almost completely. This did not matter as far as the portion of the beach that was to be a beach was concerned, but the portion that was to hold Moses' buildings and parking lots and parkway would have to be built up to a mean height of fourteen feet if they were not to be submerged in every storm, and if he wanted a road seventeen miles long along it, it would have to be built up to that height for seventeen miles. The job could be done, of course--floating dredges, huge pumps mounted on barges, could suck up hydraulic fill, which would become sand when dry, from the bay bottom, and pipelines could spill it out over the strand. But approximately forty million cubic yards of fill would be required. The job would take months--and it would be expensive. Even Smith quailed at this, and Moses had to talk fast and hard to persuade the Governor to go along. But he did persuade him, and the largest floating dredges in the United States were brought to the bay as soon as spring cleared the ice off it in 1927. The job could not be completed by the time ice set in again in the winter, but Moses refused to let the dredges leave. Their crews could live on them, he said, and all through the winter of 1927-28 they did, and the pumps kept working. "Night after night," Shapiro recalls, "they kept working to midnight."

When the sand from the bay bottom was spread on the barrier beach, it proved to be the worst problem of all. It was beautiful to look at, dazzling white and fine-grained, but the fineness meant that when it dried, it blew. Even the lightest breeze stirred it into the air in swirls so thick that the strand looked like a desert during a violent sandstorm.

"It was always blowing in your face when you worked," Shapiro remembers. "When it got bad, it would fill your eyes, your ears and your nose as fast as you could clean them out. You'd be choking and coughing. You'd be talking to somebody not three feet away and an especially violent gust would come, and you couldn't even see him any more. You couldn't even see the hand in front of your face." During the day, workmen would dig an excavation. At night, the sand would fill it in--so completely that the workmen couldn't even find its edges. One workman who left his car with its rear end turned into the wind for a few days came back to find the numerals--and all other color--completely erased from its rear license plate.

Moses dispatched landscape architects to other Long Island beaches to find out why the sand on older, natural dunes was more stable. They reported that it was because of the presence on these dunes of a form of beach grass (*Ammophilia arenaria*) whose roots, seeking water in the dry sand, spread horizontally rather than vertically and thus held sand around it in place. But to be effective, they reported, the grass had to be planted thickly --hundreds of thousands, even millions, of clumps would be required to hold down the new dunes on Jones Beach--and it could be planted only by hand. In the summer of 1928, on the desolate sand bar on the edge of the ocean, amid half-completed building skeletons that looked like ancient ruins, was a panorama out of the dynasties of the Pharaohs: hundreds, thousands of men, spread out over miles of sand, kneeling on the ground digging little holes and planting in them tiny bundles of grass.

The peculiarities of man seemed sometimes to join with those of nature to thwart Moses. But he would not be thwarted.

The contract to build the causeway had been awarded to an Atlantic City, New Jersey, firm. In the autumn of 1927, the president of the firm coolly informed Moses that its money had run out and it couldn't meet its payroll. It's laborers wouldn't work without pay, and work would have to stop unless $20,000 was found immediately. Moses drove to Albany, but Smith told him that every cent that could be squeezed out of the Highway Bureau budget--in fact every cent that could be squeezed out of the budget of any department--had already been squeezed. More money would be available in the 1928 appropriations, of course, but they wouldn't be available until after January 1. Promising to pay her back as soon as the appropriations came through--he did--Moses borrowed the $20,000 from his mother and paid the laborers himself. And they kept working.

To the east of the portion of Jones Beach ceded to him by Hempstead Town, directly in the path of his proposed Ocean Parkway, lay five hundred acres of meadowland owned by Oyster Bay Town and leased as a duckhunting preserve to a group of wealthy sportsmen headed by Solomon Guggenheim. When Moses appropriated the tract, the sportsmen obtained an injunction and served it on the contractor. But Moses met the contractor on the dredge and, as Cleveland Rodgers put it: "The legal papers somehow slipped off the deck into the swirling waters. . . . When the hunters, in their furlined cloaks and escorted by their formidable legal advisors, arrived a few days later to shoot ducks, under full protection of the law, they found the meadow" slashed through and filled in for the parkway.

And always, day after day, summer and winter, Moses was out on the job, encouraging the men in the field. "People will work harder for you if they have a good time," he told the Morses, and he urged the laborers working on Jones Beach to go swimming during their lunch hour. Organizing softball games on the beach, he umpired them himself. Joking with the men, he made fun about his hat. And he told them what a great project they were building. "He had a gift for leading men," Shapiro recalls. "those men idolized him. You'd see him walk up to a pickax gang that was tired and talk to them awhile and when he walked away, you could see those pickaxes swing faster."

The sands were running out on Al Smith's Governorship. Moses' construction crews did not quit for the winter of 1927-28; their grading machines pushed aside snow as well as earth as they smoothed the path for the Southern State Parkway. The Great South Bay froze solid; the men stretching the Jones Beach causeway across it pitched tents on the ice and lived on it.

Early in 1928, an astonished Hutchinson and Hewitt[*] realized that the completion of the Southern State Parkway and Jones Beach causeway, on which they had hinged the beginning of the Northern State [Parkway] and which they had assumed was years, if not decades, away, was rapidly approaching. Frantically, they tried to delay it. In another year, Al Smith wouldn't be Governor. Out of Moses' $4,500,000 1928 budget request, they slashed

----

[*] Eberly Hutchison and Charles Hewitt were state legislators from upstate New York who had led the fight against what they considered extravagant park developments undertaken by Robert Moses' Long Island State Park Commission.

$626,000. But Smith found the money in the departmental budgets, and the construction crews kept working.

Moses had not been given funds for his Long Island parks and parkways until the spring of 1926. By the end of the summer of 1928, in a period of less than three years, every foot of right-of-way for the Southern State Parkway was in his hands, and a seven-mile stretch, from near the New York City line to and around the Hempstead reservoir, was completed. Long rows of newly planted elms and maples lined it and stone-faced bridges, every one different, were carrying crossroads over it so that nothing should interrupt the swift passage of its users. A second seven-mile stretch, from the reservoir to Wantagh, was completed except for the landscaping. A third seven-mile stretch, from Wantagh to Babylon, was graded and ready for paving. And the fill for the Wantagh Parkway had been laid, a pavement placed on top of the fill and three of the four bridges that would carry the causeway across the bay completed.

When Moses had become president of the Long Island State Park Commission on April 18, 1924, there had been one state park on Long Island, the almost worthless 200-acre tract on Fire Island. By the end of the summer of 1928, there were fourteen parks totaling 9,700 acres. Because 6,775 of those acres had been acquired--from Hempstead, Oyster Bay and Babylon towns, the U.S. Department of Commerce, New York City and private individuals--as gifts, the Long Island parks had cost the state a total of about a million dollars. At 1928 land values, they were worth more than fifteen million.

By the end of the summer of 1928, the watershed properties off Merrick Road had been filled with bathhouses, baseball fields and bridle paths. Picnic areas with thousands of tables sat under their trees. Slides, swings, and jungle gyms spotted their clearings. Their lakes were decorated with floats, diving boards, sliding ponds, rowboats and canoes. Heckscher State Park contained miles of paved roads for cars and dirt roads for horseback riders, acres of athletic fields, bathhouses holding five thousand lockers, a boardwalk, a bathing pavilion with restaurants and snack bars, an inland canal for rowboating, and a marina at which sailboats could be moored. There were more bathhouses, more boardwalks, more playing fields, more snack bars, more picnic areas, more campsites at Sunken Meadow, Wildwood, Orient Beach, Montauk Point and Hither Hills state parks. On Jones Beach, two years before a desolate sand bar, there stood now, awaiting only the finishing touches that would be added in 1929, a bathhouse like a medieval castle, a water tower like the campanile of Venice, a boardwalk, a restaurant and parking fields that held ten thousand cars each. In the history of public works in America, it is probable that never had so much been built so fast.

During the summer of 1928, park-seeking families heading out of New York City began to feel Long Island open up to them. Week by week, word spread. At the beginning of the summer, the bathhouse at Valley Stream State Park contained a thousand lockers. For a few weekends, these were sufficient. Then they were not. And, even so, by the end of the summer, thousands of would-be bathers were being turned away every weekend. By the end of the summer, attendance at Long Island's state parks had passed half a million.

New Yorkers knew who was primarily responsible for the boon they had been given. It would have been difficult for them not to know. For the press was turning Robert Moses into a hero.

Getting Sued

Chapter 9, Getting Sued and Other Tales of the Engineering Life

Richard L.Meehan

Richard Meehan's account of his involvement in a lawsuit resulting from his professional work as a geotechnical engineer will send a chill down the spine of anyone who has ever had any reservations about a professional judgment or recommendation. However, it describes a fact of life that consulting engineers have to be prepared to live with.

I thought when it all began that it was going to be like any other business or engineering problem, something to tackle with logic and flow diagrams and Bayes theorem. And so I was surprised to find myself sitting in that chair, a lawyer circling my flank like an unfriendly dog on a dark street, and the flat look of the judge telling me my credibility had just slipped a notch; surprised that it was more like saying good-bye to a brave child in a hospital or the first week in boot camp. It just snaps through the polished surfaces of one's professional armor with the ease of an electrical charge, sets the most private of the smooth muscles to dancing. There is no refuge; at three o'clock in the morning, fermented bits and pieces of the day hiss from the recesses of your mind like gases from buried geologic strata.

They came after me when I was in bed with the flu.

The long arm of the law, they call it. To me it seemed an elephant's proboscis, the exploratory organ of a lawsuit, its delicate pink tip, mucoid and hairy, searching for an unshelled peanut. Backed by an irreversible force, feeding an innocent but insatiable appetite, it glided expectant and trembling into my office, and touched my secretary. "Out with the flu? Oh that's too bad. I suppose he's home, then?"

It was a hot afternoon in August, and I was motionless, impressing with my back a steamy dampness onto a white sheet. The fever was a brass wafer on my tongue, my skin as fine tuned as a seismometer. I had two symmetrical but distinct headaches, one behind each eye. I confess to taking secret pleasure in the flu: it is one of the few safe and legitimate refuges of middle age, a bright temple within a grove sickly sweet with blossoms of orange and almond, grounds of raked sand forbidden to salesmen, business associates, and children; friendly spirits visit me there like schools of tropical fish. It was four o'clock in the afternoon. Only the most exquisite sounds came to the room where I, the patient, lay; birds twittering in the happily spattering lawn sprinkler; the bland conversation of the mailman; a butterfly in the garden of women and children; Elton John, faintly, from my daughter Kathleen's room. My wife had gone to the marketplace. She would return with offerings, fruit peeled and cut in a china bowl, clean sheets, oatmeal, a fresh <u>National Enquirer</u>, I lack the nerve, at my age, to ask for a Superman comic book.

The doorbell rang, and Kathleen thumped down the stairs, yelling, "I'll get it." Paper boy collecting for the <u>Times</u>, I supposed. Or one of my neighbors petitioning against the affront of nuclear power--probably the bearded one two doors down with the dog that howls all night. I'd gladly trade both him and his dog for a few extra rems of radiation. Or a child, speaking a language I barely understand, selling inedible candy wrapped in magenta foil. Kathleen will handle it, whatever it is beating at my door.

There were mumbled words, and Kathleen came to my room. "There's a man who says he has to see you. He says it's important," she said.

"Tell him I'm sick in bed, asleep; ask him what he wants," I told her. Not a clear reply. Am I not leaving the door open just a crack here? Perhaps I am finding the solitude of the temple a little boring, succumbing to a dangerous curiosity about the world. Or am I seduced by "it's important"? Meaning I am important?

Kathleen returned to the front door. There was more subdued conversation. A moment later she came back, and there was a man with her. Slight, balding, with glasses, clean shaven, a short-sleeved shirt open at the collar, he stood nervously at the threshold of my bedroom. He looked like an electrical engineer. He held a folded piece of paper.

"Mr. Meehan?"

"Yes."

"This is a subpoena."

The man inspected his conscience, which he evidently kept beneath his fingernails. "Hey, I'm sorry about this; this isn't my usual job, delivering these things; I'm a detective. But the regular guy was out today."

Hard times have come to the detective business. What with the no-fault divorce laws in California, there's no romance in it any more; a man can't earn a living following wayward husbands into seedy hotels. Sam Spade reduced to delivering subpoenas to the sickroom. Decay of a proud profession.

I accepted the paper, feeling sorry for him. For a moment, I wondered whether I should have offered a tip.

The construction industry had slumped badly in the autumn of 1969, a few months after my partners and I had started our little consulting engineering firm. In those days, even the better established consultants were scratching for work. Like many other engineers going into business for themselves, we had underestimated the persistence and flair that it took to bring in new clients, at least ones who pay their bills. My father-in-law, a successful surgeon and banker, had loaned me $5,000 to start up this business, and the money was beginning to run out. It was a rainy gray autumn, and I was spending my days in San Francisco, visiting prospective clients who had no need for our services but who welcomed me politely in their steamy offices and listened patiently to my nervous pitch. I had one good pair of shoes; they grew sodden and heavy from the wet sidewalks, and I developed a bad case of athlete's foot. My wife devotedly ironed a clean button-down shirt and pressed the crease back into the rain-soaked, shapeless lower part of my trousers every day. She never asked about "new jobs" unless I brought that matter up.

At the office, we were making ends meet, but just barely. A new project, especially when it was a new client, was cause for celebration.

One of my partners, Irwin Sprague, was active in Republican politics and through those connections had been introduced to the building director of the San Ignacio School District. The town of San Ignacio occupies a valley of the same name, a wrinkle in the California coast ranges south of San Francisco. Until the mid 1960s, the floor of this valley was carpeted by matted bunch grass and desiccated cow pies. Now it contains an inland sea of pastel tract houses, warping and fading in the sun. The district had just acquired a site for a new high school, and the director contacted us about performing the standard soils investigation of the site and advising their structural engineer and architect on matters of foundation design. We were delighted and worked up a program to perform the necessary borings, soil tests, and analyses. We would charge them in accordance with our standard hourly fee schedule, we agreed--twenty-five dollars an hour for our time, plus expenses, with a guarantee that we would not exceed three thousand dollars.

Looking back on it, we were offering a real bargain, even for those days. Jobs like this were usually assigned to junior staff people in firms like ours, but we didn't have any of those, so Irwin and I, with twenty years of experience between us, handled the field and lab work and report writing ourselves. We were eager to do a good job--perhaps overeager, it occurs to me now. There is such a thing as trying too hard, and it affects one's judgment. It takes experience, and perhaps age too, to understand that professional skill is a peculiar alloy of strict standards and callous indifference.

We knew we had trouble the first day on the project when Irwin returned from drilling holes at the site. "It's a fat clay," he said, placing a lump of malevolent black adobe on my desk. I ran some tests on the soil, which confirmed that it was a fat clay, the kind that shrinks when it dries and swells when it's wet. That means trouble for building foundations in California, with its wet and dry seasons.

We sent in our report and then did not hear much about the high school for several months. One day the following spring, someone told me that the architect and structural engineer were preparing final plans and specifications. This got me thinking about the job again, especially about the two alternative solutions I had suggested. One of these was to dig out the fat clay beneath the school and replace it with gravel. The other was to mix exactly the right amount of water with the clay so that it would neither shrink nor swell.

Then I had some second thoughts.

True, the special moisture-conditioning solution would save the owner some money (we engineers are steeped in an aesthetic imperative that demands functional adequacy at lowest cost). But it also assumed that the contractor would know what he was doing and that his work would be closely supervised. How did I know whether the district would get a contractor who had any experience in this type of work? Would they have someone competent supervising the work in the field? No one had ever bothered to discuss the plans and specs with me; could I count on the designers having incorporated all of the special provisions that I had recommended for using the on-site soils? Obviously I could not assure myself on any of these points; they were all out of

569

control. Maybe, all things considered, it would be better to stick to the simpler solution that called for replacing the soils at the site.

I put a call through to Frank Schultz, the structural engineer, telling him of my concerns, that I was inclined to go for the soil-replacement solution, the safer one. Schultz disagreed. He was an old-timer, and he liked to run his jobs. Once he had made up his mind on a point, Schultz had an effective, rhetorical, direct-mail manner of presenting his case. He said that the district had just about exceeded the budget on the job already, and every penny counted. He had a lot of precautions built into his design, extra heavy reinforcing in the concrete and special protective barriers to keep the water out of the foundations. He had designed hundreds of buildings on soils just like this, or worse, and never had problems before. He wasn't worried about it. Schultz was a well-known designer, with a lot of experience, including many structures in areas where I knew the soil conditions were even worse than at our site. O.K., I told him; I agreed that we should go ahead with the cheaper alternative solution. We chatted for a few more minutes about design details, then rang off.

That was the last I thought about the high school until late that summer, when Irwin came by my office one afternoon and told me he had got a call from his friend, the building director at the district. They were starting construction in a few days and would be calling on us from time to time to perform soil tests to check the moisture conditioning operations. They had their own man on the job to supervise the work from day to day, so all they needed was a few tests, at his direction. Irwin had discussed the frequency of testing with them a bit and concluded that one of our technicians could perform the work in about thirty hours' time. He suggested they budget six hundred dollars for our testing services on the job.

Matters went downhill from there. The district had reluctantly awarded the job to a local contractor, Mullins, an affable charmer with a spotty record of past performance. The district's man on the job was a beery construction stiff hired from some other public agency.

I assigned a recent graduate with only a few months' experience as our on-call inspector. The site work started, then stopped for a few months because of a strike, then had to be partially redone, then started up again. For one long wet winter, the uncompleted work was lashed by rainstorms, with no effort being made to protect the sensitive foundation soils. Half-completed precautionary measures that I had specified, drains and moisture barriers, perversely became the conduits and baffles through which storm waters gushed into the foundations. The contractor's men appeared to be moonlighters with little experience in the delicate concrete work required for the job. Schultz, the structural engineer, raged over unacceptable workmanship from time to time, and some of the concrete work was ripped out, redone, and then ripped out again. Anything that could have gone wrong did. In the end, the district threw Mullins off the job and demanded that his bonding company finish it in time for the school year. They did, but it came out looking patched and second rate. A month before the dedication, a huge crack appeared in the floor at the main entrance. All of us who had been involved with the project were ashamed of it.

Three years later.

The date of my pretrial deposition loomed before me, piquant and dreadful, like a final exam in organic chemistry. Irwin and I had the idea that we should hire a lawyer. Our consulting business had enjoyed a modest success since that time four years before, when we had worked on the San Ignacio High School project. The project had turned out a mess, with cracked slabs and warped walls. The school district was blaming the contractor, Mullins. Mullins was blaming the district's consulting engineers, Frank Schultz on structures and us on soils. Now the issue had become a $150,000 lawsuit, in which we were named. We had no malpractice insurance, so we were on our own, had to pay for our own defense. I had never been involved in a lawsuit before. They said that an experienced trial lawyer charged fifty dollars an hour. And up.

I had a lawyer friend, a bright, droll Harvard man, Henry Worthington. Henry had just finished making some mid-life adjustments--getting a divorce, going on the wagon, and quitting his job with a big city firm. Putting myself in his hands would probably cost a few hundred, I figured, but it would be worth it if he could make it all go away.

Henry, the new Henry, had a regular schedule on Monday, Wednesday, and Friday that began with meditation and granola at 5:30 in the morning, followed by a session of vegetable gardening using the French intensive method. Then he spent an hour reloading shotgun shells, in preparation for his afternoon skeet shooting. Henry had elevated skeet shooting to the level of a spiritual activity, the transubstantiation of clay discs to puffs of black smoke. I met him at the range one windy afternoon, shot seven out of twenty-five, then, unenlightened but with my ears ringing asked him if he would review my file and come with me to this deposition.

Henry was neatly packing empty shells into a box, and he paused and looked at me. He is one of those people who can pause in a conversation for thirty seconds or so, create with perfect calmness a frightening void during which you fear that your heart will stop and you feel an urgency to explain or confess something. No, he could not handle the matter, he finally told me without explanation. Could he recommend someone else, I asked. Henry thought about that for what seemed another geologic age, then began to critique my shooting style. I was responding to the wrong kinesthetic cues. "Maladapted sternomastoid response is where the major problems begin," he began to explain.

"I know, Henry," I said, exasperated. "I think about all those things too, but I'd really appreciate it if you could just give me a name."

"Try Duncan Macaulay. Duncan's a superb litigator, just superb. A lone operator. You don't find many of those around any more."

I made an appointment to see Macaulay, and a few days later Irwin and I went to his office, a small building he shared with some other lawyers, tucked inconspicuously into a leafy residential street. No stable of young legal wizards to keep busy at forty dollars an hour, I thought. A good sign.

Macaulay proved to be a tweedy bearded chap, fiftyish, with a cagey opening manner that immediately struck me as potentially effective against our adversaries. He made no effort to convince us that he knew anything about our business or the various technical matters relating to the case, which I thought reflected favorably on him. In my experience, playing up credentials in or knowledge of the client's business is always a sign of weakness.

I explained our background on the project, that we were not too concerned about losing the case because there was no question of any negligence on our part, but that we thought we might have some exposure here, that we didn't carry any liability insurance against this kind of thing, that the insurance wasn't even available anymore.

Macaulay shook his head in the socially sympathetic manner of one unmoved by others' troubles. "You guys really land up carrying an unfair burden in these situations," he said.

This annoyed me. I didn't want professional sympathy. We were interested in making the lawsuit go away, or failing that, going for the jugular veins of the contractor, the district, whoever might be after us. But Macaulay looked so perfectly insincere in this remark that I didn't hold it against him. We agreed to retain him to work on the case, and he marked his calendar for the day of my deposition.

A couple of days later, Macaulay gave me his first piece of advice, which proved, in the end, to be bad. Or perhaps the mistake was in my asking his advice on the matter in the first place, not making my own decision.

For the past year or so, I had kept the fat manila project file for the high school job at home. Not that there was anything incriminating in the disorderly pile of pink telephone messages, photocopied letters, fold-softened ozalids* and mud-splattered calculation sheets contained in it. But I had an anxious vision of a mob of detectives bursting into my office and cleaning out my files, then something turning up in them later that would prove embarrassing to me. I make mistakes and overreact to situations and use bad language just like everyone else. It might well be that the law says that some lawyer has the right of access to all that, but that's someone else's law, not mine.

I read through the file a few times, then threw out a couple of items that made us look more foolish than criminal. There remained pages and pages of field notes and calculations, and I almost threw those out too. But the subpoena instructed me to bring all my project files and notes to the deposition. I called Macaulay and asked him what to do.

"You don't have anything to hide, do you?" he asked me.

I wasn't aware of anything. So in the end, I followed his advice and brought the files (which the other lawyers photocopied) even though it ran against my own intuition. Lawyers are skilled at explaining the rules. But there are

----

* An ozalid is a type of reproduction, usually of a drawing.

certain decisions that you had better make yourself, at midnight, alone in your office. And the point to keep in mind at that time, and at all other times, is the more you give the opposition, the more they have to work with, the bigger the ore body they have to mine for the rich veins that occur in every barely perceptible crack in your professional performance.

The day of my deposition arrived in the middle of a late August heat wave. I woke up at seven o'clock that morning to find the heat seeping through my bedroom window, displacing the glade of moist air in which the filaments of my mind had finally disentangled in the gray hour before sunrise. By the time I had reached the San Ignacio Valley, an hour's drive from my home, the morning had grown sullen and hot. Mists that during the night stroked the feverish hills like a mother's cool hand had now drawn away, and the air itself felt thin and burned by the orange sun.

I found the attorney's office, where my deposition was to be taken, in an abandoned orchard on the edge of the town. The building was a new California missionary-style stucco and chicken wire tax write-off with an imitation tile roof, partially screened by parched olive trees and surrounded by a sticky petroliferous parking lot.

Inside a receptionist with carmine lipstick and a rococo tower of hair told me that it would be a while and suggested I join a large plastic plant that seemed to have blundered into the waiting room. I flipped through a copy of <u>Audubon</u> magazine. An hour passed. An hour and a half. The day was emptying itself of otherwise paying work. I could feel whatever organs of rage exist in the region of the solar plexus inflating. Aware of the hazards of anger, I ransacked the files of my mind for consolation. I told myself jokes. I visualized an advisory panel consisting of William James, Norman Vincent Peale, Marcus Aurelius, and Doris Day.

An hour later the session recessed for lunch.

The morning's victim had been Frank Schultz, the structural engineer on the project. Frank and his lawyer (actually his insurance company's lawyer), my lawyer Macaulay, who had sat in on Frank's deposition, and I walked across the street to a Sambo's restaurant for lunch.

We ordered our burgers, and the insurance company lawyer, Don Smith, up from Los Angeles for the day, coached Frank on his morning performance. Smith wore a polyester suit that contained, under some tension, the bulk of an ex-athlete gone soft. I tried to read the inscription on a gold and ruby college ring he wore on one fleshy finger of his big left hand. I was curious because I couldn't determine to my own satisfaction whether his diffident commentary on the morning's proceedings were the gropings of a sluggish intellect or the faint output of the teaspoonful of brain matter that he was willing to assign to the conversation, holding the rest perhaps for a current love affair or a faltering investment.

Frank was vibrant with emotion and was not listening very closely to his lawyer's advice. He was in fact repeating, to my benefit, some errors he had made during the morning's session. "When I said that that contractor Mullins was incompetent, I meant it in the full sense of the word," he blurted.

573

Contractor incompetence was how Frank saw the whole issue. It was Frank who had advised the district to discharge Mullins from the job.

Smith asked him if he really had authority to direct the contractor's work in the field.

"This was my job, and I had full authority to tell him what to do," Frank said angrily, as if challenged by his lawyer.

Smith frowned and said that he didn't want to tell Frank what to say on the stand or advise him to speak untruthfully, but couldn't he soften that a bit?

It was apparent to me that Frank was blindly digging his own grave deeper, by claiming to hold responsibilities on the job that he actually did not have at all. Under fire, Frank was taking hits and going down with pride, the captain of the ship. What he should have been saying was that he was only the galley cook, that on the morning the battle was lost he dished out the chipped beef on toast just the same as on any other morning. Perhaps we engineers tend to make that mistake, put too much of ourselves into our work, I reflected. Maybe Smith had the right idea. But Smith had now done what he could. His eyes glazed at Frank's further protestations, his mind seemed to depart to Los Angeles, to whatever other arena was preoccupying him.

Smith's abandonment of the issue left the field open, and there was a silence that we filled with baconburgers tasting of substances half-prepared in distant cities. I decided that I would see what I could do to deflect Frank's self-destructive momentum.

"Frank," I said, "You've been in this business a lot longer than I have, but I'd like to say that in the last thirteen years or so I've worked as project engineer on quite a few projects, and I've never on one of them considered that I had direct authority to tell the contractor or his men what to do in the field. My role on the job, as I see it, is to advise the owner if something is going wrong; then he takes action if he chooses to."

Frank had relaxed and grown reflective during this speech and had evidently begun to listen; both lawyers were nodding vigorously in agreement.

"Right, Frank," said Smith. "Aren't you, as a consultant, really just an adviser?"

Frank thought for a moment, then began to nod. "Well, I guess you're right," he said. "I can't run everything."

I thought that even if, in my own deposition, I set our cause backward with my own mistakes, I had made one small contribution; my day would not be a total waste.

But in fact my deposition seemed to go well. I had never been deposed before and was surprised at how easy it was. In fact, it was fun in some ways, matching wits with the attorneys, always trying to say the thing that peels away that papery veneer of strutting indignation, exposing the underlying venal whine. I remember thinking that I would have to be careful during the trial, if it ever came to that, to avoid enjoying this sport; a man at pleasure is a man off his guard. But when I read my deposition a few weeks later, it seemed clean. It was typed with IBM executive boldface on crinkly white rag bond, flatteringly bound in a plastic cover. The day had cost us about a thousand dollars. Legal larceny is carried out in gentlemanly style; the victim's vanity is enhanced even as he is being robbed.

Summer passed, and in the bright crisp beginnings of fall my attention moved to other things. From time to time faint vibrations reached me as each new specimen of witness hit the web. The sullen ready-mix contractor. A retired engineer who reportedly held advanced degrees in concrete and offered expert witness services at a discount. It swayed and shuddered when struck with the district's three hundred pound field superintendent, but his struggles were brief and his submission to all interrogators so complete that Macaulay told me with satisfaction that he would not be of any use to anyone.

There was sporadic but diffident talk of settlement. "If this thing goes to trial," Macaulay told me one afternoon when he dropped by the office, "it could end up costing you five or six thousand dollars. You really shouldn't be carrying any of the blame here, but just from a practical standpoint, we're thinking about getting up a pot to get this contractor off our backs." I told him he could consider himself authorized to bargain up to three thousand. No, he couldn't see that, Macaulay said; fifteen hundred seemed like a reasonable maximum to him, for our share. But Mullins was claiming $150,000 from the district, and Macaulay told me a few days later that he had not shown any interest in the meager six thousand that our reluctant alliance--the district, Schultz's insurance company, and my firm--had put together. It looked as if the case was going to trial.

There was one other thing, Macaulay said. The district had decided to file a cross-complaint against Schultz and us. That meant that if the contractor won his case against the district, the district, our client, would go after us.

The trial began one Monday morning in February at Oakland County Courtroom 5C, a solemn, walnut-paneled room that reminded me of the Boston Harvard Club, which I used to visit with my father when I was a boy. I wore my best tweed jacket and regimental necktie.

I found Frank Schultz sitting in the visitors' gallery. Both of us greeted Mullins with funereal nods when he entered. Mullins was chewing a toothpick; he jammed his bulk into a seat on the other side of the aisle, returning our greeting with an affable grin.

The three of us were joined by one of those bleary old men who find courtroom drama preferable to television in some forlorn residential hotel lobby. That made four spectators. Not exactly the trial of the century. But that morning it seemed important enough to me; I had already poured out $5,000 in legal fees, which had done little to stem the unmistakable and ominous concentration of attention on my role as project foundation engineer. Soil and foundation engineering is supposed to be a rational business; in reality it is a soft and vulnerable nexus of structure and earth, where design objectives and the whims of nature, the responsibility of consultants, and the risks of ownership all blur together.

The lawyers sat at a long table before us. At one end was the district's lawyer, Tony DeLuca. Tony had entered a cross-complaint against my firm, putting me in the difficult tactical position of having to side with the district against the contractor, and at the same time defend myself against my own client, the district, by pointing out their contribution to the failures. "I really hated to do that," Tony told me one day, adding that the district might sue him if he had not targeted all possible pockets. Lawyers never mean any harm, of course. Nothing personal in taking away your eight-year-old Dodge Dart and the kids' college fund. Tony drove a new Alfa Romeo and wore a fresh, crisp suit every day.

Schultz's insurance company had a lawyer on the case, and he sat next to Tony, and next to him was someone representing the contractor's bonding company. Then Mullins's lawyer, a nervous, bearded villain named Lyons. With Lyons was his young apprentice, Vincent Moore, a quiet and conservatively dressed recent law graduate with a direct gaze that by the end of the trial would pierce the privacy of my dreams. And my lawyer, Duncan Macaulay, pulling at his beard and fumbling ursinely through his papers. Six lawyers. Three grand a day.

Lyons opened the trial with his kick-off statement of Mullins's position, an intense finger-thrusting speech packed with imitation outrage. The other lawyers followed, each speaking with a kind of impassioned rhetoric that was supposed to convey conviction in their client's cause. To an observer peering through the small glass window of the courtroom door, there was a ghostly semblance of brilliance, conviction, and wit, but to those of us in the courtroom, including Judge Bradford, a florid, balding man, fiftyish, with hard intelligent eyes and a dry manner of asking blunt questions, the arguments seemed unplanned, adventitious, and rhetorically untidy. In his opener, my lawyer Macaulay made a speech about how His Honor was going to have to be cautious lest he become confused by the complex technical issues involved in the case. He emphasized the critical importance of technical terms, some ill-chosen and irrelevant examples of which he then proceeded to define in a sort of sonorous and highly erroneous glossary. About a third of the way through his benumbing discourse, Judge Bradford began to fall asleep.

I was horrified, and later embarrassed, when Macaulay asked me on the way home whether I thought his opening statement had been effective. "I don't know, Duncan; you know what you're doing here," I said. "I've never been to a trial before. I don't have a television, so I don't even know what's supposed to happen in trials."

Mullins was the first witness. Lyons put him on the stand the next morning, Tuesday. By Thursday he was looking pale and addled, and Friday he called in sick. They finished with him the following Monday. "At this rate, this trial might go on for a month," Tony DeLuca said to me Monday afternoon. I was astonished, having once been found guilty of a speeding charge in thirty seconds. I had reckoned on two or three days at the outside. But I did not have time to brood about my mounting legal costs. Lyons announced that he was putting me on the stand the next morning.

There are, it should be understood by the inexperienced, certain dangerous delights in litigation, pleasures not unlike those that can be obtained at less expense in a game of seven card stud. It should have been clear by now that there was no way for me to win the case. Winning is beside the point when you are paying legal fees. I had already lost; the question to be resolved in the trial being only how badly I had lost. The drama of the trial obscures the simple fact that the real contest is not between the various litigants but rather between the litigants and the attorneys and that the attorneys always win. And yet there is a certain moral passion and euphoric stimulation in a courtroom that makes the entire proceeding a fascinating adult entertainment, a sort of puppet show directed by the legal profession, in which the litigants are offered fantastical rewards (the judge, in a ringing voice, "The court finds this defendant NOT GUILTY," cheers, tears, joyous laughter). And so it was that I went to bed that evening before my first day of testimony pleasantly intoxicated with a sort of precombat, psychic tumescence.

The next morning I learned to my surprise that I was to be examined as an unfriendly witness by Vincent Moore, Lyons's young apprentice, rather than by Lyons himself. It began well, I felt an easy confidence, and Judge Bradford seemed to be placing a high degree of credibility on my testimony. From time to time, to the evident annoyance of both lawyers, he would break in and ask me a few questions himself, my opinion on this or that. Moore got bogged down and confused several times and almost had his entire line of inquiry, which was obviously aimed at discrediting my work on the project, cut off by the judge when he failed to establish its relevance to his case. "This is his first trial," Macaulay remarked during lunch break. "He's still wet behind the ears."

But Moore, with a little help from Judge Bradford (whom I suspected of musing paternally on his first trial), recovered quickly. Moore wore a dark suit, a white shirt with that chiaroscuric gleam that new shirts have the first time back from the laundry. Earnest and spare in his manner, he seemed to have stepped not from the streets of Oakland but from the glazed shadows of a fine old oil painting. When he slipped in the course of his examination or his way was suddenly blocked by a sustained objection, I could follow and empathize with that flicker of uncertainty, the brief and dignified silence in which he walked back, head bowed in thought, the delicate rearrangement of papers, then the smooth recovery. "Mr. Meehan, is it not true that ..." he would begin on his new tack in a clear voice that sprang, it seemed, from no other source than the purest desire to get to the truth of the matter. I could not help but admire his style, as he turned back with that question, one foot advanced before the other, chin raised, fingertips on the table, fixing me with that solemn Rembrandtesque gaze. Lyons, slight and black bearded, lurked in the background with the nervous shiftiness of a Shakespearean villain. He would have been easier for me; he had judged well in giving the examination to Moore.

But yet other reasons for Moore's assignment surfaced later that afternoon when he came to the matter of our field soil test calculations. Evidently Moore had an undergraduate background in mathematics, and he had exhaustively checked every one of the hundreds of calculations that my inspector had made in the course of our construction soil testing. And he had, it appeared, found errors. I managed to stave off the impact of this by remaining calm and seemingly indifferent to these disconcerting exposures, but I was worried, and when court recessed for the afternoon, I called Roger, who had been the field technician on the project, and told him to meet me at the office that evening. It became clear to me on further examination that there were quite a few mistakes in the arithmetic, and I told Roger to run the whole batch of calculations through a computer. By midnight we had the answers, and I prepared an exhibit comparing all of the computer results, printed in that mechanical bank-statement style that looks so convincing to the layperson, with the original slide-rule field calculations. True, there were differences in the results, but the comparison did not look nearly so bad when we looked at all of the tests instead of just the ones with mistakes.

The next morning, Moore resumed his examination of every inaccuracy he could find in our mud-splattered and partially illegible field calculations. ("And do you consider, Mr. Meehan," with arched brows and mock astonishment, "THAT particular error to be acceptable work in your company, or the standard of your industry?") But after an hour or so of this, Judge Bradford was looking sleepy again, and the time seemed right for me to suggest we would save some time by examining my computer review of all the calculations, not just the ones selected by Moore. Moore hesitated, glanced at the judge, who seemed interested in my suggestion, and reluctantly agreed. My computer display blunted the impact of Moore's attack, and for the first time in hours I felt as if I had regained some control over the situation.

Moore moved on to other areas, showing how from time to time my inspector Roger and I had given field instructions and approvals that differed from what was required by the written specifications and how in one case I had rather arbitrarily "corrected" certain test results. I kept explaining for the judge's benefit the overriding importance of field judgment and experience, but by mid-afternoon I felt I was losing ground. Now the various parts of my mind, which the previous day had dodged and whirled with regimented precision, began to come apart, like a troop of boy scouts at the end of a long hike. While its leading elements forged ahead, Moore was on its flanks and rear, working on the weak and straggling member. But some very tired scoutmaster in me continued to patrol the line and pull back the stragglers, all the time maintaining a feigned scoffing attitude toward the attack.

Then suddenly, right at the end of the afternoon, Moore shifted his line of questioning and asked me whether any of our tests showed evidence of soil movement. After two days defending myself on the stand, I was primed for denial.

"I don't have any indication that any movement of soils occurred on the site at any time," the record shows I said.

Judge Bradford looked at me carefully and made some notes.

Afterward Macaulay and I talked about the day on the way home. I had the feeling that I had overstated my position, but by that time I couldn't remember what I had said. Perhaps we could straighten it out on cross-examination. "Up to today, you were clean," said Macaulay. "Today they grazed you."

That week of my testimony I spent a lot of time with Macaulay. In the morning I would meet him at his office. there Macaulay, backstage, would rush about talking in a distracted way, stuffing papers into a briefcase, shuffling through the pile of mail that was accumulating on his desk. Meanwhile I would foolishly dose myself with plastic cupfuls of bitter machine-made coffee, for which I would pay later on the stand with a full bladder and an annoying hum in the wiring of my nervous system. As usual I would resolve to learn to chew gum instead of drinking coffee.

Then in Macaulay's gray, coffin-like Mercedes, the two of us would make the one-hour drive to Oakland. Crossing the flat waters of San Francisco Bay on the Dumbarton Bridge, we would skirt Newark, its new subdivisions bordered with bewildered and homesick eucalyptus trees, then enter the river of northbound traffic flowing lazily through Union City, slowing to a stagnant pool at Hayward, then breaking into a grimy ripple at San Leandro. Oakland lay beneath a cloudless sky streaked with amber turpentine wash, stinking of inner tubes and mercaptins. I watched the litter-strewn streets, with their steel-caged liquor stores and abandoned stucco gasoline stations, flip past below the elevated freeway.

At those times Macaulay and I would discuss the case, and sometimes I would probe him for cost estimates. It had been my custom from the start to maintain a record of my best projection of the total cost of the suit, it being a theory of mine that a best-guess cost estimate and estimated coefficient of variation are necessary at all times when continuing policy decisions are required. Of course, as the trial went on, I had to increase my estimate continually. Macaulay alone was costing $500 per day. And by the third day of my testimony, I was thinking in terms of a six-week trial. My working cost estimate, including legal fees, my own time, and an allowance for a most probable settlement cost, had gone from $5,000 before the trial, to $15,000 at the beginning of the trial. Now, in the third day of the trial I was figuring $17,000 in legal costs, $5,000 in lost work time, and $15,000 in settlement costs, realizing that in the worst case the last figure could be much worse.

In this way, my best-guess estimate had gone from $5,000 to $37,000. It was an ominous trend, what would appear to be a deterioration of our position, or to put it another way, evidence of a very bad initial estimate, a failure to be realistic about the situation in the first place. With some annoyance I recalled Macaulay's early advice that we should offer no more than $1,500 in settlement. Couldn't he have foreseen that there was a significant probability of this $37,000 outcome?

The next day, Thursday, was my third on the stand. Moore, having got out of me late in the afternoon the denial that any of the tests had indicated soil movement, decided to quit while he was ahead. So the next morning I was examined briefly by the other attorneys, then cross-examined by Macaulay. He made a number of false starts, was blocked successfully and to his great annoyance several times by Moore's clever objections, then finally had me run through a demonstration of field test techniques that seemed to me irrelevant and again put the judge to sleep.

Earlier I had told Macaulay that I wanted to restate my position on the soil tests because I felt I had made an incorrect statement, and during the afternoon session he got around to asking me about what the tests really showed. Had any of them indicated soil swell? Yes, some of them had indicated that, I answered.

Judge Bradford came alive. "That's not what you said yesterday. That's not what my notes show." He gave me a long look; his eyes were like two small aluminum discs.

I couldn't remember exactly what I had said. "Your honor, I hope I didn't say that; my position is that the tests, some of them, showed an increase in soil water content, and that is one indication of swell."

Bradford said he wanted to review my previous day's testimony on that point. The court reporter fumbled with the stacks of transcript tape. "We'll look at it next session," Bradford said. "Court dismissed."

The next day, Friday, was a court holiday. That left me until Monday to find out whether I was going to be charged with perjury.

The next day I went into the office, presented the grim outlook to my partners over lunch, then took the afternoon off. Friday evening I went to bed early and slept until Sunday morning, waking up for only a few hours on Saturday. It was that kind of tempestuous dream-filled sleep from which you arise drained and exhausted. I remember one dream: Judge Bradford changed into a woman with hard eyes and bloodless lips. I recognized her; she was an East German border guard who years before had held me up at the border because of "irregularities" in my passport.

Monday morning I was on the stand at 10 A.M. I had my files with me and my journal, in which I wrote notes whenever I had a break. The calming effect of doing two things at once. Moore was going to reexamine me.

10 A.M. Sitting on the stand, waiting for Judge Bradford to come into the courtroom. Before me the five lawyers ranged like an avenging army. Heart rate a bit higher than normal, but hands steady, not excessively nervous. Court reporter enters, everyone stands. The judge will be a minute or two, he's on the phone. So far I've been preserving a cheerful and calm demeanor. Can Moore push me to the point of cracking? I don't know. Feeling pretty alert and rested this morning. A buzz. Judge enters.

During the eleven o'clock break Moore asked me if I were writing my memoirs. "Yes, I'm writing about you," I said, hoping to unnerve him.

"(Unintelligible) sold his before he wrote them," Moore said to Mullins.

"Who?" I ask, thinking he said Nixon.

"Kissinger."

1:45 P.M. Just finished putting sketches on the board, illustrating various types of slab misalignment. The drawing looks good, and I am feeling confident regarding my testimony (the contradiction in my testimony of last week has been handled and accepted as a simple "clarification"). Macaulay and the judge are now conferring on some obscure point of law.

For the last hour of my testimony, Moore worked on me in his calm manner, with his partner Lyons occasionally leaping from his chair, a sort of Roderigo, advancing at me and my diagrams, thrusting and parrying with a wooden pointer. I had my own pointer, an aluminum one, and there were moments when a still photograph would have caught us in a duel.

2:45 P.M. Finally finished my testimony. No disaster today. Managed to fend off Moore by evasion, frustrating him. Playing stupid. This is o.k.; isn't it true that the examining attorney extracts testimony by appealing to the professional's vain need to know everything? By asking for help and cooperation? What emotions interfere with my functioning perfectly as an expert witness? For one, the fear of being exposed as inadequate, the guilt over mistakes, exposure of the fact that "my" project didn't work out adequately. Remember your advice to Schultz!

My testimony was finished, but of course the trial had just begun. There were Schultz, other parties to the suit, various experts on concrete. A month later, the trial was still going strong, five days a week, with no end in sight. Judge Bradford hinted that the six lawyers were dragging it out and pressed for a negotiated settlement.

The problem, everyone agreed, was that the district's board of directors was unwilling to negotiate a settlement. The rest of us long ago had replaced moral righteousness with cynical pragmatism, but the district consisted of elected people, and elected people do not come quickly to this point of view.

Hearing that Tony DeLuca, the district's lawyer, was meeting with the district's board to brief them on the trial, I decided a bold move was in order. Never mind the god-damn lawyers. I wrote a letter appealing to the district board members directly, telling them that we would continue to support their case vigorously against the contractor but only if they would drop their cross-complaint against us. Otherwise, I hinted darkly, we were in the untenable position of being threatened by them, our allies. Inevitably it would be in our own interest to expose the district's mistakes that contributed to the problem, which would only help the contractor. Before writing the letter, I spent an hour reading some of Thucydides' accounts of successful oratorical appeals made by various envoys during the Peloponnesian War, to get the right flavor.

My partner, who knew a couple of board members, brought my letter to the meeting and tried to sit in on the conference, but the staff closed the session, although they did agree to distribute copies of our letter to the board members. Irwin waited for two hours while the meeting took place. Afterward DeLuca told him the board had voted three for; two against, to make a settlement offer of $50,000, provided it was split three ways--between district, structural engineer, and ourselves. That was $17,000 for us. Irwin and I agreed that we should tell them to go to hell.

The next evening I called Macaulay, told him that we wanted to drop out, stop supporting the district's case at our expense.

"You mean you want me to walk out of the courtroom?" Macaulay asked, with dramatized incredulity.

"Well, that's our feeling on the matter. Is it really necessary for us to be represented while people argue about concrete for weeks? I'm just disgusted with the district's position and I don't see why we should sit there supporting their case at our expense?"

A weak speech, with that apologetic "just" and whiny "I don't see why." You wouldn't catch Pericles saying that, I thought. Macaulay began to talk in the soothing manner that one saved for raving maniacs.

"Well, Dick, I can really sympathize with the way you fellows feel about this, but just walking out of the courtroom, I don't know. Let's see, if you could represent yourself instead...but god damn it, you're a corporation; you can't do that."

Macaulay was simulating serious exploration of my idea, even though I knew what the lawyers think about self-representation. ("Guy represented himself in San Jose on a purse-snatching charge," said Tony DeLuca over lunch one day; "first thing he asked the witness was, 'Did you actually see my face when I took your purse?'")

At this point, Macaulay was presumably facing one of the oldest scenarios in law: the client who wants to quit the game, leave it all up to the judge.

I can imagine that senior law partners have standard speeches they make to their young associates on that subject: "Now we know that clients sometimes become weary and disillusioned with the due process of law, and will occasionally even want to drop out of a trial," the speech would begin. "The trial lawyer must realize that such expressions reflect a state of emotional distress that is entirely natural to the client and should treat such outbursts with sympathy but firmness, presenting the client with the serious and even catastrophic consequences of proceeding without representation."

So Macaulay gave some respectful consideration to the suggestion, the suggestion that runs against the grain of the entire system. Then he said, "Dick, as soon as you leave the courtroom, you know what's going to happen. They're all going to turn on us."

There's the rub. It doesn't seem likely that the judge is so stupid that he is going to accept anything that's said about an absent party. But if a witness should say that I was drunk on the job, is the judge going to say to himself, "that sounds like a lot of baloney," or is he going to think, "If those guys refuse to stay in the ball game, then I'm going to accept whatever is said against them. That's the way the system works around here, right or wrong, and I didn't get to sit up here by bucking the system."

"Well, okay Duncan, why don't you go over there tomorrow and find out what's going on." I said, adding that we were prepared to go after the district's jugular vein, bring out how, back in the beginning, we appealed to the district to settle the issue with the contractor by negotiation, not litigation, how they told us that they had no quarrel with us and they didn't care whether the contractor sued them or not.

"Jesus, Dick, you didn't tell me about all this. Why have you been holding this back on me?" Macaulay was excited about that news, or pretended to be. He went on about how we were going to bring all that out when we presented our case. Suddenly I was struck by the bleak thought that the weeks of trial so far had represented only the contractor's case. The rest of us hadn't even begun. The trial would go on forever.

The next evening I was having dinner at a friend's house. Macaulay called, excited. The judge was pushing settlement, was going to ram $45,000 down the contractor Mullins's throat. Surely we all could come up with that, the judge had said, hinting that no one was clean in this situation.

"I told him you'd go $7,500, no more," Macaulay said. "The judge knows that you don't have any insurance. At least we managed to get that through to him. He asked me whether you could write a note for the rest of it. I said I just didn't think you'd be willing to go fifteen. But if we could go, say, thirteen, I think that would do it."

"And fifteen?" I asked.

"Fifteen and you'd look like heroes to the judge; it would look like you were really busting ass, working Saturdays and everything else to pay this thing off."

I wondered whether the other parties had exhausted their bargaining positions. DeLuca said he had, but was he holding more cards? Maybe not. Schultz said his insurance company's limit was $15,000.

"Consider yourself authorized to bargain up to fifteen," I told Macaulay.

The case was settled the next day, with my agreement to put up $13,000 cash into the settlement pot. Add to this $13,000 in legal fees, and $5,000 for my time, and my total cost came to $31,000.

Schultz, the structural engineer on the job, and his insurance company paid $15,000 settlement, plus, I suppose, $15,000 legal fees. Schultz himself had to put up only $5,000 cash, plus his time, but the insurance company will get back their money from him in the long run. The district reportedly paid $35,000 in legal fees, plus $15,000 settlement costs, plus staff time--say $55,000 altogether.

Total litigation costs to the district and its consultants and their insurers, plus the contractor, his bonding company, and the landscape architects, were about $150,000.

The six law firms involved collected $88,000.

Mullins's net gain was $33,000, less the cost of lost business during the months of litigation and trial. I was told that he owed most of the $33,000 in back taxes. The lost business may not matter, for I understand that his contracting firm has been dissolved, and Mullins himself has taken a job as some kind of government inspector.

My father, like me, is a civil engineer. After the trial was over I told him the whole sad story, how I had collected $3,500 in fees on the job and paid out $31,000. He just shrugged, as if to say, "That's the world you people have created." My father spent his life building things. He's seventy-two now and still at it, fifty hours a week. He doesn't brood about professional liability or OSHA [Occupational Safety and Health Administration] or affirmative action. "I'm an old man. What can they do to me?" he says.

583

I used to be a designer, and once I knew the secret and the satisfaction of transforming a concept into a reality. But that's dangerous business these days, and there is not much of a market for it, either.

But if you watch the lawyers, in time you learn the trick of trafficking in words. My firm has tripled in size since we worked on the San Ignacio project. We write environmental impact reports now, some of them so big it takes more than one strong man to lift them, and we earn a good profit performing studies and analyses required (but I suspect not read) by bureaucrats. "Forensic engineering" I call it. It's the new way, and business is booming.

The Song of the Bridge and Other Poems

Poems from I Built a Bridge and Other Poems
David B. Steinman

Noted bridge designer David Steinman left a legacy not only of a host of beautiful bridges but also several collections of verse which express his love of his craft and his appreciation for the symbolic role of his life's work in the lives of his fellow men. The beauty of his bridges is widely acknowledged. His ability to express himself in verse, although not as acclaimed as his engineering talent, reveals an inner self fully conscious of the significance of his professional work as both a work of art and a tool of civilization.

THE SONG OF THE BRIDGE

With hammer-clang on steel and rock
   I sing the song of men who build.
With strength defying storm and shock
   I sing a hymn of dreams fulfilled.

I lift my span, I fling it wide,
   And stand where wind and wave contend.
I bear the load so men may ride
   Whither they will, and to what end.

The light gleams on my strands and bars
   In glory when the sun goes down.
I spread a net to hold the stars
   And wear the sunset as my crown.

BROOKLYN BRIDGE: NIGHTFALL

Against the city's gleaming spires,
   Above the ships that ply the stream,
A bridge of haunting beauty stands--
   Fulfillment of an artist's dream.

From deep beneath the tidal flow
   Two granite towers proudly rise
To hold the pendant strands aloft--
   A harp against the sunset skies.

Each pylon frames, between its shafts,
   Twin Gothic portals pierced with blue
And crowned with magic laced design
   Of lines and curves that Euclid knew.

The silver meshings of the net
   Are beaded with the stars of night
Like jeweled dewdrops that adorn
   A spiderweb in evening light.

Between the towers reaching high
  A cradle for the stars is swung;
And from this soaring cable curve
  A latticework of steel is hung.

Around the bridge in afterglow
  The city's lights like firelflies gleam,
And eyes look up to see the span--
  A poem stretched across the stream.

THE BRIDGE

In days gone by, a valiant band
With consecrated heart and hand
Set out as pilgrims, seeking ways
To clear the wilderness of fear,
To bring the distant places near,
To build new roads to brighter days.

The pilgrims built a bridge of wood;
In massive strength the great span stood.
But ere the bridge with load was strained,
A flaming spark fell on the span,
A holocaust of dread began . . .
Then naught but glowing ash remained.

The pilgrims labored to atone;
This time they built a bridge of stone.
But ere the builders' thrill had waned,
An earthquake heaved and split the ground,
Felling the bridge with crashing sound . . .
Then naught but rubble heap remained.

Yet undiminished in their zeal
The pilgrims built a bridge of steel.
But as their eyes aloft were trained,
A thing of terror, hurtling past,
Dissolved the bridge in fission blast . . .
Then naught but vapor mist remained.

But in the pit of their despair
There came an answer to their prayer:
As clouds rolled back before their gaze,
A radiant vision met their eyes--
A prismed span across the skies,
Resplendent in its glowing rays.

And as they watched with wonder high,
They heard a Voice speak from the sky:
"To span the gap from man to man
Construct a bridge not made by hands,
Not wood or stone or iron bands--
Of Human Kindness build the span!"

Civil Engineers Are People
David McCullough

Civil Engineering Magazine - December, 1978

*In September, 1978, at a joint meeting of the American Society of Civil
Engineers, the British Institution of Civil Engineers and the Canadian Society
for Civil Engineering, author and historian David McCullough was asked to
speak to the assembled engineers. McCullough was invited because his books
on the Brooklyn Bridge and the Panama Canal had popularized the
achievements and contributions of the heroic civil engineers of the late
nineteenth and early twentieth century. Not only were the books themselves well
received by the general public and literary critics alike, but both books also
served as the basis for television documentaries.*

*McCullough's speech at that meeting was one of the meeting's highlights. It
was published in Civil Engineering for the edification of ASCE members who
had not been able to attend the meeting. It appears here for the edification of
those who seek to better understand the civil engineer.*

I am a fellow who never intended to have anything to do with civil engineering
or civil engineers. I had finished writing my first book, The Johnstown Flood,
an account of the famous disaster of 1889 which, as possibly you know, was
caused chiefly by human short-sightedness. The book was well received and
so, inevitably I suppose, I was urged to follow up with another. The Chicago
Fire was suggested. I wanted to keep on writing but I wanted no part of the
Chicago Fire or anything more like it, I had a horrible premonition, a life
sentence of disaster books, and right at a moment when what I wanted more
than anything was an entirely different kind of subject, one in which man
would be seen to do things right. What I needed was a symbol of affirmation,
and it was this, rather than any calculated career plan, that led me first to the
Brooklyn Bridge and, later, to the Panama Canal.

I knew nothing about engineering. I had no physics beyond high school, my
mathematics stopped at plane geometry. But then by the same token, neither
had I had any formal training as an historian. I, I confess, was one of those
English majors in college, a man for all seasons, as we were encouraged to
believe. Like so many who set out on a new course, I did so with the Brooklyn
Bridge in good part because I didn't know how very much I didn't know.

So I come before you as an odd fellow with an odd specialized experience in an
odd business, which should help explain the oddity of much of what I'm about
to say.

Let's shift the scenery. It is January 1906; the President of the United States,
Theodore Roosevelt, is writing to a friend in England.

> "You may be amused (writes the President) at an expression
> of opinion on literary matters the other day by a man over
> here to whom I have taken a great liking. He is Stevens,
> whom I have appointed as the Chief Engineer to build the
> Panama Canal. . . . He was originally a backwoods boy from
> Maine. . . . and has done an immense amount of hard

railroad engineering work in the West. (Now comes the important part) I never supposed he had any particular literary tastes, but I happened to find out that he had read a great deal and that he has the same trick that I have of reading over and over again books for which he really cares. His favorite books are all of Macaulay, all of Scott, the poetry of Edgar Allen Poe (this I regard as astounding); and . . . Mark Twain's <u>Huckleberry Finn</u>, which he reads continually. . . ."

So: isn't that wonderful?

The chief executive of the land has found himself an engineer who actually reads books? Not only that; he reads lots of them, and all the very best by Theodore's lights. "I never supposed he had any literary tastes," says Theodore, nor, do we gather, would the friend in England ever have imagined such a thing, of an engineer. "Astounding," says Theodore. "Astounding," we can almost hear the friend respond to himself.

Engineers who read, who paint, who grow roses and collect fossils and write poetry, who fall asleep in lectures, very human-like, even civilized civil engineers are scattered all through the historical record. Civil engineers have been known to go to the theater, yes indeed; they have taken pleasure in good music and fine wine and the company of charming women. There is even historical evidence of the existence among a few civil engineers of a sense of humor.

You see what an unorthodox theme I've got here. But most of all, in this most historic setting of Williamsburg, before such a very distinguished and influential body, I want to say that it is not a so-called technical bent or proficiency, nor the pragmatist's passion for order, that have distinguished the giants in the profession in days past and given their work its special, undeniable majesty. It is rather the very breadth of their *human* nature. Mathematics has counted for less than imagination. Character has counted above all.

Those generous humanistic qualities of mind, that warmth of spirit that we of the liberal arts, we non-technical people, are so quick and so proud to claim as our very own specialty are what in fact have given use, given shape to most of the great works of civil engineering.

The prime importance of such triumphs as the Brooklyn Bridge or the Panama Canal, or the little Panama Railroad or Roebling's Niagara Suspension Bridge . . . the Eads Bridge . . . the Great Northern Railroad . . . the Eiffel Tower . . . the Ferris wheel . . . Brunel's incredible iron ship, the *Great Eastern* . . . the Chicago skyline--the reason why they are so important historically, so very appealing to all of us, has mainly to do with the lift they give to the spirit. To think that men could do such things! They were wonders worked by heroes in an age that needed both wonders and heroes, as all ages do. It is their impact on the emotions, on feeling--much more than their usefulness--that makes them human events of such major consequence, not just big things on the landscape.

And the best of their builders knew this.

John A. Roebling, on the day in 1855 when his Niagara Suspension Bridge was opened to traffic, was terribly concerned about what the effect would be when the first train rolled across. It was the question on everyone's mind since this was the first suspension bridge built to carry a railroad. He climbed to the top of one of the four stone towers of the bridge, and there he sat, to gauge the vibrations through the seat of his pants, which gives you some idea of the relative state of the art at the time.

"No vibrations whatever," he noted. "Less noise and movement than in a common truss bridge." Quite enough to satisfy an engineer, presumably--but then one finds him declaring in a letter home, "No one is afraid to cross. The passage of trains is a great sight, worth seeing."

A great sight, worth seeing; it was Roebling being effusive. A bridge was worth money, a bridge was worth painstaking effort, a bridge was worth untold amounts in the prestige it brought to those who owned it, and to him who built it, to be sure, but a bridge was also worth *seeing*.

Years after, designing the Brooklyn Bridge, he devised a wholly unique feature that had nothing to do with utility. He would build a pedestrian pathway down the center but up above the main traffic, up where people could see in every direction. No such pathway had ever been part of a bridge before. He wanted people to enjoy the bridge, to be able to get up and out of the city on fine days, to go over the river in a way they had never experienced. The bridge was to be a release, a deliverance from the crowds and confinement of the city. This elevated promenade, as he called it, would be of "incalculable value."

No one would have complained had he left it out of the plan; it never would have occurred to him in the first place, had he been the prosaic, limited sort of fellow engineers are supposed to be.

The genius of Roebling is founded on what he did with the suspension bridge form and what he did had to do with mathematics and a leap of the imagination of a kind that happens only so often. But if you are looking for a tangible example of an engineer's interest in people, of the magic that an engineering structure can work for people, go walk his elevated promenade over the East River to Brooklyn.

What sort of a man was he, John August Roebling? And what of the other celebrated figures of that exceptional era--Isambard Kingdom Brunel, James B. Eads, Gustave Eiffel, Washington Roebling, George Morison, John Stevens, General George Goethals? I think, too, of Herman Haupt and his Hoosac Tunnel, and Charles Ellet and his Wheeling Bridge, and some who are still less well known, like George Totten, the hard-bitten, leathery little American who built the Panama Railroad under the worst imaginable hardships and at a greater loss of life per mile than experienced by any railroad in history.

They belonged to a different time. It was such a vastly different world from our own, and that, of course, is one of the keys to understanding them. A Jules Verne kind of confidence was in the air. Everybody knew tomorrow was going to be better than today. Progress was on the march. Had not science and technology conferred upon mankind powers hitherto known only to the gods? It was the nineteenth century after all. "They never achieve anything who do not believe in success," preached the father of the Suez Canal, Ferdinand de Lesseps. He wasn't an engineer, as it happens, but the whole world thought he was. Testifying before a congressional committee early in this century, before the First War, John Stevens said that such was the advance of modern technology that henceforth anything was theoretically possible, given the money.

It was faith of a kind we can never appreciate as they did. And the high-blown words given to it by after dinner speakers of that day are to us, often as not, plain laughable. But our laugh rings like Merlin's laugh. Merlin, you will recall, laughed in his strange way and at strange times because he knew what lay ahead. We know what was to come so close on the heels of such faith. We know about the First War and so much else that was to follow.

But that, I insist, should not cause us to misjudge the power of the faith. Belief in tomorrow is a tonic. Belief in a progeny who will be around in days to come to enjoy the fruits of your labor and to judge you by your works is a powerful motivating force. And certainly the surviving records attest to this. In letters, diaries, in formal proposals, the engineers speak of their work standing the test of time, serving future generations, making the world of their children a better world.

It was a good life, a wonderful life. I can scarcely imagine a better life. There was so much that needed doing. The demand for talent had never been greater. The work was full of variety--railroads, bridges, locks and dams, sea walls, tunnels, and with the advent of the skyscraper in the 1880s, whole new kinds of cities to build. Bessemer steel had been introduced. Dynamite and pneumatic caissons and reinforced concrete had come into use. Intellectually, for the men in charge, there was quite literally never a dull moment.

Young men of ability advanced very quickly. The average age of the engineering staff on the Brooklyn Bridge as it got under way was about thirty. John Stevens, who was entirely self-educated, was only 42 when James J. Hill made him chief engineer of the entire Great Northern system.

They worked in the open air most of their working lives. They commanded great armies of men and there was excitement and grandeur in that. Imagine yourself in de Lesseps' shoes at Suez. Picture Goethals at Panama with upwards of 50,000 men at work. If you'd like a musical score to go with such images I would suggest the grand march from *Aida*, that, as perhaps you know, being the opera commissioned by the king of Egypt to commemorate the opening of the Suez Canal. But greatest of all, surely, was the creative satisfaction to be found in the work itself, which, in combination with phenomenal outpourings of public acclaim, must have been of a kind known to very few men ever.

It was not a life without struggle, or pain or suffering, nor was it a life for anyone dependent upon the comforts of home. They were nearly all itinerants. George Morison, one of the best of the great bridge engineers, dreamed his whole life of settling in a permanent house in his beloved New Hampshire, and at last built a house overlooking Mount Monadnock. But as things turned out, because of his work, he never spent more than a few weeks a year there. "He never married," we read in one biographical sketch. "His life was spent largely in hotels, clubs, and sleeping cars." What a fate.

John A. Roebling came to a hideous end, dying of lock jaw as a result of injuries incurred in the early work on the Brooklyn Bridge. Washington Roebling was struck down by the bends and left virtually paralyzed. Then, from the emotional strain of all that beset him, he suffered what today would be called a nervous breakdown.

Old George Totten, the Panama Railroad pioneer, battled bugs and heat and rain and mud; he survived political uprisings, malaria in several varieties, and came so close to death from yellow fever at one point that his companions had a coffin ready and waiting for him.

Often, in reading the old accounts, you stop and wonder what in the world kept them going, what made them keep at it. Then you think of what Robert Louis Stevenson wrote about his grandfather, who was an engineer. "The seas into which his labors carried the new engineer were still scarce charted, the coast still dark," wrote Stevenson. ". . . The joy of my grandfather for his career was as the love of a woman."

Romanticism? Oh my yes! Was there ever a more splendid Victorian romantic than Brunel? Little Brunel with his stovepipe hat and cigar and his mighty ship, all his mighty, audacious dreams?

But what vast differences there were between them, one from another, as people. John Stevens, who never tried to hide his humble origins, was as hard as nails physically, plain spoken, very popular with men and women, and incurably profane. George Morison was a product of Harvard, both Harvard College and the Harvard Law School, and it was only after working as a lawyer that he decided to teach himself engineering, because, as he said very formally in his diary, he wished to be useful. Morison was a big butter tub of a man and a prude. Very few could say they liked him; everyone respected him. Both Roeblings, father and son, were accomplished musicians--father on the flute, son on the violin--but where the father was vain and imperious, a *presence* in any gathering, the son seemed to go out of his way to appear inconsequential, which he assuredly was not.

William LeBaron Jenney, the man who devised what is regarded as the first skeletal system for a high rise building, the man who built the first sky scraper, was a rollicking raconteur and a great gourmet. John Welborn Root, though known as an architect--he was the Root of the brilliant Chicago firm of Burnham and Root--was an engineer by training and had a wonderfully creative engineering mind. It was Root, for example, who worked out the foundation for the Montauk Building, the so-called floating foundation. But Root can be recalled also for adoring amateur theatrics, or for his sparkling improvisations on the piano. Root was a wit, an enthusiastic sailor. Root, you see, was, in the old expression, "something of a Bohemian."

It must be pretty obvious by now how much I admire these men. There's a life in them, there's such a grand kind of energy in their work and in the way their minds worked. I am in awe of their physical courage; and there is something profoundly moving, I think you'll agree, in their steady perseverance in the face of so many adversities.

591

They weren't supermen. The peculiarities of some are memorable and not a little amusing. I offer two examples. George Morison, somewhere along the line, decided that any man who liked horses and got along with them was a man of very low morals.

General George Goethals, his devoted daughter-in-law once told me, had such a passion for cleanliness that he washed his snow-white hair every day. When I asked if there was anything he especially disliked, she said, "Oh, yes. Fat men." I said I was a little puzzled by that since, as I reminded her, it was to William Howard Taft that Goethals owed his position at Panama. It was Taft who had picked him out among all the Corps of Engineers and gave him his big chance. "Well," she said, "the General once told me that William Howard Taft was the only clean fat man he had ever known."

The human side is often found in very small things. In 1933, at the time of the fiftieth anniversary of the Brooklyn Bridge, an elderly woman knocked at the door of the house on Brooklyn Heights where Washington and Emily Roebling had lived. She told the people who then owned the house that she too had once lived there. She had been the Roeblings' maid. Naturally, she was invited right in and in the course of conversation she inquired whether Colonel Roebling's initials could still be seen in the window upstairs. The owners of the house knew nothing whatever about it. So up they all went to the top floor and sure enough there in a lower pane of a window in the room which Roebling had watched the progress of the bridge, during all the years of his confinement, were his initials, W.A.R. He had cut them with the diamond in his ring. From there he had been able to watch the bridge rise year by year, at his command, a monument dwarfing everything on the New York skyline of the day; but like a solitary child, like any of us, he had cut his mark in the corner of a piece of window glass.

And what of the dark side? These are human beings, it must be remembered. What of the good men who failed, nice men, men loved by their families? What of the careers shattered? The Quebec Bridge, the world's longest cantilever span, was to have been the capstone of one of the prominent careers of the turn of the century, that of Theodore Cooper. Cooper was in his late 60's when he did the design. In August 1907, during construction, the bridge collapsed, killing 75 men, the worst bridge-building disaster on record. A short time before, in his New York office, Cooper had received word that one of the cantilever spans had deflected downward a fraction of an inch, and he had ordered an investigation. Then, on hearing that work was proceeding as usual, he sent off a wire ordering an immediate halt. It arrived too late.

All Cooper's past successes, the good name he had worked a lifetime to attain, counted for not a thing any longer. He was finished.

As was Gustave Eiffel as a builder of bridges, once his name became linked to the payoffs and political intrigue that surrounded the last desperate years of the French attempt at a Panama canal. Eiffel, whose tower remains the most conspicuous of all testaments to nineteenth century daring, was forced to abandon civil engineering and spend the rest of his life working on aeronautical problems.

As painfully human as anything to be found in the record of that era is a notation in the private papers of John A. Roebling. Were I asked to name my personal favorites among the men mentioned this evening, I suppose I would have to say Washington Roebling and John Stevens--favorites as people that is. But no one bestrides that pioneering age as does John Roebling. To him goes the crown, it seems to me. Unquestionably he was a genius and a very great man. Look at his photograph sometime. Look at that face. It's Ahab and unforgettable.

To the outside world he was a creature of iron, never tiring, driving. He tolerated no slackers, no guesswork, no fools.

But once, when his youngest son, a boy of 15 or so, committed some wrong, transgressed morally in the eyes of the father in some way we don't know much about, John Roebling very nearly beat him to death. The boy ran away, disappeared for nearly a year. Finally, by accident, he was found to be in a prison in Philadelphia. He had had himself picked up and locked up as a common vagrant in order to keep away from his father and was, as Washington Roebling observed, "enjoying life for the first time."

But I want to read to you what John Roebling wrote at about the time young Edmund was brought back home again. It is an entry in his philosophical notes, under the heading, "Man. Conscience."

> "A man may be content with the success of an enterprise; he may have succeeded in overcoming obstacles, in vanquishing his adversaries and enemies; in achieving a great task; solving a great mental problem, or accomplishing work, which was previously pronounced impossible and impracticable. The hero is admired and proclaimed a public benefaction; observed of all observers, he feels himself elated, and in his own estimation a great man. Retiring for one calm moment within the recesses of his own inner self, he reviews his past deeds, his thought and motives of action. And before the stern judgment of his own conscience, he stands condemned, an untruth, a lie to himself. But nobody knows! Does he himself now know? Who can hide me from myself?"

As I say, engineers, too, are people.

I am by nature optimistic. And it's been my experience that a look at the lives of those who have gone before us, those in whose footsteps we trod, tends to enforce that outlook. I don't think mankind is perfectable and I certainly don't see the course of history as an upward climb. We're no nearer the angels than we were in days past. But we do learn over the long run; we try. And what we make, what we build, is often very, very good. I have little patience with those who see man only as a mindless mutt. I have little patience with those who insist that anything we do to alter the natural order is a mistake, that our man-made structures only mar the landscape. That is nonsense. The Golden Gate at San Francisco is a more thrilling natural wonder *because* the bridge is there.

Man is a problem solver. It is why we have survived as long as we have, and it is why we will survive. If we seem to have painted ourselves into a corner with our nightmare weapons, our environmental blunders, well we are not helpless--to think, to plan, to act.

Probably, everything considered, it is their zest as problem solvers that appeals to me most about those figures from the past, your professional progenitors, their conviction that by God there *is* a way.

They were builders, workers of wonders, and they were men. I want to close now, with something I found not long ago; it was written by a visitor from France who had just been over the Brooklyn Bridge for the first time, not many years after the bridge was finished.

> "You see great ships passing beneath it, and this indisputable evidence of its height confuses the mind. But walk over it, feel the quivering monstrous trellis of iron and steel. . . see the trains that pass over it in both directions, and the steamboats passing beneath your body, while carriages come and go, and foot passengers hasten along, an eager crowd, and you will feel that the engineer is the great artist of our epoch, and you will own that these people have a right to plume themselves on their audacity. . . ."

It is in that spirit that I salute you, you civilized civil engineers who carry on the good work.

Selected Additional Reading

The books listed here are just a few of the hundreds of books dealing in some way with some aspect of engineering ethics or professionalism which one might find in even a modest-sized library. These particular books have been chosen because they are either books from which selections in this collection have been drawn, because they elaborate on ideas or information in these selections, or because they themselves are collections of such readings.

Agnew, Spiro T. "**Go Quietly . . . or else**" (New York: William Morrow and Company, 1980) 288 pgs. (Mr. Agnew's rebuttal to the charges that resulted in his resignation as Vice President.)

Baum, Robert J. **Ethical Problems in Engineering, Volume II: Cases** (Troy, NY: Center for the Study of the Human Dimensions of Science and Technology, 1980) 215 pgs.

Billington, David P. **The Tower and the Bridge** (New York: Basic Books Inc. 1983) 306 pgs.

Byrne, Robert. **The Dam** (New York: Atheneum, 1981) 302 pgs. (Fiction)

Byrne, Robert. **Skyscraper** (New York: Atheneum, 1984) 274 pgs. (Fiction)

Byrne, Robert. **The Tunnel** (New York: Harcourt, Brace & Jovanovitch, 1977) 214 pgs. (Fiction)

Calhoun, Daniel H. **The American Civil Engineer: Origins and Conflicts** (Cambridge, MA: Harvard University Press, 1960) 295 pgs. (A scholarly study of the early days of the American civil engineering profession)

Caro, Robert A. **The Power Broker** (New York: Knopf, 1974) 1246 pgs. (Pulitzer Prize-winning biography of Robert Moses)

Cohen, Richard M. and Witcover, Jules. **A Heartbeat Away: The Investigation and Resignation of Vice President Spiro T. Agnew** (New York: The Viking Press, 1974) 373 pgs.

Cross, Hardy. **Engineers and Ivory Towers** (New York: McGraw-Hill Publishing Co. 1952) 141 pgs. (A collection of essays by one of America's most respected engineering educators on various professional issues)

Davenport, W. H. and Rosenthal, Daniel (Editors). **Engineering: Its Role and Function in Human Society** (New York: Pergamon Press, 1967) 284 pgs. (A collection of readings dealing with the relationship between engineering and society)

Doolaard, A. Den. **Roll Back the Sea** (New York: Simon and Schuster, 1948) 435 pgs. (A novel which focuses on the work of Dutch hydraulic engineers engaged in dike construction and maintenance)

Doolaard, A. Den. **The Land Behind God's Back** (New York: Simon and Schuster, 1958) 292 pgs. (A fictional account of a Yugoslavian bridgebuilder)

Flores, Albert. **Ethical Problems in Engineering, Volume I: Readings** (Troy, NY: Center for the Study of the Human Dimensions of Science and Technology, 1980)

Florman, Samuel C. **Blaming Technology: The Irrational Search for Scapegoats** (New York: St. Martin's Press, 1981) 207 pgs.

Florman, Samuel C. **The Civilized Engineer** (New York: St. Martin's Press, 1987) 258 pgs.

Florman, Samuel C. **Engineering and the Liberal Arts: A Technologist's Guide to History, Literature, Philosophy, Art and Music** (New York: McGraw-Hill Publishing Co., 1968) 278 pgs.

Florman, Samuel C. **The Existential Pleasures of Engineering** (New York: St. Martin's Press, 1976) 160 pgs.

Layton, Edwin T. **The Revolt of the Engineers: Social Responsibility and the American Engineering Profession** (Cleveland, OH: Press of Case Western Reserve University, 1971) 286 pgs. (A scholarly study of the evolution of the American civil engineering profession)

Martin, Mike W. and Schinzinger, R. **Ethics in Engineering** (New York: McGraw-Hill Publishing Co., 1983) 335 pgs.

Meehan, Richard L. **Getting Sued, and Other Tales of the Engineering Life** (Cambridge, MA: The MIT Press, 1981) 241 pgs. (A collection of stories based on actual experiences of a modern American civil engineer)

Petroski, Henry. **Beyond Engineering** (New York: St. Martin's Press, 1986) 256 pgs. (A collection of essays on various aspects of engineering and technology by a modern civil engineering educator)

Petroski, Henry. **To Engineer Is Human: The Role of Failure in Successful Design** (New York: St. Martin's Press, 1985) 247 pgs.

Schaub, James H. and Pavlovic, Karl. **Engineering Professionalism and Ethics** (New York: John Wiley and Sons, 1983) 560 pgs.

Schaub, James H. and Dickison, Sheila K. with Morris, M. D. **Engineering and the Humanities** (New York: John Wiley and Sons, 1982) 583 pgs. (A collection of readings dealing with and illustrating the relationship between engineering and the humanities)

Steinman, David B. **I Built a Bridge, and Other Poems** (New York: Davidson Press, 1955) 38 pgs. (A collection of poetry written by an American civil engineer)

AUTHOR INDEX

BIOGRAPHICAL INDEX

PROJECT INDEX